MOLECULAR ZOOLOGY

MOLECULAR ZOOLOGY

ADVANCES, STRATEGIES, AND PROTOCOLS

Edited by

Joan D. Ferraris
National Institutes of Health
Bethesda, Maryland

and

Stephen R. Palumbi
University of Hawaii
Honolulu, Hawaii

WILEY-LISS

A JOHN WILEY & SONS, INC., PUBLICATION
New York • Chichester • Brisbane • Toronto • Singapore

Address All Inquiries to the Publisher
Wiley-Liss, Inc., 605 Third Avenue, New York, NY 10158-0012

The text of this book is printed on acid-free paper

Copyright © 1996 Wiley-Liss, Inc.

Library of Congress Cataloging in Publication Data

Molecular zoology : advances, strategies and protocols / Joan D.
   Ferraris and Stephen R. Palumbi, editors.
            p.   cm.
      Developed from a January '95 American Society of Zoologists three-day
   symposium on "Molecular approaches to zoology & evolution."
      Includes bibliographical references and index.
      ISBN 0-471-14449-5 (cloth : alk. paper). — ISBN 0-471-14461-4
   (paper : alk. paper)
      1. Molecular biology—Methodology. 2. Zoology, Experimental.
   3. Molecular genetics—Methodology. I. Ferraris, Joan D.
   II. Palumbi, Stephen R. III. Society for Integrative and Comparative
   Biology (formerly the American Society of Zoologists).
   QH506.M6645 1996
   574.8'8'072—dc20   96-208

Printed in the United States of America

10 9 8 7 6 5 4 3 2 1

CONTRIBUTORS

Judith A. Blake, Molecular Systematics Laboratory, The Institute for Genomic Research, 932 Clopper Road, Gaithersburg, Maryland 20878.

Carol J. Bult, Molecular Systematics Laboratory, The Institute for Genomic Research, 932 Clopper Road, Gaithersburg, Maryland 20878.

Ann Campbell Burke, Department of Genetics, Harvard Medical School, 200 Longwood Avenue, Boston, Massachusetts 02115.

Jane C. Burns, Department of Pediatrics, School of Medicine, University of California–San Diego, San Diego, California 92093-0609.

R. Andrew Cameron, Division of Biology 156-29, California Institute of Technology, Pasadena, California 91125.

Brenda B. Cameron, Center for Population Biology, University of California–Davis, Davis, California 95616.

Constance L. Cepko, Department of Genetics, Harvard Medical School, 200 Longwood Avenue, Boston, Massachusetts 02115.

Nagaraj Chatakondi, Department of Fisheries Science and Allied Aquaculture, Auburn University, Auburn, Alabama 36849.

Thomas T. Chen, Biotechnology Center, University of Connecticut, 184 Auditorium Rd., U-149, Storrs, Connecticut 06269-3149.

Clara M. Cheng, Center of Marine Biotechnology, University of Maryland Biotechnology Institute, Columbus Center/701 East Pratt Street, Baltimore, Maryland 21202; and Department of Biological Sciences, University of Maryland Baltimore County, Baltimore, Maryland 21228.

James A. Coffman, Division of Biology 156-29, California Institute of Technology, Pasadena, California 91125.

Eric H. Davidson, Division of Biology 156-29, California Institute of Technology, Pasadena, California 91125.

Rex A. Dunham, Department of Fisheries Science and Allied Aquaculture, Auburn University, Auburn, Alabama 36849.

Gregor Durstewitz, Oregon Institute of Marine Biology, University of Oregon, Charleston, Oregon 97420; and Department of Biology, University of Oregon, Eugene, Oregon 97403.

Joan D. Ferraris, Laboratory of Kidney and Electrolyte Metabolism, National Heart, Lung and Blood Institute, National Institutes of Health, 10 Center Drive MSC 1598/Building 10, Rm 6N307, Bethesda, Maryland 20892-1598.

Robert C. Fleisher, Molecular Genetics Program, Department of Zoological Research, National Zoological Park, Smithsonian Institution, Washington, DC 20008.

Arlyn García-Pérez, Laboratory of Kidney and Electrolyte Metabolism, National Heart, Lung and Blood Institute, National Institutes of Health, 10 Center Drive MSC 1598/Building 10, Rm 6N307, Bethesda, Maryland 20892-1598.

Jonathan B. Geller, Department of Biological Sciences, University of North Carolina–Wilmington, Wilmington, North Carolina 28403-3297.

Jeffrey A. Golden, Department of Genetics, Harvard Medical School, 200 Longwood Avenue, Boston, Massachusetts 02115.

Richard K. Grosberg, Center for Population Biology, Department of Evolution and Ecology, University of California–Davis, Davis, California 95616.

David M. Hillis, Department of Zoology, University of Texas, Austin, Texas 78712.

Linda Z. Holland, Scripps Institution of Oceanography, University of California–San Diego, La Jolla, California 92093-0202.

Nicholas D. Holland, Scripps Institution of Oceanography, University of California–San Diego, La Jolla, California 92093-0202.

Peter W. H. Holland, School of Animal and Microbial Sciences, University of Reading, Whiteknights, P.O. Box 228, Reading, RG6 2AJ United Kingdom.

John P. Huelsenbeck, Department of Zoology, University of Texas, Austin, Texas 78712.

Rosemary Jagus, Center of Marine Biotechnology, University of Maryland Biotechnology Institute, Columbus Center/701 East Pratt Street, Baltimore, Maryland 21202.

Pancharatnam Jeyasuria, Center of Marine Biotechnology, University of Maryland Biotechnology Institute, Columbus Center/701 East Pratt Street, Baltimore, Maryland 21202.

Robert Jones, Thinking Machines Corporation, 245 First Street, Cambridge, Massachusetts 02142-1214.

Valentine Lance, Center of Reproduction of Endangered Species, Zoological Society of San Diego, P.O. Box 551, San Diego, California 92112-0551.

Donald R. Levitan, Department of Biological Sciences, Florida State University, Tallahassee, Florida 32306-2043.

Chun-Mean Lin, Center of Marine Biotechnology, University of Maryland Biotechnology Institute, Columbus Center/701 East Pratt Street, Baltimore, Maryland 21202; and Department of Biological Sciences, University of Maryland Baltimore County, Baltimore, Maryland 21202.

Sabine S. Loew, Department of Biological Sciences, Illinois State University, Normal, Illinois 61761.

Jenn-Kan Lu, Center of Marine Biotechnology, University of Maryland Biotechnology Institute, Columbus Center/701 East Pratt Street, Baltimore, Maryland 21202; and Department of Biological Sciences, University of Maryland Baltimore County, Baltimore, Maryland 21202.

Wayne Paul Maddison, Department of Ecology and Evolutionary Biology, University of Arizona, Biosciences West Building, Tucson, Arizona 85721.

Margaret M. McCarthy, Department of Physiology, University of Maryland School of Medicine, 655 West Baltimore Street, Baltimore, Maryland 21201.

Craig E. Nelson, Department of Genetics, Harvard Medical School, 200 Longwood Avenue, Boston, Massachusetts 02115.

Stephen R. Palumbi, Department of Zoology and Kewalo Marine Laboratory, University of Hawaii, 41 Ahui Street, Honolulu, Hawaii 96813.

John M. Peters, Department of Neurobiology and Behavior, Cornell University, Ithaca, New York 14853.

Allen R. Place, Center of Marine Biotechnology, University of Maryland Biotechnology Institute, Columbus Center/701 East Pratt Street, Baltimore, Maryland 21202.

Ellen M. Popodi, Indiana Institute for Molecular and Cellular Biology and Department of Biology, Indiana University, Jordan Hall, Bloomington, Indiana 47405.

Wayne K. Potts, Center for Mammalian Genetics and Department of Pathology and Laboratory Medicine, University of Florida, Gainesville, Florida 32610-0275.

Dennis A. Powers, Hopkins Marine Station, Stanford University, Oceanview Boulevard, Pacific Grove, California 93950.

David C. Queller, Department of Ecology and Evolutionary Biology, Rice University, P.O. Box 1892, Houston, Texas 77251-1892.

Rudolf A. Raff, Indiana Institute for Molecular and Cellular Biology and Department of Biology, Indiana University, Jordan Hall, Bloomington, Indiana 47405.

Renate Reimschuessel, Aquatic Pathobiology Group, Department of Pathology, University of Maryland at Baltimore, Baltimore, Maryland 21201.

Patricia M. Schulte, Hopkins Marine Station, Stanford University, Oceanview Boulevard, Pacific Grove, California 93950.

Michael J. Shamblott, Center of Marine Biotechnology, University of Maryland Biotechnology Institute, Columbus Center/701 East Pratt Street, Baltimore, Maryland 21202; and Department of Biological Sciences, University of Maryland Baltimore County, Baltimore, Maryland 21202.

Carlos R. Solís, Department of Ecology and Evolutionary Biology, Rice University, P.O. Box 1892, Houston, Texas 77251-1892.

Ralf J. Sommer, Max-Planck Institut für Entwicklungsbiologie, Spemannstrasse 35, 72076 Tübingen, Germany.

Paul W. Sternberg, Howard Hughes Medical Institute and Division of Biology 156-29, California Institute of Technology, 391 Holliston Avenue, Pasadena, California 91125.

Joan E. Strassmann, Department of Ecology and Evolutionary Biology, Rice University, P.O. Box 1892, Houston, Texas 77251-1892.

Billie J. Swalla, Biology Department, Vanderbilt University, Box 1812/Station B, Nashville, Tennessee 37235.

Nora Barclay Terwilliger, Oregon Institute of Marine Biology, University of Oregon, Charleston, Oregon 97420; and Department of Biology, University of Oregon, Eugene, Oregon 97403.

Robert W. Zeller, Department of Biology, University of California–San Diego, 9500 Gilman Drive, La Jolla, California 92093.

CONTENTS

8. Strategies for Finding and Using Highly Polymorphic DNA Microsatellite Loci for Studies of Genetic Relatedness and Pedigrees

Joan E. Strassmann, Carlos R. Solís, John M. Peters, and David C. Queller

■■■■■ FOREWORD

The Society for Integrative and Comparative Biology (formerly the American Society of Zoologists) is proud to present *Molecular Zoology: Advances, Strategies, and Protocols,* the product of a major, society-wide symposium, organized by Joan D. Ferraris and Stephen R. Palumbi to showcase the actual and potential contributions of the methods and approaches of molecular biology to virtually all zoological endeavors. The SICB, now 107 years old, has always represented an assemblage of biologists with interests in the full spectrum of approaches to research and education in biology, from systematics and morphology to physiology and biochemistry. This volume is an excellent example of the directions and initiatives underway in the Society.

The idea for the symposium that led to this volume arose during discussions between Drs. Linda H. Mantel, then Program Officer for the Division of Comparative Physiology and Biochemistry, Charlotte P. Mangum, then President, and Joan D. Ferraris. It was to be among the first symposia to transcend virtually all of the traditional divisional boundaries that framed the Society's symposia in the past, and thus set a model for annual meetings of the future. The Molecular Approaches to Zoology and Evolution Symposium took place January 5–8, 1995 at the annual meeting of the American Society of Zoologists in St. Louis, MO. It more than met its two-fold expectations to focus on molecular methods that are useful to zoologists with interests in a broad range of questions and emphasize the recent innovations that make these techniques accessible to all biologists.

Molecular Zoology is one of many initiatives taken by the SICB in recent years to reinvigorate its activities and serve current and future members. The **Society for Integrative and Comparative Biology,** formerly the Society's subtitle, became our name on January 1, 1996. Concomitantly, the Society is initiating changes in its journal, the *American Zoologist,* traditionally a venue for publication of the Society's symposia, to increase the rigor of its review process and include invited reviews and other contributions in some issues. The Society is a founding sponsor of one of the first fully electronic journals in the biological sciences, *Experimental Biology Online.* One or more additional journals that take different approaches to reporting on the biological sciences will soon be added to the benefits of SICB members. A grants-in-aid program for graduate-student members begins in 1996. Our annual meetings will feature cross-divisional symposia, satellite symposia, cutting-edge colloquia and greater collaboration with affiliate societies. As a society, SICB is clearly headed for the twenty-first century with strength and vision, and the volume in hand is a good demonstration of that status.

The Society is deeply indebted to Joan D. Ferraris and Steve R. Palumbi for the incredible talent and energy they devoted to drawing together an outstanding group of biologists to participate in the symposium and prepare the up-to-the-minute chapters for this volume. This book will be useful to SICB members and their students for many years to come. To the editors and authors, I extend the Society's deepest appreciation.

The Society for Integrative   MICHAEL G. HADFIELD, PRESIDENT
and Comparative Biology

Among classically trained zoologists, there are many whose pursuit of particular biological questions has led to the need to employ molecular tools. Zoological questions have long formed the conceptual bases for many different fields of biology, and technical advances have resulted in an increasing ability to discern fundamental biological principles. Long the domain of cell biologists and biochemists, molecular tools are now appropriate for answering important questions in many other fields such as ecology, evolution, physiology and development. As the scope for molecular tools in zoology grows, it is important to be able to provide access to them to the widest possible audience.

This book describes the use of molecular tools in several major fields in zoology and evolution and attempts to span the varied disciplines of members of the Society for Integrative and Comparative Biology (SICB; formerly the American Society of Zoologists). Our selection of contributors was intended to take advantage of a wide range of molecular tools in a wide range of systems. Most were chosen because they are primarily known as zoologists whose basic research in important biological questions has led them to adopt the versatility of molecular tools. By bringing together a wide range of "molecular" zoologists from postdoctoral fellows to established researchers, we hope to encourage this combination of problem-based research and laboratory-based data, and provide zoologists with greater access to molecular studies. Contributors were asked to describe a particular molecular approach in detail, showing not only the conceptual goals, but also, the methods by which these goals are achieved. How are particular genes recognized, isolated, and studied? How are diagnostic genetic tests designed to distinguish species or individuals of a species? How can the genetic control of physiological adaptation be investigated? How can the enormous store of genetic information be tapped and sifted to look for patterns? Who is related to whom and how? These are among the questions contributors to this volume have set out to address.

The format of this book is designed to bring together researchers currently using molecular techniques in a great variety of ways and to make this expertise accessible to students, beginning researchers, and senior researchers who would like to incorporate molecular tools in their own work. Being able to use the appropriate molecular tools requires rapid access to experimental insight and practical troubleshooting. Even more important is a feeling for the reasons behind each step in a protocol. Such reasons are important in a practical sense, as well as being important in overall appreciation of the power and limits of molecular techniques.

Published protocols often leave out the reasoning and warnings; they are sometimes staccato commands whose cryptic brevity is demanded by limited journal space. In many cases, the actual procedures used in a lab are written out in a much clearer and informative fashion than are the published versions. To alieviate this problem, we include protocol sections from each contributor. These are detailed renditions of exactly what each of our contributors does in the lab to make things work. Our goal is not to provide a comprehensive set of molecular protocols for every conceivable molecular technique. Instead, we provide examples of protocols of most general techniques used in molecular zoology. Other sources of protocols include those of Sambrook, et al., 1989 (Cold Spring Harbor Laboratory Press, NY) and a special volume of *Methods in Enzymology* devoted to collecting data for studies of molecular evolution (Zimmer, et al., 1993, Vol 224, Academic Press, New York).

Protocols are listed alphabetically by author. Protocols from a particular lab are grouped together because they often go together in a series. However, we also, characterize protocols into general application categories. We provide a Table of Contents that lists protocols by application so that the reader can examine, for example, all of the DNA extraction protocols. Note that many protocols are listed in more than one category. For example, sequencing of double stranded PCR products can be found under the PCR category and the Sequencing category. Why provide more than a single protocol for each application? Different protocols are often directed at slightly different applications. For example, making replicas of lambda libraries and plasmid libraries is similar philosophically, but different technically. Second, the inclusion of different protocols can give a reader a feeling for the variation tolerated by the protocols and the different ways a given procedure might be organized.

The goal of *Molecular Zoology: Advances, Strategies, and Protocols* is to bring together the perspectives of zoologists in a number of different fields and show how molecular tools are helping them expand the frontiers of knowledge. Specific protocols should find a welcome home on the benchtop: descriptions of advances and approaches should provide fertile ground for discussions in seminar rooms. Even for those who swear never to pick up a pipetman, we hope the volume opens the molecular approach to appreciation and understanding.

The symposium "Molecular Approaches to Zoology and Evolution," the original presentation venue, was generously sponsored by a grant from the Alfred P. Sloan Foundation and by SICB. We owe many thanks to Dr. Michael G. Hadfield, who as President-elect of SICB, supported our efforts very well.

Bethesda, Maryland   Joan D. Ferraris
Honolulu, Hawaii   Stephen R. Palumbi

◼◼◼◼ PROTOCOLS

MOLECULAR ZOOLOGY

PART I

SYSTEMATICS, PHYLOGENY, AND DATABASES

CHAPTER 1

Biological Databases on the Internet

JUDITH A. BLAKE* AND CAROL J. BULT

Molecular Systematics Laboratory, The Institute for Genomic Research, Rockville, Maryland 20850

CONTENTS

SYNOPSIS

The development of electronic databases and the linking of information through the Internet has provided new opportunities for biologists to analyze large amounts of information from diverse sources in the exploration of scientific questions. In this chapter we will review the use of the Internet for retrieval of biological information, examine the range of biological databases currently available via the Internet, and highlight the development of new database design and implementation strategies, which will provide mechanisms for linking specialized databases in the retrieval of

*Present Address: The Jackson Laboratory, Bar Harbor, Maine 04609

Molecular Zoology: Advances, Strategies, and Protocols, Edited by Joan D. Ferraris and Stephen R. Palumbi.
ISBN 0-471-14461-4   © 1996 Wiley-Liss, Inc.

information in response to complex queries. Two of the public databases developed and maintained at The Institute for Genomic Research (TIGR) are used as examples of the promise and power of emerging database federations.

INTRODUCTION

Since the early 1980s, biologists have witnessed a vast increase in the amount of biological information available to them through electronic media. Computerized biological databases now represent everything from laboratory and field data to details of protein structure and function. Molecular sequence databases (e.g., Gen-Bank, SwissProt) are one example of the computerization of biological information. The number of molecular sequences, particularly DNA sequences, posted to the GenBank has increased from a few thousand in the first releases of data in the mid-1980s to over 230,000 sequences by the end of 1994, and over 500,000 sequences by the end of 1995.

A number of advances ushered in this era of computerization of biological information. Key among them are (1) the availability of powerful personal computers, (2) the refinement and automation of molecular biological and sequencing methodologies, and (3) the emergence of the Internet (the network of computers using the same protocols that are linked through high-speed data transmission lines). One consequence of these advances has been the development of multiple, specialized biological databases that are accessible using Internet protocols.

Although the number and complexity of computerized biological databases continue to grow, the ability of researchers to retrieve and compile information from them in response to complex biological questions has yet to be fully realized. New thinking about decentralized, highly specialized databases has resulted in the development of improved database designs and new software tools for exploring and synthesizing biological data.

ACCESS TO INTERNET

The impact of the Internet can be compared to that of the Gutenburg Press; it has dramatically changed how information is distributed and managed. The Internet communications protocols were developed by the Defense Advanced Research Projects Agency (DARPA) to allow electronic data transmission between diverse computing platforms. The DARPA communication protocol is known as Transmission Control Protocol/Internet Protocol (TCP/IP). This protocol is essentially a standardized instruction set for packaging and transmitting information electronically between two computers on the same network. The National Science Foundation implemented the DARPA TCP/IP protocol to network five supercomputing centers in the early 1980s, extending the power of interconnectivity to the broader scientific community. Expanded from a network linking research institutions, the Internet has grown into a worldwide communication and data transfer resource (Krol, 1992; Schatz and Hardin, 1994).

One of the essential technologies that allows information sources to be accessed via the Internet is the client/server architecture. The "client" is the software interface resident on the user's computer, which accepts input from the user and displays the results. The "server" is the computer that processes the requests issued by the user through the client. The server can be on the same computer as the client, but more often the server is at a site remote from the user. For example, a user at a small college may log on to a computer at a national supercomputing center and perform data analysis using the programs resident on the computer at the remote site. From the user's perspective, through the use of the client software, it is as though he or she had access to the analysis programs on their own local computer. Because clients can be written for different platforms (e.g., DOS, Macintosh, VAX, UNIX), the user does not have to be operating on the same platform as the server. This important feature has contributed to the portability and flexibility of the Internet.

Although new browsers and interfaces to the Internet continue to simplify and speed entry to this amazing electronic network, the standard Internet protocols still provide the basis for powerful access to Internet resources. Standard methods for accessing and using the Internet to transfer information are described below. Table 1 presents examples of some Internet access points of particular interest to biologists using these methods.

Internet Protocols and Navigational Tools

Telnet: Telnet allows users to log on to a computer at a remote site using the Internet network to do so. Using telnet, researchers with accounts can log on to the computers at a remote computing facility and use different data analysis programs interactively.

FTP: The File Transfer Protocol, or FTP, is a way to download files from a computer at a remote site to the user's local computer or to upload files in the other direction. In a typical FTP session, the user logs on to a server at a remote site and has access to specific computer files. So-called "anonymous FTP sites" contain files that can be downloaded (copied to the local computer) by anyone who has access to the Internet. FTP is often used to receive documents or software codes that have been released into the public domain. FTP protocols have been incorporated into several other programs that are more user-friendly, such as the Macintosh program FETCH (copyright 1993 Dartmouth College; e-mail fetch@Dartmouth.edu) and the menu-driven program GOPHER (see below).

FTP_by_Mail: FTP_by_Mail is a common way to use remote Internet servers. The user issues a request to a server by writing an e-mail message in a specified server-defined format. The e-mail message is automatically parsed (sorted and identified) by the server and the results are obtained from the database. These results are automatically formatted and returned by e-mail to the user.

WAIS: Wide Area Information Servers (WAIS) provide a way for users to look

TABLE 1. Internet Protocols and Interfaces

Protocols	Sample Address	Resource Description
Telnet		
Log on to another computer and use its programs and applications.	telnet consultant.micro.umn.edu	A public gopher site. Can obtain information on downloading a gopher client. LOGIN "gopher"
Anonymous FTP	ftp.ncifcrf.gov/pub/doc	Can obtain documents using the ftp 'get' commmand. Samples:
Log on to remote computer to up-load or download files.		"network_resources_for_biologists.txt" or "una_internet.quide.txt"
	expasy.hcuge.ch/databases/	Many documents describing biological databases are available as well as public domain software.
FTP_by_Mail Servers	retrieve@ncbi.nlm.nih.gov	Returns GenBank sequences based on locus name or acces-sion number.
Use e-mail to send formatted file re-quest to server.	blast@ncbi.nlm.nih.gov	Returns results from sequence similarity search using submit-ted sequence as the query sequence.
		An e-mail message with 'help' entered into message text re-turns a document file of instructions.
WAIS	waismail@net.bio.net	'Help' in the message text will return instructions on using the biosci/bionet wais-mail server to text search e-mail dis-cussion group archives for information of interest.
Conduct a text search of archives		
Gopher	gopher sunsite.unc.edu	A starting point to explore biological resources on the Inter-net using gopher.
Menu-driven navigation tool.	gopher flybase.bio.indiana.edu	A gopher access to the *Drosophila* resources.
World Wide Web	http://www.ncsa.uiuc.edu/	Web entry sites for molecular and computational biology re-sources.
Hyper text based navigational tool.	http://expasy.hcuge.ch/	
	http://www.gdb.org/hopkins.html	

up information in a database or resource on the Internet using text search words. WAIS looks through the archive and assembles a list of matches in response to the user query. The user can subsequently refine the list to find the desired information. WAIS searching can be used, for example, to query the archives of the Biosci/Bionet e-mail discussion groups (Table 1) for any reference to a specific topic (e.g., "primer design") and receive back a compilation of messages that had been posted to the biosci mail groups that contain the words of query.

Gopher: Moving around an FTP site or between remote computer sites can be cumbersome and confusing since the navigation is based on knowing the location of a resource and a series of computer commands. Gopher clients, originally developed at the University of Minnesota, improve on FTP by providing a menu-driven system to move between directories and resources. This means the user can traverse between ftp sites (using telnet) or move down site directories (using ftp commands) by selecting a menu item. Gopher software clients can be downloaded using anonymous FTP from the University of Minnesota ftp site (boombox.micro.umn.edu). Alternatively, a user can use Gopher by first telneting to a public Gopher site and using the gopher client resident there (e.g., telnet consultant.micro.umn.edu).

World Wide Web: The World Wide Web (WWW) provides a powerful graphical user interface (GUI) for users to access multiple client architectures (e.g., FTP, Gopher). The advance over Gopher is that WWW can be used to view and transmit different media types including graphics and other nontext data. WWW documents are written in HTML (Hyper Text Markup Language), which includes information on how to connect with different clients. The documents are transferred over the net using HTTP (Hyper Text Transfer Protocol). The Internet address for a WWW site is called a Universal Resource Locator (URL). Two of the most common WWW interfaces are Mosaic and Netscape. The Mosaic browser was developed at the National Center for Supercomputing Applications (NCSA), located at the University of Illinois at Urbana–Champaign, and is available by anonymous FTP from most sites providing molecular biology information (e.g., ftp.ncsa.uiuc.edu). Netscape is a commercial product (Netscape Communications Corp., Mountain View, CA; ftp.netscape.com). Both Mosaic and Netscape work on multiple computer platforms (Fig. 1).

Users can connect from one URL to another via a hypertext link (also called a hotlink) on the Mosaic or Netscape display. A mouse click on the hotlink, which is indicated by color-coded text, moves the user to the next level of information. Linking of URLs via the WWW is one way of creating a loose federation of databases although the kinds of queries that can be launched are limited. However, many databases now have direct links between their sites by incorporating hotlinks into their HTML documents.

BIOLOGICAL DATABASES

Over a hundred specialized biological databases can now be browsed using the protocols and interfaces discussed above. These biological databases are summa-

---

SUGGESTIONS FOR THE NOVICE

This box is written for those who want to get started on the Internet using the World Wide Web. Perhaps this guide will enhance your early successes.

1. Have your systems administrator or local computer literate open up Mosaic or Netscape for you.

2. (Mosaic)   Under 'File', use the 'Open URL' box to type in the following address:
   (Netscape) Under 'File', use the 'Open Location' box to type in the following address:

   http://expasy.hcuge.ch/

   Click on the "Open" button.
   When the screen opens to the requested address, hit the 'Dismiss' button to close the dialogue box.

3. With the Expasy Home Page on the screen, note that some words are highlighted (usually in blue). Clicking on any of these words will open up another screen connecting you with further information concerning the highlighted word. For example, click on 'Swiss-Prot' to gain entry into the Swiss-Prot Annotated Protein Sequence Database.

Note that there are various buttons at the top and bottom of your window. Clicking on the 'Back' button, for instance, will take you back to the previous window.

4. With the window open to Swiss_Prot, click on 'by description or identification'. This brings up a search box into which you can type a keyword for your search. In this test example, type in 'alcohol dehydrogenase'.

5. With the list of alchohol dehydrogenses documented in Swiss-Prot now before you, click on 'Adhi_dromo' to obtain more information on a Drosophila alcohol dehydrogenase. This opens up the Swiss-Prot record for the protein sequence.

6. Note that in the record, several words are highlighted. Clicking on any of these words will take you to other databases and further information about the protein sequence.

Click on "Medline' to look at the abstract of the reference describing this sequence.

Click on 'Flybase' to look at the Drosophila database record for this sequence.

Click on 'Prosite' to obtain protein structure information about the sequence.

7. At any time, you can either hit 'Back' to go back a page and follow another path, or you can click on other buttons to return to you 'Home Page' or starting point, or to other paths.

8. (Mosaic) If you find a page that you would like to be able to return to , you can enter that URL address into your 'Hotlist' list of addresses by dropping down the Menu list under 'Navigate' and clicking on 'Adding Current to Hotlist".
   (Netscape) Drop down the 'Bookmarks' menu, and 'Add Bookmark' to save an address.

In both cases, many addresses can be stored to make for a quick return to these reference sources.

Explore and be amazed. More and more information is accessible and linked every day.

Figure 1. WWW demo with Mosaic or Netscape.

rized in Table 2. Information about these databases can often be found at FTP or WWW sites, which have collected information about biological resources on the Internet (e.g., at the University of Geneva; ftp expasy.hcuge.ch or http://expasy.hcuge.ch/).

TABLE 2. Types of Biological Databases

Types of Databases	URLs (WWW Addresses)	Resource
Molecular sequence databases	http://www.ncbi.nlm.nih.gov/	National Center for Biotechnology Information Gen-Bank molecular sequence database
	http://www.gdb.org/Dan/proteins/owl.html	OWL: a nonredundant protein sequence database
Specialized molecular information databases	http://www.gdb.org	GDB Genome Data Base
	http://www.tigr.org	EGAD Expressed Gene Anatomy Database
Organism-based databases	http://moulon.inra.fr/acedb/	*Caenorhabditis elegans* DataBase
	http://www.agron.missouri.edu/	Maize Genome Database Project
Museum collections databases	http://www.mip.berkeley.edu	Museum Informatics Project University California–Berkeley
Taxonomic databases and resources	http://muse.bio.cornell.edu/	Biodiversity and Biological Collections WWW Server
	http://phylogeny.arizona.edu/tree/phylogeny.html	Tree of Life: phylogenetic navigation system
Other biological resources	http://www.wdcm.riken.go.jp:80/	WDCM World Data Center on Micro-organisms
	http://probe.nalusda.gov:8000/	Agricultural Genome WWW Server

No single database exists that provides for the needs of all biological research. Specialized biological databases exist because different researchers and research problems require different views of the information. For example, researchers interested in the metabolic enzymes may develop a database designed to hold information about the enzymes, their structure, function, cellular roles, sequences, polymorphisms, and homology to other proteins. Another database may focus on protein structural motifs and would include information about many of the same proteins in the enzyme database, but with more details about the structural motifs relative to many other proteins. Theoretically, links could be made between two databases that have different subsets of the available biological information. A nucleotide sequence may be linked to a chromosome mapping database, or to a database representing a gene family, or to proteins that are expressed at a certain stage in development. The challenge for the informatics community, that is, those people working to organize and make available electronic information, is to provide procedures and tools to link multiple database views to facilitate answering whatever question a scientist may want to ask.

Molecular Sequence Databases: There are two different classes of molecular sequence databases. First, there are the repositories of all the molecular sequences being made available by researchers. These databases include GenBank and GSDB (Genome Sequence Data Base), both located in the United States, DDBJ (DNA DataBank of Japan), and EMBL, the DNA database maintained by the European Molecular Biology Laboratories. These major repositories regularly exchange sequences so each site holds a replicate of sequences held at the other sites. This reduces the amount of Internet traffic around the world by molecular biologists interested in depositing or retrieving sequences from these repositories.

There are also protein sequence databases [e.g., SwissProt, PIR (Protein Information Resource)] or sequence databases that hold specific subsets of sequences such as the G-protein coupled receptor database. These smaller and more restricted databases serve the needs of particular communities of scientists.

Specialized Molecular Information Databases: There are also databases that contain information beyond raw DNA sequences but related to the sequences. Specialized molecular information databases contain subsets of information, which may contain information about restriction enzymes (REBASE) or provide information about eukaryotic promoter sites (EPD). The Genome Data Base (GDB) contains information about the physical location of genes on chromosomes in the human genome. The Protein Data Bank (PDB) holds details about the three-dimensional structure of proteins. In many cases, specialized molecular information databases have links to the DNA or protein repositories.

Organism-Based Databases: Sequence, mapping, genetic, taxonomic, and other types of information about a specific organism or group of related organisms are often maintained in a single database. This centralized database structure is common for model organisms such as the fly, *Drosophila melanogaster* (FlyBase), the nematode, *Caenorhabditis elegans* (ACEDB), and many crop plant species (e.g., *Zea mays;* Maize Genome Database Project). Organism-based databases contain multiple types of data such as graphics, text, and photos that are related to that organism. To accomplish this, database developers are exploring methods for representing different types of descriptive information in the same database. This has led to some innovative approaches to database design and management.

Museum Collections Databases: Museum collections information has always been maintained as annotated records whether held on file cards or as computer records. As computerization of collections data has become commonplace, many collections curators have created databases to manage all collection information including details about loans and further use of collection material. With the Internet, some collections are choosing to make some information about their collections available for general searching. Summary statistics about collections coverage, either taxonomic or geographic, has been posted for both small and large collections.

Taxonomic Databases and Resources: Taxonomic and classification information is vital to the description and interpretation of other biological information. Confusing taxonomic designations have confounded some efforts to properly link data about the same organism. Voucher specimens have not always been deposited in curated collections, especially specimens for work in molecular biology. The use of the Internet for pursuing biological information from many points of view has increased interest in taxonomic resources as a method of grouping information. Classification reference files, checklists, phylogenetic trees, and other taxonomic resources are slowly appearing as experts for different taxonomic groups agree on representations for their group.

Linking Databases: As different specialized databases become accessible over the Internet, moving between information sources can become impractical. This problem has been addressed by the creation of linking databases that hold links between different database resources. For example, the Sequences, Sources, Taxa (SST) database (see below) links molecular sequence database accession numbers with museum collection voucher specimen numbers and provides a direct accessible link between sequence information and voucher specimen information. This link would otherwise only be available by extensive searches of the primary literature.

While most linking databases currently are prototypes, they will become increasingly common as the need and demand for interconnectivity among databases grows.

DATABASE DESIGN STRATEGIES

Database design strategies can be discussed from the perspective of database architectures, details about how to represent different types of information, or descriptions of schemas of existing databases. Understanding the purpose of the database and the structure of the data to be entered into the database leads to more intelligent decisions as to which data model to use to build the database, and how to represent the different data elements. Careful consideration of what the primary entity is and how it is related to other information in the database will help create a useful design.

The choice of database architecture has a lasting impact on all aspects of database design and implementation (Lewis, 1994). How data are structured affects the types of queries that can be supported by a database; the ease of modifying user interfaces as the types and formats of data evolve and how easily a database can be "scaled up" to accommodate large amounts and increasing complexity of data are all factors to consider. The most commonly employed database architectures are the flatfile, relational, and object-oriented data models.

A database that maintains information as a series of flatfiles is essentially a data archive. Flatfile architecture offers little support of complex queries because all the annotation about a primary entity is contained in one record. If the information in each record is formatted in the same manner, it is possible to search such a database using Boolean expressions. However, precise queries that require comparing different features among records are not possible because the relationships among different components are not defined by the data model.

In a relational database management system (RDBMS), the emphasis is on relations among entities in the database. Unlike a flatfile model, information about each primary entity (i.e., the concept or object being described) is broken down into components that are represented separately. In RDBMS terminology, primary entities are represented as tables, with all relevant information or annotation about the entity represented as fields within a table (summarized in Cuticchia, 1994). One difficulty inherent in a relational data model is that some biological concepts, especially qualitative ones, are difficult to represent in such a formalized, detailed manner (Winslett, 1993). Because they are highly structured, relational data models may not be the best choice when data or relationships among data are expected to change rapidly. The power of a relational model is that the structure allows for complex, ad hoc queries using a formal data access language such as Structured Query Language (SQL). Relational database management systems also scale well and provide sophisticated methods for dealing with data versioning, transaction tracking, and security (Winslett, 1993).

In an object-oriented data model, information is organized as hierarchical sets of entities called objects (Winslett, 1993). Information is stored on each member of an object as well as the object class as a whole. Object-oriented data models (e.g., ACEDB) are very popular in the scientific community because of their flexibility and user-friendly graphical interfaces (Cherry and Cartinhour, 1994). A major criticism of object-oriented architectures is that they do not scale well. And while they are flexible, the flexibility often means that a user must have a complete understanding of the underlying organization of data in order to formulate a meaningful query.

Currently, many database designers are choosing to implement a relational database structure in order to have the query power inherent in that approach (Fields, 1992; Fasman, 1994; Keele et al., 1994). At the same time, object-oriented style software interfaces for these databases are being developed, which make the database easier to use than they are with a highly defined query language. It is likely that future database architectures will include hybrid data models that incorporate different aspects of the flatfile, relational, and object-oriented structures.

Construction and implementation of database designs require close coordination between computer programmers, computational biologists, and the primary biologists who will use the database. The people who will be using the database on a regular basis need to be fully included in the design discussions. A critical part of the process is to consider first the use and extent of the database, what kinds of information will be retrieved, and what kinds of questions will be asked. It is far easier to change paper drawings than it is to alter the database structure after it has been populated with data. First, implementations of a database design should be tested with a small dataset. Then all the participants can be polled again as to what they hope to recover from the database and whether the current design needs to be modified. The continued involvement of programmers and computational scientists as well as biologists in the process is important since both the knowledge about the biological information and the expectations of the database are sure to change over time. For the database to continue to be useful, it will need to change too.

LINKING DATABASES INTO FEDERATIONS

As the amount of available information escalates, the drudgery of recovering data relevant to a specific research question also increases. Some Internet sites hold copies of databases developed and updated at other research centers. The number of databases holding different subsets of information and the lengthy process of searching for, downloading, and merging information from different sources are a major concern for investigators attempting to synthesize large volumes of information. These problems have escalated enormously with the advent of the various genome sequencing projects now underway, which are submitting thousands of sequences to the molecular sequence databases each week.

Concerns about the usefulness of biological databases for answering complex biological questions resulted in several meetings during 1993 and 1994 (Blake et al., 1994; Waterman et al., 1994) to discuss database access and capabilities. The response to the problem has been the development of software tools and community agreements to accelerate the interconnection of biological databases using the capabilities of the Internet. The ideal database federation is one in which complex queries are broken down into components, which, in turn, are automatically routed to the appropriate database. The responses are subsequently merged into a single report and transmitted to the user. Prototypical software for achieving this level of interoperability is currently being tested, but it will not be widely implemented for several years. An example would be a query such as "Return all sequences of alpha tubulin that have been derived from gastropods collected in the Caribbean." If the appropriate databases were interconnected, the list for sequences derived from gastropods collected in the Caribbean would be gathered from a database such as SST and joined with a list of alpha tubulin sequences from a molecular sequence database. The summary list of alpha tubulin sequences derived from gastropods collected in the Caribbean would be used to pull the actual sequences from the molecular sequence database. These sequences, with requested annotation, would be sent to the requester.

The ability to link databases through the incorporation of linking fields and the shared understanding of semantics and syntax between the managers of the databases permit information to be stored in specialized databases maintained by experts. The value of the specialized databases is that the data are curated by the experts most familiar with and in closest contact with the information the database is holding. Many managers of biological databases are moving toward participating in database federations.

While complete interoperability is some years off, initial links between databases using WWW have been implemented successfully and demonstrate the promise of this approach. The development of the WWW and Gopher certainly stimulated interest in pursuing database interconnections, and the development of hotlinks between databases has led to the creation of defacto database federations. It is likely that in the future those databases that contain the same types of information may become tightly linked while others that don't directly overlap in the type of data represented will be linked with less rigor.

Technical Issues

Two immediate issues that confront database managers wanting to link to other databases are the generation of unique identifiers for primary entities and tracking of these unique identifiers during updating of the databases (versioning).

Accession Keys: Perhaps the most important issue is the designation of unique accession keys. The term "unique accession key" refers to the assignment of a permanent and unique designator for each entity in a database so that it can be retrieved or linked with other information unambiguously (i.e., to maintain referential integrity). The accession key is equivalent to a unique catalog number in a museum collection. The complexities surrounding unique accession keys include the following. First, the database must maintain a history of the use of the accession key even when the unique key for a given entity may change during updating or revision of the database. This is particularly important in a federated system since other databases may use an "old" accession key in their linking field and should not be presumed or obligated to continually update their database with another database's versioning. Second, there needs to be a convention by which databases within a federated system can be identified so that unique keys need only be maintained within a database. For example, the accession key 1543 may refer to entities in either of two databases. By adding the database acronym or another identifier (by convention), the key can be uniquely identified. For example, the M34769 accession from GenBank can be identified as GB:M34769.

Versioning: As researchers incorporate data downloaded from Internet resources in their research programs, it is necessary to document which version of the information is used in an analysis. The questions of which version of a database is being maintained at a given site and how updates are implemented become very important not only to be able to repeat the analysis described in a research project but also to maintain correct links between the data in the old version of the database and the data in the new version. This is particularly important as databases become linked together by incorporating accession numbers from one database into another database as the unit that supplies the link between the two databases. Linking via accession number is most often done without any communication or agreement between the two databases. So the integrity of the linkage between the two databases depends on the support of each database for each and every accession number it has ever assigned. Again, the analogy to museum collection data is apparent.

Community Issues

Internet access to database information results in many researchers noting that some information is missing from or needs to be linked with the database information. As well, errors in data entry may be noted. Thus the availability of the information creates the possibility of community data curation. This is not meant to assume that any but an author or designated representative would change or update an original data entry. Rather, members of the research community may want to note additional information, which could be linked to the original data, or connect the data entry with other studies with which it shares some component. Recognition of this exciting possibility has already generated some efforts of community curation such as data entry by authors of annotation for molecular sequences. Issues relating to community curation include (1) proper attribution for the information added, (2) mechanisms to ensure data integrity and security, (3) guidelines as to what data go public and when, and (4) local or global agreements concerning data access and distribution.

The TDB Example

The Institute for Genomic Research (TIGR) is a not-for-profit research institute dedicated to the nucleotide sequence-based characterization of gene expression, development, physiology, and the evolution of genes and organisms. Research at TIGR is supported by a large-scale high-throughput sequencing facility with extensive computational and genome informatics core facilities (Adams et al., 1994). As part of the TIGR information management system, several public databases have been developed to facilitate access to molecular information obtained and curated at TIGR. Two of these databases, the Expressed Gene Anatomy Database (EGAD) and the Sources, Sequences, Taxa (SST) database, provide an example of the potential for linked databases to provide rapid recovery of information to an interesting biological question. The TIGR databases are constructed using the commercial relational database system SYBASE (Sybase, Inc., Emeryville, California), which can be queried using Structured Query Language (SQL). SST and EGAD can be accessed over the Internet. The TIGR WWW URL is http://www.tigr.org; the anonymous FTP site address is ftp.tigr.org.

The Expressed Gene Anatomy Database supports dynamic curation of information related to protein expression and cellular role of the expressed gene (Adams et al., 1995). It links sequence information from public sequence databases with expression information derived from the TIGR human EST project (Adams et al., 1991). [ESTs, or Expressed Sequence Tags, are short DNA sequences derived from genes that are expressed (processed into proteins) in particular tissues or cells.]

Annotation relative to sequence structure, such as alternative splice coordinates and protein feature information, is incorporated into relevant tables and linked to isology (sequence similarity), cellular role, and phylogenetic tree tables.

The Sequences, Sources, Taxa database provides information about the source material for molecular sequences such as voucher specimen information or clone identity information. It records details about collection events and locations, provides taxonomic information, and provides data on the current location and contact person for the source material. It is directly linked to EGAD through sequence accession numbers or through gene names.

A particular feature of the EGAD database is that it holds a nonredundant set of human transcripts derived from the curation of human sequences from GenBank. This Human Transcript (HT) set provides a mechanism by which one can ask complex questions such as "What proportion of known human transcripts are involved in cellular signaling?" where the designation that the sequences are derived from human samples comes from a link to SST and the information about transcripts and cellular role comes from EGAD.

EGAD and SST are built using relational database architecture, they provide unique keys to the objects that they describe, and they can provide direct links to other databases. Using direct HTML links, sequences from non-TIGR, molecular sequence databases such as GSBD can be retrieved in response to questions such as "Show me all gene sequences that have been derived from toad (*Bufo*) samples." EGAD and SST together provide a model of how specialized databases can be linked to provide answers to interesting questions that span different domains of expertise.

SUMMARY

With new technologies and methods the pace of data acquisition only quickens. But simultaneously, now, there is a vastly increased effort to improve data integration mechanisms by creating database structures and software tools to support rapid access to and interactive use of molecular and related biological information. Biological databases existed long before the advent of computers and the Internet. We are just now, however, beginning to realize the capacity that computers give us to use the databases not just as archives, but as research tools. The future of computerized scientific databases will be in their ability to rapidly retrieve and manipulate data. Then the full value of the information they contain can be recognized as it is used to answer scientific inquiries.

ACKNOWLEDGMENTS

This work is supported in part by National Science Foundation grant DEB-9400861 to C. Bult, J. Blake, A. Kerlavage, and C. Fields. The authors thank J. Kelley, O. White, and A. Kerlavage for reviewing early drafts of the manuscript for this chapter.

REFERENCES

Adams MD, Kelley JM, Gocayne JD, Dubnick M, Polymeropoulos MH, Xiao H, Merril CR, Wu A, Olde B, Moreno RF, Kerlavage AR, McCombie WR, Venter JC (1991): Complementary DNA sequencing: expressed sequence tags and human genome project. *Science* 252:1651–1656.

Adams MD, Kerlavage AR, Kelley JM, Gocayne JD, Fields C, Fraser CM, Venter JC (1994): A model for high-throughput automated DNA sequencing and analysis core facilities. *Nature* 368:474–475.

Adams MD, Kerlavage AR, Fleischmann RD, Fuldner RA, Bult CJ, Lee NH, Kirkness EF, Weinstock KG, Gocayne JD, White O, Sutton G, Blake JA, Brandon RC, Chiu M-W, Clayton RA, Cline RT, Cotton MD, Earle-Hughes J, Fine LD, FitzGerald LM, FitzHugh WM, Fritchman JL, Geoghagen NSM, Glodek A, Gnehm CL, Hanna MC, Hedblom E, Hinkle PS, Jr, Kelley JM, Klimek KM, Kelley JC, Liu L-I, Marmaros SM, Merrick JM, Moreno-Palanques RM, McDonald LA, Nguyen DT, Pellegrino SM, Phillips CA, Ryder SE, Scott JL, Saudek DM, Shirley R, Small KV, Spriggs TA, Utterback TR, Weidman JF, Li Y, Bednarik DP, Cao L, Cepeda MA, Coleman TA, Collins EJ, Dimke D, Feng P, Ferrie A, Fischer C, Hastings GA, He W-W, Hu J-S, Greene JM, Gruber J, Hudson P, Kim A, Kozak DL, Kunsch C, Ji H, Li H, Meissner PS, Olsen H, Raymond L, Wei Y-F, Wing J, Xu C, Yu G-L, Ruben SM, Dillon PJ, Fannon MR, Rosen CA, Haseltine WA, Fields C, Fraser CM, and Venter JC (1995): Initial assessment of human gene diversity and expression patterns based upon 83 million nucleotides of cDNA sequence. *Nature,* 37 (Suppl) 3–174.

Blake JA, Bult CJ, Donoghue MJ, Humphries J, Fields C (1994): Interconnection of biological databases: a meeting report. *Syst Biol* 43:585–589.

Cherry JM, Cartinhour SW (1994): ACEDB: a tool for biological information. In: Adams MD, Fields C, Venter CJ (eds). *Automated DNA sequencing and analysis.* San Diego: Academic Press, pp 347–356.

Cuticchia AJ (1994): A relational database primer for molecular biologists. In: Adams MD, Fields C, Venter CJ (eds). *Automated DNA sequencing and analysis.* San Diego: Academic Press, pp 339–346.

Fasman K (1994): Restructuring the Genome Data Base: a model for a federation of biological databases. *J Comput Biol* 1:165–171.

Fields C (1992): Data exchange and inter-database communication in genome projects. *Trends Biotechnol* 10:58–61.

Keele JW, Wray JE, Behrens DW, Rohrer GA, Sunden SLF, Kappes SM, Bishop MD, Stone RT, Alexander LJ, Beattie CW (1994): A conceptual database model for genomic research. *J Comput Biol* 1:65–76.

Krol E (1992): *The whole internet.* Sebastopol: O'Reilly and Associates.

Lewis S (1994): Design issues in developing laboratory information management systems. In: Adams MD, Fields C, Venter CJ (eds). *Automated DNA sequencing and analysis.* San Diego: Academic Press, pp 329–338.

Schatz BR, Hardin JB (1994): NCSA, Mosaic and the World Wide Web: global hypermedia protocols for the Internet. *Science* 265:895–901.

Waterman M, Uberbacher E, Spengler S, Smith FR, Slezak T, Robbins R, Marr T, Kingsbury DT, Gilna R, Fields C, Fasman K, Davison D, Cinkosky M, Cartwright P, Branscomb E,

Berman H (1994): Genome Informatics I: Community Databases. *J Comput Biol* 1:173–190.

Winslett M (1993): New database technology for nontraditional applications. In: Fortuner R (ed). *Advances in computer methods for systematic biology: artificial intelligence, databases, computer vision.* Baltimore: The Johns Hopkins University Press, pp 257–273.

■■■■■■ CHAPTER 2

Parametric Bootstrapping in Molecular Phylogenetics: Applications and Performance

JOHN P. HUELSENBECK AND DAVID M. HILLIS

Department of Zoology, University of Texas, Austin, Texas 78712

ROBERT JONES

Darwin Molecular Corporation, Bothell, Washington 98021

CONTENTS

Molecular Zoology: Advances, Strategies, and Protocols, Edited by Joan D. Ferraris and Stephen R. Palumbi.
ISBN 0-471-14461-4 © 1996 Wiley-Liss, Inc.

SYNOPSIS

Parametric bootstrapping is a statistical tool for producing independent replicates of a study based on parameters estimated from a unique data set. There are at least four obvious uses for parametric bootstrapping in phylogenetic studies: (1) to examine a potential source of systematic bias (e.g., long branches or base composition); (2) to evaluate the relative support of two or more competing phylogenetic hypotheses; (3) to conduct power analyses, which estimate the length of sequence that will be needed to resolve a particular phylogenetic tree; and (4) to estimate the reliability of individual clades in a phylogenetic tree. Parametric bootstrapping is well suited for the first three of these applications, but it suffers some of the same biases that exist for nonparametric bootstrapping (and T-PTP tests) when applied to the fourth problem. However, in cases in which alternative phylogenetic hypotheses have been proposed, parametric bootstrapping may be the most appropriate approach to evaluating the respective support of empirical data.

INTRODUCTION

Simulation usually is used prospectively in phylogenetic analysis. That is, the performance of phylogenetic methods is examined using simulated data with the assumption that these simulations give an idea of the general performance of the methods. Although the recent literature is full of studies that examine the statistical properties of different phylogenetic methods (see reviews by Hillis, 1995; and Huelsenbeck, 1995), these studies do not necessarily give an accurate idea of how a method will perform for any particular data set. Any particular data set has peculiarities unique to itself (e.g., an underlying tree with unique branch lengths) that have not been explored by any simulation study. An underutilized method applies simulation retrospectively; that is, the behavior of a phylogenetic method for a particular data set is explored using simulation. This application of simulation has been called the parametric bootstrap method (Efron, 1985) and may be "one of the best uses of simulation" (Felsenstein, 1988). In this chapter, we describe some applications of the parametric bootstrap in phylogenetics.

WHAT IS THE PARAMETRIC BOOTSTRAP?

A basic problem faced by many statistical phylogenetic methods is how to construct replicate data sets. Once new sample data sets have been constructed, the variability of trees (or other parameters of interest) estimated from the replicates can be assessed without difficulty. Bootstrapping refers to a general set of methods designed to create replicate data sets based on the original data. The nonparametric bootstrap creates replicate data sets (= pseudoreplicates) by randomly sampling the characters of the original character matrix with replacement, such that new character matrices of the same size are created (Efron, 1979, 1982; Felsenstein, 1985). The nonparametric bootstrap commonly is referred to as the "bootstrap" method in systematics (Felsenstein, 1985).

The parametric bootstrap, on the other hand, creates replicates using numerical simulation (Efron, 1985; Felsenstein, 1988; see also Gouy and Li, 1989; and Bull et al., 1993). In general, the parametric bootstrap works as follows: (1) assume a model of evolution; (2) estimate the parameters of the model from the data (e.g., tree topology, branch lengths, transition:transversion bias); (3) simulate new character matrices of the same size assuming this parameterized model; and (4) analyze the replicate character matrices using the method of interest. This protocol will be discussed in more detail below.

One use of parametric bootstrapping is to explore the possibility that an analysis has "branch-length problems." If the true tree has two very long branches (a long branch is one with many mutations) separated by several shorter branches, many phylogenetic methods will converge to a tree in which the long branches are put together (Felsenstein, 1978). Parametric bootstrapping can examine whether long branches on an estimated tree are long enough to attract in a phylogenetic analysis; thus parametric bootstrapping can rule out branch attraction as a possible problem. Another use of parametric bootstrapping is the extension of data to new designs (Bull et al., 1993). For example, the effect that certain evolutionary parameters have on phylogenetic estimation for a particular data set can be explored by systematically varying that parameter (e.g., to explore the effect of transition:transversion bias on phylogenetic estimation, the transition:transversion bias can be varied for the simulated data sets). This use of the parametric bootstrap allows the robustness of phylogenetic estimation to be examined. The parametric bootstrap also has uses for determining the null distribution of test statistics in phylogenetic hypothesis testing. Although simulation of null distributions can be computationally intensive, the advantage is that some of the assumptions that go into parametric null distributions, such as the χ^2 distribution, can be avoided.

Parametric bootstrapping also can be used to conduct power tests for phylogenetic studies (e.g., Hillis et al., 1994a). Many phylogenetic studies do not examine enough data to provide definitive resolution of a particular group. Parametric bootstrapping can be used to estimate how much additional data would be needed to achieve a given level of resolution. For some problems, parametric bootstrapping may indicate that more sequence data are needed than exist in the genomes of the organisms under study (see Hillis et al., 1994b). Thus the parametric bootstrap may provide a means for identifying unsolvable problems before large amounts of time and money are wasted attempting an answer. On a more positive note, parametric bootstrapping can be useful for planning a sequencing study, to ensure that a reasonable number of data are collected to address a given problem.

Several of the potential uses of the parametric bootstrap will be explored in more detail below for DNA sequence data.

SIMULATING DATA IN PHYLOGENETIC ANALYSIS
OF DNA SEQUENCES

A phylogenetic model consists of three parts: a tree, the lengths of the branches of the tree, and a stochastic model of DNA substitution. In order to simulate a data set, all three must be specified. In this section, we outline how reasonable choices can be made about what tree to simulate, branch lengths to use, and model of DNA substitution to assume.

Choosing a Model Tree

The problem of interest dictates which tree to assume. For example, to test the adequacy of a model of DNA substitution, such as the Jukes–Cantor model, the maximum likelihood tree under that model would be simulated (Goldman, 1993). Similarly, if one wanted to test whether a more complicated model provided a significant increase in the likelihood of a tree, the maximum likelihood tree under the simpler model would be simulated (Goldman, 1993). For other problems, simulating suboptimal trees is most appropriate. Consider a hypothetical case in which the estimated tree has two very long branches that are adjacent on the tree. One possible explanation for such a pattern is that convergence along the long branches has been misinterpreted as phylogenetic signal (Felsenstein, 1978). One possible test of branch attraction is to simulate model trees in which the two long branches are separated. If the two branches come together in a large portion of the analyses of the simulated data set, then branch attraction may be a problem.

Choosing a Model of DNA Substitution

All current models of DNA substitution are specific examples of what are called Markov chains. With Markov chain models, the state in the next instance of time depends only on the current state of the system. What state the system was in one day ago or 1,000,000 days ago has no influence. At the heart of Markov chains is a matrix of instantaneous rates of change. For example, for the Jukes–Cantor model of DNA substitution (Jukes and Cantor, 1969), the instantaneous matrix has the following form:

$$\mathbf{Q} = \begin{bmatrix} -3\alpha & \alpha & \alpha & \alpha \\ \alpha & -3\alpha & \alpha & \alpha \\ \alpha & \alpha & -3\alpha & \alpha \\ \alpha & \alpha & \alpha & -3\alpha \end{bmatrix} \tag{1}$$

where α is the instantaneous rate of change from one nucleotide to another. Entries in the columns and rows are in the order A, C, G, T. The first row, for example, gives the rates of change from nucleotide A to nucleotides A, C, G, and T, the second row the rate of change from nucleotide C to nucleotides A, C, G, and T, and so on. For reasons that will not be discussed here, the rows of the matrix must sum to zero, hence the "-3α" along the diagonals. To calculate the probability of a change from one nucleotide to another over an arbitrary amount of time, the following matrix operation is performed:

$$\mathbf{P}_t = e^{Qt} \tag{2}$$

$\mathbf{P}_t$ is a matrix that describes the probabilities of changing from one nucleotide state to another over a specified period of time (t).

For the matrix $\mathbf{Q}$, shown above for the Jukes–Cantor model, the transition probabilities are

$$\mathbf{P}_t = \begin{bmatrix} \dfrac{1 + 3e^{-4\alpha t}}{4} & \dfrac{1 - e^{-4\alpha t}}{4} & \dfrac{1 - e^{-4\alpha t}}{4} & \dfrac{1 - e^{-4\alpha t}}{4} \\ \dfrac{1 - e^{-4\alpha t}}{4} & \dfrac{1 + 3e^{-4\alpha t}}{4} & \dfrac{1 - e^{-4\alpha t}}{4} & \dfrac{1 - e^{-4\alpha t}}{4} \\ \dfrac{1 - e^{-4\alpha t}}{4} & \dfrac{1 - e^{-4\alpha t}}{4} & \dfrac{1 + 3e^{-4\alpha t}}{4} & \dfrac{1 - e^{-4\alpha t}}{4} \\ \dfrac{1 - e^{-4\alpha t}}{4} & \dfrac{1 - e^{-4\alpha t}}{4} & \dfrac{1 - e^{-4\alpha t}}{4} & \dfrac{1 + 3e^{-4\alpha t}}{4} \end{bmatrix} \tag{3}$$

Once again, the entries of the rows and columns are in the order A, C, G, and T. The Jukes–Cantor model of DNA substitution is the simplest of the models that are available. Other models of DNA substitution add more parameters to make the model more biologically realistic. For example, the matrix of instantaneous rates of change for the Kimura (1980) model is

$$\mathbf{Q} = \begin{bmatrix} -2\alpha - \beta & \alpha & \beta & \alpha \\ \alpha & -2\alpha - \beta & \alpha & \beta \\ \beta & \alpha & -2\alpha - \beta & \alpha \\ \alpha & \beta & \alpha & -2\alpha - \beta \end{bmatrix} \tag{4}$$

where α is the rate of transversions and β is the rate of transitions. The Kimura model relaxes the constraint that all possible nucleotide changes occur at the same rate by allowing transitions and transversions to have different rates. More parameters can be added to make the models more realistic. For example, the Hasegawa–Kishino–Yano model (1985) allows for different equilibrium nucleotide frequencies but is otherwise the same as the Kimura model. Also, the assumption of identically distributed rates of change for all sites can be relaxed by assuming that the rates of change follow a gamma distribution (Yang, 1993). Other models allow many more rate parameters in the $\mathbf{Q}$ matrix (Yang, 1994) or permit nonhomogeneous change across the tree (i.e., equilibrium nucleotide frequencies or rate parameters may change from branch to branch) (Yang and Roberts, 1995).

How does one choose among all the models of DNA substitution that are available? Goldman (1993) described likelihood ratio tests that allow (1) testing of the adequacy of the model of DNA substitution and (2) testing of one model of DNA substitution against another model. The likelihood ratio test statistic is

$$\delta = 2(\ln L_1 - \ln L_0) \tag{5}$$

where $\ln L_1$ is the log likelihood under the more parameter-rich model of DNA substitution and $\ln L_0$ is the log likelihood under the simpler model of DNA substitution. This number will be greater than 0 because the likelihood under the more parameter-rich model will always be as good as or better than the simpler model. Goldman (1993) suggested that the null distribution for the test statistic δ be determined using simulation (= parametric bootstrapping). In this case, the tree and other parameters are estimated under the null hypothesis that the simpler model of DNA substitution is correct. Many data sets of the same size as the original are simulated using these parameters estimates under the simple model of DNA substitution. The maximum likelihood is calculated for each simulated data set under the simple and parameter-rich model of DNA substitution and the difference between the log likelihoods under each model is noted. If the difference for the original data is greater than 95% of the differences calculated from the simulated data, then the simpler model of DNA substitution is rejected.

Usually, one is interested in whether a particular parameter that can be added to a model provides a significant improvement in the likelihood. For example, a transition:transversion bias is one parameter that is often assumed by models implemented in likelihood. However, it is possible that such a parameter is unnecessary (i.e., the parameter does not provide a statistically significant increase in the probability of observing the data). This can be examined by testing a Kimura model against the Jukes–Cantor model (or the HKY model against the Felsenstein, 1981, model). In this case, the log likelihood under the Kimura model is $\ln L_1$ and the log likelihood under the Jukes–Cantor model is $\ln L_0$. However, Goldman (1993) also devised a test of the adequacy of the model of DNA substitution. He suggested that the log likelihood under the multinomial distribution ($\ln L_1$) be tested against the model of DNA substitution of interest ($\ln L_0$). The expectation for this test is that the model of DNA substitution is rejected as being inadequate. This result should be taken with a grain of salt, however, because we know *a priori* that all of our models are false, at least in the details (i.e., make some unrealistic assumptions). The best we can do is choose the model that best captures the important aspects of the substitution process.

Estimating Branch Lengths and Other Parameters

Minimally, the parameters that should be estimated include the branch lengths of the tree. If more complicated models of DNA substitution are assumed, then other parameters, such as the transition:transversion bias, equilibrium nucleotide frequencies, or shape parameter for the gamma distribution, also must be estimated. Maximum likelihood is a statistically sound method for estimating these parameters. The maximum likelihood method finds those parameter values that maximize the probability of observing the data. Several programs are available to the systematist that provide maximum likelihood estimates of parameters. These programs include DNAML (Felsenstein, 1993), fastDNAML, and PAUP* (Swofford, 1995).

APPLYING THE PARAMETRIC BOOTSTRAP

Examining Potential Sources of Phylogenetic Bias

Imagine an investigator has just collected data to test a hypothesis of relationships that has been suggested in the literature. The data seem to strongly reject the previously suggested hypothesis, but there are still two possibilities: (1) the original hypothesis is incorrect; or (2) the assumptions about the data are incorrect and have biased the interpretation. For instance, perhaps long branches produced a region of inconsistency, or perhaps base-composition bias is being misinterpreted as phylogenetic signal. How can these possibilities be discounted?

One potential source of phylogenetic bias results from attraction of long branches in phylogenetic trees. The combination of evolutionary parameters for which a method will provide inconsistent estimates of phylogeny (converge to the wrong tree as more data are added) has been termed the Felsenstein Zone. Felsenstein (1978) first pointed out that the parsimony method will converge to a phylogenetic estimate in which long branches are linked together when, in reality, the long branches are separated by very short branches, hence the maxim that "long branches attract." But how can we know if the branch attraction problem applies to a particular data set of interest? Parametric bootstrapping can examine whether the branches are long enough to attract for a particular example. By varying the assumed tree (the tree underlying the simulations, or the model tree), problems of branch attraction can be addressed for a particular data set. The object is to assume trees in which the long branches of the tree are separated, assume a model of DNA substitution, estimate the parameters of the model (branch lengths, transition:transversion bias, etc.) using likelihood, simulate many data sets of the same size as the original, and analyze the simulated data sets using parsimony (or whatever method was originally used in the analysis). If the long branches are repeatedly grouped together in analysis of simulated data sets even though the branches were originally separated in the model (simulated) trees, branch attraction may be indicated. Alternatively, if the model tree is consistently found in analysis of the replicate data sets, then branch attraction is probably not a problem, at least for the model tree(s) tested. The parametric bootstrap cannot definitively indicate that the results were caused by branch attraction, although it is possible to rule out branch attraction in many cases.

As an example of the protocol described above, consider the problem of amniote phylogeny. The phylogeny of amniotes has been controversial because different data sets provide different estimates of phylogeny. One data set (18S rRNA) suggests a sister-group relationship between birds and mammals, whereas other data sets (e.g., most other genes, as well as analyses that include certain fossil taxa) suggest that

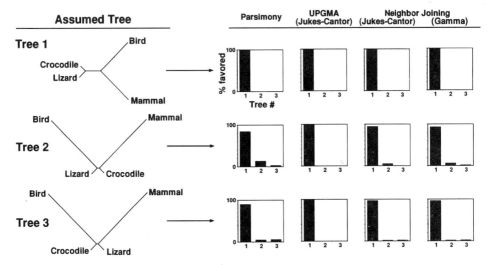

Figure 1. A parametric bootstrap analysis designed to explore long-branch attraction as an explanation for the bird–mammal grouping (tree 1) suggested by the 18S rRNA gene. All possible unrooted trees were assumed and the parameters (e.g., branch lengths, shape parameter of the gamma distribution, transition:transversion bias) for each tree were estimated using maximum likelihood. Many (1000) data sets of the same size as the original 18S rRNA gene were simulated for each model tree. The simulated data were analyzed using parsimony, UPGMA, and neighbor joining, with the distance measures indicated. The boxes show the percentage of the time the various trees were estimated using each phylogenetic method. Tree 1 (the tree for which the two long branches are linked together) was estimated a high proportion of the time regardless of which tree was used to simulate data, or which method was used to analyze the data.

birds are more closely related to crocodilians (Gauthier et al., 1988; Eernisse and Kluge, 1993; Hedges, 1994). Currently, the bird–crocodilian phylogeny (tree 3 of Fig. 1) is favored; the bird–crocodilian phylogeny is more consistent with the appearance of fossil taxa in the rock record (Gauthier et al., 1988) and most of the independent data sets support this phylogeny (Hedges, 1994). However, the question remains why some genes, especially the 18S rRNA gene, strongly support a bird–mammal phylogeny (tree 1 of Fig. 1).

One possible explanation for the bird–mammal grouping favored by the 18S rRNA data is long-branch attraction. If one considers the most parsimonious tree for the 18S rRNA data set, the branches leading to both the birds and mammals are very long and linked together in the phylogeny (tree 1, Fig. 1). Can this grouping be explained as an artifact of the long branches? To explore this possibility, we conducted a parametric bootstrap analysis of the four taxa (represented by sequences from an alligator, a lizard, a mammal, and a bird). We considered all three unrooted trees as model trees and used a parameter-rich model of DNA substitution (the HKY-Γ model, Hasegawa et al., 1985; Yang, 1993; see the end of this chapter for information about the program used for this analysis). The branch lengths and other parameters were estimated using maximum likelihood for each possible tree. The relative branch lengths and trees examined using this protocol are shown in Figure 1. For each model tree, 1000 data sets of the same size as the original 18S rRNA data were generated using simulation. These simulated data sets were then analyzed using parsimony, UPGMA, and the neighbor joining method. The distance methods were based on Jukes–Cantor and gamma distances, as indicated in Figure 1.

The results from these analyses of the simulated data are shown in the boxes of Figure 1. Interestingly, regardless of the starting tree (the tree underlying the simulations), tree 1 — the tree linking birds and mammals — is estimated a high proportion of the time. When tree 1 is in fact the true tree, all methods correctly estimate tree 1 in analyses of the simulated data. However, when the true tree is tree 2 or tree 3, tree 1 is estimated 85–95% of the time. This suggests that branch-length attraction could be a problem in this case, because none of the methods are able to distinguish among the three possible trees. No matter which tree is correct, the long branches of the bird and mammal lineages appear to lead the analysis to the bird–mammal tree. The analysis also indicates that long-branch attraction problems are not limited to parsimony analyses but affect corrected distance methods as well.

As another example, consider the controversy over bat monophyly. Pettigrew (1986, 1991) has argued, primarily on the basis of neuroanatomy, that the "megabats" (flying foxes) are more closely related to primates than they are to the "microbats" (the smaller, echo-locating bats). This hypothesis (called "the flying primate" hypothesis, because it suggests that megabats are essentially flying primates) stimulated a series of molecular studies on bat relationships (Bennet et al., 1988; Adkins and Honeycutt, 1991; Mindell et al., 1991; Ammerman and Hillis, 1992; Bailey et al., 1992; Stanhope et al., 1992). All these studies supported the monophyly of bats and, taken collectively, would seem to indicate strong and consistent rejection of Pettigrew's flying primate hypothesis. But Pettigrew (1994) has since argued that all these studies suffer from the same bias: namely, that megabats and microbats have a base compositional bias toward adenines and thymines. Pettigrew (1994) argued that this base compositional bias results from the higher metabolic rate that is required by flying, which requires higher cytosolic ATP concentrations, which biases the nucleotide precursor pool used for DNA repair and replication. Thus Pettigrew (1994:279) dismissed the six independent molecular studies with the simple statement that "these studies merely confirm what was already known—that bats share an AT bias in their DNA sequences."

Clearly, if Pettigrew is correct that bats share an AT bias of independent origins, then this could potentially bias the studies. But are the data consistent with this "flying DNA" hypothesis? This presents an ideal case for parametric bootstrapping. Van Den Bussche et al. (1996) estimated the best tree that supported Pettigrew's flying primate hypothesis (i.e., the best solution that separated the microbats and megabats and put the latter with the primates) and then varied base composition in the two bat lineages in simulated replicate data sets. They found that, even at 100% AT bias in the two bat lineages (with all other lineages at equal base composition), Pettigrew's explanation of the results was not supported. They also showed that many of the genes that have been examined show no base composition artifacts along the lines suggested by Pettigrew, and that where such biases do exist, they are much too small to account for the support for bat monophyly. This does not prove that bats are monophyletic, but it does show that Pettigrew's explanation of the results is unsupported. Thus one is left with strong support from several independent mitochondrial and nuclear genes that bats really are monophyletic.

Evaluating Alternative Phylogenetic Hypotheses

Parametric bootstrapping also can be used to test the relative support of data for two competing hypotheses. In the case presented above, Van Den Bussche et al. also looked at the relative difference between the optimality scores (tree lengths and maximum likelihood scores) for the alternative hypotheses (bat monophyly versus flying primate). For the data set examined, the bat monophyly hypothesis required a tree 23 steps shorter than the best tree required by the flying primate hypothesis. Would we expect a difference this large if the flying primate hypothesis were true? Obviously, even if the flying primate tree were true, stochastic variation (or some systematic bias generated by the tree topology) might lead us to find shorter solutions, at least on occasion. Parametric bootstrapping provided the means for generating the expected distribution of tree length differences. In 100 simulations (parametric bootstrap replicates), the optimal tree was always within 3 steps of the model (flying primate) tree, so the observed 23-step difference between the bat monophyly tree and the best flying primate tree can be taken as highly significant.

When Is the Data Set Big Enough?

A power analysis usually indicates the degree to which a given experimental design has the potential to discriminate among alternative hypotheses. In the context of estimating phylogenetic history, a power analysis estimates the number of characters (evolving under a given rate and model of evolution) that are needed to resolve a tree at a given frequency. In practice, parametric bootstrapping can be used in an early stage of a phylogenetic study to estimate the size of the data set needed, especially in the context of sequence studies. Of course, power analyses typically assume that new data collected will be roughly equivalent in rate of evolution to the original data collected, so the estimates must be taken as only a rough approximation. Nonetheless, power analyses can be extremely helpful in planning the extent or evaluating the feasibility of a study, as well as for evaluating the relative efficiency of competing methods of phylogenetic analysis for a specific problem.

Examples of simple power analyses are shown in Figures 2 and 3. In Figure 2, a simple four-taxon tree with very long branches of equal length is evaluated. Under these conditions, several commonly used phylogenetic methods differ dramatically in their efficiency of recovery of the correct tree. Thus, if the first 100 nucleotides collected were used to estimate the tree shown, this analysis could be used to predict that approximately 200 total nucleotides would be sufficient to resolve the tree nearly 100% of the time, as long as the weighted parsimony or maximum likelihood methods were chosen to analyze the data. Obviously, under these conditions, these methods would be preferable to an unweighted parsimony analysis (which would

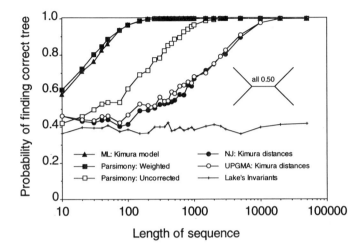

Figure 2. The accuracy of several phylogenetic methods for a simple four-taxon tree when all branches are equal in length and very long. All methods are consistent (converge to the correct tree) but differ dramatically in their efficiency.

require about 10 times as many data to reach the same level of resolution), neighbor-joining or UPGMA with Kimura distances (which would require about 100 times as many data), or Lake's method of invariants (which would require about 10 million times as many data; result is off scale of graph). However, a different initial estimate could produce markedly different results, as illustrated in Figure 3. Three conclusions have changed. First, with unequal rates of evolution across the different

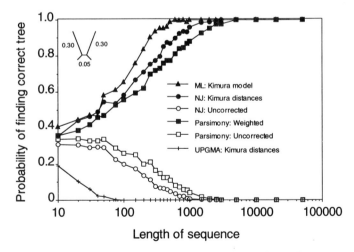

Figure 3. The accuracy of several phylogenetic methods for a simple four-taxon tree when the internal branch and two opposing peripheral branches of the tree are very short and the remaining two peripheral branches are very long. Several phylogenetic methods are inconsistent (converge to the wrong tree as more data are added) for the conditions simulated here.

branches in the tree, all the methods require more data to find the correct solution. Second, under these conditions, the most efficient method in Figure 2 (weighted parsimony) is now the third-most efficient method, and it has been surpassed by neighbor-joining with Kimura distances (which was relatively inefficient for the conditions in Fig. 2). And third, some methods are inappropriate for this problem, since they converge on the incorrect solution as more data are collected (i.e., they are inconsistent).

Parametric Bootstrapping and the Reliability of Individual Clades

The three previous applications of parametric bootstrapping are fairly straightforward. As long as a model of character evolution can be formulated that captures the relevant features of the substitution process, parametric bootstrapping can be used to generate independent replicate data sets for analysis, which can be used to compare alternative hypotheses or conduct power analyses. However, most methods for examining phylogenetic reliability focus on evaluating clades in the optimal tree, without reference to any alternative trees or sample sizes. The most common method for analyzing trees in this manner is nonparametric bootstrapping, which is known to provide biased (generally conservative) estimates of phylogenetic accuracy (Zharkikh and Li, 1992a, 1992b; Hillis and Bull, 1993). In this section, we explore the possibility of using parametric bootstrapping for this purpose.

Systematists have devised numerous methods to estimate the reliability of phylogenetic trees. In general, these methods can be divided into six classes with the following purposes: (1) to evaluate trees estimated from parsimony methods (Cavender, 1978, 1981; Felsenstein, 1985; Sneath, 1986; Templeton, 1983), (2) to test distance-derived trees (Barry and Hartigan, 1987; Felsenstein, 1984, 1986; Hasegawa et al., 1984, 1985; Nei et al., 1985; Rohlf and Sokal, 1981; Sarich and Wilson, 1967a, 1967b; Templeton, 1985), (3) to construct confidence limits using likelihood (reviewed in Felsenstein, 1988; Kishino and Hasegawa, 1989), (4) to evaluate invariants (Cavender and Felsenstein, 1987; Lake, 1987a, 1987b), (5) to resample the original character matrix according to some prescribed scheme (Felsenstein, 1985, 1988; Lanyon, 1985; Mueller and Ayala, 1982; Penny and Hendy, 1986; Wu, 1986), or (6) to incorporate randomization of the original character matrix (Archie, 1989; Faith, 1991; Faith and Cranston, 1991; Hillis, 1991; Steel et al., 1993). A few of these methods have been discussed in terms of their assumptions (Carpenter, 1992; Felsenstein, 1988; Källersjö et al., 1992) and in terms of their presumed power (Faith, 1991; Felsenstein, 1985, 1988; Fitch, 1986; Ruvolo and Smith, 1986; Saitou, 1986; Thomas et al., 1990). However, with but few exceptions (e.g., Felsenstein and Kishino, 1993; Hillis and Bull, 1993; Huelsenbeck, 1991; Zharkikh and Li, 1992a, 1992b), the performance of different methods for estimating reliability has not been tested. Furthermore, the performance of different methods for estimating the reliability of phylogenetic trees has never been directly compared. Here we compare the performance of parametric bootstrapping against three other methods for estimating the reliability of maximum parsimony trees: the *a priori* and *a posteriori* topology-dependent cladistic permutation tail

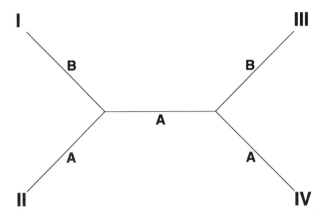

Figure 4. An unrooted four-taxon tree. The three-branch length refers to the length of the branches marked A in the figure. The two-branch length refers to the length of the branches marked B in the figure.

probability (T-PTP; Faith, 1991) tests and nonparametric bootstrapping (Felsenstein, 1985, 1988).

Methods. The general approach taken here was to construct a tree using computer simulation, estimate the maximum parsimony tree, and then estimate the reliability of the estimated tree using randomization (T-PTP test) and resampling (bootstrap) methods.

An unrooted four-taxon tree was used as the model tree in this study (Fig. 4). This tree has one internal branch and four peripheral branches. Because branch-length variation is a critical determinant for the performance of most methods of phylogenetic inference, and particularly for maximum parsimony (Felsenstein, 1978; Huelsenbeck and Hillis, 1993; Huelsenbeck, 1995), the branch lengths of the model tree were varied extensively. The length of a branch is here defined as the percentage of characters that are expected to differ between the two ends of the branch. Every simulated tree was constructed from two branch-length parameters. One parameter established the expected branch lengths of the internal branch and two opposing peripheral branches (branches A in Fig. 4, here called the "three-branch length"). The other parameter established the lengths of the remaining two opposing peripheral branches (branches B in Fig. 4, here called the "two-branch length"). The lengths of the constrained branches were varied in increments of 3%, from 3% to 75% expected internodal change. The parameter space examined in this study, then, included 625 combinations of branch lengths (Fig. 5). Some of these branch-length combinations are known to cause the parsimony method to fail consistently (Felsenstein, 1978).

A Jukes–Cantor model of molecular evolution was used as the mechanism of sequence change on the simulated phylogenies (Jukes and Cantor, 1969). The Jukes–Cantor model employs a parameter α that describes the rate at which a given

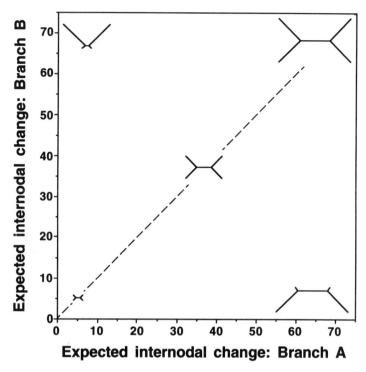

Figure 5. The simulations were performed under a wide variety of branch-length conditions. The three-branch length was plotted on the abscissa and the two-branch length was plotted on the ordinate. Different areas of the parameter space represent trees with different branch-length combinations. Change along the branches was varied from 3% expected change to 75% expected change in increments of 3%.

nucleotide changes to one of the three remaining nucleotides. The probability of observing a nucleotide substitution depends on the product of α and time (t). However, given enough time (or large α) there is a meaningful probability of back substitutions to the original nucleotide, in which case, no substitution will be observed. The substitution probabilities for the Jukes–Cantor model are given in equation 3.

For a four-taxon tree, $4^4 = 256$ combinations of nucleotides can occur at the tips of the tree. If one uses the notation of Lake (1987b), however, there are only 36 classes of nucleotide changes possible for a four-taxon tree. These classes are determined by pooling certain configurations of nucleotide assignments to the tips of the four-taxon tree. Any nucleotide for the first taxon (= taxon I or the reference taxon) is denoted by a "1." If the nucleotides for taxa II to IV are the same as the reference taxon, they are denoted "1"; if they differ by a transition, they are denoted "2"; and if they differ by a transversion, they are denoted "3" or "4." Of the 36 classes, maximum parsimony considers only six as informative: 1122, 1133, 1212, 1313, 1221, and 1331. Once branch lengths were specified, the probability of each of these 36 combinations was calculated for the model tree.

Simulated trees were then constructed in several steps. (1) Using the probabilities of each class of character change, 36 intervals between 0 and 1 were determined. (2) A pseudorandom number was chosen using a multiplicative linear congruential generator with multiplier 16,807 and prime modulus $2^{31} - 1$ (Park and Miller, 1988). This number was in the closed interval [0, 1]. (3) The pseudorandom number was used to determine which combination of nucleotides occurred at the nth site in the sequence string. (4) This process was repeated for all n sites. In this study, sequence length was standardized by the number of variable sites (i.e., for a simulation of 100 characters, sites were generated until 100 were variable). Invariant sites were recorded because of their effect on parametric bootstrapping.

Simulations were performed on a Connection Machine™ (CM-5). Twenty thousand independently constructed trees were examined for each of the 625 combinations of two-branch and three-branch lengths, resulting in a total of 1.25×10^7 trees.

As a first step, maximum parsimony was used to estimate the phylogeny from each set of simulated data (Fitch, 1971). The reliability of each estimated tree was then determined using the *a priori* and *a posteriori* T-PTP tests (Faith, 1991) and the parametric and nonparametric bootstrap (Felsenstein, 1985, 1988).

The *a priori* T-PTP test uses as a test statistic the difference in the length (in number of character steps) between the shortest tree in which a particular clade is monophyletic and the shortest tree in which a particular clade is nonmonophyletic. The null distribution of this test statistic is determined by randomly permuting the character states of each character. For example, if the states assigned to taxa I, II, III, and IV were originally G, A, A, T, respectively, one possible permutation of the character states is G, A, T, A. In this case, the assignment of states to taxa I and II did not change. Also note that the number of each character state for a character does not change; only the assignment of these character states to taxa changes. After all n sites have been permuted, the test statistic is calculated again. That is, the difference in the length between the shortest tree in which the clade of interest is monophyletic and the shortest tree in which the clade is nonmonophyletic is calculated for the randomized matrix. This difference is noted and the procedure is repeated many times. The proportion of time that the original length difference is exceeded or met in the randomizations represents the *a priori* T-PTP proportion.

The *a posteriori* T-PTP uses the same test statistic as the *a priori* T-PTP test but differs from the *a priori* T-PTP test in that it determines the null distribution of the test statistic differently. As with the *a priori* T-PTP test, the original character matrix is randomly permuted many times. However, with the *a posteriori* T-PTP test, one calculates the length difference of all clades of the same size as the clade of interest and picks the greatest length difference between the shortest tree in which the clade of interest is monophyletic and the shortest tree in which the clade is nonmonophyletic.

For an unrooted four-taxon tree, three clades were examined for each permuted matrix. The maximum difference in length was noted and the process was repeated many times. The proportion of the time that the original length difference is exceeded or met in the randomizations represented the *a posteriori* T-PTP proportion.

In nonparametric bootstrapping, the characters of the original character matrix are randomly sampled with replacement until a new replicate data set of the same dimensions as the original character matrix is constructed. The optimal tree(s) for this replicate data matrix is determined and the procedure is repeated many times. The proportion of the time that the original clade appears in analysis of the replicate data sets is often (misleadingly) called the bootstrap p-value (Hillis and Bull, 1993). In this chapter, this value is called the nonparametric bootstrap proportion.

The parametric bootstrap first assumes a model of sequence evolution and a model tree. The parameters of the model are estimated from the data. For example, for the Jukes–Cantor (1969) model, the length of the branches in terms of the product of mutation rate and time could be estimated using likelihood or a least-squares criterion. Replicate data sets are constructed by numerically simulating matrices of the same size as the original character matrix using the parameterized model of evolution. The optimal tree for each replicate data set is then calculated and the procedure repeated many times. The proportion of the time that the original clade appears in analysis of the replicate data sets is called the parametric bootstrap proportion in this chapter.

We performed 100 randomizations of the original character matrix and con- structed 100 nonparametric as well as 100 parametric replicate data sets of the original character matrix for each simulated tree. For the parametric bootstrap, we used a Jukes–Cantor (1969) model of sequence evolution and the estimated tree(s) as the model tree(s). The lengths of the branches were estimated using a least-squares criterion (Kidd and Sgaramella-Zonta, 1971).

Results. The incidence of rejecting or accepting the correct tree is plotted in three-dimensional graphs. Figure 6 depicts the axes of Figures 7 and 8. Figure 7 shows the incidence of rejecting the correct tree when the level of Type I error is set to 0.05 for the four methods examined in this study. The probability of rejecting the correct

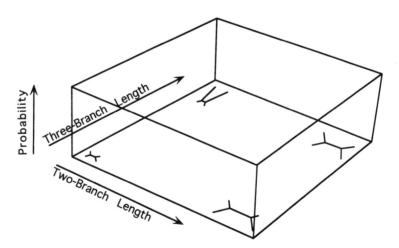

Figure 6. Key to the parameter space depicted in Figures 7 and 8. The parameter space is turned at an angle.

Nonparametric Bootstrap   **Parametric Bootstrap**

A Priori T-PTP   **A Posteriori T-PTP**

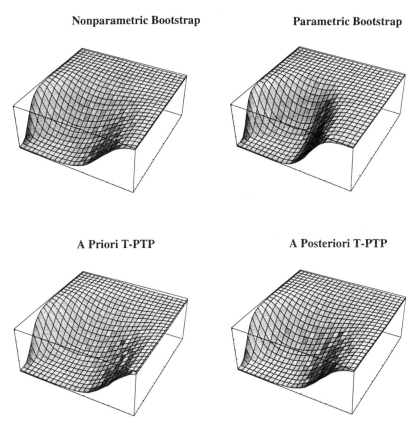

Figure 7. The incidence of rejecting the correct tree in different areas of the parameter space when the level of Type I error is set to 0.05. In general, the correct tree is rejected most often for high rates of evolution or under conditions in which the parsimony method produces inconsistent estimates of phylogeny.

tree differs in various parts of the parameter space. In general, the correct tree is rejected most often in areas where parsimony is inconsistent (the upper left region of the parameter space) and in areas of the parameter space in which rates of evolution are high. On the other hand, all methods have a very low incidence of accepting the incorrect tree over most of the parameter space (Fig. 8). However, in areas of the parameter space where parsimony is inconsistent, the incidence of accepting the incorrect tree is quite high. This means that not only is maximum parsimony converging on the wrong tree but that parsimony converges on the incorrect phylogeny with a high indication of reliability.

Although Figures 7 and 8 give a good indication of the failure of the methods in different areas of the parameter space for a given level of Type I error, they do not indicate what a given proportion means in terms of the probability of the clade being correct. Figure 9 shows the probability of correctly estimating the clade as a function of the proportion returned by different methods of estimating reliability for one

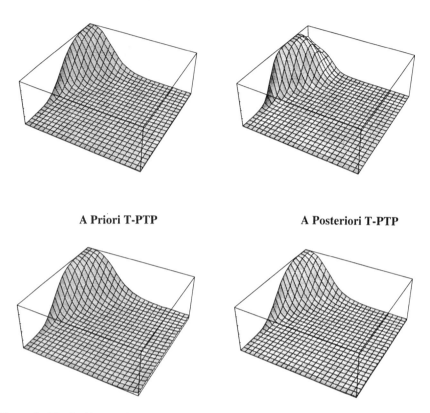

A Priori T-PTP   **A Posteriori T-PTP**

Figure 8. The incidence of accepting the incorrect tree for the four methods examined in this study. The level of Type I error is 0.05. The incorrect tree is accepted most often in areas of the parameter space in which the parsimony method is inconsistent.

point in the parameter space. Ideally, all simulated points in this graph would fall along the dashed diagonal line; if this were the case, the proportion returned by the methods would truly indicate the probability that a given group represents a true historical group. This, however, is not the case. Instead, most methods are highly conservative for proportional values above about 50% (the plotted lines for the methods fall above the dashed diagonal line) for this point in the parameter space. This means that the probability of a clade representing a true historical group is actually higher than the proportion returned by the method of estimating reliability (a point noted previously for nonparametric bootstrapping by Hillis and Bull, 1993). Conversely, for proportional values below about 50%, most methods are not conservative; for proportions below approximately 50%, the probability that the clade is correct is lower than the proportion.

The proportional values returned by the bootstrap and randomization methods differ in their meaning. For example, the *a posteriori* T-PTP test is the most

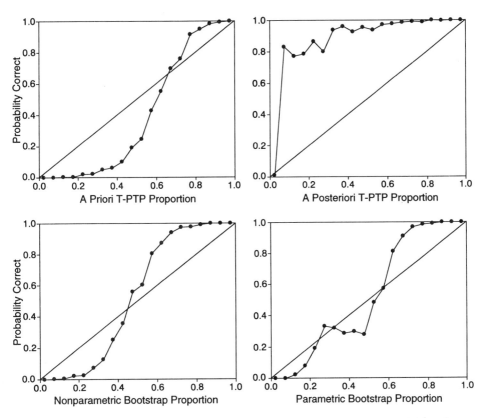

Figure 9. Relationship between the proportion returned by various methods and the probability of the clade representing a true historical group. The graph is based on 20,000 trees simulated with all branches equal in length (30% expected change).

conservative of all the methods for the conditions examined in Figure 9; for even very low *a posteriori* T-PTP proportion values, the probability that the clade represents an actual historical group is quite high (the probability is >75% for proportion values above 10%). The *a priori* T-PTP test is the least conservative of the methods examined. The probability of the clade representing a true historical group is much lower for the *a priori* T-PTP test than for the other methods. The performance of the bootstrap methods is intermediate to the *a posteriori* and *a priori* T-PTP tests. Furthermore, the nonparametric and parametric bootstrap methods are similar in their performance for proportional values above about 60% for the conditions examined in Figure 9. For values below about 60%, the bootstrap methods differ. However, this difference may be an artifact of how the performance of the parametric bootstrap was calculated in the simulations; for low parametric bootstrap proportions, an incorrect tree is often estimated using the parsimony method. The parameters for the simulation were then estimated from this incorrect tree, which appears to affect the performance of the parametric bootstrap. Figure 10 depicts the probability

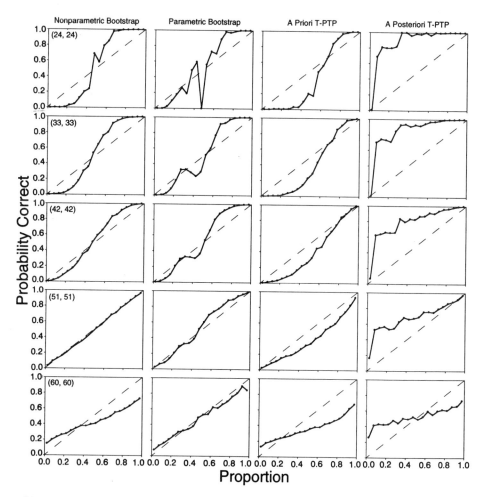

Figure 10. The probability of a clade representing a true historical group as a function of the proportion returned by a method. The graphs are from the dashed diagonal line of Figure 5. The branch-length conditions are expressed in the parentheses next to each row of graphs.

of correctly estimating a clade as a function of the proportion value returned by methods for simulations of equal branch lengths and from low to high rates.

For the conditions examined in Figure 9, the maximum parsimony method is expected to converge on the correct tree given sequences of sufficient length. Figure 11 shows the performance of the bootstrap and randomization methods under conditions in which parsimony is inconsistent (i.e., maximum parsimony is expected to converge on an incorrect tree given sequences of sufficient length). In Figure 11 the probability of correctly estimating the tree is plotted as a function of bootstrap or T-PTP proportions. When parsimony is strongly converging on an incorrect tree, the highest proportion of values returned by methods for estimating reliability are

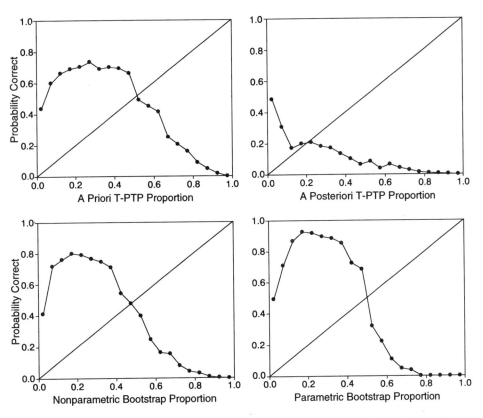

Figure 11. Relationship between the proportion returned by various methods and the probability of the clade representing a true historical group. The graph is based on 20,000 trees simulated under conditions in which the parsimony method converges on an incorrect estimate of phylogeny (three-branch length = 3%; two-branch length = 42%).

for an incorrect tree. Typically, the probability of correctly estimating the tree for proportional values of 0 is 50%. The probability of correctly estimating the tree increases above 50% for intermediate proportions and then the probability decreases to about 0 for high proportion values.

For the conditions simulated in Figure 11, the bootstrap methods and the *a priori* T-PTP test perform very similarly. However, the probability of correctly estimating a clade for a given proportion is much lower for the *a posteriori* T-PTP test than for the other methods examined. This means that under evolutionary conditions in which parsimony is inconsistent, the *a posteriori* T-PTP will underestimate the probability of a clade representing a true historical group compared with the other methods. Figure 12 shows the performance of the bootstrap and randomization methods taken along a transect of Figure 5 [two-branch and three-branch lengths are from (three-branch length, two-branch length): (3, 3) to (3, 75)].

How does this relationship between the probability of correctly estimating a

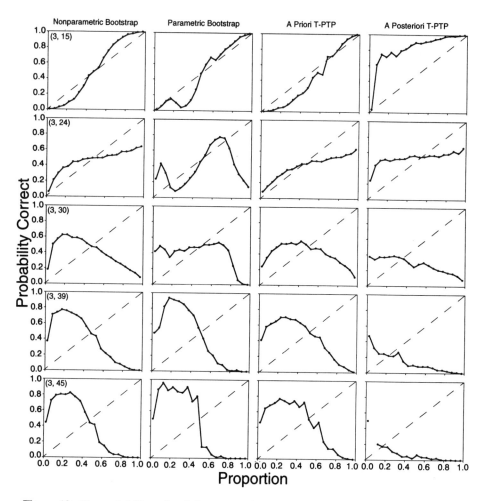

Figure 12. The probability of a clade representing a true historical group as a function of the proportion returned by a method. The graphs explore the behavior of the methods when the parsimony method is expected to converge on an incorrect estimate of phylogeny. The branch-length conditions are expressed in the parentheses next to each row of graphs.

clade for a given proportion change throughout the parameter space? Figure 13 shows the average deviation from the dashed diagonal line of Figure 9 for the parameter space examined. The deviation is plotted only for proportions greater than 90%. The deviation is smallest for areas in which rates of change are moderate and becomes greatest in areas of the parameter space in which the parsimony method is inconsistent. All the methods can be expected to be quite conservative for proportions above 90% over most of the parameter space. That is, in most areas of the parameter space, the probability of correctly estimating the clade is higher than the bootstrap or T-PTP proportions.

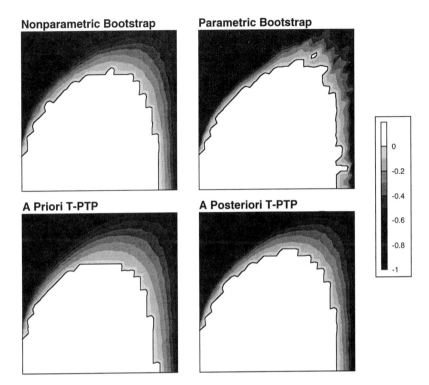

Figure 13. The maximum deviation above the diagonal of Figures 8 and 10 for proportions above 90%. White areas of the graphs represent areas in which the proportions represent conservative estimates of the probability of a clade being correct. Gray and black areas represent situations in which the probability of the clade being correct is actually lower than the proportion value.

CONCLUSIONS

The chief advantage of parametric bootstrapping over nonparametric bootstrapping is that the parametric approach generates independent stochastic replicates, whereas any bias present in the original data set is preserved in the "pseudoreplicates" generated by the nonparametric approach. The chief disadvantages of parametric bootstrapping are the dependence on a detailed model of character evolution and the greater computational requirements. Nonparametric bootstrapping is widely used because it provides a relatively simple way of evaluating relative support for each of the branches in an estimated tree without making special assumptions about the distribution of the data. Despite the fact that the numerical values of bootstrap proportions have no simple interpretation, they at least represent conservative estimates of phylogenetic accuracy under a wide range of conditions (Hillis and Bull, 1993). Our results here indicate that parametric bootstrapping also provides biased estimates when used to assess accuracy of individual branches on a tree, although perhaps to a lesser degree than some other methods (especially T-PTP tests). However, we believe the method has greater value in some other applications, such as evaluating the possibility of method inconsistency, comparing the relative support for alternative hypotheses, or in conducting power analyses for predicting the size of the data set that needs to be collected to resolve a given problem. Perhaps by framing questions of phylogenetic reliability in these latter contexts, we can avoid the problems of bias associated with estimating phylogenetic reliability of individual clades in a phylogenetic estimate.

COMPUTER PROGRAM

A program, written in C, is available on request from JPH. The program simulates nucleotide data for user-defined trees under the Hasegawa–Kishino–Yano model of DNA substitution with or without rate heterogeneity.

ACKNOWLEDGMENTS

We thank Danny Hillis and Thinking Machines Corporation for access to the Connection Machine™ (CM-5), which was used for many of the simulations. Jim Bull, Cliff Cunningham, Daniel Faith, and Jim Archie provided helpful discussions. This work was supported by NSF grant DEB-9221052 awarded to David Hillis.

REFERENCES

Adkins RM, Honeycutt RL (1991): Molecular phylogeny of the superorder Archonta. *Proc Natl Acad Sci USA* 88:10317–10321.

Ammerman LK, Hillis DM (1992): A molecular test of bat relationships: monophyly or diphyly? *Syst Biol* 41:222–232.

Archie J (1989): A randomization test for phylogenetic information in systematic data. *Syst Zool* 38:239–252.

Barry D, Hartigan JA (1987): Statistical analysis of hominoid molecular evolution. *Stat Sci* 2:191–210.

Bailey WJ, Slightom JL, Goodman M (1992): Rejection of the "flying primate" hypothesis by phylogenetic evidence from the e-globin gene. *Science* 256:86–89.

Bennet S, Alexander LJ, Crozier RH, MacKinlay AG (1988): Are megabats flying primates? Contrary evidence from a mitochondrial DNA sequence. *Aust J Biol Sci* 41:327–332.

Bull JJ, Cunningham CW, Molineux IJ, Badgett MR, Hillis DM (1993): Experimental molecular evolution of bacteriophage T7. *Evolution* 47:993–1007.

Carpenter JM (1992): Random cladistics. *Cladistics* 8:147–154.

Cavender JA (1978): Taxonomy with confidence. *Math Biosci* 40:271–280.

Cavender JA (1981): Tests of phylogenetic alternatives under generalized models. *Math Biosci* 54:217–229.

Cavender JA, Felsenstein J (1987): Invariants of phylogenies in a simple case with discrete states. *J Class* 4:57–71.

Eernisse DJ, Kluge AG (1993): Taxonomic congruence versus total evidence, and Amniote phylogeny inferred from fossils, molecules, and morphology. *Mol Biol Evol* 10:1170–1195.

Efron B (1979): Bootstrapping methods: another look at the jackknife. *Ann Statist* 7:1–26.

Efron B (1982): The jackknife, the bootstrap, and other resampling plans. *Conf Board Math Sci Soc Ind Appl Math* 38:1–92.

Efron B (1985): Bootstrap confidence intervals for a class of parametric problems. *Biometrika* 72:45–58.

Faith DP (1991): Cladistic permutation tests for monophyly and nonmonophyly. *Syst Zool* 40:366–375.

Faith DP, Cranston P (1991): Could a cladogram this short have arisen by chance alone? On permutation tests for cladistic structure. *Cladistics* 7:1–28.

Felsenstein J (1978): Cases in which parsimony or compatibility methods will be positively misleading. *Syst Zool* 27:401–410.

Felsenstein J (1981): Evolutionary trees from DNA sequences: a maximum likelihood approach. *J Mol Evol* 17:368–376.

Felsenstein J (1984): Distance methods for inferring phylogenies: a justification. *Evolution* 39:16–24.

Felsenstein J (1985): Confidence limits on phylogenies: an approach using the bootstrap. *Evolution* 39:783–791.

Felsenstein J (1986): Distance methods: reply to Farris. *Cladistics* 2:130–143.

Felsenstein J (1988): Phylogenies from molecular sequences: inference and reliability. *Annu Rev Genet* 22:521–565.

Felsenstein J (1993): *PHYLIP (Phylogeny Inference Package)*, v. 3.5c. Seattle: Distributed by author, Department of Genetics, University of Washington.

Felsenstein J, Kishino H (1993): Is there something wrong with the bootstrap on phylogenies? A reply to Hillis and Bull. *Syst Biol* 42:193–200.

Fitch WM (1971): Toward defining the course of evolution: minimum change for a specific tree topology. *Syst Zool* 20:406–416.

Fitch WM (1986): Commentary (on papers by Ruvolo and Smith, Saitou, and Templeton). *Mol Biol Evol* 3:296–298.

Gauthier J, Kluge AG, Rowe T (1988): Amniote phylogeny and the importance of fossils. *Cladistics* 4:105–209.

Goldman N (1993): Statistical tests of models of DNA substitution. *J Mol Evol* 36:182–198.

Gouy M, Li W-H (1989): Phylogenetic analysis based on rRNA sequences supports the archaebacterial rather than the eocyte tree. *Science* 339:145–147.

Hasegawa M, Yano T, Kishino H (1984): A new molecular clock of mitochondrial DNA and the evolution of hominoids. *Proc Jpn Acad* 60B:95–98.

Hasegawa M, Kishino H, Yano T (1985): Dating of the human–ape splitting by a molecular clock of mitochondrial DNA. *J Mol Evol* 22:160–174.

Hedges SB (1994): Molecular evidence for the origin of birds. *Proc Natl Acad Sci USA* 91:2621–2624.

Hillis DM (1991): Discriminating between phylogenetic signal and random noise in DNA sequences. In: Miyamoto M, Cracraft J (eds): *Phylogenetic analysis of DNA sequences.* New York: Oxford University Press, pp 278–294.

Hillis DM (1995): Approaches for assessing phylogenetic accuracy. *Syst Biol* 44:3–16.

Hillis DM, Bull JJ (1993): An empirical test of bootstrapping as a method for assessing confidence in phylogenetic analysis. *Syst Biol* 42:182–192.

Hillis DM, Huelsenbeck JP, Cunningham CW (1994a): Application and accuracy of molecular phylogenies. *Science* 264:671–677.

Hillis DM, Huelsenbeck JP, Swofford DL (1994b): Hobgobblin of phylogenetics? *Nature* 369:363–364.

Huelsenbeck JP (1991): Tree-length distribution skewness: an indicator of phylogenetic information. *Syst Zool* 40:257–270.

Huelsenbeck JP (1995): Performance of phylogenetic methods in simulation. *Syst Biol* 44:17–48.

Huelsenbeck JP, Hillis DM (1993): Success of phylogenetic methods in the four-taxon case. *Syst Biol* 42:247–264.

Jukes TH, Cantor CR (1969): Evolution of protein molecules. In: Munro H (ed): *Mammalian protein metabolism.* New York: Academic Press, pp 21–132.

Källersjö M, Farris JS, Kluge AG, Bult C (1992): Skewness and permutation. *Cladistics* 8:275–288.

Kidd KK, Sgaramella-Zonta LA (1971): Phylogenetic analysis: concepts and methods. *Am J Hum Genet* 23:235–252.

Kimura M (1980): A simple method for estimating evolutionary rates of base substitution through comparative studies of nucleotide sequences. *J Mol Evol* 16:111–120.

Kishino H, Hasegawa M (1989): Evaluation of the maximum likelihood estimate of the evolutionary tree topologies from DNA sequence data, and the branching order in Hominoidea. *J Mol Evol* 29:170–179.

Lake JA (1987a): Determining evolutionary distances from highly diverged nucleic acid sequences: operator metrics. *J Mol Evol* 26:59–73.

Lake JA (1987b): A rate-independent technique for analysis of nucleic acid sequences: operator metrics. *Mol Biol Evol* 4:167–191.

Lanyon S (1985): Detecting internal inconsistencies in distance data. *Syst Zool* 34:397–403.

Mindell DP, Dick CW, Baker RJ (1991): Phylogenetic relationships among megabats, microbats and primates. *Proc Natl Acad Sci USA* 88:10322–10326.

Mueller LD, Ayala FJ (1982): Estimation and interpretation of genetic distance in empirical studies. *Genet Res* 40:127–137.

Nei M, Stephens JC, Saitou N (1985): Methods for computing the standard errors of branching points in an evolutionary tree and their application to molecular data from humans and apes. *Mol Biol Evol* 2:66–85.

Park SK, Miller KW (1988): Random number generators: good ones are hard to find. *Commun ACM* 31:1192–1201.

Penny D, Hendy M (1986): Estimating the reliability of evolutionary trees. *Mol Biol Evol* 3:403–417.

Pettigrew JD (1986): Flying primates? Megabats have the advanced pathway from eye to midbrain. *Science* 231:1304–1306.

Pettigrew JD (1991): Wings or brain? Convergent evolution in the origins of bats. *Syst Zool* 40:199–216.

Pettigrew JD (1994): Flying DNA. *Curr Biol* 4:277–280.

Rohlf FJ, Sokal RR (1981): Comparing numerical taxonomic studies. *Syst Zool* 30:459–490.

Ruvolo M, Smith TF (1986): Phylogeny and DNA–DNA hybridization. *Mol Biol Evol* 3:285–289.

Saitou N (1986): On the delta Q-test of Templeton. *Mol Biol Evol* 3:282–284.

Sarich VM, Wilson AC (1967a): Rates of albumin evolution in primates. *Proc Natl Acad Sci USA* 58:142–148.

Sarich VM, Wilson AC (1967b): Immunological time scale for hominoid evolution. *Science* 158:1200–1203.

Sneath PHA (1986): Estimating uncertainty in evolutionary triads from Manhattan-distance triads. *Syst Zool* 35:470–488.

Stanhope MJ, Czelusniak J, Si J-S, Nickerson J, Goodman M (1992): A molecular perspective on mammalian evolution from the gene encoding interphotoreceptor Retinoid Binding Protein, with convincing evidence for bat monophyly. *Mol Phyl Evol* 1:148–160.

Steel MA, Lockhart PJ, Penny D (1993): Confidence in evolutionary trees from biological sequence data. *Nature* 364:440–442.

Swofford DL (1995): *PAUP*: Phylogenetic Analysis Using Parsimony,* version 4.0. Sunderland, MA: Sinauer.

Templeton AR (1983): Phylogenetic inference from restriction endonuclease cleavage site maps with particular reference to the evolution of humans and the apes. *Evolution* 37:221–244.

Templeton AR (1985): The phylogeny of the hominoid primates: a statistical analysis of the DNA–DNA hybridization data. *Mol Biol Evol* 2:420–433.

Thomas RH, Paabo S, Wilson AC. 1990. Chance marsupial relationships—reply. *Nature* 345:394.

Van Den Bussche RA, Hillis DM, Huelsenbeck JP, Baker RJ (1996): Base compositional bias and phylogenetic analyses: a test of the "flying DNA" hypothesis. *Mol Biol Evol (submitted).*

Wu CFJ (1986): Jackknife, bootstrap, and other resampling plans in regression analysis. *Ann Statist* 14:1261–1295.

Yang Z (1993): Maximum-likelihood estimation of phylogeny from DNA sequences when substitution rates differ over sites. *Mol Biol Evol* 10:1396–1401.

Yang Z (1994): Estimating the pattern of nucleotide substitution. *J Mol Evol* 39:105–111.

Yang Z, Roberts D (1995): On the use of nucleic acid sequences to infer early branching in the tree of life. *Mol Biol Evol* 12:451–458.

Zharkikh A, Li W.-H (1992a): Statistical properties of bootstrap estimation of phylogenetic variability from nucleotide sequences: I. Four taxa with a molecular clock. *Mol Biol Evol* 9:1119–1147.

Zharkikh A, Li W-H (1992b): Statistical properties of bootstrap estimation of phylogenetic variability from nucleotide sequences: II. Four taxa without a molecular clock. *J Mol Evol* 35:356–366.

■■■■■ **CHAPTER 3**

Molecular Approaches and the Growth of Phylogenetic Biology

WAYNE PAUL MADDISON

Department of Ecology and Evolutionary Biology, University of Arizona, Tucson, Arizona 85721

CONTENTS

Molecular Zoology: Advances, Strategies, and Protocols, Edited by Joan D. Ferraris and Stephen R. Palumbi.
ISBN 0-471-14461-4   © 1996 Wiley-Liss, Inc.

SYNOPSIS

The large-scale structure of genetic descent is the phylogenetic tree of life. Phylogenetic biology focuses on this phylogenetic tree and the evolutionary processes it elucidates. These include the processes of character evolution, which are reflected in the history and patterns of character change, species selection, which yields asymmetries in clade sizes in the tree, and interactions between clades, which may lead to parallel phylogenesis. The fine-scale structure of phylogeny is now accessible through molecular data, which are leading to a new appreciation of the relationship between gene histories and species histories. The many avenues of research available, new analytical tools, and accessibility of molecular data have led to rapid growth of phylogenetic biology. Because molecular approaches allow phylogeny reconstruction by nonsystematists, and because of phylogenetic biology's focus on evolutionary processes, the growth of phylogenetic biology could lead, paradoxically, to a decrease in support for systematics in general, for morphological systematics in particular, and for organism-centered research. Morphological systematics cannot be abandoned, because strong morphological systematics is what gives a molecular phylogeny its value, by allowing it to be linked to the vast data we have on organisms. Organism-centered research is one of our best means of biological exploration. In pursuing the exciting new questions of phylogenetic biology, it will be important not to abandon the traditional approaches vital to it.

INTRODUCTION

In the past two decades a focus in evolutionary biology has emerged and grown. It concerns phylogenetic trees, and how these phylogenetic trees can elucidate the processes of evolutionary change. A typical study will reconstruct a phylogenetic tree for a group of species, map the evolutionary history of a character of interest on the tree, then make conclusions about the factors behind observed patterns in the character's evolution. Studies of this sort, and other phylogenetic studies such as those on relative diversification rates, have become common, even infiltrating disciplines that formerly saw few phylogenetic trees in the pages of their journals.

These studies don't precisely fit the mold of traditional phylogenetic systematics, because there is little concern for classification. They are concerned entirely with genetic descent and the processes of diversification and change. Phylogenetic trees are reconstructed to answer specific questions about evolutionary processes, not for their own sake, nor as a means to set up a general-purpose classification. Nor do these studies fit the mold of evolutionary theory of a few decades ago, because they are so attentive to phylogeny. They represent a new focus that has arisen from the integration of phylogenetic systematics with evolutionary theory. If one were to grace the new focus represented by such studies with a name, it could be "phylogenetic biology."

The growth of phylogenetic biology might be attributed both to theoretical advances, such as improved analytical tools in estimating and interpreting phylogeny

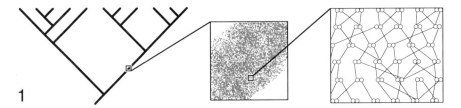

Figure 1. Phylogeny, from the broad scale to the fine structure. At the broad scale is the phylogenetic tree of species branching. At the fine scale are the breeding relationships of individual organisms within a population.

(for reviews, see Swofford and Olsen, 1990; Maddison and Maddison, 1992; Hillis et al., 1994), and to advances in gathering data, most notably molecular data (Hillis and Moritz, 1990). Here I will discuss the scope of phylogenetic biology, why it is so exciting, and the role that molecular approaches have played in its growth and independence. Then I will discuss some of the consequences of the growth of phylogenetic biology. Not all of these consequences are positive, and it will be important for the field, in its exuberant growth, not to discard all of the traditions from which it has grown.

WHAT IS PHYLOGENETIC BIOLOGY?

Before considering a definition of phylogenetic biology, it would be appropriate to review what is meant herein by "phylogeny." Phylogenetic history is the history of genetic connections through evolutionary time. At a fine scale, the genetic connections happening around us are not continuous across all organisms from bacteria through humans, but instead (in sexual organisms) are grouped into bundles of genetic descent that we call "species." These bundles, these species lineages, descend through time, occasionally branching to yield new species (Fig. 1). Thus the large-scale structure of genetic connections in evolutionary history is, in general, that of an ever-branching tree.

With that concept of phylogeny in mind, we can define phylogenetic biology as *the study of the phylogenetic history of organisms, and of its relationships to the processes of evolution.*

THE QUESTIONS OF PHYLOGENETIC BIOLOGY

It is not surprising that such hope is being invested in phylogeny; genetic descent, after all, has a powerful place in our explanations of why organisms are the way they are. The scope of biological processes that can be related to phylogeny is great (Brooks and McLennan, 1991; Harvey and Pagel, 1991; Maddison and Maddison, 1992), and here I will give some examples of how various evolutionary processes can be viewed from a phylogenetic perspective.

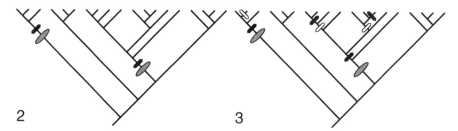

Figures 2 and 3. Two examples of the correlated evolution of two characters. 2: When the feature marked by the gray oval arises, there is immediate adaptive evolution of the other feature marked by the black line. 3: When the feature marked by the gray oval arises, the evolution of another character becomes unconstrained, leading to an increase in the rate of its evolution, resulting in frequent changes in state as marked by the white and black lines.

Character Evolution

The branches of the phylogenetic tree were populated by organisms living, reproducing, and dying. By selection and drift, the characteristics of the organisms in a species lineage change from generation to generation, and thus there is change along the branches of the phylogenetic tree. This evolutionary change may be affected by many other features of the organisms, including their developmental system and the ecological context in which they place themselves. Depending on the exact nature of the interactions, many possible evolutionary patterns could result.

Imagine several species of organisms whose surfaces could be either hairy or smooth, and the phylogenetic tree depicting their interrelationships (Fig. 2). If a feature evolved (marked by the gray ovals in the figure) that causes them to select a new environment, it may be immediately adaptive for them to become hairy (marked by the black lines in the figure). We may see, therefore, a pattern of evolution of the two characters in which their changes are coincident on the same branches of the tree (Fig. 2).

However, there are different processes that cause correlated evolution of two characters, and the phylogenetic patterns they yield can be different. An evolutionary change in the folding of a protein (marked by the gray ovals in Fig. 3) may place some amino acids in a position in which their nature is no longer so important to the protein's functioning. These amino acid sites find their evolution no longer so constrained, and they may suffer a higher rate of evolution—not evolution to any particular amino acid, but just greater freedom to change (marked by the black and white lines in Fig. 3). Alternatively, for morphological characters, the release of a constraint may allow a specific response on the part of the affected character, allowing it to move to some adaptive peak that had been developmentally inaccessible to it.

Detecting these patterns, and the causes underlying them, is often not easy. Phylogeny is critical, for it supplies information about how many replicates there have been in the natural experiments of correlated evolution. Much work has been accomplished concerning methods to reconstruct the history of character evolution (Farris, 1970; Fitch, 1971; Swofford and Maddison, 1992; Maddison, 1994) and to

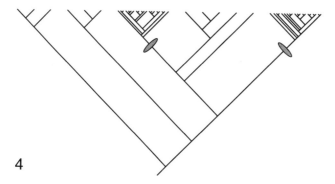

Figure 4. Species selection. When the feature marked by the gray oval arises, speciation rate increases and/or extinction rate decreases. The clades possessing it are therefore more species-rich.

test for phylogenetic character correlations (e.g., Felsenstein, 1985; Maddison, 1990; Pagel, 1994), but much work remains to be done to deal with the multitude of types of data and biological processes that could be investigated. Reviews of this topic are given by Harvey and Pagel (1991) and Maddison and Maddison (1992).

Species Selection

The probability that a species will speciate or go extinct depends to a certain extent on its traits, such as habitat specificity and dispersal modes (Stanley, 1979; Vrba, 1984). Insofar as its traits will tend to be passed on to its daughter species, there will be natural selection at the species level (Lewontin, 1970). Clades with diversity-favoring traits will be more successful over the long term than others, and this will be reflected in the shape of the phylogenetic tree, with these clades being larger than their sisters. They might achieve their success by raising speciation rates, or by lowering extinction rates.

An example is given in Figure 4. In each instance that the feature marked by a gray oval arose, the clade possessing it achieved greater diversity than its sister clade. Mitter et al. (1988) give an example of such a pattern in insects, among which herbivorous clades are consistently more species-rich than their carnivorous sisters. The use of phylogeny is clearly crucial to this inference, for it defines what are the clades and what comparisons to make.

Fine Structure of Phylogenetic Trees

The branches of the phylogenetic tree of life, which are the species lineages descending through time, are not simple, single lines. They have a fine-scale structure, for, in sexual organisms, they are bundles of genetic descent. The bundles are made of very fine fibers indeed, each representing the descent of an individual copy of a gene. With the advent of DNA sequence data, phylogenetic biology is beginning to pay serious attention to this fine structure.

Within Species. Within a sexually reproducing species, the relationships of individual organisms form a reticulate network (Fig. 1). However, when we look closely, we see that the fine structure of genetic descent can be characterized as a series of overlain gene trees (Hein, 1990; Maddison, 1995). The reason for this is simple. Suppose we follow the descent of copies of a small section of the genome, such as a single gene or part of a gene, within which there has been no recombination in the time span considered. We would see this section's history of descent to take the form of a tree, because a copy in one generation can leave one or more copies in the next generation, and yet it came from only one copy in the previous generation. Such trees have been called gene trees or gene genealogies (for the moment using the word "gene" to refer to one of these small unrecombined sections of the genome). The history of another section of the genome would follow a different tree, because recombination between this section and the first would mean that some individuals would find the sources of their copies in the two genes routed through different grandparents. Together, all of these sections of the genome descending through organisms in tangled tree-like histories make up the genetic history of the species.

All the forces of population genetics — mutation, recombination, selection, migration, drift — operate by affecting the forms of gene trees and the information flowing along them. A branch of population genetics, coalescence theory, is taking advantage of the tree-like structure of gene descent to make inferences about processes within populations (Kingman, 1982; Slatkin and Maddison, 1989; Hudson, 1990; Felsenstein, 1992).

Between Species. The importance of thinking in terms of gene trees and the fine structure of genetic descent extends beyond population genetics to phylogenetic biology. It has been realized for a number of years that species that are each other's closest relatives may not be each other's closest relatives in terms of some of their gene copies (Goodman et al., 1979; Avise et al., 1983; Tajima, 1983; Pamilo and Nei, 1988; Takahata, 1989). For instance, even if humans and chimpanzees are each other's closest relatives, there is a good chance that, for at least some genes, human copies of the gene are more closely related to gorilla copies than they are to chimp copies (Fig. 5). This would seem to contradict the species relationships, yet it can happen. If we follow the ancestry of the human copies back to the speciation event between chimps and humans, the ancestor of a human copy would find itself coexisting in the common ancestral species with the ancestor of a chimp copy (open arrow in Fig. 5). Chances are, the ancestors of the human and chimp copies would not find their own common ancestral copy right at the point of speciation, but at some point deeper in the history of the common ancestral species, perhaps even as far back as the previous speciation event (between gorillas and the ancestor of chimps and humans). If this happens, then ancestors of copies in all three species were coexisting: human, chimps, and gorillas. Just by chance, the human copy may find a common ancestral copy with the gorilla copy (closed arrow, Fig. 5) before it finds one with the chimp copy. The descendants, the copies in humans and gorillas, would then be more closely related to each other than to the copies in chimpanzees.

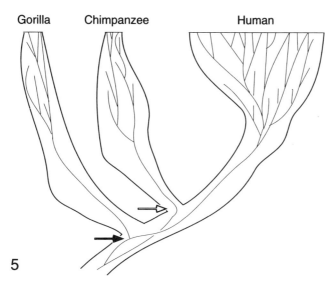

Figure 5. Hypothetical example of a gene tree whose relationships disagree with the species tree of extant hominids. The tree of descent of copies of the gene is shown by fine lines contained within the species tree. Note that the human ancestral copy and chimp ancestral copy do not have a common ancestral copy at the point of human–chimp speciation, marked by the open arrow, but rather the human and chimp copies persisted together in human–chimp ancestral species. The human copy finds a common ancestor, marked by the closed arrow, with the gorilla ancestral copy before (as one looks backward in time) it finds a common ancestor with the chimp copy.

This gene, though it may mislead us about the species relationships (Doyle, 1992), is not a mistake. It is a part of genetic history. Acceptance of the legitimate existence of such gene histories may very well be leading to a revolution in our conception of phylogeny itself. All of us learned the Bohr model of the atom in school: a central nucleus with electrons zooming about in orbits like satellites. Then we learned that it is too simplistic, that the electrons are not like satellites. Instead, there are electron clouds, with electrons occupying a broad cloud-like space but concentrated in certain areas more than others. The electron is not wandering through this cloud — it *is* this cloud, since in some sense it exists in all places throughout the cloud simultaneously. What gene trees tell us is that phylogeny is similarly fuzzy. A phylogeny is like a gene-history cloud (Fig. 6), with the greatest density of gene histories forming a tree-like shape (the human and chimp copies of most genes are each other's closest relatives), but with a potentially broad cloud of other histories completing the picture. Just like the electron, phylogenetic history is spread out, composed simultaneously of different relationships for different portions of the genome. When we see a phylogenetic tree of simple lines, and assume it informs us about the history of genetic descent, we must realize that it tells us only part of the story. What it tells us is something like the modal gene history, and not the variance about it (Maddison, 1995).

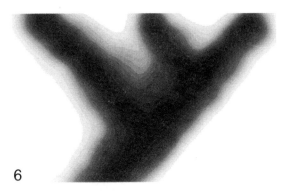

6

Figure 6. Phylogeny viewed as a gene-history cloud, analogous to an electron cloud of an atom. Because gene trees can disagree with species trees, the genetic history of a group of species can be a composite, with most genes agreeing on which species are sisters, but with some fraction of the genes disagreeing. The phylogenetic history therefore resembles a statistical cloud, and our usual representation of a phylogeny with simple thin lines is only depicting something like a modal gene history.

Our new awareness of phylogeny's fine structure will change phylogenetic biology. For instance, most phylogeneticists usually treat a character as evolving passively along the branches of a phylogenetic tree, not affecting the tree or, at most, affecting only speciation and extinction rates. This model is implicit in the widespread use of methods that reconstruct the evolution of a character on a phylogenetic tree: the tree is taken as a given, and the character's states are mapped onto the tree. It is as if the tree's existence is independent of the characters, which just flow along its branches as an afterthought. However, a gene-tree perspective tells us that the fixation of a gene in an ancestral species, which to a phylogeneticist is simply a change in character state along the branch of the tree, obviously affects the gene-tree history of that part of the genome. As such, changes in character state are *part* of the fine-scale structure of phylogeny, not just passive passengers along it. For many genes, we will still be able to treat the shape of phylogeny as independent of the particular character's history, for that character's genes are involved in only a small part of the total genetic connections. For some characters, such as those affecting mating systems, it may be far harder to untangle phylogeny and character evolution.

Associated Trees

Though I have suggested that the relationship between a gene tree and the species tree is one of part and whole, for most genes it is probably a good approximation to view the gene tree as passively flowing within a containing species tree. This relationship is one of the possible types of associations between trees: one tree is contained within another.

Gene trees can disagree with their containing species tree for the reason outlined above and in Figure 7, that common ancestry of gene copies at a single locus can

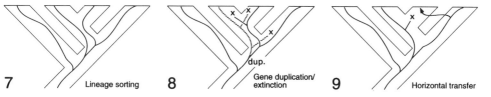

Figures 7–9. Three processes that can cause a gene tree and species tree to disagree. In lineage sorting, an ancestral polymorphism was maintained through more than one speciation event, and therefore the common ancestors of sampled gene copies are found deep in ancestral species. In gene duplication and extinction, a gene duplication event is followed by extinction (or failure to sample) in some lineages. "X" marks extinction of the gene. In horizontal transfer, a gene jumps across species lineages, possibly via a virus or other vector.

extend deeper than speciation events. That is, separate lineages of genes coexisted in ancestral species, then sorted themselves out differentially into the descendant species' sampled copies (Avise et al., 1983). Lineage sorting is not the only possible source of discordance, however. If a gene is duplicated so as to place a copy of itself at some other locus in the genome, the two loci will coexist through phylogenetic time, the history of each being independently contained within the species tree. If, in different descendant species, one or the other locus is lost, then the tree of surviving copies might very well disagree with the species tree (Fig. 8; Goodman et al., 1979). A third possible cause of discordance is horizontal transfer (Kidwell, 1993; Cummings, 1994), in which the gene copy breaks the bounds of the species tree and is transferred directly from one branch to another (Fig. 9), either via hybridization or a vector such as a virus.

Which of the three sources is a more likely explanation of discordance depends on various issues (Clark et al., 1994; Cummings, 1994). Hidden duplication and extinction are superficially similar to lineage sorting, in that multiple gene copies coexist and are sorted among descendant species, but the dynamics of the process are different, since duplicated genes are not competing for the same locus. Deep coalescence (lineage sorting) for copies at the same locus is unlikely when branches on the tree are long relative to population sizes (Pamilo and Nei, 1988), unless balancing selection is occurring; hidden duplication and extinction depend on other factors. Horizontal transfer would seem less likely if there seem to be no appropriate means of transfer, for instance, if phylogenetic distance rules out hybridization but appropriate vectors and means of incorporation are not known to exist (Cummings, 1994).

There are other situations in phylogenetic biology in which one tree is contained within another. In historical biogeography, the successive breakup of a habitat (as of Pangea in the Mesozoic) can take the form of a tree of areas (Nelson and Platnick, 1981), contained within which is the descent of species inhabiting these areas. In closely associated organisms, such as a host and parasite, speciation of one (say, the host) may lead to speciation in the other, and we can model their relationship as contained (parasite) and containing (host) trees (Brooks and McLennan, 1991; Farrell et al., 1992). In both of these cases, there are alternative causes for a discordance between contained and containing trees. Hidden speciation and extinction can lead to discordance, just as in the case of gene trees and species trees (Page, 1993). The equivalent of horizontal transfer is, in the case of biogeography, dispersal from one area to another, and in the case of host–parasite associations, a parasite's shifting from one host to another (Futuyma and McCafferty, 1991).

The contained–containing relationship between trees is only one of many conceivable. Associated organisms may have a reciprocal effect on each other's speciation. Interactions among organisms can of course be of varying degrees of specificity and intensity, and there will be correspondingly varying degrees of effect on each other's phylogeny.

ROLE OF MOLECULAR APPROACHES IN PHYLOGENETIC BIOLOGY

The rise of phylogenetic biology is attributable not only to the exciting questions it addresses, but also to two developments that are allowing faster and more confident phylogeny reconstruction: the development of new analytical tools, and the ease of acquisition of nucleotide and protein sequence data (Hillis et al., 1994). The molecular approaches in particular have brought phylogenetic reconstruction to branches of biology that previously have had little to do with evolutionary history. Two reasons might be cited.

First, molecular approaches can now yield data comparing diverse organisms in more abundance than morphological and other phenotypic approaches, and perhaps with less effort (though not necessarily less expensively). The fact that molecular techniques are generally not organism-specific means that only minimal learning is needed to obtain data from unfamiliar organisms. The data achieved are not only abundant, but they may be better behaved statistically than morphological data. Any given molecular character may not be any better than a morphological character, but the uniformity of behavior within a class of molecular characters (e.g., third positions in coding regions) may allow application of powerful statistical techniques like maximum likelihood.

Second, molecular approaches yield a high-resolution view on genetic history. Because the histories of different genes can be resolved separately, DNA sequence data are beginning to allow us to peek into the fine structure of phylogeny and population histories. This will give us an unprecedented ability to explore processes over phylogenetic time.

From what I have said, it would seem that phylogenetic biology is heading toward limitless and unqualified success, and that part of its success is due to molecular approaches. It is therefore appropriate to introduce two notes of caution.

WHAT IS KNOWABLE?

My first note of caution concerns the limits of phylogenetic biology. Phylogenetic biology is full of optimism, with many theoretical papers beginning with the premise that the phylogenetic tree is known precisely, even the length of its branches. For trees of species, such precision may someday be approximated, but for trees of genes within species, there may simply be insufficient information left in a short DNA sequence. As large as the genome is, history is even larger, and for many of our attempts to understand it, we will be left with too few and too ambiguous data. In fact, much of the progress in phylogenetic biology in the next decade might actually focus on clearly delimiting what we can hope and can't hope to know. If we discover that phylogeny cannot help us understand certain processes, then phylogenetic skeptics will not necessarily be able to celebrate phylogeny's failure, because in many cases the lack of historical trace may render the evolutionary process unresolvable by any means, phylogenetic or otherwise.

PHYLOGENETIC BIOLOGY VERSUS SYSTEMATICS

My second note of caution is that not all of the consequences of the growth of phylogenetic biology may be desirable. One possible conflict is hinted in the title of this section, which might seem ridiculous: How could phylogenetic biology be against systematics? I will suggest, however, that the manner of phylogenetic biology's current growth could actually, over the long term, be damaging to systematics and eventually to phylogenetic biology itself.

In what follows, please don't take my use of the word "will" too seriously. What I present is one hypothetical scenario for the future of phylogenetic biology, one which I hope does not come to pass. In fact, some parts of the scenario are simply not sustainable, as I will argue later.

Two characteristics of our current phylogenetic biology are worth recalling: it tends to be question-oriented, using phylogenies to test general hypotheses about evolutionary processes, and it is using molecular data more and more prominently. Both of these characteristics will shift the center of gravity of phylogenetic reconstruction away from its traditional home—systematists in museums and herbaria—toward evolutionary and molecular biologists in universities and other academic institutions. The question-oriented evolutionary biologists will pull phylogeny reconstruction into their labs to answer their questions, and the generality and success of molecular techniques will allow them to reconstruct phylogenies without extensive apprenticeship in the peculiarities of the morphological systematics of their group. Nonsystematists will incorporate sequencing and phylogeny reconstruction as a necessary but *peripheral* part of a larger project. An evolutionary biologist will be able to deal quickly with the phylogeny and spend more time focusing on questions of more interest. Phylogeny reconstruction will become just an incidental tool used in the course of a study, like measuring sucrose concentrations or videotaping behavior.

The ease of phylogeny reconstruction will devalue systematics as an independent discipline. No longer will anyone devote one's life to the systematics of a group, because everyone can reconstruct phylogeny themselves, and all the funding will go toward questions answered using phylogeny, not to the reconstruction of phylogeny as an end in itself. (The taxonomic side of systematics, which recognizes species, will either move into conservation biology, and be performed by "parataxonomists" working on biodiversity studies, or will focus on the bundling of life's genetic connections, thus becoming part of phylogenetic and population biology.) Empirical systematics will therefore disintegrate as a recognizable discipline.

Expertise in morphological systematics, except as already discovered and codified in identification keys and guides, will start to fade as molecular data begin to take over completely, and there are no longer dedicated systematists from whom to learn the intricacies of a group of organisms. For that matter, the premium put on being "question-oriented" will mean that fewer and fewer young biologists will dedicate a career to a particular group of organisms, whether in systematics or any other discipline. Instead they will take their toolbox of molecular and theoretical techniques and their focus will bounce from clade to clade, stopping only long enough in each to examine their favorite general question there.

MOLECULES ARE NOT ENOUGH

This scenario may be viewed by many as highly unlikely, because there will always be some value placed on systematics, on morphological systematics, and on organism-centered research. But there are already trends leading toward the scenario I have given. Phylogenetic research is judged more interesting the more it seeks to answer general questions instead of reconstructing particular histories. More and more students are being trained not by specialists in the systematics of their group of organisms, but in mixed-taxon labs that have the molecular techniques or theoretical orientation they need. Even if our future is not as extreme as I have suggested, a continued devaluation and decrease of systematics in general, morphological systematics in particular, and organism-centered research seem very real possibilities.

Instead of asking how likely is the scenario, a more appropriate question at this point is perhaps, are these trends undesirable? Or is it simply a matter of progress in efficiency of pursuing our goals? I will argue that many of the trends I described in my scenario are undesirable, for two simple reasons: molecules aren't enough, and question-oriented researchers aren't enough.

The literature on the use of morphological versus molecular characters in reconstructing phylogeny has primarily dealt with the question of which is more likely to give us a correct branching pattern for the taxa sampled (Hillis, 1987). This issue is not yet resolved, but perhaps most systematists would agree (myself included) that if they could have both morphological and molecular data, they would be more confident of their results than if they had only one of the two. This is one argument as to why we should keep morphological systematics active. But there is a much more potent one.

Let us imagine that in the future we discover a way to analyze molecular data such that it always gives us the exactly correct phylogenetic tree for a set of taxa sampled, but that no such claim can be made for morphological data. Even were this

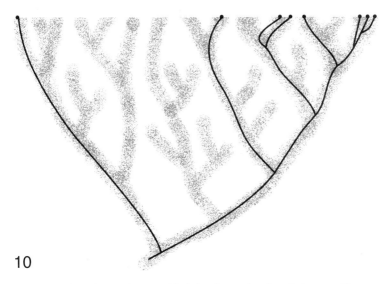

10

Figure 10. Sampling of organisms and species for molecular phylogeny. The gray fuzzy lines depict the species phylogeny; they are wide to imply many individual organisms within each species. Within some of the extant species, a few gene copies have been sampled and are represented by black spots. The black lines follow the gene tree for the sample copies. Because relatively few individuals and few species are sampled, a molecular phylogeny represented by the black gene tree at best is a thin skeleton. The vast majority of organisms, the gray portions, must be linked to it via phenotypically based systematics.

to happen, we still would need to keep morphological systematics active for many years to come.

The reason arises from the limitations of technology. The vast majority of the millions of species on earth will not be sequenced any time soon (Fig. 10). This means that a molecular phylogeny can at best be a thin skeleton of what we need to know, and the majority of species will be placed on that skeleton by their phenotypic (including morphological) characteristics (Maddison and Maddison, 1992: 72). The framework from morphological systematics is needed not only after a molecular study is completed, but also beforehand, to ensure that there is efficient sampling of relevant species. Molecular sampling may become dense in familiar groups like higher primates, but it may never be in groups like spiders, where thousands of species receive the attention of a handful of specialists. And molecular sampling will be close to nonexistent for the vast portions of biological diversity preserved as fossils. Similarly, the vast majority of the countless individual organisms within species that we encounter and gather data on will not be sequenced any time soon (Fig. 10). This includes the individuals flying overhead in a migration study, those we monitor for physiology of respiration, those we observe scurrying under a rock in a behavioral study, those we count in remote sensing studies of tropical forests, and those we uncover in Cambrian strata.

Almost all of our biological information came and will continue to come from specimens that were and must be identified and placed phylogenetically by phenotype. If we did not have an understanding of morphological systematics, then we might as well throw away all our biological data that came from specimens that were not sequenced, or from species none of whose individuals was sequenced. The value of those data depends critically on the extent to which we can accurately place a specimen on the phylogenetic tree by phenotypic data alone. At present, we cannot do that as well as we need, because much remains to be achieved in morphological systematics. If we let morphological systematics decay, we will find ourselves unable to link accurately a molecular phylogeny to the organisms we encounter. *Strong morphological systematics is what gives a molecular phylogeny its value,* by allowing it to be linked to the vast data we have on organisms.

THE IMPORTANCE OF EXPLORATION

The second claim I made was that question-oriented researchers aren't enough. I don't mean to say we need researchers who don't ask questions. Rather, I mean we also need researchers who focus their work on an organism or group of organisms, regardless of what questions those organisms might lead them to ask. This is in contrast with researchers who focus on general questions, and use whatever organisms are convenient to answer them. I take some of my inspiration for the following discussion from D. Futuyma's presidential address to the American Society of Naturalists in 1994, which was a compelling presentation of the importance of, but prevailing lack of respect for, organism-centered research.

An organism-centered career is often belittled. I have seen workers praised for not dedicating themselves to a group of organisms, as if such dedication were necessarily a bad thing. I have heard some biologists, who in fact are dedicated to working on a particular group of organisms, carefully hide the fact by pretending they chose their group because it is useful for testing some general hypothesis. We want to be known as "molecular evolutionists" or "life history theorists" and not "arachnologists" or "mammalogists." (Strangely enough, it is perfectly legitimate to maintain a technique-centered career, as long as the technique is expensive or new enough.)

Of course, all of us want our work to be of broad interest, and science values generalities. But generalities are built on particulars that someone had to discover, and some particulars are hard-won indeed. There is a need for biologists who discover facts and histories about organisms, and who specialize in this. If there is a value to becoming intimately familiar with a topic, this value applies to organisms as well as to questions.

Perhaps the difficulty in explaining the value of an organism-centered approach is that it depends on a long-term perspective. In the short term, a question-centered biologist can tell you precisely what important general hypotheses will be addressed. An organism-centered biologist would express his or her orientation as an exploration, one that might lead to an interesting phenomenon relevant to a question not yet identified. Wilson (1985) eloquently explained the value of such exploration of biological diversity and noted that whenever an interesting phenomenon is discovered by a systematist or organism-focused researcher, its study is usually taken over by a question-oriented biologist, whose field then gets sole credit for the progress in our understanding.

Organisms chosen solely to test a question may in fact be those most likely to hold a biased answer to the question. If our organisms drive our research, we are perhaps more likely to find the inconvenient facts that challenge our old questions, or for that matter to stumble on questions we hadn't yet thought of, than if we approach selected organisms with questions already in hand. Biology needs to maintain a mix of exploration and intensely focused research, and organism-centered research is one of our best sources of exploration.

CONCLUSIONS

However much we may celebrate yet another new synthesis, the reintegration of phylogenetic systematics into evolutionary biology, we need to recognize that the echos of life's history cannot be interpreted easily, nor casually. We will always need people who are intimately familiar with reconstructing phylogeny, with morphological systematics, and with a group of organisms. This will usually take the dedication of a career, not just dabbling. Focus on history itself is as legitimate a scientific pursuit as is research involving the most controlled experiment, although, with the exception of the Big Bang and dinosaurs, reconstruction of history is usually not accorded an exhalted place in science. But biologists are realizing that life's rich history plays a central role in their discipline, and the attention being paid to phylogenetic biology bodes well for its future. I have expressed fears that in our excitement we will neglect some traditional but vital approaches in phylogenetic biology. I hope that my fears are unwarranted, and that through appropriate collaborations, scientists of all persuasions—historical, experimental, molecular, morphological, organism-centered, question-driven, technique-centered—can all manage to hang on together and prosper.

ACKNOWLEDGMENTS

I am grateful to Joan Ferraris and Steve Palumbi for the opportunity to participate in this symposium, and to some of its participants for encouraging me to publish this chapter despite its pretensions and its only qualified endorsement of molecular approaches. Alan de Queiroz, John Lundberg, and David Maddison provided helpful discussion and comments on the manuscript. This work was supported in part by a fellowship from the David and Lucile Packard Foundation.

REFERENCES

Avise JC, Shapiro JF, Daniel SW, Aquadro CF, Lansman RA (1983): Mitochondrial DNA differentiation during the speciation process in *Peromyscus*. *Mol Biol Evol* 1:38–56.

Brooks DR, McLennan DA (1991): *Phylogeny, ecology and behavior: a research program in comparative biology.* Chicago: University of Chicago Press.

Clark JB, Maddison WP, Kidwell MG (1994): Phylogenetic analysis supports horizontal transfer of P-transposable elements. *Mol Biol Evol* 11:40–50.

Cummings MP (1994): Transmission patterns of eukaryotic transposable elements: arguments for and against horizontal transfer. *Trends Ecol Evol* 9:141–145.

Doyle JJ (1992): Gene trees and species trees: molecular systematics as one-character taxonomy. *Syst Bot* 17:144–163.

Farrell BD, Mitter C, Futuyma DJ (1992): Diversification at the insect–plant interface: insights from phylogenetics. *Bioscience* 42:34–42.

Farris JS (1970): Methods for computing Wagner trees. *Syst Zool* 19:83–92.

Felsenstein J (1985): Phylogenies and the comparative method. *Am Nat* 125:1–15.

Felsenstein J (1992): Estimating effective population size from samples of sequences: a bootstrap Monte Carlo integration approach. *Genet Res* 60:209–220.

Fitch WM (1971): Toward defining the course of evolution: minimal change for a specific tree topology. *Syst Zool* 20:406–416.

Futuyma DJ, McCafferty SS (1991): Phylogeny and the evolution of host plant associations in the leaf beetle genus *Ophraella* (Coleoptera, Chrysomelidae). *Evolution* 44:1885–1913.

Goodman M, Czelusniak J, Moore GW, Romero-Herrera AE, Matsuda G (1979): Fitting the gene lineage into its species lineage, a parsimony strategy illustrated by cladograms constructed from globin sequences. *Syst Zool* 28:132–163.

Harvey PH, Pagel MD (1991): *The comparative method in evolutionary biology.* Oxford: Oxford University Press.

Hein J (1990): Reconstructing evolution of sequences subject to recombination using parsimony. *Math Biosci* 98:185–200.

Hillis DM (1987): Molecular versus morphological approaches to systematics. *Annu Rev Ecol Syst* 18:23–42.

Hillis DM, Moritz G (eds) (1990): *Molecular systematics.* Sunderland, MA: Sinauer.

Hillis DM, Huelsenbeck JP, Cunningham CW (1994): Application and accuracy of molecular phylogenies. *Science* 264:671–677.

Hudson RR (1990): Gene genealogies and the coalescent process. *Oxford Surv Evol Biol* 7: 1–44.

Kidwell MG (1993): Lateral transfer in natural populations of eukaryotes. *Annu Rev Genetics* 27:235–256.

Kingman JFC (1982): The coalescent. *Stochast Proc Appl* 13:235–248.

Lewontin RC (1970): The units of selection. *Annu Rev Ecol Syst* 1:1–18.

Maddison DR (1994): Phylogenetic methods for inferring the evolutionary history and processes of change in discretely valued characters. *Annu Rev Entomol* 39:267–292.

Maddison WP (1990): A method for testing the correlated evolution of two binary characters: Are gains or losses concentrated on certain branches of a phylogenetic tree? *Evolution* 44:539–557.

Maddison WP (1995): Phylogenetic histories within and among species. In: Hoch PC, Stevenson AG, Schaal BA (eds). *Experimental and molecular approaches to plant biosystematics.* Monographs in Systematics. St Louis: Missouri Botanical Garden.

Maddison WP, Maddison DR (1992): *MacClade 3: analysis of phylogeny and character evolution*. Sunderland, MA: Sinauer.

Mitter C, Farrell B, Wiegemann B (1988): The phylogenetic study of adaptive zones: Has phytophagy promoted insect diversification? *Am Nat* 132:107–128.

Page RDM (1993): Genes, organisms and areas: the problem of multiple lineages. *Syst Biol* 42:77–84.

Pagel M (1994): Detecting correlated evolution on phylogenies: a general-method for the comparative analysis of discrete characters. *Proc R Soc Lond B Biol Sci* 255:37–45.

Pamilo P, Nei M (1988): Relationships between gene trees and species trees. *Mol Biol Evol* 5:568–583.

Nelson G, Platnick NI (1981): *Systematics and biogeography: cladistics and vicariance*. New York: Columbia University Press.

Slatkin M, Maddison WP (1989): A cladistic measure of gene flow inferred from the phylogeny of alleles. *Genetics* 123:603–613.

Stanley SM (1979): *Macroevolution: pattern and process*. San Francisco: WH Freeman.

Swofford DL, Maddison WP (1992): Parsimony, character-state reconstructions, and evolutionary inferences. In: Mayden RL (ed). *Systematics, historical ecology, and North American freshwater fishes*. Stanford: Stanford University Press, pp 187–223.

Swofford DL, Olsen GJ (1990): Phylogeny reconstruction. In: Hillis DM, Moritz G (eds). *Molecular systematics*. Sunderland, MA: Sinauer, pp 411–501.

Tajima F (1983): Evolutionary relationships of DNA sequences in finite populations. *Genetics* 123:229–240.

Takahata N (1989): Gene genealogy in three related populations: consistency probability between gene and population trees. *Genetics* 122:957–966.

Vrba E (1984): What is species selection? *Syst Zool* 33:318–328.

Wilson EO (1985): The biological diversity crisis: a challenge to science. *Issues Sci Technol* 2:20–29.

PART II

POPULATION STRUCTURE AND MOLECULAR MARKERS

Characterization of Genetic Structure and Genealogies Using RAPD-PCR Markers: A Random Primer for the Novice and Nervous

RICHARD K. GROSBERG

Center for Population Biology, Section of Evolution and Ecology,
University of California–Davis, Davis, California 95616

DONALD R. LEVITAN

Department of Biological Sciences, Florida State University, Tallahassee, Florida 32306-2403

BRENDA B. CAMERON

Center for Population Biology, University of California, Davis, California 95616

CONTENTS

Molecular Zoology: Advances, Strategies, and Protocols, Edited by Joan D. Ferraris and Stephen R. Palumbi.
ISBN 0-471-14461-4   © 1996 Wiley-Liss, Inc.

SYNOPSIS

The genetic and genealogical structure of natural populations affords both a reflection of evolutionary history, as well as a set of constraints and possibilities for future evolutionary change. Evolutionary biologists, armed with a variety of recently developed, high-resolution genetic markers, now possess unprecedented capacity to portray structure and reconstruct genealogies. Unfortunately, fashion and novelty, rather than utility, often dictate the popularity of a given technique. The truth is that no single type of marker yet offers the capacity to reveal cheaply and powerfully genetic structure and genealogical relationships across a full spectrum of temporal and spatial scales. In this chapter, we evaluate the strengths (cheap, fast, easy, and polymorphic) and weaknesses (repeatability, comigration of bands, detectability, dominance, and nonmendelian inheritance) of randomly amplified polymorphic DNA (RAPD) markers for depicting population structure and genealogies. Overall, RAPD markers, because they are anonymous and expressed as dominant alleles potentially concealing substantial amounts of hidden variation, generally rate poorly for their ability to portray breeding systems and phylogenetic relationships (especially above the level of populations). However, the available data suggest that in many taxa RAPD-PCR can easily generate hundreds of highly polymorphic, independent markers, the majority of which behave as mendelian alleles. Many of the markers may exhibit substantial variation in evolutionary rates. As such, RAPD markers offer evolutionary biologists still unmatched opportunities to evaluate intraspecific genetic structure and genealogical relationships in large samples that exhibit structure across diverse spatial scales.

INTRODUCTION

Evolutionary ecology rests on a foundation of understanding how variation in morphology, behavior, and resource allocation relates to fitness in different environments. The first step in building this foundation involves measuring phenotypic traits and analyzing the contributions that environmental, ontogenetic, and genetic processes make to the expression of these traits. The next, and still more challenging, step entails portraying the spatial and temporal scales over which environmental and genetic structure vary, so that character evolution can be set into a relevant selective context (Bell, 1992). The last step—one that has proved the most technically elusive—requires establishing how phenotypic variation correlates with reproductive success.

The measurement of reproductive success, achieved through either sexual reproduction or asexual propagation, fundamentally relies on the ability to assign offspring to parents. When females brood offspring or exhibit some form of parental care, establishing maternity can be relatively easy, even without genetic markers (but see Westneat et al., 1987). However, when females (or males) show no parental care, or communally care for young, direct assignment of parentage becomes substantially more difficult. In the extreme, when male and female gametes are released into the environment and motile propagules can disperse extensively (as in many sessile invertebrates and plants), assigning parentage and inferring population structure by direct observation are essentially impossible.

The development of allozyme markers in the 1960s opened the door to genetic characterization of population structure and reproductive success. Numerous individuals can be analyzed through allozyme electrophoresis at relatively low cost and with few technical impediments: that is one of the reasons why a number of us have stuck by what many consider an antiquated technique. However, 30 years of experience have shown that, in many taxa, the number of polymorphic loci and alleles per locus is often too low to characterize all but larger scale genetic patterns or to assign parentage with high confidence, especially when the parental pool is large or inbred. In addition, destructive sampling is usually obligatory, precluding analysis of reproductive success and genetic structure beyond a single generation.

Over the last decade, the development of new molecular tools that provide access to highly variable DNA markers, especially mini- and microsatellite VNTR (variable number of tandom repeat) loci (Burke et al., 1991) and loci of the major histocompatibility complex, has made high-resolution analysis of genealogies and genetic structure possible, but still not easy or risk free (reviewed in Avise, 1994). Most of these methods involve complex and expensive protocols and demand expertise in the tools of molecular genetics; nearly all are time-consuming and expensive. Some of these techniques, such as single-locus and multilocus DNA fingerprinting, depend on the availability of large amounts of undegraded DNA (Weatherhead and Montgomery 1991) and may require destructive sampling. For population biologists, the recent development of microsatellite markers offers several distinct advantages over multilocus fingerprinting (Queller et al., 1993; also see Strassmann et al., Chapter 8, and Fleischer, Chapter 7, in this volume).

As long as cost is an object, time a constraint, or technical ability and motivation limiting, population ecologists must always face a trade-off between gathering high-resolution genetic information and sampling large numbers of individuals (Lessa and Applebaum, 1993). This trade-off is especially intrusive where associations among particular traits (or suites of traits), selective regimes, and fitness may be weak, but nonetheless highly significant, both in a statistical and evolutionary sense. The accurate portrayal of such associations necessarily involves large samples. Thus, although mini- and microsatellite loci, and nuclear introns, offer access to highly polymorphic genetic markers, in terms of both time and money, their expense highlights, rather than ameliorates, the trade-off between genetic resolution and large sample sizes.

The development five years ago of randomly amplified polymorphic DNA (RAPD) markers by Williams et al. (1990), along with arbitrarily primed (AP) markers by Welsh and McClelland (1990) and DAF (DNA amplification fingerprinting) by Caetono-Anollés et al. (1991), appeared to provide an escape from the choice between *Scylla* (high genetic resolution) and *Charybdis* (large sample size) for molecular ecologists (Hedrick, 1992): hundreds of apparently neutral polymorphic markers throughout the genome can be generated quickly (Bowditch et al., 1993); large numbers of individuals can be processed relatively inexpensively and without any prior DNA sequence information; radioisotopes are unnecessary; and— because the polymerase chain reaction is used to amplify the RAPD markers—little (but nevertheless undegraded) DNA is needed, and nondestructive sampling is feasible (Hadrys et al., 1992; Black, 1993). Moreover, different loci exhibit different degrees of polymorphism, hence RAPD markers should be useful for deciphering population structure across a range of spatial and genealogical levels of resolution.

A number of technical and analytical problems soon tempered the initial enthusiasm for RAPD markers (reviewed by Hadrys et al., 1992; Black, 1993), leading some workers to question their utility in population ecology and phylogeny reconstruction (e.g., Black, 1993). Some of the negative reactions toward RAPD-PCR are undoubtedly justified; some, however, are hyperbole. In this chapter, we critically assess the strengths and weaknesses of RAPD markers in the context of characterizing parentage and genetic structure in populations of sessile organisms. We avoid discussion of the use of RAPD markers for the reconstruction of higher level spatial and genealogical/phylogenetic relationships, in part because our own research concerns intraspecific patterns and processes, but also because we, like others (Clark and Lanigan, 1993; Smith et al., 1994), doubt that these markers can generally yield useful information about transpecific relationships. Our approach is first to outline the basic protocols for generating, scoring, and interpreting RAPD markers. We then reexamine some of the problems encountered by us, and many others, in using RAPD markers and offer remedies to some of these problems. Finally, we evaluate the prospects of RAPD-PCR for characterizing population-level parameters in sessile and sedentary organisms with motile propagules by chronicling our own experiences with the technique.

WHAT ARE RAPDs?

Introduction

Randomly amplified polymorphic DNA markers consist of relatively short DNA fragments (about 200–2000 base pairs long), amplified via PCR by small (usually 10 bases in length) arbitrary (with a G + C content > 50%) primers. Typically, primers are used singly and must anneal to priming sites in opposite orientations in order for amplification to occur. Following convention, we term the pair of inverted priming sites, plus the intervening sequence of nucleotides, a RAPD locus, and the amplified product from a particular locus a RAPD marker. The resulting amplification product(s) can be size-separated electrophoretically on an agarose gel and visualized by ethidium bromide staining.

Depending on the number of inverted complementary priming sites in an individual's genome, and the lengths of the intervening DNA sequence, a given primer may amplify from 0 to 30 products. Hundreds of RAPD primers are commercially available (from Operon Technologies Inc., Alameda, CA), and when several primers are used independently, hundreds of polymorphic markers potentially can be identified. Nevertheless, different species (and even populations, e.g., Dawson et al., 1993; Huff et al., 1993) exhibit dramatically different degrees of polymorphism, both in terms of the proportion of RAPD loci that vary and the number of loci that amplify (Caetano-Anollés et al., 1991). Reasons for this variation in levels of polymorphism remain unclear.

Because RAPD-PCR primers are not designed to amplify a specific target sequence, the amplified loci are anonymous and presumably scattered throughout the genome (Williams et al., 1990, 1991; Tinker et al., 1993). RAPD loci carry the advantages that (1) there is no need for prior nucleotide sequence data for the taxa under study and (2) many of the loci may be acting as neutral markers. On the other hand, in the absence of information concerning the nature of the DNA being amplified by RAPD-PCR, there is no guarantee that any single primer will produce usable markers or that the variation is, in fact, neutral.

Most workers assume that RAPD loci amplify only if the priming sites perfectly match the oligonucleotide used (Hadrys et al., 1992), based on Williams et al. (1990), who showed that single nucleotide substitutions at a priming site can preclude amplification in tests with human, soybean, corn, and yeast DNA. However, annealing temperatures in RAPD-PCR are typically quite low (35–45°C) and primers are relatively short (8–15 bp); consequently, primer–template complementarity could be less than perfect and still result in amplification of template DNA. Fortunately, it is possible to choose PCR conditions that minimize imperfect priming (see below).

Scoring RAPD Polymorphisms

The usual gel assay for genetic variation at RAPD loci is the presence or absence of a band of a specific molecular weight, amplified by a given primer, in an individual sample. Bands (i.e., putative loci) that appear in all individuals are considered monomorphic; those present in some individuals but not others are polymorphic in the sampled population. Bands of different size are usually considered to represent independent loci and are scored as independent traits. This cannot be the case in every instance. For example, both Martin et al. (1991) and Smith et al. (1994) showed through Southern blot hybridizations that different bands amplified by a single primer contain homologous sequences, suggesting that different bands are not necessarily independent traits. Such bands may represent homologous alleles at the same locus or gene duplications. In addition, there should be some concern that bands amplified by different primers may be homologous, once again due to duplications or the presence of multiple priming sites within putative loci.

In some cases judgment of the presence or absence of a band has a subjective component, as band intensity can quantitatively vary over a wide range (see "Problems with RAPD-PCR and Some Remedies: Variation in Band Intensity and Limits to Detection"). Although band intensity itself can be heritable (Hunt and Page, 1992; Heun and Helentjaris, 1993; Levitan and Grosberg, 1993), nonheritable factors, including PCR conditions, DNA concentrations, and associations with other loci (Hunt and Page, 1992; Heun and Helentjaris, 1993; Wilkerson et al., 1993; Smith et al., 1994), may cause considerable brightness variation (reviewed in Black, 1993).

As a matter of convenience, most workers assume that when a given primer amplifies bands of the same apparent molecular weight, the bands represent the same allelic state (the dominant allele); conversely, the absence of a band represents a single alternative allelic state (the null, or recessive, allele). Because RAPD bands are usually expressed as phenotypically dominant markers in diploids, it is generally impossible to distinguish between the genotypes of individuals homozygous and heterozygous for the dominant allele (see "Problems with RAPD-PCR and Some Remedies: Dominance"). Segregation analysis, however, can be performed in organisms such as hymenopterans, in which males are haploid (Fondrk et al., 1993; Shoemaker et al., 1994), in those plants with a macrogametophytic stage (Bucci and Menozzi, 1993), or in individual gametes.

Sources of Polymorphism and Amplification Artifacts

Not all variation at RAPD loci should be detectable simply in terms of the presence or absence of a given band, and not all mutations will necessarily lead to a shift from presence to absence (or *vice versa*). Polymorphism, detectable or not, can arise in numerous ways. For example, an amplified band can be turned into a null in one of at least three ways: (1) either one or both priming sites can be lost due to mutations that reduce or eliminate priming; (2) an insertion may occur between priming sites that is so long that the fragment between sites is no longer amplifiable by PCR; or (3) sequences may be rearranged so that the priming sites are no longer complementary. A null can be transformed into an amplifiable marker when a mutation occurs at one or both inverted priming sites that makes those sites recognizable by a complementary primer, a deletion occurs between sites that shortens the intervening fragment so that it can be amplified by PCR, or by sequence rearrangements. Finally, an insertion or deletion between inverted priming sites of an amplifiable locus may alter the length of the amplified region between the priming sites, producing a length polymorphism of codominant alleles, rather than a presence/absence polymorphism.

Classes of mutations causing a gain or loss of a RAPD band appear straightforward to detect, but the mutational events themselves will be difficult to distinguish, at least without sequencing a locus or Southern blotting using amplified product as a genomic probe (Hadrys et al., 1992; see Smith et al., 1994 for an exemplary study). Mutations that produce length polymorphisms are far more difficult to detect without formal genetic analysis, largely because most RAPD primers amplify multiple loci, most of which are assumed to produce markers of different molecular weights (but see Hunt and Page, 1992, and "Problems with RAPD-PCR and Some Remedies: Comigration of Bands"). In practice, as with multilocus DNA fingerprinting, it is often impossible to distinguish bands of different molecular weights representing alleles at the same locus from bands representing alleles at nonhomologous loci (Lynch, 1988). For this reason, most workers simply ignore length polymorphisms ascribable to a single locus and score presence *versus* absence. Nevertheless, thanks to the haplo-diploid genetics of Hymenoptera, Hunt and Page (1992), as well as Shoemaker et al. (1994), identified a number of codominant alleles from single loci (also see Williams et al., 1990; Martin et al., 1991; Roehrdanz et al., 1993).

There are other potential problems in scoring RAPD variation solely in terms of presence or absence of a band. Southern blot analysis (using probes derived from amplified RAPD bands) of amplification products demonstrates that although some comigrating bands in different conspecifics are, indeed, homologous (Hadrys et al., 1992; Smith et al., 1994), some clearly are not (Smith et al., 1994). Overall, few studies have undertaken such time-consuming steps; hence, for the time being, we simply lack the data to judge whether this form of cryptic variation is widespread within and among populations.

In principle, therefore, conventional presence/absence scoring of RAPD markers likely underestimates levels of genetic variation (Apostol et al., 1993; Rossetto et al., 1995) and may provide, at best, a hazy reflection of underlying genotypic frequencies. For these reasons alone, our view is that RAPD markers should be scored as biallelic systems only when appropriate sequencing, blotting, or inheritance studies support such an assumption. Whether each RAPD marker should be scored as an independent character remains an open question.

There is an additional nongenetic concern when using RAPD markers to analyze population genetic parameters and phylogenies: presence/absence variation may be due to amplification artifacts (Ellsworth et al., 1993; "Problems with RAPD-PCR and Some Remedies: Repeatability of Banding Patterns Within and Among Individuals") or the insensitivity of detection methods ("Problems with RAPD-PCR and Some Remedies: Variation in Band Intensity and Limits to Detection"). These amplification and detection artifacts may be difficult to distinguish, in part because they may often have similar causes (including imperfect primer–template annealing, interlocus interactions, and effects of copy number). The existence of such variation, if pervasive, can confound parent–offspring relationships and lead to overestimation of levels of allelic diversity and genetic distances. However, unless particular populations or families express nongenetic variation differentially, this "noise" should be of similar magnitude in different populations or families.

Population Genetic, Genealogical, and Phylogenetic Implications of Dominance and Cryptic Variation: A Preliminary Overview

If a population is in Hardy–Weinberg equilibrium, and allelic variation at a locus comprises two, and only two, allelic states, then the inability to distinguish dominant homozygotes from heterozygotes does not pose a serious barrier to the analysis of parentage and population structure. This is because the frequency of the recessive allele, particularly if it is common (>0.1) and the sample size is large, can be reckoned from the frequency of null/null homozygotes (Lynch and Milligan, 1994). Allelic and genotypic frequencies can then be analyzed by conventional software to calculate F-statistics and their analogues [e.g., BIOSYS-1 (Swofford and Selander, 1989), Genetic Data Analysis (Lewis and Zaykin, 1996), or AMOVA (Excoffier et al., 1992)] or relatedness (e.g., Reeve et al., 1992). If, however, there is more than one null allele, then it will be risky to calculate allelic and genotypic frequencies. To the extent that cryptic variation represents a general phenomenon, the implications for RAPD-based population genetic studies are clear: indices of genetic structure, gene flow, and mating patterns based on RAPD markers scored as biallelic character states in Hardy–Weinberg equilibrium should be interpreted cautiously.

The potential for cryptic variation in the null allelic class, and comigration of nonhomologous alleles, should also temper enthusiasm for using RAPD markers for transpecific comparisons of population genetic parameters, and even for intraspecific comparisons of variation exhibited by RAPD markers to other nuclear or mitochondrial markers (Liu and Furnier, 1993; Haig et al., 1994; but see Peakall et al., 1995). Applications of cladistic phylogenetic reconstruction (e.g., parsimony) that treat shared bands (or nulls) as synapomorphies (e.g., Welsh et al., 1992; Landry et al., 1993) are invalid if there are more than two allelic states (Smith et al., 1994). Even a purely phenetic approach to reconstructing genealogies, based on overall similarity in band-sharing patterns, can be confounded by the problem of cryptic variation, depending on the similarity metric and whether there is bias in the number of cryptic alleles in the presence *versus* absence (e.g., Rossetto et al., 1995). However, on balance, there are more ways to lose a RAPD band than to gain one, and shared presence alleles are more likely to be homologous than shared nulls. Thus, as we discuss in "Problems with RAPD-PCR and Some Remedies: Nonmendelian Variation," similarity indices based on sharing of expressed bands alone (e.g., Nei and Li, 1979) should be more robust than those that weigh shared presence alleles equally with shared nulls (e.g., Black, 1993; Rossetto et al., 1995), especially when used to build phenetic trees.

PROBLEMS WITH RAPD-PCR AND SOME REMEDIES

Repeatability of Banding Patterns Within and Among Individuals

Background. A high-resolution genetic marker is only useful when each can be reliably and repeatably amplified from a given individual, and each element is heritable (see "Problems with RAPD-PCR and Some Remedies: Nonmendelian Variation"). As with any high-resolution, PCR-based technique (including direct sequencing of PCR products), different amplification conditions can yield different amplification products (Palumbi et al., 1991). Susceptibility to such amplification artifacts is potentially a very serious problem for RAPD-PCR, because the amplified loci are generally anonymous, the primer–template annealing conditions are comparatively permissive, undegraded DNA must be used, and a wide range of amplification parameters can affect banding patterns (reviewed in Schweder et al., 1995).

Not surprisingly, many RAPD-based studies report some difficulties in obtaining repeatable results in terms of band number, molecular weight, and brightness (reviewed in Ellsworth et al., 1993; Bielawski et al., 1995; Schweder et al., 1995). Most problems seem to involve (1) different banding patterns arising from replicate samples of the same individual or (2) bands appearing and disappearing haphazardly in a series of supposedly identical amplifications.

Schweder et al. (1995) recently garnered a fearsome list of potential causes of unreliable and inconsistent amplifications. These confounding variables act in part by influencing the specificity and efficiency of primer–template interactions during the initial phases of PCR amplification. The most frequently cited causes include (1) the identity or model of thermocycler (Klein-Lankhorst et al., 1991; Williams et al., 1991; Fani et al., 1993; Penner et al., 1993; but see Haig et al., 1994); (2) annealing temperatures (Welsh and McClelland, 1990; Welsh et al., 1991; Ellsworth et al., 1993; Levi et al., 1993); (3) denaturation, priming, and extension times and ramping profiles (Yu and Pauls, 1992; Levi et al., 1993; systematically examined by Schweder et al., 1995); (4) primer and template combinations *and* concentrations (Welsh and McClelland, 1990; Hadrys et al., 1992; Ellsworth et al., 1993; Levi et al., 1993; Micheli et al., 1994; Muralidharan and Wakeland, 1993; Williams et al., 1993; Smith et al., 1994); (5) dNTP concentration (Levi et al., 1993; Williams et al., 1993); (6) Mg^{2+} concentration (Ellsworth et al., 1993; Levi et al., 1993; Williams et al., 1993); and (7) *Taq* polymerase concentration (Levi et al., 1993; Williams et al., 1993) and source (Aldrich and Cullis, 1993; Fani et al., 1993; Levi et al., 1993; Williams et al., 1993). Additionally, some primers give notoriously inconsistent results—perhaps because of nonspecific annealing to the template DNA (e.g., Bielawski et al., 1995)—whereas others are extremely reliable. Finally, the quality of a DNA sample may change over time because of contamination or degradation (Black et al., 1992), a problem we have faced with unextracted tissue stored at −80°C, and with apparently clean DNA stored at −20°C.

Inconsistency at the second level can arise because each extraction starts from a different piece of tissue and then follows an independent extraction (and perhaps amplification) trajectory. Variation in (1) the presence (or concentration) of pathogens, symbionts, or other contaminants (e.g., Micheli et al., 1994) in different tissues; (2) somatic mutations; or (3) extraction procedures and reagents can all produce initially different DNA samples at different concentrations. Subsequent differences in sample processing and storage (e.g., number of freeze–thaw cycles) can also introduce variation. For the reasons noted in the previous paragraph, such variation can affect banding patterns.

Some Remedies. We initially encountered many problems with consistency of amplifications at all the levels identified above. The first step toward success with RAPD-PCR is ensuring that the template DNA is consistently undegraded and uncontaminated by endonucleases, polysaccharides, and other garbage that would interfere with amplification by PCR. Our initial analyses of our extracted DNA on mini-agarose gels showed that it varied in quality, at times being quite degraded. Sometimes it could be fully digested by restriction endonucleases; other times it resisted digestion. We finally settled on the extraction procedures presented in Grosberg.1, using 2% hexadecyltrimethyl ammonium bromide (CTAB). CTAB is a buffer/detergent that seems to work extremely well in combination with standard proteinase-K plus phenol:chloroform extraction protocols, especially with mucusy organisms like many invertebrates and plants (Rogers and Bendich, 1988).

The next step is to optimize the relative concentrations and components of the PCR reaction mixture and the amplification conditions for your thermocycler. An essential step is to quantify the amount of DNA in a sample (preferably using a fluorometer, rather than a spectrophotometer), so that primer–template ratios can be optimized and consistently maintained. In setting up our protocols, we manipulated virtually all the variables listed in "Problems with RAPD-PCR and Some Remedies: Repeatability of Banding Patterns Within and Among Individuals"; however, we always used the same thermocycler and source of purified water. We ultimately found a reaction formula and amplification profile that allowed a wide range of DNA concentrations (three orders of magnitude) to be used without altering banding patterns or intensities for all the primers we used to generate data sets.

Others have not been so fortunate. To them, we first recommend taking whatever time is necessary to develop collection, extraction, and DNA quantification protocols that are simple and reliable. Once this step is overcome, it is worth consulting Williams et al. (1993), Micheli et al. (1994), and Bielawski et al. (1995), for suggestions on optimizing consistency in the amplification step of RAPD-PCR. Although we have not tried it ourselves, among the more interesting recent suggestions is the use of a single-stranded binding protein [gene protein (Gp) 32] in the PCR cocktail to enhance the specificity of primer annealing, and thereby increase the intensity and repeatability of RAPD amplification products (Bielawski et al., 1995).

The bottom line concerning repeatability *within* individual genotypes is simple: at the outset, develop procedures ensuring that replicate tissue extractions provide identical banding patterns for a particular primer–template combination. If the same set of protocols gives reliable amplifications across a set of primer–template combinations, then do not change any extraction and amplification protocols, including ramping times and PCR machine (unless you are interested in adding to the list of potential woes, or are a molecular biologist who actually seeks to understand the causes of the problems). Use positive (i.e., samples with known banding patterns) and negative (i.e., blanks) controls throughout a study. In other words, don't ever gloat, don't cut corners, and don't underestimate the fickleness of PCR. When consistency problems arise, be prepared to start over, first checking your reagents, your DNA, your thermocycler, and your horoscope.

To our knowledge, there is only one way to *ensure* that bands are both repeatable and reliable mendelian markers *among* genotypes: perform crosses to ensure that each and every band is transmitted in a manner consistent with mendelian inheritance. Clearly, such a task would be impossible for large numbers of bands. In "Problems with RAPD-PCR and Some Remedies: Nonmendelian Variation," we offer a statistical approach to this problem, although it will not satisfy those requiring a guarantee of absolute heritability.

Comigration of Bands

Background. RAPD bands can be numerous and tightly spaced on an agarose gel (Fig. 1) and can be very difficult to distinguish. In principle, the amplification products of distinct RAPD loci, or even different alleles at the same locus, could be of indistinguishable molecular weight (see "What are RAPDs: Sources of Polymorphism and Amplification Artifacts"). The probability of this problem increases as the number and density of bands and the number of sampled individuals increase.

Comigration of homologous (but different) alleles, or nonhomologous alleles, *in the same individual* (i.e., the same lane on a gel) will produce a composite band that comprises two, or more, amplification products. Such composites will yield an underestimation of the number of alleles per locus, heterozygosity, and the frequency of polymorphic loci and will diminish the power to discriminate among individual genotypes and populations. If extensive, comigration of both homologous and nonhomologous products will also confound cladistic analyses (Smith et al., 1994).

Some Remedies. If the goal is simply to increase the resolving power of RAPD-PCR, then there are at least four courses of action. The first three involve attacking the problem at the level of the gel and minimizing scoring errors: agarose gel electrophoresis, in combination with ethidium bromide staining, is notoriously insensitive to revealing small differences (5–10 base pairs, at best) in molecular weight of amplified DNA, especially when bands are slightly blurry. An expensive, time-consuming, and toxic remedy is to use acrylamide (because of its higher

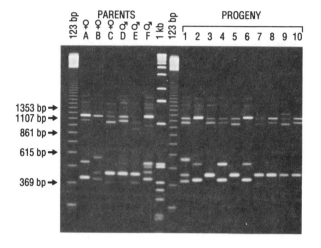

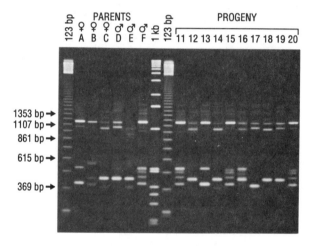

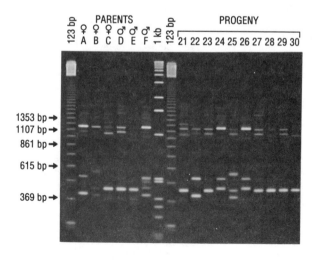

resolving power) and silver-staining or radioactive labeling to visualize PCR products. Alternatively, and much less expensively, one can simply run longer gels (Smith et al., 1994). We (Levitan and Grosberg, 1993), like Hunt and Page (1992), also found that the use of relatively low concentration agarose gels (0.6%), coupled with the addition of a cross-linker (1.0% Synergel; Diversified Biotech, Newton Center, MA), greatly reduces blurriness, allowing us to resolve bands that differ by as little as 10 base pairs. The fourth course of attack is to increase the number of primers used to screen a sample, and thereby reveal more polymorphic loci. This brute force strategy has the advantage of simplicity; it should also increase statistical power for the estimation of relatedness, as it is more likely (although not guaranteed) to reveal polymorphism at independent loci (Queller and Goodnight, 1989; Reeve et al., 1992).

If, on the other hand, the goal is to characterize accurately levels of polymorphism, breeding systems, or phylogenetic relationships, then it is essential to establish whether comigrating bands in different individuals are homologous. There are at least three approaches to addressing this question. Smith et al. (1994) suggest digesting comigrating RAPD amplification products with one (or more) restriction endonucleases, then re-running the digested products on an agarose gel. Digestion of homologous bands should produce the same number and sizes of products. Two additional approaches, both considerably more powerful, expensive, and time-consuming, involve sequencing amplified bands, or use of amplified products as probes in blots of genomic DNA (Hadrys et al., 1992; Smith et al., 1994). These approaches should be most successful when comigration occurs across samples. However, as discussed above, if comigration occurs within lanes too, then neither probing nor sequencing will necessarily clarify just how much cryptic variation there is. To date, Southern hybridizations suggest that at least some comigrating bands represent amplification products from nonhomologous loci (Smith et al., 1994; but see Martin et al., 1991; Hadrys et al., 1992).

Variation in Band Intensity and Limits to Detection

Background. RAPD bands are usually scored as discrete presence/absence characters; yet, (1) some bands consistently amplify more strongly than others and (2) a given band in different individuals may quantitatively vary in intensity from barely visible to unmistakably brilliant. Furthermore, bright bands tend to amplify more

---

Figure 1. RAPD-PCR products amplified from *Hydractinia symbiolongicarpus* by Operon Technologies primer F-07 on a 0.6% agarose (plus 1.0% Synergel), ethidium bromide-stained gel. The number, intensity, resolution, and spacing of amplification products depend on the primer–template combination (compare lanes labeled Progeny 1 to 10). The amplification products are from the three matings described in the text (♀A × ♂F; ♀B × ♂D; ♀C × ♂E) and their offspring (♀A × ♂F: 2, 4, 6, 11, 13, 15, 20, 22, 24, 26; ♀B × ♂D: 1, 3, 8, 10, 17, 19, 23, 25, 27, 29; ♀C × ♂E: 5, 7, 9, 12, 14, 16, 18, 21, 28, 30). Comparison of banding patterns in parents and offspring shows that all the scoreable bands (see text) on this gel behave as mendelian markers. DNA ladders (123 bp and 1 kb) and base pair sizes provide standardized reference points.

reliably than faint ones (*personal observations;* also see Black, 1993). Because it may be difficult to detect a weakly amplified product on an ethidium bromide-stained agarose gel, a locus could be scored as a null/null homozygote when, in fact, a weakly amplified allele is present. For example, in five of 47 strains of *Xanthomonas,* Southern analysis revealed an amplification product undetectable on ethidium bromide-stained agarose gels (Smith et al., 1994).

Some variation in band intensity appears to be both repeatable and heritable (Hunt and Page, 1992; Heun and Helentjaris, 1993; Levitan and Grosberg, 1993). Heritable variation in band intensity may be due to variation in (1) the number of copies of loci amplifiable by a particular primer–template combination, (2) copy number of tandemly repeated amplifiable sequences within a locus, or (3) additive interactions among loci (Hunt and Page, 1992; Smith et al., 1994). However, due to priming site competition and other nonadditive interactions among loci (Smith et al., 1994), as well as the effects of amplification conditions on primer–template affinity, not all variation in band intensity is likely to be repeatable and heritable.

Finally, even when the intensity of a specific RAPD band is both repeatable and heritable, the question remains of how to compare two individuals, one of which expresses a band strongly, the other of which expresses the same band weakly. One problem with scoring such intensity differences is that intensity itself is a notoriously difficult trait to quantify, especially on ethidium bromide-stained gels. Second, in no case is it known how differences in intensity relate to differences in underlying genotype.

Some Remedies.   One could classify bands according to their brightness, analyze the inheritance of brightness classes of a particular band, and score different brightness categories as different alleles at the same locus. This option is worth pursuing in haplo-diploid taxa (e.g., Hunt and Page, 1992; Shoemaker et al., 1994), because the association between genotype and phenotype can be discerned in the haploid gender, or perhaps when relatively few loci are being analyzed. However, if it is desirable to score numerous loci, then this approach will be impractical, at best.

There are at least two other realistic options. The first makes use of every polymorphic band: provided it is detectable, a band is scored as present, regardless of its intensity and absent if it is undetectable. This procedure does not discriminate between genetic and nongenetic causes of variation in banding intensity and assumes that there are no population-specific biases in the expression of brightness variation, including the pattern of false negatives (i.e., undetectable, but nevertheless amplified, products).

The alternative is to score only those bands that are bright when present, with little or no quantitative variation from this state to absence (i.e., undetectability). This scoring procedure appears to lack some of the subjective bias of classifying all bands, regardless of intensity, as present or absent. Unfortunately, as the number of sampled individuals increases, so too does the likelihood that brightness polymorphism for a band will be exhibited in the sample. In our own studies, we were in several instances quite confident that we were scoring a clear presence/absence polymorphism for a given band, only to discover that as we expanded our sample to include individuals from other populations, the band exhibited quantitative variation in brightness. One then must decide whether to move on to other loci, or to develop an unbiased scoring protocol. This approach may severely underestimate overall levels of genetic variation, because it discards bands (loci?) exhibiting quantitative variation for band intensity. However, unless there is population-specific variation in the number of loci exhibiting such variation, comparisons of genetic similarity may not be severely biased.

Dominance

Background. RAPD alleles are expressed as dominant markers, complicating the inference of underlying genotypes. If either allele at a locus carries an amplifiable fragment, then the phenotype will be a visible band, but—assuming the locus is biallelic—the underlying genotype could be either $+/+$ or $+/-$. If the absence of a band represents a single null allele, individuals that lack a particular band should be homozygous (i.e., $-/-$) for the null allele at that putative locus. The frequency of that allele in a sample (in Hardy–Weinberg equilibrium) can then be calculated from the square root of the proportion of individuals that lack a particular band. It should therefore be possible to estimate allelic and genotypic frequencies reasonably accurately at a RAPD locus, provided that (1) the sample is large (thereby minimizing error in the estimate of the true allelic frequencies); (2) the frequency of the null category exceeds 0.1; (3) the population is in Hardy–Weinberg equilibrium; and (4) the locus in question is biallelic (Lynch and Milligan, 1994). [If the sampled population is in Hardy–Weinberg equilibrium, it is possible to calculate the frequency (and variance) of the expressed marker allele (if it represents a single allele), even if the null category conceals multiple alleles.] If all of these conditions are met, then these frequencies can be used to calculate F-statistics (and their corresponding errors; see Lynch and Milligan, 1994) and to characterize likelihoods of parentage. If they are not, then it will be virtually impossible to estimate allelic and genotypic frequencies.

Satisfaction of the first two conditions is simply a matter of scoring a large sample, and using only those loci with reasonably high frequencies of nulls. Although most studies assume Hardy–Weinberg equilibrium, they fail to demonstrate it. In our view, but without independent knowledge of whether sampled populations and loci conform to Hardy–Weinberg equilibrium, it seems risky to assume that they are.

The available theoretical (see "What are RAPDs: Sources of Polymorphism and Amplification Artifacts") and empirical evidence suggests that the final condition may not be generally realized. Southern blots have shown that a presence band can conceal more than a single allelic variant (Smith et al., 1994). Our own data (Levitan and Grosberg, 1993) suggest that the same is true for the null allelic category at numerous loci in the hydrozoan *Hydractinia symbiolongicarpus*.

We reached this conclusion through simple-minded probabilistic reasoning, rather than sequencing null alleles. Assume that RAPD loci are truly neutral, and that each locus is represented by two alleles, one expressed, the other null. At some loci, one of the alleles would be more common than the other, whereas the reverse would be true at other loci. However, for a large sample of loci, the distribution of allelic frequencies across loci should approximate a binomial, with a mean of 0.5, because there is no reason to believe that one allele should be systematically more common than the other (i.e., the ratio of presence to absence alleles should be approximately 50:50). In *H. symbiolongicarpus,* we estimated the mean frequency of the null allele, averaged across 133 polymorphic loci, to be 0.809 (lower 95% CI = 0.785; upper 95% CI = 0.831) in the parental generation (*n* = 6 individuals), and 0.818 (lower 95% CI = 0.789; upper 95% CI = 0.845) in their 30 offspring. This implies that null alleles actually outnumber presence alleles by several times, and that the absence phenotype comprises a pool of alleles that all share the characteristic of not being amplifiable by RAPD-PCR. If this is generally true, and presence phenotypes also encompass numerous alleles, we see no obvious way to estimate RAPD allelic frequencies from presence/absence data.

Some Remedies. There appear to be three alternatives, all of which entail substantial compromises. The "see-no-evil" approach is to hope that there really are only two alleles per locus and that the population sample is in Hardy–Weinberg equilibrium. A slightly better "hear-no-evil" variant would be to use the binomial test we discussed above, with the caveat that the approach only gives a general picture for many loci, not for a specific locus. For the connoisseur of molecular genetics and evolution, the "speak-no-evil" choice would be to use Southern blotting (using probes derived from amplified markers) to reveal all cryptic variation in both scorable classes at all loci, then use this information rather than presence/absence data to estimate allelic and genotypic frequencies. This converts RAPD loci to anonymous single-copy loci (*sensu* Karl and Avise, 1993), thereby losing the major advantages of RAPDs.

The third option is to accept that cryptic variation lies concealed in both the presence and absence categories and acknowledge that RAPD markers are generally ill-suited to the calculation of allelic and genotypic frequencies. Despite this, it is still possible to estimate frequencies of different bands and use those frequencies in either an AMOVA (Excoffier et al., 1992) or a G-test to analyze genetic structure. In this approach, different band frequencies indicate different gene frequencies in different populations. Similarly, although high-resolution analysis of parentage requires considerably more RAPD markers than would be the case for codominant markers (Milligan and McMurry, 1993), this is unlikely to be a limiting feature of RAPD-PCR, except perhaps in highly inbred populations.

Nonmendelian Variation

Background. We refer to repeatable variation in RAPD-PCR products that does not correspond to underlying genetic variation as nonmendelian variation. We exclude from this designation those sources of banding variation that are not repeatable (see "Problems with RAPD-PCR and Some Remedies: Repeatability of Banding Patterns"). Nonmendelian inheritance of RAPD markers, whatever its source, potentially confounds analyses of parentage based on inclusion or exclusion techniques and, at the population level, leads to overestimates of the amount of allelic variation and heterozygosity at RAPD loci. Because RAPD loci are anonymous, and RAPD banding patterns reflect the inheritance of multiple loci and only dominant alleles, nonmendelian inheritance of RAPD markers can be very difficult to detect. Fortunately, most analyses of the transmission genetics of RAPD markers suggest that the majority of bands behave as mendelian alleles (Williams et al., 1990; Carlson et al., 1991; Martin et al., 1991; Welsh et al., 1991; Hunt and Page, 1992; Bucci and Menozzi, 1993; Fondrk et al., 1993; Levitan and Grosberg, 1993). Several studies, however, report the amplification of bands from offspring DNA that did not appear in either putative parent (Riedy et al., 1992; Hunt and Page, 1992; Fondrk et al., 1993).

There are at least three major sources of nonparental bands in offspring, which are often confounded (e.g., Riedy et al., 1992). The first and most obvious (but rarely admitted) is that the putative parents are not the true genetic parents, an option best tested through remating. The second (and, in our opinion, the most common) source comprises laboratory or field artifacts, such as contamination, degradation of template DNA, and inconsistent technique, that promote the spurious appearance (or disappearance) of bands, at least some of which may be repeatably and reliably amplified. Finally, because RAPD-PCR uses fairly permissive annealing conditions, it can produce heteroduplexes (Hadrys et al., 1992; Hunt and Page, 1992), which form when homologous alleles also contain a nonhomologous region. This yields a hybrid fragment that differs in gel migration rate from the actual length of either allele at a locus. RAPD bands may also disappear due to varying levels of competition among primer binding sites, such that efficient amplification depends on the identity and complexity of the remainder of the genome (Hunt and Page, 1992). Indeed, Smith et al. (1994) identified several cases in the phytobacterium genus *Xanthomonas* in which the presence of one fragment appeared to preclude efficient amplification of another fragment.

Some Remedies. The only definitive remedy is to conduct segregation analyses on all bands, separate those markers that behave as mendelian alleles from those that do not, and eliminate all nonsegregating bands from subsequent analyses. In theory, if all scored bands behave as mendelian markers, then it should be possible to analyze parentage using any combination of exclusion, inclusion, or likelihood algorithms (Meagher, 1986; Thompson, 1986; Meagher and Thompson, 1987; Devlin et al., 1988). Similarly, all such markers would provide useful information for the calculation of higher-order genealogical relationships (reviewed in Reeve et al., 1992). Nevertheless, there are several significant drawbacks to this approach. First, it is not clear that establishing the fidelity of markers with a subset of individuals guarantees that a nonhomologous comigrating band will never appear. Second, the number of polymorphic markers that can be assayed, the frequencies of those markers, and the sampling variances of the frequencies fundamentally constrain the power to reconstruct genealogies (Westneat et al., 1987; Lynch, 1988; Burke, 1989; Reeve et al., 1992). Even if it were possible to conduct the necessary matings and identify the subset of markers that behave as mendelian alleles, it would be especially burdensome to establish such behavior for large numbers of dominant markers in diploid individuals.

The less palatable (but more practical) option is to accept that a comprehensive segregation analysis is either impossible, or not worth the trouble, and devise analytical techniques that reduce (or ignore) errors caused by nonmendelian bands. The virtue of this option is that it makes use of essentially all bands, the high-resolving power of numerous markers being one of the strengths of RAPD-PCR. However, nonmendelian bands can wreak havoc with parentage analyses based on inclusion (the search for parent-specific diagnostic bands in progeny) or exclusion (the search for marker bands that can eliminate one or more individuals from a pool of parents).

A straightforward alternative for characterizing genealogical relationships and genetic structure in a sampled population is first to use all data to build a similarity (or dissimilarity) matrix for all individuals based on the degree of band-sharing, and then to analyze the similarity matrix with a cluster analysis or other tree-building algorithm. This yields a hierarchical depiction of patterns of similarity across all loci, which should mirror genealogical relationships, if band-sharing indices reflect underlying genetic distances. Cluster analyses have the advantage of pooling information that alone would be insufficient to establish genealogical relationships. For example, a marker shared by all members of a sibship, but not private to those individuals in that sibship alone, cannot alone definitively establish membership in that group; however, it can add to the weight of data supporting a particular set of genetic relationships (Apostol et al., 1993). Likewise, if a few nonparental bands appear in offspring, clustering and other tree-building (e.g., neighbor-joining) algorithms will weigh those bands against the high percentage of bands inherited as mendelian markers.

On the other hand, the clustering approach lacks the ability to discriminate classes of relatives of similar relatedness (e.g., full-sibs *versus* parent–offspring), and (especially with overlapping generations) it may be very difficult to discern discrete classes of relatedness. It is possible to plot indices of band-sharing against relatedness inferred from known pedigrees, to calibrate the relationship between band-sharing and relatedness (Piper and Rabenold, 1992; Apostol et al., 1993).

Although clustering based on overall similarity should be relatively insensitive to the occasional appearance of nonmendelian bands, particularly with numerous (say, >100) polymorphic markers, detection of lower-order groupings may be most sensitive to nonmendelian "noise." For this reason, and to appease purists, it is always imperative to minimize contamination, degradation, and amplification artifacts. It is also worth carefully inspecting the raw data matrix to determine whether the presence (or absence) of one band, or a few bands, defines a particular grouping. This can be done by observing how PAUP (Swofford, 1991), or some other cladistic method, assigns character state transitions on the tree generated by distance methods. However, most cladistic tree-building algorithms become glacially mired if the data set includes numerous (say, >20–25) individuals or higher level taxa.

The choice of a similarity metric is also not trivial; in particular, several commonly used indices make very different assumptions about weighting of presence *versus* absence information. For example, Nei and Li's (1979) similarity index considers only band presence, not absence; whereas many other commonly used indices incorporate both presence and absence data (e.g., Apostol et al., 1993; Black, 1993; Rossetto et al., 1995). The former index ignores band absence when estimating similarity; we prefer it because we suspect that on average shared bands are more likely than nulls to represent homologous alleles. In the case of rare or endangered organisms, others (e.g., Rossetto et al., 1995) oddly advocate the use of indices that include both the presence and absence of bands, arguing that the risk of overestimating genetic similarity acts as a "safety margin" in genetic conservation strategies. We suggest trying a variety of indices, many of which are conveniently available in the RAPDistance software package (Armstrong et al., 1994), and comparing distance matrices using programs such as DIPLOMO (DIstance PLot MOnitor) analysis (Weiller and Gibbs, 1995).

Finally, different phenetic tree-building algorithms, because they use different agglomeration procedures and make different assumptions about rooting and evolutionary rates, may yield different tree topologies from identical similarity matrices (Felsenstein, 1993). We recommend using a variety of phenetic tree-building methods (e.g., UPGMA, Ward's minimum variance, and neighbor-joining), analyzing the statistical robustness of the resulting trees with some sort of randomization test (e.g., bootstrapping), and comparing the topologies of the trees generated by different algorithms. If the trees differ beyond the details, you should carefully consider why.

TWO CASE STUDIES WITH COLONIAL HYDROZOANS

Background

Against this problem-laden background, critics, sceptics, and population biologists may wonder why anyone would step into the breach of RAPD-PCR. One trivial answer is that all genetic markers have their problems, but their advocates are generally not terribly forthright. For us, the answer is fairly simple: RAPD-PCR gives ready access to literally hundreds of characters that appear to evolve at rather different rates and that can be assayed in hundreds or thousands of individuals, all at relatively low cost in time and materials.

In this section, we act as advocates rather than critics of RAPD-PCR and show how RAPD markers can be used to characterize parent–offspring and higher-order genetic relationships, as well as population structure. We focus on our own studies of two congeneric species of colonial marine hydrozoans, in which the temporal association between adults and their gametes, or adults and offspring, is so brief that genetic methods must be used to deduce parentage and to analyze population structure.

The first study centers on *Hydractinia symbiolongicarpus,* which lives in protected nearshore waters along the New England coast of the United States. We considered three questions in this species: Can we use RAPD markers to (1) associate parents with their offspring; (2) identify sibships correctly; and (3) characterize the genetic composition of chimerical colonies? The second study concerns two questions that involve portraying genetic structure in natural populations of *H. milleri,* an inhabitant of the Pacific Northwest (from northern California, northward at least to Vancouver Island): (1) at what level of resolution can we identify genetic structure, and (2) to what extent is our ability to discern structure limited by variation in the marker?

In addition to living on opposite coasts and distinct habitats, these two species of gonochoric hydroids have life histories that differ in several key aspects. *Hydractinia symbiolongicarpus* colonizes gastropod shells occupied by the hermit crab *Pagurus longicarpus* (Buss and Yund, 1989). Males release their sperm into the water, whereas females release their sinking eggs onto the benthos (Bunting, 1894; Ballard, 1942). Fertilized eggs rapidly develop into demersal larvae (Müller, 1969; Weis and Buss, 1987; Yund et al., 1987; Shenk and Buss, 1991). These larvae settle onto passing hermit crab shells, metamorphose, and develop into colonies consisting of hundreds to thousands of polyps, linked to one another by a common gastrovascular system (reviewed in Yund et al., 1987). Two or more larvae frequently colonize a single shell, and when they do, there is the potential for intense intraspecific competition for space (Yund et al., 1987). As the colonies come into contact, one of two outcomes generally ensues. Either the colonies fuse via their gastrovascular systems, forming a physiologically and morphologically unified genetic chimera; or one or both colonies produce highly modified extensions of the gastrovascular system, termed hyperplastic stolons (Ivker, 1972, and references therein). Large batteries of nematocysts cover these specialized stolons, which can inflict fatal damage to an opponent (Buss et al., 1984). The hyperplastic response can involve a substantial investment of somatic tissue, decreasing somatic growth and reproductive output by up to 50% (R.K. Grosberg and L.W. Buss, *unpublished observations*). Although *H. symbiolongicarpus* colonies themselves are sessile, the hermit crabs that bear them appear to mix extensively; thus populations ought to exhibit little genetic structure. In fact, allozyme markers revealed a high degree of polymorphism, but no evidence of genetic structure or nonrandom mating, among *H. symbiolongicarpus* colonies from different hermit crabs at the same geographic locality.

Colonies of *Hydractinia milleri,* in contrast to *H. symbiolongicarpus,* inhabit large boulders on wave exposed shores, low in the intertidal. In terms of expected genetic structure, their reproductive biology differs from *H. symbiolongicarpus* in one significant respect: although males release sperm into the water, females brood their embryos (*personal observations*). The brooded larvae eventually take up residence among the polyps of their mother, before crawling away to found a new colony. Colonies are typically 5 (about 10 polyps) to 100 mm across (several thousand polyps) but occasionally reach 200 mm or more. Colonies generally form aggregations consisting of from 10 to 100 colonies, with discrete aggregations separated by anywhere from 50 cm to hundreds of meters of apparently habitable, but uncolonized, substrate.

Unlike *H. symbiolongicarpus,* natural interactions between conspecific *H. mil-*

leri do not apparently elicit the production of hyperplastic stolons, or any other obvious sign of aggression. Colonies seem to coexist with stable boundaries for several years. One explanation for this lack of aggression is that neighboring colonies are so closely related that they lack sufficient genetic disparity to evoke such behavior. Indeed, interactions between colonies taken from different localities do elicit the production of nonfeeding, hypertrophied polyps (tentaculozooids) along colony margins. Taken together, the reproductive biology, dispersion, and lack of aggression exhibited by *H. milleri* colonies suggest that aggregations consist of a small number of founding individuals and their offspring, which, if persistent, may become genetically differentiated and inbred with respect to nearby aggregations. If both sperm and larval dispersals are primarily restricted to the bounds of an aggregation, then an aggregation provides a conveniently small subpopulation for investigating parent–offspring relations and correlates of reproductive success.

For both species, we needed a genetic marker that would distinguish closely related individuals, yet allow us to assay thousands of individuals. Allozymes revealed a rich store of genetic variation in *Hydractinia symbiolongicarpus,* allowing us to distinguish all sampled individuals based on their multilocus genotypes. Similar attempts using allozyme markers to distinguish individuals of *H. milleri* failed. Moreover, we became interested in analyzing the parentage of larvae, and in nondestructively sampling natural populations through time. This required that we employ not only a highly polymorphic marker, but one that could be amplified by PCR. At about this time, the first papers describing RAPD-PCR and its variations began to appear. We jumped on the bandwagon. In the next two sections we describe how we developed RAPD markers in these two species, and we summarize some preliminary results. Levitan and Grosberg (1993) fully report the data summarized here for *Hydractinia symbiolongicarpus.*

Study I: Parentage in *Hydractinia symbiolongicarpus*

DNA Extraction. At the time we began these studies, there were no protocols for extracting high molecular weight DNA from hydroids. Conventional proteinase-K/phenol:chloroform extraction techniques with various detergents (e.g., SDS and Triton-X-100) were undependable, and the DNA was often severely degraded. Finally, we followed botanical advice (Milligan, 1992) and added CTAB (hexadecyltrimethyl ammonium bromide) to the extraction buffer. Electrophoresis on minigels, as well as fluorometric analysis, revealed consistent yields of high-quality DNA.

Optimization of PCR-Amplification Conditions. The first step is to develop a PCR cocktail that gives repeatable amplification of mendelian bands using the minimum amount of template DNA (often in short supply) and DNA polymerase (the most expensive reagent, by far). This involves careful adjustment of the amount of template DNA in each extraction, followed by optimization of the concentrations of template DNA, RAPD primer, dNTPs, DNA polymerase, and salts ($MgCl_2$ and KCl).

The second step is the amplification process itself. A typical PCR reaction involves about 30–50 amplification cycles, each consisting of three phases: denaturation of template DNA, primer annealing, and polymerization/extension. RAPD-PCR differs from other types of amplification procedures in that only a single primer is generally used to amplify template DNA, and primer annealing takes place at a relatively low temperature (35–50°C). The duration and temperature of each phase can and should be adjusted so as to minimize the probability of nonspecific priming (the highest feasible annealing temperature) and maximize repeatability. If possible, the effects of transition profiles (i.e., "ramping times") between phases should also be explored (see references in "Problems with RAPD-PCR and Some Remedies: Repeatability of Banding Patterns Within and Among Individuals"). Finally, we found it worthwhile in terms of repeatability to start the reaction with a single cycle of prolonged denaturation.

Electrophoresis and Detection of Amplification Products. Of the various electrophoretic media currently available, agarose gels, at concentrations between 0.5% and 2.0%, are by far the most popular for RAPD-PCR. An alternative medium is polyacrylamide, preferred by some for RAPD-PCR because of its apparently superior ability to resolve differences in molecular weight. However, acrylamide is more expensive, cumbersome, and more toxic (in its nonpolymerized form) than agarose.

In our experience, agarose concentration can strongly influence the migration patterns and resolution of RAPD bands. We were often disappointed by the thickness and blurriness of bands we obtained across a range of agarose concentrations, making it difficult to resolve differences among bands migrating at similar rates. We then tried adding a cross-linker (1% Synergel) to our gels and decreased the agarose concentration to 0.6%. This gave us narrow, sharp bands. On the other hand, gel thickness and size had little effect on our ability to detect RAPD variation.

Screening for RAPD Polymorphism. In our study of *H. symbiolongicarpus* we screened 56 primers on three male and three female individuals, which we then mated in three pairs to generate offspring for an inheritance study. We identified 13 primers that exhibited polymorphism, produced sharp banding patterns, and amplified between five (enough to provide genetic information) and 15 bands (the greatest number that we were comfortable scoring). Of the other primers we screened, 19 failed to reveal polymorphism; 14 amplified less than 5 polymorphic bands; two produced more than 15 polymorphic bands; and eight amplified only blurred bands. All told, we used 23% of the screened primers for further analyses, a greater success rate than for *Hydractinia milleri*.

Scoring Protocol. We first tried using a state-of-the art optical imaging program to analyze banding patterns from the gel photographs. For several reasons, it quickly became apparent that using these programs would be more time-consuming than hand scoring the gels. The primary problem is that RAPD bands are often tightly clustered. To avoid having the program pool narrowly separated bands into the same category, the range of sizes used to classify any single band must be very small.

However, as we reduced this "binning" range, the small and inevitable variation in band migration introduced by otherwise trivial amounts of gel warping caused the program to score identical bands as different. Even with three or four DNA ladders per gel as standards, slight gel warping caused the misidentification of many bands; eventually, we gave up on this kind of automation. Instead, we created separate enlarged templates depicting the positions (i.e., molecular weights) in a "synthetic" lane of all bands amplified by each primer in the whole sample. We could then quickly score each gel by recording the subset of bands found in each lane. Because some bands consistently amplified strongly in some individuals or subpopulations, but weakly in others, we initially scored all bands that repeatably amplified (see next section), regardless of intensity.

Analysis of Repeatability. We excised duplicate tissue samples from each of the six parental colonies and 30 offspring used in the inheritance study (described in the next section) and extracted the DNA independently in each sample. We amplified these samples using the full set of primers and compared banding patterns between duplicates and from the same sample run on different days. Once we had optimized extraction and amplification conditions, we found no variation for a given primer–template between independent extractions or days (Levitan and Grosberg, 1993; Fig. 1). It is also worth noting that in our initial screening of the six parental genotypes using 56 primers, DNA concentration in the amplification cocktail ranged from 1 to 30 ng/μl in duplicate extractions (largely because we started with different amounts of tissue). Over this range of variation in DNA concentration, we found no discernible effect on banding pattern in the duplicated extractions.

Inheritance of Markers. To explore whether RAPD markers behave as mendelian alleles in *Hydractinia*, we mated three randomly paired male and three female colonies collected haphazardly from a large population at Barnstable Harbor, MA. We then analyzed ten randomly selected offspring from each mating, for a total of 30 offspring (Fig. 1). We coded and scrambled the 30 F_1 DNA samples so that we could not discern their relationships to each other or to their parents through the time of band scoring and analysis. We screened the six adults and 30 offspring with 13 primers, generating a total of 156 markers of which 133 were polymorphic. Only four nonparental markers appeared in the offspring.

Rather than attempting to determine whether every polymorphic marker behaved as a mendelian allele, we once again adopted a statistical approach to characterize the overall transmission properties of the majority of bands. The analysis is based on binomial expectations for the transmission of a dominant allele to offspring in a sibship. From a phenotypic perspective, matings can involve parents in which both express a band, or only one expresses a particular band. From a genetic perspective, a parent expressing a particular band can either be $+/+$ or $+/-$ at that locus, whereas a parent lacking a band will be $-/-$. So when both parents in a mating express a marker, the mating could be one of three types: (1) $+/+ \times +/-$; (2) $+/+ \times +/+$; or (3) $+/- \times +/-$. Offspring from the first two types of mating should all inherit the band, whereas on average 75% of the offspring from the second type of mating should inherit the band. When one parent expresses a band, but the other does not, the mating could be one of two types: (1) $+/+ \times -/-$ or (2) $+/- \times -/-$. In the first case, all offspring would inherit the band; in the second, on average 50% of the offspring should inherit the band.

When accumulated across many loci, the distribution of numbers of offspring in our sibships of ten that inherited a band when both parents expressed the band should have two modes: one at seven or eight, reflecting $+/- \times +/-$ matings, and one at ten, reflecting $+/+ \times +/-$ and $+/+ \times +/+$ matings. Similarly, for matings in which only one parent expressed the band, there should be modes at five and ten offspring, the former reflecting $+/- \times -/-$ matings, and the latter reflecting $+/+ \times -/-$ matings. These are the distributional patterns we found empirically (Levitan and Grosberg, 1993). This outcome by no means guarantees that each and every band will be transmitted as a mendelian allele; but it does support the assumption that the majority of bands must be transmitted as mendelian alleles.

Genealogical Analysis of Banding Patterns. We analyzed the banding pattern data in parents and offspring with three methods: exclusion, inclusion, and phenetic clustering. Exclusion involves searching for markers in offspring that are absent from some, but not all, individuals in the population of potential parents. This protocol is most effective when one parent (usually the mother) is known. For example, if an offspring carries a particular marker, and the mother lacks that marker, then all males who also lack the marker can be excluded from the pool of potential fathers. However, if nonparental bands occur in the offspring, then true genetic parents may be incorrectly excluded from consideration. In the best of circumstances, only a single potential father will carry the marker and remain in the pool. More commonly, several prospective fathers will remain (Lewis and Snow, 1992): likelihood methods can then be used to assign relative probabilities of paternity to these individuals (Meagher, 1986; Devlin et al., 1988; Roeder et al., 1989).

Inclusion requires identification of a marker (or better, markers) unique to a single parent, and then assigning offspring carrying those markers to that parent. Unlike exclusion, parentage assignment by inclusion can at least partially overcome the appearance of nonparental bands in offspring, as long as several markers can be used to associate an offspring with a particular parent. However, inclusion suffers from the danger that an unsampled potential parent outside the hypothetical parental pool has a set of markers identical to a sampled prospective parent within the pool. Thus parentage assignment by inclusion will be most powerful when there is a high degree of polymorphism and when the pool of potential parents can be circumscribed.

Because it is relatively easy to survey potential parents and offspring for hundreds of polymorphic markers using RAPD-PCR, phenetic tree-building methods (given the assumptions discussed in "Problems with RAPD-PCR and Some Remedies: Nonmendelian Inheritance") offer an alternative approach to the analysis of genetic relationships. Tree-building methods group individuals in a hierarchical framework based on overall similarity of banding patterns. Even if there are some nonparental bands in offspring, the "signal" of well-behaved mendelian bands should overwhelm the "noise" of nonmendelian bands, provided enough mendelian characters (i.e., markers) are considered. We constructed a presence/absence matrix for all 156 bands in the six parents and 30 offspring, and then calculated Nei and Li's (1979) similarity index for all pairwise combinations. We analyzed this similarity matrix using a variety of tree-building algorithms, all of which yielded concordant topologies.

In comparing inclusion and exclusion approaches to parentage analysis in this simple situation, we found that inclusion was slightly more powerful than exclusion, but both accurately assigned parents to offspring using the 13 polymorphic primers (Levitan and Grosberg, 1993). However, with both methods the presence of nonparental markers made some assignments ambiguous. The clustering approach was far and away the most powerful technique (also see Apostol et al., 1993). It correctly and significantly associated all offspring with their true parents and grouped together each sibship without error or ambiguity. We plan to explore the power of clustering further, as we expand the number of prospective parents and offspring that we include in the analysis.

Analysis of Chimeric Colonies. Fusion between *Hydractinia symbiolongicarpus* colonies may be a reasonably common event (Yund et al., 1987), especially when sibling larvae cosettle on the same shell. Fusion potentially confers substantial fitness benefits, as well as costs, which together influence the evolution of specificity in self/nonself, or allorecognition, systems (reviewed in Grosberg, 1988). For example, fusion produces a larger chimeric individual, which may enjoy increased resistance to predation and enhanced competitive ability (Buss, 1990). On the other hand, because most clonal invertebrates do not sequester their germ lines, fusion opens several avenues to intergenotypic competition between cell lines inhabiting the same soma (Buss, 1990). Thus, in order minimally to assay the fitness costs and benefits of intergenotypic competition, it is necessary to characterize the representation of both genotypes in somatic and reproductive tissues.

Unfortunately, when two colonies fuse and form a genetically chimeric individual, after a short period of time it is often impossible visually to distinguish the two genotypes. In the case of *Hydractinia,* because the extensive polymorphism in the self/nonself recognition system almost necessarily limits fusion to closely related individuals (our own data), the marker must be highly polymorphic, so that tissues of close relatives can be genetically distinguished. We therefore began to explore the utility of RAPD-PCR markers for characterizing the genetic composition of somatically fused colonies of *Hydractinia symbiolongicarpus.*

In these preliminary studies, we first clonally replicated sibling colonies and paired them to determine which sibling combinations fused. At the same time, we began an extensive RAPD-PCR screening of these siblings to identify markers that would unambiguously diagnose each genotype in a fusible pair. We then established chimeric colonies by allowing compatible siblings to fuse. At roughly monthly intervals postfusion, we clipped one polyp (or a few polyps) from beneath 20–40 random positions on a grid placed over the colony. We then extracted DNA in the manner described in Grosberg.1 for small tissue samples and amplified the DNA using a primer that would allow us to discern to which component genotype the sampled polyp(s) belonged. Figure 2 shows a diagram of one such chimera some 540 days postfusion, with the corresponding gel. This figure clearly illustrates that RAPD markers can be used to identify the genotypic identity of components of a chimera, and that intergenotypic chimeras can remain fairly stable over sustained periods.

We also are beginning to use RAPD markers to portray the gametic composition of chimeras. This is, however, a more complex undertaking than characterizing somatic composition, both because RAPD markers are dominant (thus only half the gametes from a heterozygous parent will receive the marker) and gametes usually carry only a single copy of the parental DNA. Although several protocols exist for extracting and amplifying very small amounts of DNA (e.g., Landry et al., 1993; Grosberg.1), at this point, we are not convinced we can reliably amplify the DNA from a single gamete. We therefore fertilize ova of unknown genotypic identity with sperm from a male carrying a diagnostic marker and assay the resulting (multicellular) larvae. Conversely, we use sperm of unknown genotypic identity to fertilize ova with a diagnostic marker and assay the larvae for the presence of one or the other diagnostic bands of the chimera. This procedure is most efficient when both components of the chimera are homozygous for the presence of their diagnostic marker band; but even when they are not, 50% of the progeny should carry the marker, and the analysis should be unbiased.

Study II: Population Structure in *Hydractinia milleri*

Screening for RAPD Polymorphism. Using the same procedures described above for *Hydractinia symbiolongicarpus,* we extracted DNA from five *H. milleri* colonies collected from a single aggregation from Tatoosh Island, Washington. We screened these samples with 101 primers. Of these primers, nine failed to amplify any loci; 35 gave bands that were too smeared to score; 23 amplified clear but invariant bands; and 34 gave clear polymorphic banding patterns. We chose the 14 most promising primers (using the criteria in "Parentage in *Hydractinia symbiolongicarpus:* Screening for RAPD Polymorphism") and screened those on 57 individuals from an additional site at Waadah Island, Washington (approximately 10 km from Tatoosh Island). We repeated this screening three times and chose the best eight primers that produced repeatable, polymorphic bands. Thus, of the original 101 primers that we screened, slightly more than 8% proved useful. We then extended the geographic scope of our sampling to include two sites in northern California and ultimately screened over 200 colonies, carrying 278 polymorphic bands.

Population Sampling Design. Because *H. milleri* females brood demersal, crawl-away larvae, it is likely that most sibling larvae will settle close to their mother. Consequently, populations of *H. milleri* ought to be more genetically structured than, say, *H. symbiolongicarpus* (as well as free-spawning invertebrates with feeding larvae, or plants with wind-dispersed propagules). But it is also possible that waves carry some larvae a considerable distance, where they successfully

CLONES3/4 --565 DAYS POSTFUSION

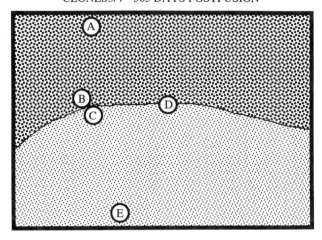

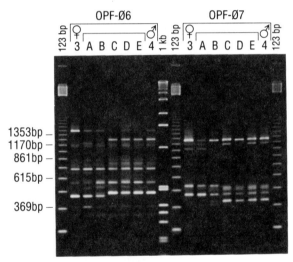

Figure 2. *Top panel:* Diagrammatic representation of a chimeric colony of *Hydractinia symbiolongicarpus* 565 days postfusion. The different shading patterns denote the apparent extent of the two strains (♀3 in the upper section and ♂4 in the lower section) that form the chimera, inferred from the phenotypic differences in the structure of their corresponding gastrovascular networks. At each of the lettered positions (A–E), we sampled two polyps and analyzed them for the presence of strain-specific, diagnostic RAPD markers. *Bottom panel:* Amplification products from two RAPD primers (OPF-06 and OPF-07) applied to DNA extracted from positions A–E. Lanes denoted ♀3 and ♂4 show amplification products from pure cultures of the constituent strains. The amplification products show that samples A and B have profiles characteristic of ♀3, whereas samples C–E have profiles like that of ♂4.

metamorphose. Thus the spatial scale over which *H. milleri* populations ought to exhibit genetic structure is difficult to predict.

We used a stratified sampling design in which we mapped the positions of over 200 colonies in 17 aggregations at three localities, two about 10 km apart in northern California (Doran Rocks: 73 colonies in seven aggregations; Coleman Beach: 73 colonies in six aggregations), and one about 1000 km northward near the mouth of the Straight of San Juan de Fuca in Washington State (Waadah Island: 54 colonies in four aggregations). For analysis of genetic structure, we carefully removed 3–5 polyps from each colony; we also removed brooded embryos from females for analysis of individual reproductive success. At the Waadah Island site, we mapped the colonies over a 3-year period, monitoring patterns of recruitment, mortality, fission, and fusion. This allowed us to compare the genetically based picture with an independent set of demographic observations, and thereby assess the power of the RAPD markers to discriminate clonemates from close relatives.

Data Analysis and Population Structure. We assayed each colony with the battery of eight primers. We then computed Nei and Li's (1979) similarity index for all pairs of colonies and analyzed the resulting similarity matrix using a variety of tree-building algorithms: all gave virtually, but not perfectly, identical tree topologies.

At the broadest sampling scale, the cluster analysis accurately assigns each individual colony to its correct site. As expected, individuals from the Doran Rocks and Coleman Beach sites are far more similar to each other than either is to samples from the Waadah Island site. At the level of aggregations within sites, the cluster analysis correctly grouped >90% of colonies from the same aggregation together. In most cases, "misassigned" colonies grouped with members of the nearest aggregation. However, about 2–3% of individuals grouped by themselves, suggesting that they immigrated from outside the area we sampled. Taken together, these data suggest that most dispersal occurs on scales of millimeters to centimeters (within aggregations). Some dispersal occurs on the order of tens of meters (among aggregations within a site), and perhaps farther, but not so often as to overwhelm the persistence of substantial genetic structure at very fine spatial scales.

There is an additional noteworthy pattern in the relationships among colonies in the Waadah Island aggregations: we could not distinguish two pairs of individuals. Either the 278 RAPD markers could not resolve these genotypes, perhaps because they represent pairs of close relatives, or the two pairs represent clonemates derived by asexual fragmentation. We addressed this question by comparing the genetically based inference of clonality to the time-series maps of the colonies in the Waadah Island aggregations. The chronologies of the two pairs of colonies show that both pairs fragmented 3 years before we sampled the aggregations for genetic analysis. Thus the array of markers could distinguish clonemates from close kin and further showed that cloning, although rare in *H. milleri,* does occur.

Overall, this analysis of genetic structure shows that genetic similarity declines with distance separating colonies, even at very fine spatial scales. The picture reassuringly reflects the reproductive ecology of *H. milleri,* with its limited potential for gene flow.

CONCLUSIONS

Our principal aim in this chapter was to assess some of the positive and negative attributes of RAPD markers in terms of their power to characterize genetic relationships and structure. In our view, RAPD markers, because they are anonymous and expressed as dominant alleles that may conceal substantial amounts of cryptic genetic variation, are generally ill-suited to the characterization of breeding systems, the calculation of population genetic parameters such as F-statistics, or the inference of phylogenetic relationships above the species level. Granted, with a lot of work, it could be possible to render a small number of markers useful for these purposes; but it seems far more profitable not to fight the shortcomings of RAPD-PCR.

On the other hand, RAPD-PCR can easily generate hundreds of potentially independent markers, the majority of which appear to behave as neutral mendelian alleles. In many cases, RAPD loci harbor sufficient polymorphism to reveal genetic structure at very fine spatial and temporal scales, but not so much as to obscure structure at broader geographic (and presumably temporal) scales. With some caution in both the generation and interpretation of these markers, they hold as yet unmatched promise for characterizing hierarchical genetic structure and close genealogical relationships in large samples, especially when population structure spans a range of spatial scales. For those willing to wait with bated breath, there can be little doubt that high-resolution markers—newer, more fashionable, and almost certainly better than RAPDs—will soon be developed and publicized as the final solution to the daunting problems of characterizing genetic structure and reproductive success. Until that time actually arrives, we urge the novice, nervous, or antagonistic population ecologist to look critically and fairly at all potentially useful genetic markers, including RAPDs, to assess their value for the problem at hand, and to employ the technique that works.

ACKNOWLEDGMENTS

NSF grants to RK Grosberg and DR Levitan allowed us to develop RAPD-PCR markers and supported our research on the population biology of colonial marine invertebrates. Funds from the Agricultural Experiment Station of the University of California also supported this research. J Ferraris and S Palumbi goaded us into writing this chapter; the reader can judge if they were right. We thank several reviewers, who prefer to remain anonymous, for their useful comments. The American Society of Zoologists and the Alfred P Sloan Foundation funded the symposium that was the original venue for this work.

REFERENCES

Aldrich J, Cullis CA (1993): RAPD analysis in flax: optimization of yield and reproducibility using Klen*Taq* 1 DNA polymerase, Chelex 100 and gel purification of genomic DNA. *Plant Mol Biol Rep* 11:128–141.

Apostol BL, Black WC IV, Miller BR, Reiter P, Beaty BJ (1993): Estimation of the number of full sibling families at an oviposition site using RAPD-PCR markers: applications to the mosquito *Aedes aegypti*. *Theor Appl Genet* 86:991–1000.

Armstrong JS, Gibbs AJ, Weiller G (1994): The RAPDistance package. Obtainable via Anonymous ftp or WWW. ftp://life.anu.edu.au/pub/RAPDistance; http://life.anu.edu.au/molecular/software/rapd.html.

Avise JC (1994): *Molecular markers, natural history, and evolution*. New York: Chapman and Hall.

Ballard WW (1942): The mechanism of synchronous spawning in *Hydractinia* and *Pennaria*. *Biol Bull* 82:329–339.

Bell G (1992): Five properties of environments. In: Grant PR, Horn HS (eds). *Molds, molecules, and Metazoa*. Princeton, NJ: Princeton University Press, pp 33–56.

Bielawski JP, Noack K, Pumo DE (1995): Reproducible amplification of RAPD markers from vertebrate DNA. *Bio techniques* 18:856–860.

Black WC (1993): PCR with arbitrary primers: approach with care. *Insect Mol Biol* 2:1–6.

Black WC, DuTeau NM, Puterka GJ, Nechols JR, Pettorini JM (1992): Use of the random amplified polymorphic DNA polymerase chain reaction (RAPD-PCR) to detect DNA polymorphisms in aphids (Homoptera: Aphididae). *Bull Entomol Res* 82:151–159.

Bowditch BM, Albright DG, Williams JGK, Braun MJ (1993): Use of randomly amplified polymorphic DNA markers in comparative genome studies. In: Zimmer EA, White TJ, Cann RL, Wilson AC (eds). *Methods in enzymology, Volume 224. Molecular evolution: producing the biochemical data*. San Diego, CA: Academic Press, pp 294–309.

Bucci G, Menozzi P (1993): Segregation analysis of random amplified polymorphic DNA (RAPD) markers in *Picea abies*. *Mol Ecol* 2:227–232.

Bunting M (1894): The origin of sex-cells in *Hydractinia echinata* and *Podocoryne*. *J Morphol* 9:203–236.

Burke T (1989): DNA fingerprinting and other methods for the study of mating success. *Trends Ecol Evol* 4:139–144.

Burke T, Hanotte O, Bruford MW, Cairns E (1991): Multilocus and single locus multisatellite analysis in populations biological studies. In: Burke T, Dolf G, Jeffreys AJ, Wolff R (eds): *DNA fingerprinting: approaches and applications*. Basel: Birkhäuser, pp 154–168.

Buss LW (1990): Competition within and between encrusting invertebrates. *Trends Ecol Evol* 5:352–356.

Buss LW, Yund PO (1989): A sibling species group of *Hydractinia* in the northeastern United States. *J Mar Biol Assoc UK* 69:857–875.

Buss LW, McFadden CS, Keene DR (1984): Biology of hydractiniid hydroids. 2. Histocompatibility effector system mediated by nematocyst discharge. *Biol Bull* 167:139–158.

Caetano-Anollés G, Bassam BJ, Gresshoff PM (1991): DNA amplification fingerprinting using very short arbitrary oligonucleotide primers. *Biotechnology* 9:553–557.

Carlson JE, Tulsieram LK, Glaubitz JC, Luk VWK, Kauffeldt C, Rutledge R (1991): Segregation of random amplified DNA markers in F_1 progeny of conifers. *Theor Appl Genet* 83:194–200.

Clark AG, Lanigan CMS (1993): Prospects for estimating nucleotide divergence with RAPDs. *Mol Biol Evol* 10:1096–1111.

Dawson IK, Chalmers KJ, Waugh R, Powell W (1993): Detection and analysis of genetic variation in *Hordeum spontaneum* populations from Israel using RAPD markers. *Mol Ecol* 2:151–160.

Devlin B, Roeder K, Ellstrand NC (1988): Fractional paternity assignment: theoretical development and comparison to other methods. *Theor Appl Genet* 76:369–380.

Ellsworth DL, Rittenhouse KD, Honeycutt RL (1993): Artifactual variation in randomly amplified polymorphic DNA banding patterns. *Biotechniques* 14:214–216.

Excoffier L, Smouse PE, Quattro JM (1992): Analysis of molecular variance inferred from metric distances among DNA haplotypes: application to human mitochondrial DNA restriction data. *Genetics* 131:479–491.

Fani R, Damiani G, Di Serio C, Gallori E, Grifoni A, Bazzicalup M (1993): Use of random amplified polymorphic DNA (RAPD) for generating DNA probes for microorganisms. *Mol Ecol* 2:243–250.

Felsenstein J (1993): *PHYLIP (Phylogeny Inference Package). Version 3.5C*. Seattle: Department of Genetics, University of Washington.

Fondrk MK, Page RE, Hunt GJ (1993): Paternity analysis of worker honey bees using random amplified DNA (RAPD). *Naturwissenschaften* 80:226–231.

Grosberg RK (1988): The evolution of allorecognition specificity in clonal invertebrates. *Q Rev Biol* 63:377–412.

Hadrys H, Balick M, Schierwater B (1992): Applications of random amplified polymorphic DNA (RAPD) in molecular ecology. *Mol Ecol* 1:55–63.

Haig SM, Rhymer JM, Heckel DG (1994): Population differentiation in randomly amplified polymorphic DNA of red-cockaded woodpeckers *Picoides borealis*. *Mol Ecol* 3:581–595.

Hedrick P (1992): Shooting the RAPD's. *Nature* 355:679–680.

Heun M, Helentjaris T (1993): Inheritance of RAPDs in F_1 hybrids of corn. *Theor Appl Genet* 85:961–968.

Huff DR, Peakall R, Smouse PE (1993): RAPD variation within and among natural populations of outcrossing buffalograss [*Buchloe dactyloides* (Nutt.) Engelm.]. *Theor Appl Genet* 86:927–934.

Hunt GJ, Page RE (1992): Patterns of inheritance with RAPD molecular markers reveal novel types of polymorphism in the honey bee. *Theor Appl Genet* 85:15–20.

Ivker FB (1972): A hierarchy of histo-incompatibility in *Hydractinia echinata*. *Biol Bull* 143:162–174.

Karl SA, Avise JC (1993): PCR-based assays of Mendelian polymorphisms from anonymous single-copy nuclear DNA: techniques and applications for population genetics. *Mol Biol Evol* 10:342–361.

Klein-Lankhorst RM, Vermunt A, Weide R, Liharska T, Zabel P (1991): Isolation of molecular markers from tomato (*L. esculentum*) using random amplified polymorphic DNA (RAPD). *Theor Appl Genet* 83:108–114.

Landry BS, Dextraze L, Boivin G (1993): Random amplified polymorphic DNA markers for DNA fingerprinting and genetic variability assessment of minute parasitic wasp species (Hymenoptera: Mymaridae and Trichigrammatidae) used in biological control programs of phytophagous insects. *Genome* 36:580–587.

Lewis PO, Snow AA (1992): Deterministic paternity exclusion using RAPD markers. *Mol Ecol* 1:155–160.

Lewis PO, Zaykin D (1996): *Genetic data analysis* (computer software). Sunderland, MA: Sinauer.

Lessa EP, Applebaum G (1993): Screening techniques for detecting allelic variation in DNA sequences. *Mol Ecol* 2:119–129.

Levi A, Rowland LJ, Hartung JS (1993): Production of reliable randomly amplified polymorphic DNA (RAPD) markers from DNA of woody plants. *HortScience* 28:1188–1190.

Levitan DR, Grosberg RK (1993): The analysis of paternity and maternity in the marine hydrozoan *Hydractinia symbiolongicarpus* using randomly amplified polymorphic DNA (RAPD) markers. *Mol Ecol* 2:315–326.

Liu Z, Furnier GR (1993): Comparison of allozyme, RFLP, and RAPD markers for revealing genetic variation within and between trembling aspen and bigtooth aspen. *Theor Appl Genet* 87:97–105.

Lynch M (1988): Estimation of relatedness by DNA fingerprinting. *Mol Biol Evol* 5:584–599.

Lynch M (1990): The similarity index and DNA fingerprinting. *Mol Biol Evol* 7:478–484.

Lynch M, Milligan B (1994): Analysis of population genetic structure with RAPD markers. *Mol Ecol* 3:91–100.

Martin GB, Williams JGK, Tanksley SD (1991): Rapid identification of markers linked to a *Pseudomonas* resistance gene in tomato by using random primers and near-isogenic lines. *Proc Nat Acad Sci USA* 88:2336–2340.

Meagher TR (1986): Analysis of paternity within a natural population of *Chamaelirium luteum*. 1. Identification of most-likely male parents. *Am Nat* 128:199–215.

Meagher TR, Thompson EA (1987): Analysis of parentage for naturally established seedlings of *Chamaelirium luteum* (Liliaceae). *Ecology* 68:803–812.

Micheli MR, Bova R, Pascale E, D'Ambrosio E (1994): Reproducible DNA fingerprinting with the random amplified polymorphic DNA (RAPD) method. *Nucleic Acids Res* 22:1921–1922.

Milligan BG (1992): Plant DNA isolation. In: Hoelzel AR (ed). *Molecular genetic analysis of populations. A practical approach.* New York: IRL Press, pp 59–88.

Milligan BG, McMurry CK (1993): Dominant vs. codominant genetic markers in the estimation of male mating success. *Mol Ecol* 2:195–284.

Müller W (1969): Auslosung der Metamorphose durch Bakkterien bei den Larven von *Hydractina echinata*. *Zool Jahrb Abt Anat Ontog Tiere* 86:84–95.

Muralidharan K, Wakeland EK (1993): Concentration of primer and template qualitatively affects products in random amplified polymorphic DNA PCR. *Bio techniques* 14:362–363.

Nei M, Li WH (1979): Mathematical model for studying genetic variation in terms of restriction endonucleases. *Proc Nat Acad Sci USA* 74:5267–5273.

Palumbi SR, Martin A, Romano S, McMillan WO, Stice L, Grabowski G (1991): *The simple fool's guide to PCR (Version 2.0).* Honolulu: Department of Zoology and Kewalo Marine Laboratory, University of Hawaii.

Peakall R, Smouse PE, Huff DR (1995): Evolutionary implications of allozyme and RAPD variation in diploid populations of dioecious buffalograss *Buchloë dactyloides*. *Mol Ecol* 4:135–147.

Penner GA, Bush A, Wise R, Kim W, Domier L, Kasha K, Laroche A, Scoles G, Molnar SJ, Fedak G (1993): Reproducibility of random amplified polymorphic DNA (RAPD) analysis among laboratories. *PCR Methods Applications* 2:341–345.

Piper WH, Rabenold PP (1992): Use of fragment-sharing estimates from DNA fingerprinting to determine relatedness. *Mol Ecol* 1:69–78.

Queller DC, Goodnight KF (1989): Estimating relatedness using genetic markers. *Evolution* 43:258–275.

Queller DC, Strassmann JE, Hughes CR (1993): Microsatellites and kinship. *Trends Ecol Evol* 8:285–288.

Reeve HK, Westneat DF, Queller DC (1992): Estimating average within-group relatedness from DNA fingerprints. *Mol Ecol* 1:223–232.

Riedy MF, Hamilton WJ, Aquadro CF (1992): Excess non-parental bands in offspring from known pedigrees assayed using RAPD PCR. *Nucleic Acids Res* 20:918.

Roeder K, Devlin B, Lindsay BG (1989): Application of maximum likelihood methods to population genetic data for estimation of individual fertilities. *Biometrics* 45:363–379.

Roehrdanz RL, Reed DK, Burton RL (1993): Use of polymerase chain reaction and arbitrary primers to distinguish laboratory-raised colonies of parasitic Hymenoptera. *Biol Control* 3:199–206.

Rogers SO, Bendich AJ (1988): Extraction of DNA from plant tissues. *Plant Mol Biol Manual* A6:1–10.

Rossetto M, Weaver PK, Dixon KW (1995): Use of RAPD analysis in devising conservation strategies for the rare and endangered *Grevillea scapigera* (Proteaceae). *Mol Ecol* 4:321–329.

Schweder ME, Shatters RG Jr, West SH, Smith RL (1995): Effect of transition interval between melting and annealing temperatures on RAPD analyses. *Biotechniques* 19:38–42.

Shoemaker DD, Ross KG, Arnold ML (1994): Development of RAPD markers in two introduced fire ants, *Solenopsis invicta* and *S. richteri,* and their application to the study of a hybrid zone. *Mol Ecol* 3:531–539.

Shenk MA, Buss LW (1991): Ontogenetic changes in fusibility in the colonial hydroid *Hydractinia symbiolongicarpus. J Exp Zool* 257:80–86.

Smith JL, Scott-Craig JS, Leadbetter JR, Bush GL, Roberts DL, Fulbright DW (1994): Characterization of random amplified polymorphic DNA (RAPD) products from *Xanthomonas campestris* and some comments on the use of RAPD products in phylogenetic analysis. *Mol Phylogenet Evol* 3:135–145.

Swofford DL (1991): *Phylogenetic Analysis Using Parsimony (PAUP). Version 3.0s.* Champaign, IL: Illinois Natural History Survey.

Swofford DL, Selander RB (1989): *BIOSYS-1. Release 1.7.* Champaign, IL: Illinois Natural History Survey.

Thompson EA (1986): *Pedigree analysis in human genetics.* Baltimore: Johns Hopkins University Press.

Tinker NA, Fortin MG, Mather DE (1993): Random amplified polymorphic DNA and pedigree analysis in spring barley. *Theor Appl Genet* 85:976–984.

Weatherhead PJ, Montgomery RD (1991): Good news and bad news about DNA fingerprinting. *Trends Ecol Evol* 6:173–174.

Weiller G and AJ Gibbs (1995): DIPLOMO: The tool for a new type of evolutionary analysis. *CABIOS* 11: in press.

Weis VM, Buss LW (1987): Biology of hydractiniid hydroids. 5. Ultrastructure of metamorphosis in *Hydractinia echinata. Postilla* 200:1–24.

Welsh J, McClelland M (1990): Fingerprinting genomes using PCR with arbitrary primers. *Nucleic Acids Res* 18:7213–7218.

Welsh J, Honeycutt RJ, McClelland M, Sobral BWS (1991): Parentage determination in maize hybrids using arbitrarily primed polymerase chain reaction (AP-PCR). *Theor Appl Genet* 82:473–476.

Welsh J, Pretzman C, Postic D, Saint Girons I, Baranton G, McClelland M (1992): Genomic fingerprinting by arbitrarily primed polymerase chain reaction resolves *Borrelia burgdorferi* into three distinct phyletic groups. *Int J Syst Bacteriol* 42:370–377.

Westneat DF, Frederick PC, Wiley RH (1987): The use of genetic markers to estimate the frequency of successful alternative reproductive tactics. *Behav Ecol Sociobiol* 21:35–45.

Wilkerson RC, Parsons TJ, Albright DG, Klein TA, Braun MJ (1993): Random amplified polymorphic DNA markers readily distinguish cryptic mosquito species (Diptera: Culicidae: Anopholes). *Insect Mol Biol* 1:205–211.

Williams JGK, Kubelik AR, Livak KJ, Rafalski JA, Tingey SV (1990): DNA polymorphisms amplified by arbitrary primers are useful as genetic markers. *Nucleic Acids Res* 18:6531–6535.

Williams JGK, Kubelik AR, Rafalski JA, Tingey SV (1991): Genetic data analysis with RAPD markers. In: Bennett JW, Lasure LL (eds). *More gene manipulations in fungi.* San Diego, CA: Academic Press, pp 433–439.

Williams JGK, Hanafey MK, Rafalski JA, Tingey SV (1993): Genetic analysis using random amplified polymorphic DNA markers. *Methods Enzymol* 218:704–740.

Yu K, Pauls KP (1992): Optimization of the PCR program for RAPD analysis. *Nucleic Acids Res* 20:2606.

Yund PO, Cunningham CW, Buss LW (1987): Recruitment and post-recruitment interactions in a colonial hydroid. *Ecology* 68:971–982.

Weiller GF, Gibbs A (1993): *DIPLOMO: Distance Plot Monitor version 1.* Computer Program distributed by the Australian National University. Available via WWW: http://life.anu.edu.au/molecular/software/rapd.html

Macrospatial Genetic Structure and Speciation in Marine Taxa with High Dispersal Abilities

STEPHEN R. PALUMBI

Department of Zoology and Kewalo Marine Laboratory, University of Hawaii, Honolulu, Hawaii 96813

CONTENTS

SYNOPSIS

In general, high gene flow among populations is thought to limit opportunities for genetic divergence and to inhibit species formation. Nevertheless, recent fossil and genetic studies have shown that the Pleistocene has been an active period of marine

Molecular Zoology: Advances, Strategies, and Protocols, Edited by Joan D. Ferraris and Stephen R. Palumbi.
ISBN 0-471-14461-4 © 1996 Wiley-Liss, Inc.

species appearance. We examined the spatial scale of genetic differentiation in four species of congeneric species in the tropical sea urchin genus *Echinometra* using mtDNA sequences of a part of the cytochrome oxidase I gene. Despite a larval planktonic stage that persists for 6–8 weeks, *Echinometra* spp. have strong genetic structure throughout the Indo-West Pacific. Localities separated by approximately 2500 km have strong differences in frequency of major mitochondrial haplotypes, although some haplotypes occur across the ranges of each species. The combination of widespread haplotypes and significant mtDNA frequency differences suggests a recent period of colonization followed by local genetic differentiation. These genetic differences may be the signature of repeated Pleistocene changes in climate and sea level and may provide the genetic differentiation on which speciation is based.

INTRODUCTION

High dispersal potential is a common feature of many marine organisms. Whether it is due to gametes shed into the water, to larvae released from brooding females, or to highly vagile adults, movement of hundreds or thousands of kilometers is a key life history feature of many marine taxa. However, it has been difficult to document precisely the movement of most marine species throughout the world's oceans. Most marine larvae cannot be tracked individually (see Olson, 1985; Stoner, 1990 for exceptions using low dispersal tunicate larvae), and chemical means of labeling large numbers of larvae (reviewed by Levin, 1990) have yet to be generalizable to many different systems. Direct observation of the larvae of coastal species in plankton tows far from shore (Scheltema, 1986), via satellite observation of gamete slicks (Willis and Oliver, 1990), or the demonstration of potentially long planktonic periods (e.g., Richmond, 1987) show the potential for planktonic larvae to move large distances but provides little information about average dispersal or the chance that such teleplanic larvae will successfully integrate into a breeding population.

Indirect Measures of Gene Flow

Successful larval transport implies gene flow, and this link allows an indirect measurement of dispersal in marine systems. Attempts to correlate genetic differentiation with life history attributes like dispersal ability have shown that, in general, high dispersal potential is associated with high gene flow over scales of hundreds or thousands of kilometers (reviewed by Palumbi, 1992, 1994). For example, taxa as diverse as molluscs, echinoderms, and corals show low genetic differentiation in species with high dispersal potential but high differentiation in species with larvae that disperse poorly (Berger, 1973; McMillan et al., 1992; Helberg, 1994).

However, there are many exceptions to this simple rule (Burton and Feldman, 1982; Palumbi, 1994). The reasons for these exceptions are diverse and are known to include (1) behavioral mechanisms that limit random dispersal (Burton and Feldman, 1982; Bowen et al., 1989; Baker et al., 1990), (2) selection against immigrants (Koehn et al., 1980) or selection for balanced polymorphisms (Karl and

Avise, 1992), (3) complex oceanographic circulation patterns (e.g., one-way current flow) (Benzie and Stoddart, 1992), and (4) historic barriers to gene flow (Bert and Harrison, 1988; Avise, 1992; Burton and Lee, 1994). These exceptions can lead to strong patterns of genetic differentiation over spatial scales far shorter than average dispersal distance (e.g., Koehn et al., 1980). At the other extreme, they can lead to genetic homogeneity over scales far larger than average dispersal distance (Reeb and Avise, 1990; Karl and Avise, 1992).

Gene Flow and Speciation

Because species formation is traditionally linked to the buildup of genetic differences between isolated or semi-isolated demes (Mayr, 1940), the relationship between gene flow, geographic distance, and speciation has been of interest to marine biogeographers for many years. For example, Hansen (1983) described a link between developmental mode and rates of speciation and extinction in marine invertebrates. Gastropod species with low dispersal potential tended to give rise to other species more often and to become extinct faster than did species with high dispersal potential.

However, speciation can be rapid in even high dispersal species. The Indo-West Pacific is home to the world's highest marine diversity, yet it is dominated by taxa with high dispersal potential. Recent molecular investigations of species relationships of sea urchins and coral reef fish (Palumbi and Metz, 1990; McMillan and Palumbi, 1995) suggest that at least some sibling species in this region have arisen in the Pleistocene. For such taxa, mechanisms by which species form are poorly understood (Palumbi, 1992, 1994).

Genetic Measurement of Gene Flow

As a result of the complexity of factors that can affect gene flow in marine environments, it is important to be able to assess empirically the degree of genetic differentiation between particular marine populations. Using population genetic models that assume neutrality and particular modes of gene flow (e.g., the island model of Sewall Wright, 1978, in which a population is made up of a set of demes that are interconnected equally by gene flow), it is possible to estimate long-term gene flow from population genetic data. Wright's F_{ST} is a widely used statistic of population differentiation. Weir's theta and Nei's G_{ST} are related statistics with slightly different properties (Weir and Cockerham, 1984; Nei, 1987).

An important caveat in these analyses is that gene flow estimates are based on the assumption that the populations studied are at genetic equilibrium. In other words, the models assume that a balance has already been reached between local genetic drift (which tends to homogenize local demes and allow their differentiation from other demes) and gene flow (which tends to add heterogeneity to demes and blur the genetic distinctions between them). Such equilibrium conditions are reached eventually in a population, but the approach to equilibrium may take a long time.

Simulations of the approach of Nei's G_{ST} to equilibrium suggest that it takes on the order of N generations for G_{ST} to reach half its eventual equilibrium value (Crow and Aoki, 1984), where N is the number of breeding adults in a deme. For many marine species, local population size is large and N may be on the order of 10^5 to 10^6. Thus genetic equilibrium will take up to millions of generations to become established.

Against this backdrop of slow genetic approach to equilibrium, there have been periodic global changes in marine habitats in the Pleistocene. During the past 2 million years, marine environments around the world have been affected by a series of glacially induced changes that have had profound effects on species distributions. Low sea level stands of the Pleistocene have exposed reefs, dried out lagoons (Paulay, 1990), drained continental estuaries (Reeb and Avise, 1990), contributed to productivity shifts (Lyle et al., 1992), changed coastal and oceanic circulation patterns (Schnitker, 1974; Duplessy, 1982), altered species distributions (Briggs, 1970), and created peninsular land masses from archipelagoes (Quinn, 1970; Potts, 1983). Sea water temperatures dropped 6–10°C in temperate areas, and the tropic and subtropic zones became substantially compressed (CLIMAP, 1976). The last major glacial maximum occurred 8000–12,000 years ago, and sea level did not return to its current position until about 6000 years ago. Distributions of marine species during glacial maxima are largely unknown, but in temperate zones changes in water temperature and coastal sediments probably altered the ranges of many species (Briggs, 1970). In the tropical Pacific, species that are ecological specialists on inner lagoon habitats (especially bivalves) became extinct much more commonly than other molluscs (Paulay, 1990), suggesting that the effects of climatic cycles are taxon-specific.

These lessons from paleoclimatology suggest that many marine populations were affected by Pleistocene climate changes, and that in many cases, genetic equilibrium may not have been reestablished.

What are the effects of genetic disequilibrium on efforts to measure marine gene flow? Some empirical genetic patterns may be due to modern-day genetic processes like gene flow and drift. Others may be relic patterns left over from past oceanographic conditions. Reeb and Avise (1990) suggest this as an explanation for distinct spatial patterns in mtDNA haplotypes in the oyster *Crassostrea virginica* on the East Coast of North America. This species inhabits coastal estuaries that may have drained during low sea stands of the Pleistocene, thereby fragmenting its habitat. Another East Coast example is the discovery of alleles from the Gulf of Mexico stone crab (genus *Menippe*) in local populations of its Atlantic coast congener (Bert and Harrison, 1988). Bert and Harrison (1988) suggested that this injection of alleles occurred during a transient high sea level period during the Pleistocene, a time when a seaway connection between the northern Gulf of Mexico and the Atlantic existed across the Florida peninsula. The persistence of these alleles near the position of the temporary seaway is thus best explained as an historic artifact rather than a feature of equilibrium genetic processes.

The occurrence of historic alterations in gene flow patterns and the ubiquitous nature of marine habitat alterations in the recent past suggest that genetic techniques used to identify and interpret gene flow patterns should be of the highest possible resolution. High-resolution DNA sequence techniques allow the description of small differences in alleles on the basis of one or a few nucleotide differences, as well as allowing characterization of more distantly related alleles. This focus on allele phylogeny has great impact on the power of population structure analyses (Avise et al., 1987; Avise, 1994, Palumbi and Baker, 1995). In particular, more recently diverged alleles (represented by sequences that are very similar) may have been affected by fewer changes in climate and habitat than those that have persisted in a species for millions of years. As an example, Baker et al. (1993) showed that two major mtDNA clades of humpback whales arose 3–5 million years ago. Both clades occur in southern hemisphere populations around Antarctica and in the North Atlantic. As a result, an analysis of population structure using only clade identification (a fairly low-resolution approach) shows only frequency differences among ocean populations. However, high-resolution DNA sequence data from the mitochondrial control region of humpback whales shows that, although both clades are in both oceans, oceans never share specific alleles within clades. Thus our perception of the degree of population division in this species is greatly enhanced by a higher resolution analysis.

High-Resolution Genetic Tools in Population Biology

Use of different molecular techniques to obtain information about DNA variation within and between populations allows a wide range of choices of resolution. High-resolution techniques allow us to distinguish recently diverged alleles. Both DNA sequence analysis and size variation analysis of microsatellite loci (see Queller et al., 1993) can provide such data. For DNA sequences, the resolution of the data set depends on the amount of information gathered per allele, either by direct sequencing, RFLP analysis (Avise et al., 1989), gradient denaturation (Lessa, 1992; Lessa and Applebaum, 1993), or other techniques.

For microsatellite loci, resolution depends on the ability to discern small changes in length of amplified products. This resolution is probably most severely limited by uncertainty about whether alleles with the same size are in fact closely related, or whether they have evolved to be the same size from different ancestors. Thus the relationship between DNA similarity of alleles and their relation by evolutionary descent is more muddied than for DNA sequences.

We have concentrated on collecting DNA sequence data from highly variable mitochondrial and nuclear loci in order to discern patterns in marine population structure. Mitochondrial sequences are becoming more and more common in analyses of population structure (e.g., Thomas et al., 1990; Vigilant et al., 1991; Baker et al., 1993) and are quickly replacing RFLP analyses. Sequences of nuclear genes, like highly variable introns, provide high-resolution data with which to compare mitochondrial patterns (e.g., Palumbi and Baker, 1994). However, DNA sequence data sets are time consuming and expensive to produce (Lessa, 1992), especially for nuclear introns, which must be cloned (after polymerase chain reaction amplification) in order to separate alleles within diploid individuals (Slade et al., 1993; Palumbi and Baker, 1994, 1995).

Recent advances in PCR and sequencing technology, however, simplify the collection of DNA sequence data sets. Although an exhaustive review is not possible here, there have been a number of recent publications of new methods in molecular population biology that provide an entry into the literature (e.g., Erlich, 1989; Zimmer et al., 1993; Hillis and Moritz, 1996). Here, I will describe the procedures we use to obtain sequence data from mitochondrial and nuclear DNA using PCR followed by solid-phase sequencing with streptavidin-coated magnetic beads. Using this approach, we are characterizing the population structure of a suite of closely related tropical sea urchins whose ranges include the scattered archipelagoes of the Indo-West Pacific.

These four species, in the echinoid genus *Echinometra*, are very similar ecologically (Nishihira et al., 1991) and remain reproductively isolated because of inability of eggs and sperm from different species to fuse (Metz et al., 1994). The Pacific species are also closely related evolutionarily (Palumbi and Metz, 1990). Measurement of mitochondrial DNA differences between them ranges from 2% to 4%, compared to the 6% differences between species in this genus that have been separated by the Isthmus of Panama about 3.5 million years ago (Bermingham and Lessios, 1993). Thus the Pacific *Echinometra* species probably arose within the past 1–3 million years.

Because all *Echinometra* species have a normal pluteus larva that spends 4–8 weeks in planktonic development (Arakaki and Uehara, 1991), all species have similar potential for long-distance dispersal. Other sea urchin species that have been examined using mtDNA (Palumbi and Wilson, 1989; Palumbi and Kessing, 1991; McMillan et al., 1992; Bermingham and Lessios, 1993; Palumbi, 1995) or allozymes (Marcus, 1977) have shown low genetic differentiation over large spatial scales. Is rapid speciation in *Echinometra* associated with low genetic dispersal or are these species an exception to the patterns seen in other genera?

To address this question, we have examined mitochondrial DNA variability in a section of the cytochrome oxidase I gene of all four Pacific *Echinometra* (S. Palumbi, G. Grabowski, T. Duda, and N. Tachino, *in preparation*). Sequence data have been generated using the polymerase chain reaction followed by direct sequencing of the double-strand product. Data analysis is based on the equations for F_{ST} developed by Hudson et al. (1992) for use with DNA sequence data. The results show an unexpected level of genetic distinction between populations in different archipelagoes in the Pacific. Nevertheless, the most common haplotypes in every species are widespread. These two results (frequency differences between populations, along with widespread haplotypes) suggest a dual role for recent changes in gene flow and local genetic drift in Pacific populations of *Echinometra*.

A COMPARISON OF METHODS

Use of conserved primers has made PCR amplification of the DNA of many species relatively straightforward (see Zimmer et al., 1993; Palumbi, 1996a). This has opened the door to population-level analysis of DNA variation in diverse species.

However, there are important questions in population biology that can only be answered by sequencing PCR products from hundreds of individuals collected from several localities. In these cases, even mild technical problems can represent a logistic nightmare. It is our experience that most of these nightmares arise when trying to sequence a PCR amplified gene product. The single most important factor contributing to "nice" sequences is generating abundant amounts of high-quality DNA template.

Four Methods of Sequencing PCR Products

Basic methods of amplifying DNA using PCR have been detailed in a number of places (Erlich, 1989; Innis et al., 1990; Zimmer et al., 1993; Palumbi, 1996a). DNA sequencing from these PCR products can be accomplished in several contrasting fashions. The major methods are (1) double-strand sequencing from purified PCR products, (2) solid-phase sequencing using a biotinylated primer and streptavidin beads, (3) single-strand sequencing from asymmetric PCR or single-strand digestion, and (4) automated sequencing.

In all these cases, the sequencing reaction depends on single-strand DNA. The difference is in how such single-strand products are generated. Double-strand PCR products can be denatured by heat into single-strand components. Alternatively, one strand of the double-strand DNA can be "tagged" with biotin. The "tagged" strand can then be peeled away, leaving ample pure single-strand DNA. The earliest (but least reproducible) of these methods was to let the PCR machine generate single-strand DNA for you. This is done by limiting one of the two PCR primers and sequencing directly from this reaction. Automated DNA sequencers take advantage of a technique termed "cycle sequencing," which uses a thermal cycler to denature double-strand DNA into single-strand components. A thermostable polymerase and dideoxy nucleotides are used to generate the sequencing products.

The protocols section of this book includes methods for obtaining DNA sequence from PCR products using asymmetric PCR, double-strand sequencing, and solid-phase sequencing. We have used all three to generate sequence data for population or systematic analyses.

Asymmetric amplifications have been supplanted largely by the other methods. This is because asymmetric sequencing depends on the quality of single-strand template produced by asymmetric PCR, and this varies from template to template or even from day to day with the same template. Double-strand sequencing is more reliable, but the sequencing reactions often have a high background and can be read for only 200–300 base pairs in most cases. As an alternative to both these techniques, we have used solid-phase DNA stripping to make single-strand DNA directly from double-strand templates.

Solid-Phase DNA Sequencing

This method combines the power and sensitivity of double-strand PCR reactions with the clean sequencing results of asymmetric amplifications. The techniques have been described (see Bowman and Palumbi, 1993), and the necessary reagents are commercially available.

The key to solid-phase sequencing is the attachment of one of the strands of a DNA duplex to a solid support. To accomplish this, we synthesize a PCR primer with a biotin molecule attached to the 5' end. Such biotinylated primers can be purchased from many biotechnology companies (Operon, Synthetic Genetics, DNAgency).

A normal double-strand PCR reaction then produces a product that is "tagged" on the 5' end of one of the strands. Biotin is used as a tag because it binds to the protein streptavidin in a nearly irreversible reaction. Therefore biotin-tagged DNA can be attached to any support that can be coated with streptavidin. Most of our work has been with small magnetic beads coated with streptavidin, although streptavidin-coated agarose beads and streptavidin-coated microcolumns are also available.

Once attached to the streptavidin on the solid support, double-strand DNA is dissociated with heat or high pH. This strips off the DNA strand that is not bound to the support. (Remember that only one of the strands has a biotin attached.) This bound strand remains attached in high pH or for brief periods at high temperatures, and the other strand drifts off into the liquid phase. The bound strand can be washed and sequenced using normal dideoxy methods (this is called solid-phase sequencing because the reaction takes place while attached to a solid support). The single-strand DNA (called the "stripped strand") can be collected, concentrated, and also used as a sequencing template.

The primary advantages of this method are that (1) any DNA segment that can be double-strand amplified can be single-strand sequenced; (2) only one round of amplification need be done, thereby limiting both costs and opportunity for contamination; and (3) a highly conserved primer that will amplify a DNA segment from a wide variety of taxa (e.g., Kocher et al., 1989) can be biotinylated, and therefore used to rapidly collect sequence data from many species.

DATA ANALYSIS

Sequence data collected from stripped and bound strands of amplified DNA are combined into a single file containing data from a single individual. These files are aligned to one another and compared. Sequence analysis to detect population structure can be accomplished in several contrasting ways.

Phylogenetic analysis of gene flow (Slatkin and Maddison, 1989) treats the locality from which a sample was collected as a phylogenetic characteristic. Gene flow is observed as a series of shifts in locality of a lineage throughout the phylogenetic tree. Thus if two indistinguishable sequences are found in two separate localities, then there must have been gene flow between those places. This method relies on the phylogenetic relationships among alleles, which can be described using distance methods (e.g., neighbor-joining; Saitou and Nei, 1987), parsimony (e.g., PAUP; Swofford, 1993), or maximum likelihood (PHYLIP; Felsenstein, 1993).

Hudson et al. (1992) suggested a method to use DNA sequence data to estimate Wright's F_{ST}. In this approach, the average pairwise nucleotide difference among sequences from within a locality (H_w) is compared to the average between different localities (H_b):

$$F_{ST} = 1 - (H_w/H_b)$$

If DNA sequences within a locality are more similar to each other than they are to sequences from other localities, then $H_b > H_w$ and $F_{ST} > 0$. But how much more than zero should F_{ST} be in order to indicate significant differences among demes? In an ideal island model, with infinite populations that are at genetic equilibrium, significant gene flow will drop F_{ST} to very low values. But in real data sets, this is not necessarily the case. One approach is to perform a Monte Carlo simulation of the data set, randomly assigning a new location to each sequence while keeping the sample sizes per location the same as the original data set. Then F_{ST} can be recalculated for the permuted data set and compared to the actual value. We have written a PASCAL program that accomplishes this, as well as providing a way to manage large DNA data sets (Heap Big Alignment, available from SRP).

A third method of DNA analysis is based on the partitioning of sequence variation within and between several hierarchical levels of organization using MANOVA (Escoffier et al., 1992). This analysis uses a matrix of distance values for each pair of sequences and provides an indication of the proportion of the sequence variation explained by, say, localities within an ocean versus localities between oceans (see Baker et al., 1993, for a marine example). Escoffier et al. (1992) also suggest the use of a Mantel test to compare different matrices. Thus the genetic distance matrix could be compared with a matrix of geographical distance between sequences considered pairwise. This type of analysis might be done at the population level instead of the individual level. For example, a matrix of 100 sequences from five localities might be distilled down to average within and between deme nucleotide differences, and this distilled 5×5 matrix could be compared to other deme properties like geographic distance and language groups.

SPECIATION AND POPULATION STRUCTURE
IN TROPICAL SEA URCHINS

Cytochrome oxidase sequences from positions 6451–7039 [relative to the *Strongylocentrotus purpuratus* sequence (Jacobs et al., 1988)] show from 2% to 13% differentiation among Pacific and Panamanian species of *Echinometra* (Fig. 1). The two Panamanian species we analyzed are clear outgroups to all the Indo-Pacific species: the two species groups differ by about 11–13%. Examination of silent positions suggests saturation of substitutions: 43–76% of the fourfold sites are different between groups. Thus it appears that Panamanian and Indo-Pacific species of the genus *Echinometra* have been separated for a long time. This genus has a fossil record that extends back to the early Miocene, and so deep clades within *Echinometra* are possible.

Within the Indo-Pacific species, differentiation is much smaller. Average percent

Timing of diversification in Pacific *Echinometra*

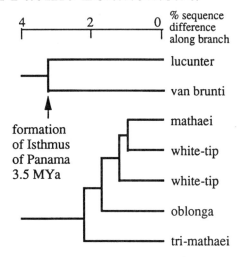

Figure 1. Calibration of rate of cytochrome oxidase I (COI) sequence evolution in Panamanian *Echinometra* and comparison to Pacific species. *Echinometra lucunter* and *E. van brunti* (samples courtesy of H. Lessios) diverged 3.5 million years ago at the rise of the Isthmus of Panama. COI diverges at an overall rate of about 2% per million years in this comparison. Sequence difference between Pacific species ranges from 2% to 5%, suggesting divergence dates of between 1 and 3 million years. Note that there are two clades of mtDNA sequences in the white-tipped species.

divergence between individuals of different species ranges from 2% to 5%. Fourfold sites vary from about 8% between *E. mathaei* and the white-tipped species (nominally *E.* species A *sensu* Uehara and Shingaki 1985), to 19% between *E.* sp. C *sensu* Uehara and Shingaki 1985 (the species with triradiate gonad spicules; see also Palumbi, 1996b) and the other three species. To place this degree of differentiation in a temporal perspective, we compared differentiation of *E. van brunti* and *E. lucunter,* which were separated by the rise of the Isthmus of Panama approximately 3.5 million years ago. Trans-Isthmian species are about 6.2% different in this gene region, corresponding to an overall rate of mitochondrial sequence differentiation of about 2% per million years. Most of the observed changes in these sequences are at silent positions. Comparing *E. van brunti* and *E. lucunter* yields an estimate of divergence at silent sites of 19%. Thus the rate of divergence at silent sites in these taxa is about 5.4% per million years. Using this rate to approximate the divergence of the Indo-Pacific species suggests an origin 1.5–3 million years ago in all cases. These results are similar to those previously reported for RFLP variation of purified mtDNA within and between species (Palumbi and Metz, 1990).

Comparison of 40–60 sequences from each of the four Indo-Pacific species (S. Palumbi, G. Grabowsky, T. Duda, and N. Tachino, *in preparation*) shows that three of the four form monophyletic clades (Fig. 1), with species C as the outgroup. Species A (with white-tipped spines) has individuals with two distinct mtDNA clades, one of which is more closely related to *E. mathaei* (*sensu* Edmondson 1934).

There is currently no evidence that the two clades of *E. sp.* A represent two distinct species. Both mitochondrial clades are present in most local populations, and the strong blocks to cross-fertilization that are seen between species (e.g., Metz et al., 1994) are not seen within species (although milder patterns of assortative mating occur; S. Palumbi, *unpublished data*). Thus it appears that *E. mathaei* arose from an early lineage of the white-tipped *E. sp.* A, and that two ancestral clades have persisted in the latter species.

Population structure in Indo-Pacific *Echinometra* is apparent for all the species that we have examined (Palumbi, 1995; S. Palumbi, G. Grabowski, T. Duda, and N. Tachino, *in preparation*). Because the white-tipped species harbors two distinct mtDNA clades, population structure is evident in this species at both high and low levels of resolution. For example, the proportion of individuals in each of the two major lineages varies from archipelago to archipelago (Fig. 2). In the Society Islands only one clade occurs (denoted clade A in Fig. 2), whereas in Guam, this clade makes up only about 15% of the population. Fiji, Bali, and New Guinea populations have intermediate haplotype frequencies (Palumbi, 1996b).

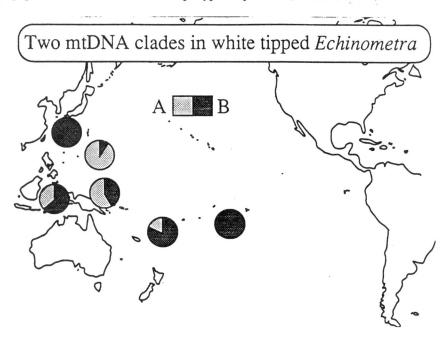

Figure 2. In the white-tipped species of *Echinometra* (*E. sp.* A *sensu* Uehara and Shingaki 1985), two clades of mtDNA sequences are distributed unevenly among populations. Clade A occurs exclusively or predominantly in Okinawa and Tahiti, whereas clade B is most common in Guam and New Guinea.

Low-resolution comparisons show very similar frequencies of the two clades in several of the populations (Fig. 2). However, high-resolution analysis of the entire 450-bp sequences shows that even in these populations there are substantial genetic differences. For example, comparing data from Fiji, Bali, and New Guinea, average within-deme nucleotide diversity is 1.5%, whereas individuals between demes show on average a 1.8% nucleotide difference ($F_{ST} = 0.15$, $p < 0.01$). In this case, genetic differences are not the result of different frequencies of the major allele clades but rather because of differences in the alleles within each clade.

Even in analyses showing strong population structure, there are mitochondrial haplotypes that occur across large geographic areas. Fiji and New Guinea share two haplotypes, and Guam and Okinawa share one. Although it is possible that further sequence information from these individuals would reveal some differences, it is clear from the 450 base pairs of data we have that these sequences are closely related and diverged recently. These results suggest recent gene flow across the Pacific on the scale of at least thousands of kilometers. How can populations accrue genetic differences if there is long-distance gene flow?

There are at least three possible explanations for the simultaneous observation of long-distance gene flow and large genetic differences between localities. First is that long-distance gene flow may occur at a steady but very low rate, and that local genetic drift (which allows mitochondrial haplotype frequencies to shift) occurs more quickly. In general, if gene flow is much less than one migrant per locality per generation, then local genetic drift (and eventually mutation) is more important than gene flow in determining gene frequencies (Slatkin, 1987; Avise, 1994). In this case, we should observe only a small number of identical sequences from individuals separated by large distances.

The second explanation is that gene flow may be episodic, with long-distance dispersal during some time periods and local genetic drift during others. In this scenario, immigrants to isolated habitats (e.g., scattered Pacific archipelagoes) might be broadly transported during some types of oceanic conditions, but subsequent gene flow between localities is curtailed by shifting climate. Observations of the invasion of the Galapagos Islands by Indo-Pacific species during El Niño events (see Richmond, 1990) suggests that changes in Pacific circulation patterns may affect gene flow. Climatic changes associated with low sea levels during glacial maxima may also have contributed to altered patterns of gene flow.

The third explanation for genetic differentiation in the face of long-distance gene flow is the action of selection on local gene frequencies. Mitochondrial haplotypes have long been assumed to be selectively equivalent because many of the changes observed between different haplotypes are silent. In our data set, there is no amino acid differentiation between any of the Pacific species of *Echinometra* in the 450-bp section of cytochrome oxidase I that we examined. However, the mitochondrial genome is a linked series of 37 genes, and selection on a variant of any one of those genes will affect the frequency of the whole linked haplotype in a population. Tests for the action of selection on gene sequences generally depend on the ratio of silent to replacement changes within and between species (e.g., McDonald and Kreitman,

1991). Because we have no amino acid variation within species in this data set, such a test is impossible.

CONCLUSIONS

Analysis of mtDNA sequences from four closely related species of the tropical sea urchin genus *Echinometra* allows a replicated view of marine population genetics across large geographic distances. Speciation during the Pleistocene in *Echinometra* is associated with strong genetic structure in these potentially high-dispersal species. Mitochondrial haplotypes vary in frequency in different archipelagoes of the south and central Pacific, yet some haplotypes are widely distributed. Furthermore, distance between localities is a very poor predictor of the genetic distance between those localities. These results suggest the dual action of episodic gene flow (moving haplotypes large distances) and local genetic drift during periods of low larval exchange (allowing regional differentiation of haplotype frequencies). The climatic and sea level changes that occurred during the Pleistocene may have been responsible for such changes in gene flow and may have set the stage for species diversification during transient periods of genetic isolation.

ACKNOWLEDGMENTS

I thank T. Duda, G. Grabowsky, and N. Tachino for help with some of the sequencing, and F. Cipriano, W. O. McMillan, and S. Romano for general and specific discussions about marine population genetics. This work has been supported by grants from the NSF.

REFERENCES

Arakaki Y, Uehara T (1991): Physiological adaptations and reproduction of the four types of *Echinometra mathaei*. In: Yanagisawa T, Yasumasu I, Oguro C, Suzuki N, Motokawa T (eds). *Biology of Echinodermata*. Rotterdam: AA Balkema, pp 105–112.

Avise JC (1992): Molecular population structure and biogeographic history of a regional fauna: a case history with lessons for conservation and biology. *Oikos* 63:62–76.

Avise JC (1994): *Molecular markers, natural history, and evolution*. New York: Chapman and Hall.

Avise JC, Arnold J, Ball RM, Bermingham E, Lamb T, Neigel JE, Reeb CA, Saunders NC (1987): Intraspecific phylogeography: the mitochondrial DNA bridge between population genetics and systematics. *Annu Rev Ecol Syst* 18:489–522.

Avise JC, Bowen B, Lamb T (1989): DNA fingerprints from hypervariable mitochondrial genotypes. *Mol Biol Ecol* 6:457–473.

Bachmann B, Luke W, Hunsmann G (1990): Improvement of PCR amplified DNA sequencing with the aid of detergents. *Nucleic Acids Res* 18:1309.

Baker CS, Palumbi SR, Lambertsen RH, Weinrich MT, Calambokidis J, O'Brien SJ (1990): The influence of seasonal migration on the distribution of mitochondrial DNA haplotypes in humpback whales. *Nature* 344:238–240.

Baker CS, Perry A, Abernethy B, Alling A, Bannister J, Calambokidis J, Clapham P, Lambertsen RH, Lien J, O'Brien SJ, Urban J, Vasquez O, Weinrich M, Palumbi SR (1993): Abundant mitochondrial DNA variation and world-wide population structure in humpback whales. *Proc Natl Acad Sci USA* 90:8239–8243.

Benzie JA, Stoddart JA (1992): Genetic structure of crown-of-thorns starfish (*Acanthaster planci*) in Australia. *Mar Biol* 112:631–639.

Berger EM (1973): Gene-enzyme variation in three sympatric species of *Littorina*. *Biol Bull* 145:83–90.

Bermingham E, Lessios HA (1993): Rate variation of protein and mitochondrial DNA evolution as revealed by sea urchins separated by the Isthmus of Panama. *Proc Natl Acad Sci USA* 90:2734–2738.

Bert TM, Harrison RG (1988): Hybridization in western Atlantic stone crabs (genus *Menippe*): evolutionary history and ecological context influence species interactions. *Evolution* 42:528–544.

Bowen BW, Meylan AB, Avise JC (1989): An odyssey of the green sea turtle: Ascension Island revisited. *Proc Natl Acad Sci USA* 86:573–576.

Bowman B, Palumbi SR (1993): Rapid production of single stranded sequencing template from amplified DNA using magnetic beads. In: Zimmer E, White T, Cann R, Wilson AC (eds). *Molecular evolution: producing the biochemical data*. Methods in Enzymology, Vol 224. New York: Academic Press, pp 399–405.

Briggs JC (1970): A faunal history of the North Atlantic Ocean. *Syst Zool* 19:19–34.

Burton RS, Feldman MW (1982): Population genetics of coastal and estuarine invertebrates: does larval behavior influence population structure? In: Kennedy S (ed). *Estuarine comparisons*. New York: Academic Press, pp 537–551.

Burton RS, Lee B-G (1994): Nuclear and mitochondrial gene genealogies and allozyme polymorphisms across a major phylogeographic break in the copepod *Tigriopus californicus*. *Proc Natl Acad Sci USA* 91:5197–5201.

CLIMAP Project Members (1976): The surface of the ice age earth. *Science* 191:1131–1137.

Crow JF, Aoki K (1984): Group selection for a polygeneic behavioral trait: estimating the degree of population subdivision. *Proc Natl Acad Sci USA* 81:6073–6077.

Duplessy JC (1982): Glacial to interglacial contrasts in the northern Indian Ocean. *Nature* 295:494–498.

Edmondson CH (1934): *Reef and shore fauna of Hawaii*. Honolulu: Bishop Museum Press.

Erlich HA (1989): *PCR technology*. New York: Stockton Press.

Excoffier L, Smouse PE, Quattro JM (1992): Analysis of molecular variance inferred from metric differences among DNA haplotypes: application to human mitochondrial DNA restriction data. *Genetics* 131:479–491.

Felsenstein J (1993): PHYLIP: Phylogeny Inference Package. Seattle: University of Washington.

Gaillard C, Strauss F (1990): Ethanol precipitation of DNA with linear polyacrylamide to carrier. *Nucleic Acids Res* 18:378.

Gyllensten UB, Erlich HA (1988): Generation of single stranded DNA by the polymerase chain reaction and its application to direct sequencing of the *HLA-DQA* locus. *Proc Natl Acad Sci USA* 85:7652–7656.

Hansen TA (1983): Modes of larval development and rates of speciation in early tertiary neogastropods. *Science* 220:501–502.

Helberg ME (1994): Relationships between inferred levels of gene flow and geographic distance in a philopatric coral. *Balanophyllia elegans Evol* 48:1829–1854.

Hillis D, Moritz C (1995): *Molecular systematics,* 2nd ed. Sunderland, MA: Sinauer.

Hudson RR, Slatkin M, Maddison WP (1992): Estimation of gene flow from DNA sequence data. *Genetics* 132:583–589.

Innis MA, Gelfand DH, Sninsky JJ, White TJ (1990): *PCR protocols: a guide to methods and applications.* New York: Academic Press.

Jacobs HT, Elliot DJ, Math VB, Farquharson A, (1988): Nucleotide sequence and gene organization of sea urchin mitochondrial DNA. *J Mol Biol* 202:185–217.

Karl SA, Avise JC (1992): Balancing selection at allozyme loci in oysters: implications from nuclear RFLPs. *Science* 256:100–102.

Kocher TD, Thomas WK, Meyer A, Edwards SV, Paabo S, Villablanca FX, Wilson AC (1989): Dynamics of mitochondrial DNA evolution in animals: amplification and sequencing with conserved primers. *Proc Natl Acad Sci USA* 86:6196–6200.

Koehn RK, Newell RI, Immerman F (1980): Maintenance of an aminopeptidase allele frequency cline by natural selection. *Proc Natl Acad Sci USA* 77:5385–5389.

Lessa E (1992): Rapid surveying of DNA sequence variation in natural populations. *Mol Biol Evol* 9:323–330.

Lessa E, Applebaum G (1993): Screening techniques for detecting allelic variation in DNA sequences. *Mol Ecol* 2:119–129.

Levin L (1990): A review of methods used for labeling and tracking marine invertebrate larvae. *Ophelia* 32:115–144.

Lyle MW, Prahl FG, Sparrow MA (1992): Upwelling and productivity inferred from a temperature record in the central equatorial Pacific. *Nature* 355:812–815.

Marcus N (1977): Genetic variation within and between geographically separated populations of the sea urchin *Arbacia punctulata. Biol Bull* 153:560–576.

Mayr E (1940): *Systematics and the origins of species.* New York: Columbia University Press.

McDonald JH, Kreitman M (1991): Adaptive protein evolution at the *Adh* locus in *Drosophila. Nature* 351:652–654.

McMillan WO, Palumbi SR (1995): Concordant evolutionary patterns among Indo-West Pacific butterflyfishes. *Proc R Soc Lond B* 260:229–236.

McMillan WO, Raff RA, Palumbi SR (1992): Population genetic consequences of developmental evolution and reduced dispersal in sea urchins (genus *Heliocidaris*). *Evol* 46:1299–1312.

Metz EC, Kane RE, Yanagimachi H, Palumbi SR (1994): Fertilization between closely related sea urchins is blocked by incompatibilities during sperm–egg attachment and early stages of fusion. *Biol Bull* 187:23–34.

Nei M (1987): *Molecular evolutionary genetics.* New York: Columbia University Press.

Nishihira M, Sato Y, Arakaki Y, Tsuchiya M (1991): Ecological distribution and habitat preference of four types of sea urchin *Echinometra mathaei* on the Okinawan coral reefs.

In: Yanagisawa T, Yasumasu I, Oguro C, Suzuki N, Motokawa T (eds). *Biology of Echinodermata*. Rotterdam: AA Balkema, pp 91–104.

Olson RR (1985): The consequences of short distance larval dispersal in a sessile marine invertebrate. *Ecology* 66:30–39.

Palumbi SR (1992): Marine speciation on a small planet. *Trends Ecol Evol* 7:114–118.

Palumbi SR (1994): Reproductive isolation, genetic divergence, and speciation in the sea. *Annu Rev Ecol Syst* 25:547–572.

Palumbi SR (1995): Using genetics as an indirect estimator of larval dispersal. In: McEdward L (ed). *Ecology of marine invertebrate larvae*. Boca Raton, FL: CRC Press, pp 369–387.

Palumbi SR (1996a): PCR and molecular systematics. In: Hillis D, Moritz C (eds). *Molecular systematics*, 2nd ed. Sunderland MA: Sinauer Press.

Palumbi SR (1996b): What can molecular genetics contribute to marine biogeography? An urchin's tale. submitted *J Exp Mar Biol Ecol*.

Palumbi SR, Baker CS (1994): Contrasting population structure from nuclear intron sequences and mtDNA of humpback whales. *Mol Biol Evol* 11:426–435.

Palumbi SR, Baker CS (1995): Nuclear genetic analysis of population structure and genetic variation using intron primers. In: Wayne R, Smith T (eds). *Conservation genetics*. Oxford Surveys in Evolutionary Biology. New York: Oxford University Press.

Palumbi SR, Kessing B (1991): Population biology of the Trans-Arctic exchange: mtDNA sequence similarity between Pacific and Atlantic sea urchins. *Evolution* 45:1790–1805.

Palumbi SR, Metz E (1990): Strong reproductive isolation between closely related tropical sea urchins (genus *Echinometra*). *Mol Biol Evol* 8:227–239.

Palumbi SR, Wilson AC (1989): Mitochondrial DNA diversity in the sea urchins *Strongylocentrotus purpuratus* and *S. droebachiensis. Evolution* 44:403–415.

Paulay G (1990): Effects of late Cenozoic sea-level fluctuations on the bivalve faunas of tropical oceanic islands. *Paleobiology* 16:415–434.

Potts DC (1983): Evolutionary disequilibrium among Indo-Pacific corals. *Bull Mar Sci* 33:619–632.

Queller DC, Strassmann JE, Hughes CR (1993): Microsatellites and kinship. *TREE* 8:285–288.

Quinn WH (1970): Late Quaternary meteorological and oceanic developments in the equatorial Pacific. *Nature* 229:330–331.

Reeb CA, Avise JC (1990): A genetic discontinuity in a continuously distributed species: mitochondrial DNA in the American oyster, *Crassostrea virginica. Genetics* 124:397–406.

Richmond RH (1987): Energetics, competency, and long distance dispersal of planula larvae of the coral *Pocillipora damicornis. Mar Biol* 93:527–533.

Richmond RH (1990): The effects of the El Niño/Southern oscillation on the dispersal of corals and other marine organisms. In: Glynn PW (ed). *Global ecological consequences of the 1982–83 El Niño–southern oscillation*. Amsterdam: Elsevier, pp 127–140.

Saitou N, Nei M (1987): The neighbor-joining method: a new method for reconstructing phylogenetic trees. *Mol Biol Evol* 4:406–425.

Sambrook J, Fritsch EF, Maniatis T (1989): *Molecular cloning: a laboratory manual*. Cold Spring Harbor: Cold Spring Harbor Laboratory Press.

Scheltema RS (1986): On dispersal and planktonic larvae of benthic invertebrates: an eclectic overview and summary of problems. *Bull Mar Sci* 39:290–322.

Schnitker D (1974): West Atlantic abyssal circulation during the past 120,000 years. *Nature* 248:385–387.

Slade RW, Moritz C, Heideman A, Hale PT (1993): Rapid assessment of single copy nuclear DNA variation in diverse species. *Mol Ecol* 2:359–373.

Slatkin M (1987): Gene flow and the geographic structure of natural populations. *Science* 236:787–792.

Slatkin M, Maddison WP (1989): A cladistic measure of gene flow inferred from the phylogenies of alleles. *Genetics* 123:603–613.

Stoner DS (1990): Recruitment of a tropical ascidian: relative importance of pre-settlement vs post-settlement processes. *Ecology* 71:1682–1690.

Swofford D (1993): PAUP release 3.1.1. Washington, DC: Smithsonian Institution.

Thomas WK, Paabo S, Villablanca F, Wilson AC (1990): Spatial and temporal continuity of kangaroo rat populations shown by sequencing mitochondrial DNA from museum specimens. *J Mol Evol* 31:101–112.

Uehara T, Shingaki M (1985): Taxonomic studies in the four types of the sea urchin, *Echinometra mathaei*, from Okinawa, Japan. *Zool Sci* 2:1009.

Vigilant L, Stoneking M, Harpending H, Hawkes K, Wilson AC (1991): African populations and the evolution of human mitochondrial DNA. *Science* 253:1503–1507.

Weir B, Cockerham CC (1984): Estimating F-statistics for the analysis of population structure. *Evolution* 38:1358–1370.

Willis BL, Oliver JK (1990): Direct tracking of coral larvae: implications for dispersal studies of planktonic larvae in topographically complex environments. *Ophelia* 32:145–162.

Wright S (1978): *Evolution and the genetics of populations: Volume 4. Variability within and among populations.* Chicago: University of Chicago Press.

Zimmer EA, White TJ, Cann RL, Wilson AC (1993): *Molecular evolution: producing the biochemical data.* Methods in Enzymology, Vol 224. New York: Academic Press.

CHAPTER 6

Molecular Approaches to the Study of Marine Biological Invasions

JONATHAN GELLER

Department of Biological Sciences, University of North Carolina at Wilmington, Wilmington, North Carolina 28403-3297

CONTENTS

SYNOPSIS

Marine invasions sometimes involve large, conspicuous, and easily identified species. However, it is likely that most invasions are cryptic for one of several reasons, all linked to the incompleteness of marine systematics: invading species may not be properly identified; invading species may be undescribed; invading species may be erroneously redescribed as native species; and invading species may resemble native species. The tools of molecular systematics are well suited to approach these

Molecular Zoology: Advances, Strategies, and Protocols, Edited by Joan D. Ferraris and Stephen R. Palumbi.
ISBN 0-471-14461-4   © 1996 Wiley-Liss, Inc.

problems. Problems of identification also hamper process-oriented studies of marine invasions. Most modern invasions are caused by the transport of plankton in the seawater ballast of ships, and much of this plankton, especially meroplankton, cannot be identified at the species level. However, nucleotide sequence variation can be used to identify species-specific molecular markers. Such markers can be used to identify larvae in ballast water and to correlate larvae with newly discovered invasive populations. While molecular markers can be detected in many ways, one technique discussed in detail uses site-directed mutagenesis with PCR to create diagnostic restriction sites. Other applications of molecular information to invasion biology discussed here include investigations of evolutionary change in invading populations, testing the naturalness of suspect cosmopolitan distributions, and reconstructing the chronology of invasions.

INTRODUCTION

Biological invasions are the establishment of populations of species in regions where they were formerly absent. In only the past few years, invasions have been recognized as a major force in altering the structure and composition of nearshore marine communities (Carlton, 1989; Paine, 1993). To a great extent, this is due to several recent and highly visible aquatic invasions: examples are the zebra mussel *Dreissena polymorpha* in the Great Lakes (Hebert et al., 1989), the Asian clam *Potomocorbula amurensis* (Carlton et al., 1990), and the ctenophore *Mnemiopsis leidyi* in the Black Sea (Vinogradov et al., 1989). However, Carlton and Geller (1993) have argued that because of the highly incomplete status of the systematics of marine organisms, marine invasions are likely to be far more pervasive than the list of high-profile invasions suggests.

Biological invasions often become environmental and conservation crises, and this has motivated much ongoing research (e.g., Nalepa and Schlosser, 1992). However, we should recognize that invasions have always been important in shaping the species composition of biogeographic units and deserve study for fundamental as well as applied reasons. For example, when the Bering Straits opened about 3.5 million years ago, the North Pacific and North Atlantic regions experienced a period of biotic interchange, with species from both regions invading the other. Vermeij's (1991a) analysis of molluscan participants in this event has shown that this exchange was strikingly asymmetric, with Pacific species being about tenfold more successful in invading the Atlantic, and Vermeij (1991a, 1991b) has proposed a hypothesis of ecological opportunity to explain this asymmetry. This hypothesis holds that, because the Atlantic region had suffered a more recent history of species extinction, more ecological space was available for invaders to fill. This scenario implies an equilibrium view of biotas at the biogeographic level, at odds with a more nonequilibrium view of communities that largely prevails today. These views at different scales need to be reconciled, and this illustrates the contribution that the study of invasions can make to ecological theory. An understanding of marine invasions poses conceptual challenges at many levels, and no one approach will be sufficient to produce a predictive model. Here, after discussing some important issues for the marine invasion studies, I focus on ways in which molecular approaches can make a contribution toward such an understanding.

Despite their importance as a natural phenomenon, most marine invasions of recent times are human-mediated, and while various mechanisms for transport of marine organisms have been important in the past, today the primary agent of transport is the use of ballast water in ocean-going ships (Carlton, 1985; Carlton and Geller, 1993). Ships take up seawater, without treatment or effective filtration against plankton, for stability and maneuverability. Thus any planktonic organisms in the water column in a port can be captured in the ballast tanks of these ships. After transport across an ocean, these organisms may be released into a bay or estuary that is environmentally similar to the one they left behind only 10 or 14 days prior. Truly massive amounts of water are involved. Typical cargo and tanker ships may carry 15–50 million liters of seawater. Jim Carlton and co-workers have estimated that, in 1991, about 76 billion liters of foreign seawater were released into U.S. waters by commercial shipping (J. T. Carlton, *personal communication*).

The potential for invasion from ballast water movement is very high. For example, we (Carlton and Geller, 1993) sampled plankton from 150 ships arriving in Coos Bay, Oregon, from various ports in Japan and identified all organisms as narrowly as possible. For most organisms, and especially larvae, species names could not be assigned, but we recognized and counted morphologically distinguishable forms as morphospecies. Altogether, we found 367 morphospecies from all major phyla and plant divisions, with the exception of sponges and ctenophores, which have now been found in other studies of ballast transport (J. T. Carlton, *personal communication*). Indeed, ballast water transport has been implicated in the introduction of the North American ctenophore *Mnemiopsis* to the Black Sea, where it threatens to destroy important fisheries (Shuskina and Musayeva, 1990).

With a fleet of tens of thousands of ships connecting all the continents, ballast water invasions pose an unprecedented threat to native communities. In fact, as Paine (1993) recently commented, we should be much more concerned about the immediate effects of invasions than about the long-term effects of climate change. By the time climate change is relevant, the issue of biotic change may be moot.

The foregoing introduction is meant to convince the reader that marine biological invasion is an issue of considerable importance for theoretical, conservation, and economic reasons. Ultimately, we would like to be able to predict which species will successfully invade a particular region if released there and what may be the ecological consequences. We are a long way from this level of understanding.

If we break down the steps that occur in a ballast water-mediated invasion (Carlton, 1985), we see that there are a series of "filters" that an aspiring invader must pass through to ultimately be successful (Fig. 1). To understand what allows plankton to pass through each of these filters requires a multidisciplinary effort to study, for example, how turbulence affects larvae during intake, how darkness or starvation affects viability, how plankton respond to acute temperature or salinity change at the time of release, how the initial population size of invaders affects chances of establishing a self-sustaining population, and so on. A starting point, at

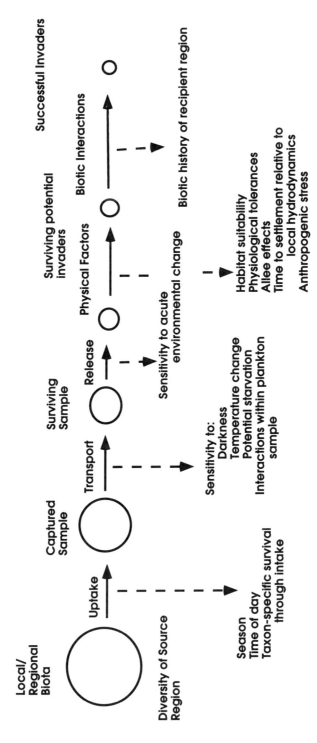

Figure 1. Filters (biological and physical factors) that attenuate the number of potential marine invaders during transport in ballast water. (Modified from Carlton, 1985.)

least, would be to compare the properties of successful invaders with those of failed invaders. Unfortunately, we generally have information only about successful invaders, since these are the only cases that come to our attention. We really do not know if the observed characteristics of successful invaders predispose them for invasion, or if successful invaders are a representative subset of all species transported by ballast.

Moreover, I suspect that we have greatly underestimated the number of invasions that have already occurred. I call these cryptic invasions, and they may occur because (1) invading species are inconspicuous and have not been noticed yet; (2) the invader is an undescribed species and when found is considered a new addition to the local biota; (3) while a species may be described in its native range, it may not be recognized in the new region and may therefore be given a new (invalid) name; or (4) the invader resembles a native species morphologically and so is not recognized as exotic (see Chapman and Carlton, 1991; Taylor and Hebert, 1993; Carlton, 1995).

In Figure 2, using data from Carlton and Geller (1993), I illustrate the number of morphospecies found in ballast water samples for taxonomic groups that I judge to be inconspicuous or taxonomically difficult. It must be remembered that these

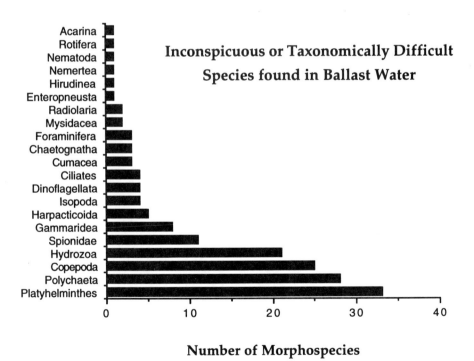

Number of Morphospecies

Figure 2. Numbers of morphospecies found in ballast water samples collected from ships releasing Japanese seawater into Coos Bay, Oregon, that fall into groups where field identification is difficult. (Data from Carlton and Geller, 1993.)

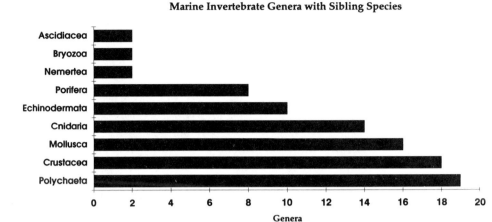

Figure 3. Numbers of genera in several marine invertebrate phyla that contain multispecies complexes. (Data from Knowlton, 1993.)

morphospecies are based mostly on larval forms, and so the actual number of possible cryptic invasions is probably higher than shown here. While some systematists might debate my assignments here, these are species that would be difficult for a field ecologist to identify or confirm as an invasion if they have in fact invaded Coos Bay. With the deemphasis of alpha taxonomists in academia, it might take an ecologist years to obtain an identification from a systematic specialist. This problem is compounded by the diversity of higher taxa in ballast samples.

The introduction of species morphologically similar to native species may be more common than we realize. Many marine genera are characterized by sibling species complexes, as depicted in Figure 3, which shows the number of genera with sibling species groups in several higher taxa (extracted from Knowlton, 1993). It is worrisome that most of the genera on this chart are from well-studied groups; the absence of many phyla from this list is probably more indicative of our ignorance of the systematics of those missing groups than of reality. It is also worrisome that sibling species continue to be discovered in even extremely well-studied or economically important genera, such as *Mytilus, Penaeus, Acropora,* and *Capitella* (references in Knowlton, 1993). How many poorly studied "species" are actually sibling species complexes? Knowlton (1993) estimated that the total number of described species would increase by an order of magnitude if all sibling species among them were known.

APPLICATIONS OF MOLECULAR TECHNIQUES TO MARINE INVASIONS

The preceding analysis of ballast water biota is meant to underscore that one of the major obstacles in the study of marine invasions is simply our inability to identify the players. This is an area where molecular tools for identifying and discovering marine species can contribute much. The number of genetic loci and techniques that can be used are large and, generally speaking, are selected depending on the nature and scope of the problem at hand. The purpose of this contribution is not to review all loci and techniques that might be used, but to give some examples of problems that can be addressed with these methods. I present some examples of work by others that relate to these questions, and also some examples, in more detail, from my own work. Applications of molecular tools that I discuss are: identifying potential invaders in ballast water, correlating putative invaders with released larvae, describing disjunct populations, investigations of evolutionary change in invading populations, revealing pseudocosmopolitan distributions as cryptic invasions, and reconstructing the chronology of invasions.

Identifying Potential Invaders in Ballast Water

Many of the organisms found in ballast water are larvae for which no identification guides exist. Molecular markers could be developed by sequencing appropriate loci in common species in potential source geographic regions. Larvae in ballast samples could then be analyzed using the polymerase chain reaction for matches against these known sequences. Though not applied to an invasion study, this approach was taken by Olson et al. (1991), who sequenced a portion of the 16S rRNA gene from two species of sea cucumbers, *Psolus fabricii* and *Cucumaria frondosa*. The larvae of these species are not morphologically distinguishable but are readily identified at the sequence level. It is interesting to note that these larvae were thought to be from a single species until DNA analysis was performed, emphasizing again the problem of identification in any studies involving marine larvae. Sequence data can be used to develop many detection systems, and I return to this example in a discussion of one such approach involving site-directed mutagenesis.

Correlating Invading Adults and Larvae

Because many organisms are released as larvae and most invertebrate life cycles have not been fully described, it may be difficult to link a newly discovered species in that taxon to the released larvae. Molecular species markers would allow one to make the life cycle connection between larva and adult. One way to achieve this would be to obtain DNA sequences from putative invaders and from larvae of the proper taxon in ballast samples and make a direct comparison. A more rapid approach would be heteroduplex mobility assay, in which amplification products of putative invaders are hybridized with amplification products from larvae. Mismatches in hybrid duplexes retard electrophoretic mobility and are thereby distinguishable from homoduplexes (W. K. Potts, Chapter 9 in this volume).

Disjunct Distributions

One important line of evidence for an invasion is a disjunct species distribution (Chapman and Carlton, 1991). Molecular evidence can show that a putatively invading population belongs phylogenetically within a group that occupies a different geographic region. This of course will be most useful when morphological characters are of little use, as in the case of sibling species. Using allozymes, Woodruff et al. (1986) were able to show that an invading epibiotic gastropod in Hong Kong was actually the Californian snail *Crepidula onyx*. Meehan et al. (1989) showed that clams called *Macoma balthica* in North American Pacific and Atlantic sites, as well as in Europe, are actually a complex of species, and that a San Francisco Bay population is more closely related to western Atlantic populations than to another Pacific population in Oregon.

Goff et al. (1992) used a molecular approach to show that a population of the green alga *Codium fragile tomentosoides* in San Francisco Bay is an invader. They isolated chloroplast DNA and cut it with several restriction enzymes. The resulting restriction fragment patterns matched with populations from the U.S. Atlantic coast and not with a native open coast species *Codium fragile fragile*. Thus the hypothesis that this new population is an ecological variant of the native species is rejected. While *Codium fragile tomentosoides* in San Francisco is invasive, this study could not determine whether the source of this invasion was from the Asian natural range or from populations already introduced to the European or American Atlantic regions or New Zealand.

Evolutionary Change in Invaders

Invasions can be opportunities to test ideas about evolution in a changing environment and about the roles of colonization in influencing patterns of genetic variation and, ultimately, speciation. Once a source region is identified, genotypic and phenotypic characteristics can be compared. The above-mentioned allozyme studies and others (Duda, 1994) are examples of such research, but genetic resolution has not been sufficient to identify point sources for invaders. Molecular methods with higher resolving power—microsatellite analysis, for example (R. C. Fleischer, Chapter 7; J. E. Strassman et al., Chapter 8 in this volume)—may be able to achieve this precision. These few studies comparing invading marine populations to native populations have shown that most of the within-species genetic variation is retained and that disjunct invading populations are undifferentiated from core populations, implying a minimal role for founder effects or drift. Presumably, this is due to the size of colonizing populations in marine invasions or to maintained gene flow, though neither have been demonstrated. It is possible that genetic change could be demonstrated if a source population could be accurately pinpointed.

Cryptic Invasions

As described above, cryptic invasions can occur when a species in a multispecies complex invades its sibling's range. If sibling species are identified at the molecular level, then these markers can be used to detect cryptic invasions. I suspect that molecular ecologists investigating suspect cosmopolitan species will uncover sibling species groups faster than expert systematists can formally describe them. In the meantime, the molecular markers can serve as temporary taxonomic labels to allow ecological investigation to proceed. My work on sibling species of mussels serves as a model of molecular detection of a cryptic invasion and builds on previous work by McDonald and Koehn (1988).

The *Mytilus edulis* species complex is a group of three species that are, for practical purposes, morphologically indistinguishable because of a strong environmental influence on shell shape. *Mytilus edulis* is native to the North Atlantic, *M. galloprovincialis* is from southern Europe and the Mediterranean Sea, and *M. trossulus* is from the North Pacific. *Mytilus galloprovincialis* has also been introduced to many regions around the globe, including Japan and California. McDonald and Koehn (1988), using allozymes, have shown that *M. galloprovincialis* has invaded southern California while the native *M. trossulus* occurs from about San Francisco and further north. However, our study of Japanese ballast water revealed *Mytilus* larvae to be one of the most frequent and abundant bivalves in our samples (Carlton and Geller, 1993). The estimated number of larvae released from a ship often exceeded 15 million, and in many cases larvae were released ready to settle and metamorphose. We also noted that in prior allozyme studies, mussels in ports of the Pacific Northwest were not sampled, and we reasoned that invasions of *M. galloprovincialis* may be most likely there.

My approach was to sequence a portion of the mitochondrial genome to identify sequence variability that would sort to *M. trossulus,* the native species, and *M. galloprovincialis,* the putative invader. DNA was prepared from mussels from populations already characterized by allozyme analysis. Ovarian tissue was used because it is enriched with mitochondria and homogenizes easily. For Japanese mussels, whole juveniles cultured from ballast water were used. For analysis, I chose the mitochondrial 16S ribosomal RNA gene, largely because universal primers that worked with *Mytilus* were already available; later, *Mytilus*-specific primers were designed (Geller et al., 1994). PCR products were cloned into a plasmid, and single-strand DNA was generated by infecting transformed *Escherichia coli* cells with a helper phage. Single-strand templates were then sequenced using standard dideoxy-chain terminating reactions. Details of routine procedures are given in Geller et al. (1993, 1994).

DNA sequences, reported in Geller et al. (1993), showed, first, that there is great divergence between two haplotypes of *M. trossulus,* in a single population in Tillamook Bay, Oregon. In fact, there were 37 differences between the two haplotypes in 350 bases pairs (bp) sequenced. This has now been shown to reflect male-specific and female-specific mitochondrial genomes (Geller, 1994). Despite this variability in *M. trossulus,* there were five nucleotide (nt) sites with fixed differences separating *M. trossulus* from either *M. galloprovincialis* or *M. edulis,* which could serve as species markers to detect the invader *M. galloprovincialis.* The next problem was how to detect these nucleotide differences in larger samples. It would have been too time, labor, and money intensive to sequence all individuals. A faster and cheaper approach would be restriction digestion of PCR products if some of these nucleotide differences caused restriction site differences. Unfortunately, they did not. However, as an alternative, I found that I could create a restriction site difference using a variation of PCR called site-directed mutagenesis.

Site-Directed Mutagenesis

The idea here is that a mismatch in a primer will be incorporated into all PCR products. Consider any aligned pair of sequences where a single variable site is flanked in the 5' direction by at least five conserved nucleotides: 5' 123456 3' (6 = variable site). One of these 6 nt sites may by chance be recognized as a restriction site. If not, altering nucleotide sites 1, 2, and/or 3 in this site can create a 6-bp palindrome, which, with luck, may be a restriction site. For robust PCR, it is preferable for the 3' end of a primer to be a close match to the template, so in creating mutagenic primers, it will be best to choose sites requiring one or two changes in sites 1–3. There are 64 possible 3-bp sequences, but only one (1.5%) will by chance be a palindrome with sites 4–6. However, nine of the 64 sequences can be changed into a palindrome with one change, and 27 with two changes. Thus 36 of 63 (57%) of nonpalindromic sequences can be converted into a possible restriction site. It should be noted that the robustness of primers with one or two mismatches is likely to be sequence dependent and PCR conditions for each new primer will need to be optimized.

Figure 4 illustrates this approach to the design of a specific primer to create a diagnostic restriction site for *Mytilus* (for details see Geller and Powers, 1994). The primer is made unusually long (51 bp) so that, after incubation with the enzyme *EcoRI*, the size difference between cut PCR products from *M. galloprovincialis* and uncut products from *M. trossulus* can easily be seen after electrophoretic separation on agarose gels. If polyacrylamide gels are used, a mutagenic primer of typical length for PCR (25–35 bp) can be used.

To illustrate the generality of this approach, I analyzed the data of Olson et al.

TEMPLATE SEQUENCE

Mytilus galloprovincialis	5'----------------------------------**AAATTC** 3'
Mytilus trossulus	5'----------------------------------**AAATTG** 3'
Primer MYT 16SA-RI	5'----------------------------------**GAATT** 3'

PCR PRODUCT SEQUENCE

Mytilus galloprovincialis	5'----------------------------------**G**/AATTC 3'
Mytilus trossulus	5'----------------------------------**GAATTG** 3'

Figure 4. *Top panel:* The 3' region of the DNA sequence of templates from *Mytilus trossulus* and *M. galloprovincialis* (see Geller et al., 1993) aligned with primer MYT16SA-R1. Position in boldface indicates mismatch in primer–template complementarity, resulting in sequence alteration in PCR products. *Bottom panel:* Sequence of PCR product showing altered nucleotide in boldface and cleavage site for the *EcoRI* site created in the *M. galloprovincialis* product.

PSOLUS   ACT-GAA<u>ACAGAT</u>ATAGTTACCGCAGGGATAAC<u>AGCGTC</u>ATCTCCTTTAAGAG

CUCUMARIA C..G.T......A.A.......................A.............

Figure 5. A portion of the 16S rDNA sequence of two species of holothurians, *Psolus frabricii* and *Cucumaria frondosa* (Olson et al., 1991). Potential regions in which site-directed mutagenesis could create a restriction site are underlined.

(1991), who showed nucleotide substitutions that discriminate two species of holothurians (Fig. 5). Consider the sequence 5′ . . . ACAGAN 3′, where N is T for *Psolus* and A for *Cucumaria*. If the 3′ end of a primer is 5′ . . . ATCGA 3′, a *ClaI* site (ATCGAT) is created in amplification products only of *Psolus*. Another variable site is AGCGTN, where N is C for *Psolus* and A for *Cucumaria*. A primer ending with 5′ . . . TACGT 3′ will uniquely create an *AatII* site for *Psolus*, while a primer ending in 5′ . . . GACGT 3′ will only create a *SnaBI* site in *Cucumaria*. Because not all restriction enzymes are commercially available, some mutagenic primers will be preferred (e.g., the *ClaI* primer above).

Using this method to detect *M. galloprovincialis*, I surveyed mussels from populations between the Aleutian Islands, Alaska, and San Diego, California. Amplification products from mussels in populations north of San Francisco never showed the mutagenic restriction site, even in heavily used ports such as Coos Bay and Seattle. Therefore we concluded that *M. galloprovincialis* has not invaded Northwest ports despite decades of continuous release of larvae (Geller et al., 1994). Populations in sites south of San Francisco were either *M. galloprovincialis* or heteroplasmic, with mitochondrial genomes with and without the mutagenic restriction site. Studies published after this analysis have shown that male mussels are often heteroplasmic (Geller, 1994; Skibinski et al., 1994; Zouros et al., 1994) and that there may be more variation in the 16S rDNA gene than revealed by my initial sequencing (J. B. Geller, *unpublished data*). Both of these findings complicate the use of this particular mtDNA marker in southern California. As allozyme studies do not show significant hybridization far south of San Francisco Bay, it appears that the native mussel, *M. trossulus,* is completely absent in southern California.

Reconstructing the History of Invasion

This brings up an intriguing question: If all bay mussels in southern California are recent introductions, to what species belong mussels found as fossils, in archaeological sites, or found by early naturalists of the last century? This question can now be answered, because DNA analysis of many preserved tissues is feasible. I have obtained material from museum collections from central and southern California from 1870 up to 1947. Some of these collections are dry while others are now in alcohol and have had an unknown history of exposure to fixatives. All attempted DNA extractions produced small amounts of low molecular weight DNA. All efforts to amplify sufficient product for cloning or sequencing failed. However, reamplification using primers internal to those used in the first-round PCR produced high yields. In all reamplifications, it is especially important to run negative controls, including the use of the first-round negative control in the second-round amplification. By using a biotin-labeled primer, I was able to use solid-phase methods to sequence these PCR products (S. R. Palumbi, Chapter 5 in this volume).

With this approach, I was able to sequence up to 120 bp from several old mussel collections from Monterey Bay in 1870, Santa Catalina Island off Los Angeles from around 1900, and from Los Angeles in 1947. Preliminary analysis (J. B. Geller, *in preparation*) indicates that mussels from Monterey Bay from 1870 and Santa Catalina Island from 1900 cluster either with female *M. trossulus* haplotypes or with male *M. trossulus* haplotypes. However, the 1947 sequence from Los Angeles clusters with present day female *M. galloprovincialis*. It therefore appears that the invasion of *M. galloprovincialis* occurred between 1900 and 1947. More interestingly, these data imply that a mussel similar to *M. trossulus* was found in southern California as late as 1900 and cannot now be found. It is possible that this native southern Californian mussel is in fact *M. trossulus* whose southern range has been displaced to the north. Alternatively, this mussel could have been an endemic southern California species, related to but distinct from *M. trossulus*. If this is so, then this species appears now to be extinct. Either interpretation implicates, but does not prove, the role of an invader in a regional or global extinction.

SUMMARY

In conclusion, I have attempted to demonstrate with examples from my work and others that molecular methods have many applications to the study of marine invasions and should be implemented in a multidisciplinary effort to deal with this growing problem. As more species markers are developed, we will better be able to know what species are being transported and which among them succeed and which fail in actually invading new areas. These markers, along with the study of species with suspect cosmopolitan distributions, will uncover more cryptic invasions. Molecular invasion studies can contribute to an understanding of basic population genetic and evolutionary processes, as ballast water transport acts as a grand scale, though inadvertent, transplant experiment. Molecular studies of populations may even uncover unsuspected processes at the cellular and molecular levels, as illustrated by mitochondrial inheritance in *Mytilus*. Finally, molecular techniques provide an opportunity to study the history of invasions by turning museum collections into genetic depositories.

ACKNOWLEDGMENTS

I wish to thank my collaborators on various parts of this research, Drs. J. T. Carlton and D. A. Powers. Thanks to Drs. J. D. Ferraris and S. R. Palumbi for organizing and inviting me to this symposium. This work was supported by the National Science Foundation through a Marine Biotechnology Fellowship and award OCE-9458350 to JBG, and a Cahill Faculty Development Award from the University of North Carolina at Wilmington.

REFERENCES

Carlton JT (1985): Transoceanic and interoceanic dispersal of coastal marine organisms: the biology of ballast water. *Ocean Mar Biol Annu Rev* 23:313–371.

Carlton JT (1987): Mechanisms and patterns of transoceanic marine biological invasions in the Pacific Ocean. *Bull Mar Sci* 41:467–499.

Carlton JT (1989): Man's role in changing the face of the ocean: marine biological invasions and the implications for the conservation of near-shore environments. *Conserv Biol* 3:265–273.

Carlton JT (1995): Biological invasions and cryptogenic species. *Ecology (in press).*

Carlton JT, Geller JB (1993): Ecological roulette: the global transport and invasion of nonindigenous marine organisms. *Science* 261:78–82.

Carlton JT, Thompson JK, Schemel LE, Nichols FH (1990): Remarkable invasion of San Francisco Bay, California, U.S.A. by the Asian clam *Potamocorbula amurensis.* I. Introduction and dispersal. *Mar Ecol Prog Ser* 66:81–94.

Chapman JW, Carlton JT (1991): A test of criteria for introduced species: the global invasion of the isopod *Synidotea laevidorsalis* Miers, 1881. *J Crustacean Biol* 11:386–400.

Duda TF (1994): Genetic population structure of the recently introduced Asian clam, *Potamocorbula amurensis,* in San Francisco Bay. *Mar Biol* 119:235–241.

Geller JB (1994): Sex-specific mitochondrial DNA haplotypes and heterplasmy in *Mytilus trossulus* and *Mytilus galloprovincialis* populations. *Mol Mar Biol Biotechnol* 3:334–337.

Geller JB, Powers DA (1994): Site-directed mutagenesis with the polymerase chain reaction for identification of sibling species of *Mytilus. Nautilus Suppl* 2:141–144.

Geller JB, Carlton JT, Powers DA (1993): Interspecific and intrapopulation variation in mitochondrial ribosomal DNA sequences of *Mytilus* spp. (Bivalvia: Mollusca). *Mol Mar Biol Biotechnol* 2:44–50.

Geller JB, Carlton JT, Powers DA (1994): PCR-based detection of mtDNA haplotypes of native and invading mussels on the northeastern Pacific coast: latitudinal patterns of invasion. *Mar Biol* 119:243–249.

Goff LJ, Liddle L, Silva PC, Voyek K, Coleman AW (1992): Tracing species invasion in *Codium,* a siphonous green alga, using molecular tools. *Am J Bot* 79:1279–1285.

Hebert PDN, Muncaster BW, Mackie GL (1989): Ecological and genetic studies on *Dreissena polymorpha* Pallas: a new mollusc in the Great Lakes. *Can J Fish Aquat Sci* 46: 1587–1591.

Knowlton N (1993): Sibling species in the sea. *Annu Rev Ecol Syst* 24:189–216.

McDonald JH, Koehn RK (1988): The mussels *Mytilus galloprovincialis* and *M. trossulus* on the Pacific coast of North America. *Mar Biol* 99:111–118.

Meehan BW, Carlton JT, Wenne R (1989): Genetic affinities of the bivalve *Macoma balthica* from the Pacific coast of North America: evidence for recent introduction and historical distribution. *Mar Biol* 102:235–241.

Nalepa TF, Schlosser DW (1992): *Zebra mussels: biology, impacts, and control*. Boca Raton, FL: Lewis Publishers (CRC Press).

Olson RR, Runstadler JA, Kocher TD (1991): Whose larvae? *Nature* 351:357–358.

Paine RT (1993): A salty and salutary perspective on global change. In: Kareiva PM, Kingsolver JG, Huey RB (eds). *Biotic interactions and global change*. Sunderland, MA: Sinauer, pp 347–355.

Skibinski DOF, Gallager C, Beynon CM (1994): Mitochondrial DNA inheritance. *Nature* 368:817–818.

Shuskina EA, Musayeva EI (1990): Structure of the planktic community of the Black Sea epipelagic zone and its variation caused by invasion of a new ctenophore species. *Oceanology* 30:225–228.

Taylor DJ, Hebert PDN (1993): Cryptic intercontinental hybridization in *Daphnia* (Crustacea): the ghost of introductions past. *Proc R Soc Lond B* 254:163–168.

Vermeij GJ (1991a): Anatomy of an invasion: the trans-Arctic interchange. *Paleobiology* 17:281–307.

Vermeij GJ (1991b): When biotas meet: understanding biotic interchange. *Science* 253:1099–1104.

Vinogradov MYe, Shuskina EA, Musayeva EI, Sorokin PYu (1989): A newly acclimated species in the Black Sea: the ctenophore *Mnemiopsis leidyi* (Ctenophora: Lobata). *Oceanology* 29:220–224.

Woodruff DS, McMeekin LL, Mulvey M, Carpenter MP (1986): Population genetics of *Crepidula onyx:* variation in a Californian slipper snail recently established in China. *Veliger* 29:53–63.

Zouros E, Ball AO, Saavedra C, Freeman KR (1994): An unusual type of mitochondrial DNA inheritance in the blue mussel *Mytilus*. *Proc Natl Acad Sci USA* 91:7463–7467.

CHAPTER 7

Application of Molecular Methods to the Assessment of Genetic Mating Systems in Vertebrates

ROBERT C. FLEISCHER

Molecular Genetics Program, Department of Zoological Research, National Zoological Park, Smithsonian Institution, Washington, DC 20008

CONTENTS

Molecular Zoology: Advances, Strategies, and Protocols, Edited by Joan D. Ferraris and Stephen R. Palumbi.
ISBN 0-471-14461-4 © 1996 Wiley-Liss, Inc.

SYNOPSIS

Verification of vertebrate mating systems is important for evaluating mating system and kin selection theory, and for estimating demographic and population genetic variables. Until recently, such assessments of "genetic" mating systems relative to "behavioral" or "social" ones in natural populations have been very difficult. Modern molecular genetics has provided a panoply of potentially useful methods to document parentage, and these vary in power, appropriateness, degree of difficulty, and expense. In this chapter I describe and evaluate the relative merits of a number of these potential genetic markers. I assess and compare the use of morphology, allozymes, single-copy nuclear DNA [scnDNA; assayed by restriction fragment length polymorphism [RFLP] or DNA sequence analyses], randomly amplified polymorphic DNA (RAPD), and variable number of tandem repeat DNA (VNTRs; i.e., minisatellites and microsatellites assayed by single-locus or multilocus probes, or by polymerase chain reaction amplification). In general, morphological, allozyme, and scnDNA variants provide low-resolution exclusion but lack the variability needed for easy assignment of parentage. RAPD protocols are simple, but dominance and sometimes low replicability of variable bands can hinder their use. Hypervariability of the many loci assayed in multilocus DNA fingerprinting provides much greater resolution, but inabilities to assign alleles to loci and to compare individuals on different gels limit this method mostly to small family groups. Single-locus VNTR methods generally render adequate variability and allow designation of alleles and among-gel comparisons. Microsatellite amplifications are superior in that very small and even degraded DNA samples are sufficient and alleles can be sized exactly, but they usually require considerable time, effort, and technical skill to develop for a limited set of related taxa. I conclude with a brief review and comparison of studies that have used these approaches to assess genetic mating systems in behaviorally monogamous, polygynous/promiscuous, and polyandrous vertebrates.

INTRODUCTION

Biologists have long been interested in the behaviors associated with mating and sexual selection (e.g., Darwin, 1871), but it is not until recently that we have had much more than a suspicion that the mating systems we infer from behavioral observations (i.e., the "social" or "behavioral" mating systems) are not necessarily the same as the ones discernible from genetic analyses (i.e., the "genetic" mating systems; Westneat et al., 1990; Dunn and Lifjeld, 1994). From the behavioral side this may be because it is often very difficult to follow the "private lives" of organisms in nature: copulations can be rapid and extra-pair mating may often be conducted covertly, thus requiring considerable effort to observe. In addition, it would be nearly impossible to observe the complete mating history of even a single individual free-roaming animal let alone a reasonable sample size of individuals.

Lastly, even if copulations are observed, there is almost no way to apportion or identify reproductive success if more than one male is involved.

From the genetic side of the equation, our ignorance has largely been because we have not had methodologies until recently that allow us to accurately evaluate genetic mating systems. The past decade has seen an incredible surge in the availability of molecular and other methods for the determination of parentage and kinship and, correspondingly, in the number of studies that document and compare genetic mating systems with social ones. It is my purpose here to review the various methods that have been developed to assess parentage, and to describe how and how well they have been applied to studies of mating systems in natural populations of vertebrates. I begin by briefly addressing the importance of documenting genetic mating systems. I then survey and compare the methods that have been used over the past two decades and discuss their strengths and weaknesses. I conclude with a general literature review and several specific case studies to demonstrate applications of the methods to the assessment of a variety of mating and unique social systems (e.g., cooperative breeders, brood parasites). The protocol by Sabine Loew and Robert Fleischer in Loew and Fleischer.1 contains a detailed methodology for multilocus DNA fingerprinting; another protocol by Robert Fleischer and Sabine Loew (Fleischer and Loew.2) provides a comprehensive protocol for constructing microsatellite-enriched plasmid libraries using hybridization selection (modified from Armour et al., 1994).

Rationale

Why is it important to know the genetic mating system? First, theories concerning the evolution of mating systems depend on accurate representations of the mating system and these may be particularly difficult to determine by observation in studies of nocturnal, fossorial, or otherwise secretive species. Models for mating system evolution require knowledge of the ratio of male reproductive success to female reproductive success and of the relative success of alternative reproductive tactics (Emlen and Oring, 1977; Payne, 1979; Trivers, 1985; Westneat et al., 1990), and strictly behaviorally based studies may provide biased and inaccurate estimates of these variables (e.g., Gibbs et al., 1990; Morton et al., 1990; Pemberton et al., 1992; Boness et al., 1993; Dunn and Lifjeld, 1994). In addition, in species where sperm storage or sperm competition may be important aspects of the mating system, molecular or other determinations of parentage are often essential (Smith, 1984; Birkhead and Møller, 1992; Oring et al., 1992). Many models of kin selection also require accurate assessments of relatedness (Trivers, 1985).

A correct determination of parentage is also needed to accurately estimate a number of variables of importance to studies of population dynamics and population and quantitative genetics. These include demographic variables such as lifetime reproductive success, effective population size, and the level of inbreeding in populations. Reliance on estimates of variance in reproductive success from behavioral methods alone could result in biased estimates of these variables. Lastly, estimates of heritability of morphological or other characters may be significantly modified by unrevealed extra-pair parentage (e.g., Alatalo et al., 1984).

METHODS FOR DETERMINING PARENTAGE

A wide variety of methods exist for determining parentage in natural populations. I first provide a brief overview of some nongenetic methods, then present and discuss in greater detail several genetic approaches (Table 1). The goal of the genetic methods is to resolve variable, replicable, and easily assayed sets of markers that can be used to identify, with high probability, the parents of individual offspring. In some cases such markers only have the resolving power to exclude offspring from putative parents without also identifying extra-pair parents. Each of the methods also differs in their expense and degree of difficulty. I attempt to summarize these strengths and costs as equitably as possible and to recommend which methods I feel are best used for particular questions and/or systems. However, I would advise researchers not to choose a method because it is the most advanced, difficult, novel, or expensive if an easier, cheaper, or older method can just as effectively answer the questions of their study.

Nongenetic Methods

In the past, researchers have used nongenetic methods such as vasectomies (Bray et al., 1975), removal experiments, radioactive tracers (e.g., Wolff and Holleman, 1978; Tamarin et al., 1983), colored glass microspheres (Quay, 1988), fluorescent pigments (Ribble, 1991), and even antibody response to rare antigens (Glass et al., 1990) to exclude or identify one or both parents of an individual. Some of these methods allow only exclusion of paternity from a putative father (e.g., vasectomies, removals), while others involve difficult procedures or only determine matrilines

TABLE 1. Types of Genetic Markers/Methods that Have Been, or Could Potentially Be, Used to Document Parentage in Studies of Genetic Mating Systems

Morphology
Allozyme or protein variants
    Protein electrophoresis / isoelectric focusing
Mitochondrial DNA / single-copy nuclear DNA / introns
    Restriction fragment length polymorphism (RFLP)—random probes
    Sequencing—control region / introns
Variable number tandem repeats (VNTRs)
    Multilocus DNA fingerprinting
    Single-locus VNTR probes
    Single-locus amplified VNTRs/microsatellite
Randomly amplified polymorphic DNA (RAPD)

(e.g., antibody response). Few approach DNA methods in their level of resolution and none have been used extensively.

Phenotypic Marker Methods

Morphological Markers. Most studies that have taken advantage of heritable morphological variants for identification of extra-pair young have been conducted in captivity (where potential partners can be controlled; e.g., Burns et al., 1980). However, a small number of studies of wild birds (e.g., Mineau and Cooke, 1979; Alatalo et al., 1984; Norris and Blakey, 1989; Payne and Payne, 1989; Møller and Birkhead, 1992) have used analyses of inherited color polymorphism or heritability of quantitative morphological traits (e.g., tarsus length) to determine whether given nestlings were not likely the true offspring of one or both of their putative parents. While such studies often allow exclusion of offspring from putative parents, they rarely allow identification of extra-pair parents.

Protein Variant Markers. With the emergence of inexpensive protein gel electro-phoresis in the 1970s, a number of researchers began to take advantage of allozyme variants to document genetic mating systems (e.g., Tilley and Hansman, 1976; Han-ken and Sherman, 1981; Westneat, 1987; Pope, 1990; Bollinger and Gavin, 1991; Gowaty and Bridges, 1991; Xia and Millar, 1991) and/or brood parasitism in verte-brates (e.g., Fleischer, 1985; Brown and Brown, 1988; Smyth et al., 1993). Proteins can be analyzed from blood or other tissue samples, or from growing feathers (Marsden and May, 1984) or eggs (Fleischer, 1985; Smyth et al., 1993). Blood and feather samples are often the chosen tissue because they can be taken nondestruc-tively. This, however, can limit the number and types of loci that can be screened, as other tissues (e.g., liver, muscle) generally have more expressed loci than plasma and erythrocytes. A number of researchers (e.g., Westneat, 1987) have used breast muscle biopsies to obtain sufficient numbers of allozyme loci for parentage analyses.

In allozyme methods, tissue extracts are wicked into starch gels or loaded into wells of polyacrylamide or cellulose acetate gels. Protein variants that differ in charge, conformation, or size are separated in the gel by electrophoresis using one of a vari-ety of buffer systems varying in pH and ionic strength. Proteins are visualized by enzyme-specific or general protein stains applied to the whole gel or to a number of horizontal slices in the case of starch gels. The resultant bands are scored and inter-preted as genotypes at mendelian-inherited loci (e.g., Fig. 1). There are several comprehensive protocols available for analysis of allozyme variability (e.g., Selander et al., 1971; Harris and Hopkinson, 1976; Evans, 1987; Murphy et al., 1990).

Researchers have found that the proportion of polymorphic allozyme loci is relatively low in many groups of vertebrates (e.g., Selander, 1976; Barrowclough et al., 1985; Nei, 1987). In addition, the few loci that are polymorphic often have low allelic diversity, with rarely more than 2–3 alleles per locus, and low hetero-zygosity. The low heterozygosity is in part caused by one allele having very high frequency (>80–90%), while others have very low frequencies. Such low levels of variation make it difficult for biochemical methods, along with any other low-

PGM / Brown-headed Cowbird

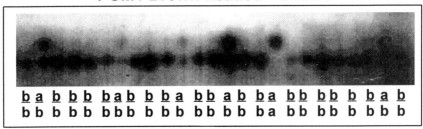

b a	b	b b	b a b	b	b a	b b	a b	b a	b b	b b	b	b a	b
b b	b	b b	b b b	b	b b	b b	b b	b a	b b	b b	b	b b	b

19bb/5ab/1aa; a = 7/50 = 0.14, b = 43/50 = 0.86

Figure 1. Typical allozyme result: diallelic phosphoglucomutase locus of the brown-headed cowbird. Genotypes are noted below each lane.

resolution method, to reliably exclude offspring from parents (Westneat et al., 1987; Burke et al., 1991). In addition, low variability makes identification of the individuals responsible for extra-pair offspring even less likely (Lewis and Snow, 1992).

Single-Copy DNA Methods

RFLP and Sequence Analyses. As methods of analyzing variation in DNA became less complicated and less expensive, it became apparent that molecular methods could also be applied to studies of genetic mating systems. Two common methods for documenting variation in both single-copy nuclear (scnDNA) and mitochondrial DNA (mtDNA) are restriction fragment length polymorphism (RFLP) analysis and DNA sequencing (see Hillis and Moritz, 1990; Hoelzel, 1992; and Avise, 1994, for applications and protocols). RFLP analyses use restriction endonucleases that cleave DNA at particular short sequences (usually 4–6 bp). If the sequences vary, the enzymes will not cut and variation can be revealed by electrophoretic separation of fragments in gel media. DNA sequencing is usually accomplished by electrophoretic separation of dideoxynucleotide-terminated, isotope-labeled, complementary strand DNA that is polymerase-synthesized from a primer annealed to single-strand template DNA (Sanger et al., 1977). Recently, additional methods involving PCR and a variety of screening methods have been developed (Karl and Avise, 1993; Lessa and Applebaum, 1993; Slade et al., 1994). None of these methods have yet been used extensively in parentage analyses, mostly because levels of intrapopulation variation in both exons and introns of nuclear genes appear to be too low for easy resolution of parentage. Both mtDNA and scnDNA have been used in parentage analyses: Quinn et al. (1987) used RFLP analysis with random single-copy nuclear probes to assess parentage in snow goose (*Anser caerulescens*) families; while Morin and Ryder (1991) used mtDNA RFLP, along with DNA fingerprinting, to reconstruct a pedigree of captive lion-tailed macaques (*Macaca silenus*). Avise et al. (1989) found a hypervariable region in mouse mtDNA and suggested it could be used to identify matrilines.

Variable Number of Tandem Repeat (VNTR) Markers

These methods take advantage of relatively small tandemly repeated sequences that are dispersed throughout the genome. Some VNTRs (referred to as minisatellites) have repeat lengths of 7–65 bp (Jeffreys et al., 1985; Shin et al., 1985; Nakamura et al., 1987; Vassart et al., 1987); microsatellites (sometimes called STRs for short tandem repeats or SSRs for simple sequence repeats; Edwards et al., 1992; Goff et al., 1992) are considered to have repeat lengths of 1–6 bp (Ali et al., 1986; Epplen, 1988). For both size classes of VNTRs, the primary source of variation is the difference in the number of repeats within an array. Also, for both types, replication slippage is thought to play a major role in generating additional repeats and variation in repeat number (Levinson and Gutman, 1987; Jeffreys et al., 1991; Schlötterer and Tautz, 1992). Minisatellite variation may also be produced by unequal sister chromatid exchange during mitosis or, less likely, unequal crossing-over between homologous chromosomes during meiosis (Smith, 1976; Jarman and Wells, 1989; Stephan, 1989; Wolff et al., 1991). Both classes of VNTR are perhaps arbitrary sections of a continuum, and it has been suggested that minisatellites originate from microsatellite duplications (Wright, 1994).

Multilocus DNA Fingerprinting. Multilocus DNA fingerprinting has been the primary method of choice for studies of genetic mating systems since its discovery in the mid-1980s until very recently (Jeffreys et al., 1985; Burke, 1989; Burke et al., 1991; Amos and Pemberton, 1992; Bruford et al., 1992). This has been because the extreme mutability and concomitant high allelic diversity and heterozygosity at minisatellite (and, to some extent, microsatellite) loci invariably permits exclusion of offspring from putative parents and usually also allows, with high probability, the assignment of excluded offspring to extra-pair parents (e.g., Burke et al., 1989; Gibbs et al., 1990; Westneat, 1990, 1993; Rabenold et al., 1990; Ribble, 1991; Smith et al., 1991; Oring et al., 1992; Stutchbury et al., 1994). In addition to high variability of markers, DNA fingerprinting is greatly simplified by the use of a large number of nearly "universal" minisatellite and microsatellite probes. Thus a major advantage of this method over others is that little or no preliminary work is required when one switches to a new species.

Multilocus DNA fingerprinting involves restriction digestion of samples of clean, high-molecular-weight genomic DNA with a tetranucleotide-recognizing restriction endonuclease followed by size fractionation in an agarose gel (Fig. 2; see Loew and Fleischer.1 for explicit protocol; also Jeffreys et al., 1985; Bruford et al., 1992). The digested DNA is denatured and transferred (i.e., capillary, vacuum, or pressure blotted) to a nylon or nitrocellulose membrane and then bound to the membrane by UV crosslinking and/or baking. Cloned or synthesized segments of DNA that contain the minisatellite or microsatellite sequences cited above are labeled (isotopically or nonisotopically) by random priming, nick translation, polymerase chain reaction, or 5′ or 3′ end-labeling. The probe is cleaned of excess labeled nucleotides, heat denatured, and hybridized to the DNA on the membrane. Some oligonucleotide probe protocols involve hybridization within a partially dried

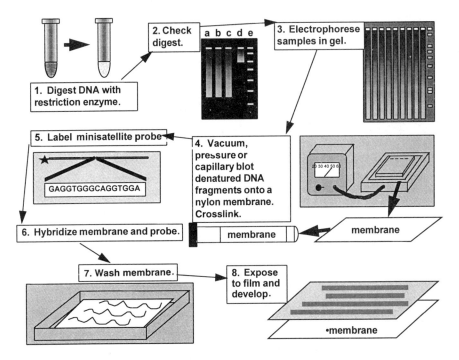

Figure 2. Abbreviated flowchart showing steps of multilocus DNA fingerprinting. In-depth protocols for multilocus DNA fingerprinting are in Loew and Fleischer.1.

gel (i.e., without transfer). X-ray film is exposed to the filter or gel for isotope and chemiluminescent methods and its development produces an autoradiograph in which 10–30 highly variable fragments (representing up to 25 or more separate loci) are usually visualized per lane (e.g., Fig. 3).

When controlled crosses or families with known pedigrees are subjected to multilocus DNA fingerprinting, the band distributions of parents and offspring generally conform to mendelian expectation (Fig. 3; e.g., Jeffreys et al., 1985, 1986, 1987; Nakamura et al., 1987; Longmire et al., 1988; Kuhnlein et al., 1989; Lang et al., 1993) with the occasional exception of a small percentage (usually $\leqslant 1\%$) of "excluding" fragments that are presumably shifted by mutation. Mutation rates for vertebrates have generally been observed in the range of 10^{-4} to 10^{-2}/gamete/generation for minisatellites and 10^{-4} to 10^{-3}/gamete/generation for microsatellites (Jeffreys et al., 1987, 1988a, 1991; Burke et al., 1989; Nürnberg et al., 1989; Kuhnlein et al., 1990; Westneat, 1990; Dallas, 1992; Dietrich et al., 1992; Kwiatkowski et al., 1992; Fleischer et al., 1994; Verheyen et al., 1994). One hypermutable human minisatellite locus (MS1; Jeffreys et al., 1988a), however, shows an astounding mutation rate of 0.05/gamete/generation. In a number of studies of vertebrates, one or more fragments have been found to be sex-linked (e.g., Rabenold et al., 1991; Ellegren et al., 1994) and have proved useful for identifying the sex of individuals in sexually monomorphic taxa.

a. toque macaques | b. palila | c. common loon

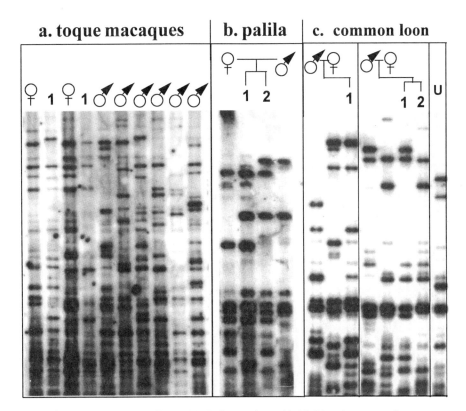

Figure 3. Examples of multilocus DNA fingerprints. (a) Multi-male group of toque macaques (S. Loew et al., *unpublished data*). Shown are two females and offspring followed by six potential fathers. (b) Nestlings (1,2) and putative parents of palila (Fleischer et al., 1994). (c) Two common loon families and an extra individual (W. Piper, *unpublished data*). Note that offspring profiles contain only fragments that are also found in parental ones.

In addition, with a few exceptions (e.g., Brock and White, 1991), most studies have revealed independent assortment of fingerprint fragments (e.g., Jeffreys et al., 1986, 1987; Jeffreys and Morton, 1987; Burke et al., 1991; Lang et al., 1993). The assumption of linkage can be assessed by a number of analyses (Burke et al., 1991; Amos et al., 1992; Bruford et al., 1992), including direct inspection of family pedigrees for co-segregation of fragments, statistical testing for nonrandom association of fragments, comparison of the widths or variances of the distributions of band-sharing among first-order relatives versus among unrelated individuals (Amos et al., 1992; Fleischer et al., 1994), and comparison of the predicted and actual mean band-sharing coefficient of first-order relatives. However, even if there is a fair level of linkage (e.g., 25% of fragments on average), there appears to be only a minor problem of incorrect parentage assignment (Amos et al., 1992).

Multilocus DNA fingerprints of putative family groups are usually scored for parentage analyses in three stages. First, extra-pair offspring are "excluded" from a pair of putative parents if there are fragments in offspring profiles that are not present in one or both parental profiles. A single extra fragment in a profile is likely to have resulted from mutation or artifact rather than nonparentage based on the fit of novel fragment frequencies to a Poisson distribution (Westneat, 1990), but this likelihood should be worked out for each population under study. In families from normal populations it is typical to have five or more novel fragments per probe in extra-pair young. The power of the data to exclude can be further tested by comparing unexcluded offspring profiles to those of "sham" parents (i.e., adults in adjacent lanes) and producing a histogram of novel fragments. The number of unattributable fragments for the actual attending males is then compared to the histogram to indicate the likelihood of exclusion (Fleischer et al., 1994).

Second, coefficients of band-sharing (S of Lynch, 1990, or "unweighted" x of Jeffreys et al., 1985) are calculated between each putative parent and each offspring. These values are compared to a distribution of "background band-sharing" from comparisons of randomly sampled, putatively unrelated individuals. In normal outbred taxa, S usually ranges between 0.0 and 0.5 and averages 0.2–0.3 for randomly sampled, presumably unrelated individuals (Burke et al., 1991; Amos et al., 1992), although it has been found to be higher for some groups of organisms regardless of current inbreeding levels. If S for the male of an excluded pair falls within this range, and S for the female is greater than the range, the excluded offspring represents an extra-pair fertilization (EPF). If S is low for the female as well, the offspring is likely the result of intraspecific brood parasitism (ISBP).

Third, males from neighboring territories, from the same group in social breeders, or floaters are run on the same gel adjacent to the excluded offspring and female. These extra males should be included on the original gel if extra-pair fertilizations are expected or if no male was dominant for group breeders; or they can be run later on a second gel if the behaviorally assigned males are excluded. Each male is then subjected to stages 1 and 2, as above, by comparing his profile to those of the offspring and female. If a male cannot be excluded and has high band-sharing with the offspring, it is concluded that the individual is the actual father. It is very important to include equal amounts of digested DNA in each lane of a fingerprint. If any deviation must be made, it should be for offspring to have less DNA than parents so that extra fragments are not mistaken for nonparentage when they are actually artifacts of light fingerprints.

Additional advantages to multilocus DNA fingerprinting over other methods are that multilocus VNTR probes assay for as many as 10–25 highly variable loci with a single probing, and that filters can easily be reprobed with several independent (usually, but see Armour et al., 1990) probes. Thus a good deal of data can be accumulated rapidly. Several disadvantages include (1) the amount of DNA that is required (minimum of about 1 μg per individual, but usually 3–10 μg; however, these amounts are not as big a problem for birds or lower vertebrates because of their nucleated erythrocytes as they may be for small mammals and invertebrates), (2) the methodology involves more steps and can be more problematic than allozyme (or even than VNTR amplification) methods, (3) specific loci and alleles usually cannot be determined, and, perhaps most important, (4) all individuals being compared should be run on a single gel (even if internal or between lane size standards are included). This is because intergel differences in fragment mobility make it nearly impossible to obtain accurate sizes and to standardize between gels. Thus, in contrast to prior optimism (Burke et al., 1991; Galbraith et al., 1991), I do not advocate scoring multilocus fingerprint similarity among individuals on different gels. [*Note:* This may be acceptable if similarity is high and a few individuals have all the variable fragments and can be used as markers on each gel. (e.g., Rave et al., 1994).] For situations where a large number of offspring are being tested (often for fish, amphibian, or reptile studies), or if there are a large number of potential fathers (or mothers), multilocus fingerprinting is generally less useful or reliable than microsatellite amplification (see below). In fact, the type of study for which multilocus fingerprinting may be best suited is for assessing parentage in monogamous vertebrates with small family sizes and low predicted EPF rates.

Single-Locus VNTR Probes. These probes assay only a single minisatellite "locus" but in protocol and otherwise are very similar to multilocus probes. The procedure begins by "lifting" or replicating onto nylon filters clones from genomic libraries grown on media-containing agar plates (genomic libraries are bacterial cells with engineered plasmid or viral "vectors" that contain random "inserts" of DNA isolated from a species of interest). Multilocus VNTR probes are then labeled and hybridized to the nylon filters in order to locate clones containing vectors with minisatellite-containing inserts (Wong et al., 1986; Nakamura et al., 1987; Armour et al., 1990; Hanotte et al., 1991; Bruford et al., 1992; Verheyen et al., 1994). The single-copy regions that flank the VNTR are the part of the probe that identifies a single locus. The isolated insert from the phage or charomid is used as a probe on nylon membranes bearing restriction enzyme-digested genomic DNA, as in the multilocus method above. The hybridization can include competitor DNA from a distant relative (to "soak up" the minisatellite part of the probe), or the insert can be restriction-digested and only the flanking regions used as a probe (and subcloned).

Such probes *usually* reveal size variants at single Mendelian loci (Fig. 4), but they can require a lot of effort to obtain and process: genomic library construction is not trivial, and even after screening and obtaining positive clones a considerable amount of fine-tuning remains. For example, in one study (Armour et al., 1990), only 12.4% of 185 positive clones assayed ended up providing useful, novel polymorphic VNTR probes. Single-locus VNTR probes, however, have an obvious advantage in allowing determination of allele frequencies, which makes them more amenable to classical population genetic analysis. On the other hand, their visualization requires precisely as much effort as in multilocus DNA fingerprinting, and exact sizing and scoring between gels can still be problematic.

Single-Locus Amplified VNTRs/Microsatellites. With the advent of the polymerase chain reaction (PCR), a method of producing highly variable single-locus genetic markers was developed for humans (Litt and Luty, 1989; Tautz, 1989; Weber and May, 1989). This method takes advantage of microsatellites and, rarely, of minisatellites (because their lengths are usually beyond the abilities of standard

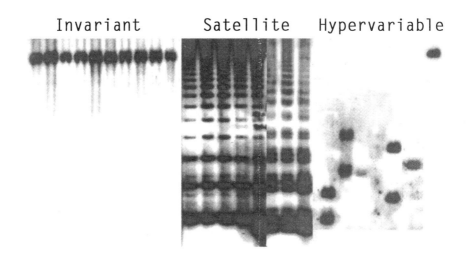

Figure 4. Potential results from hybridization with putative single-locus minisatellite probes. In some cases probes uncover no variability, satellite DNAs, or, preferably, hypervariability.

PCR and sizing is inexact; but see Jeffreys et al., 1988b; Horn et al., 1989; Decorte and Cassiman, 1991; Armour et al., 1992).

To develop primers for microsatellite amplification, microsatellite probes (the same ones used for multilocus fingerprinting; e.g., CAC_n, CA_n) are used to screen genomic libraries (usually in plasmid vectors) of the species of interest in order to locate clones that contain microsatellite and flanking sequences (Fig. 5; Rassmann et al., 1991; Ellegren et al., 1992; Hughes and Queller, 1993). A few methods for constructing libraries that are greatly enriched for microsatellite-bearing inserts have been developed (e.g., Ostrander et al., 1992; Armour et al., 1994); one such method (Armour et al., 1994) used successfully in my laboratory (Fig. 5d) is detailed in Fleischer and Loew.2. The regions that flank the repeat region are sequenced and the sequences are used to design synthetic oligonucleotide primers (Fig. 5a,b). These specific primers are then used in the polymerase chain reaction to amplify across the microsatellite to produce small products (usually <300 bp) that can be isotope-labeled and resolved on a polyacrylamide gel (Fig. 5c). Some laboratories visualize products with ethidium bromide or silver staining. The products can be sized exactly in multiples of the repeat length with a DNA sequence as a size marker and are highly variable, with up to 50 or more alleles and heterozygosities ranging up to 99% (Tautz, 1989; Amos et al., 1993).

Another way in which microsatellites have been identified is by "probing" Gen-Bank or other DNA sequence computer databases with microsatellite sequences

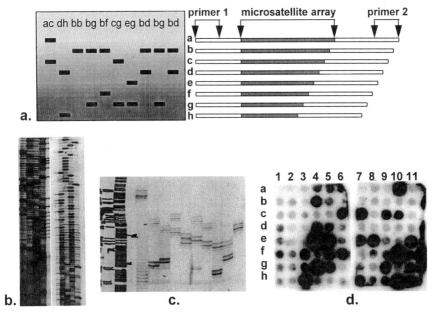

Figure 5. (a) PCR-amplified microsatellite bands (alleles) differ in length based on the number of microsatellite repeat units. Variation is assayed by amplification with two flanking region primers followed by electrophoretic separation of the products (see text). The bands in the gel correspond to the amplified products displayed on the right and differ in size by multiples of the repeat unit. (b) Examples of CA/TG_n repeat and GGA_n repeat, both isolated from a lizard genomic library (N. Zucker et al., *unpublished data*); note how the sequencing enzyme does not always read clearly through the dinucleotide repeat. Primers are designed from the flanking region sequences. (c) Example of microsatellite amplifications of blue-headed wrasses (L. Woonink et al., *unpublished data*). (d) Microtiter plate replica filters from a giant kangaroo rat (*Dipodomys ingens*) genomic library screened with labeled microsatellites (protocol of Armour et al., 1994; Fleischer and Loew.2). Positive clones increased from ≪1% in a normal library to over 30%; most of these yielded long microsatellite arrays (>10 repeats; S. Loew et al., *unpublished data*).

(e.g., Moore et al., 1991; Stallings et al., 1991). Microsatellites can often be found in introns of protein coding genes sequenced for other reasons. Also, if one is fortunate (e.g., Morin and Woodruff, 1992; Gotelli et al., 1994), microsatellites developed for a domestic or laboratory species may work with closely related species in natural populations.

Initially microsatellites were developed for humans (above) and domesticated or laboratory organisms (e.g., Love et al., 1989; Fries et al., 1990), but now amplifiable microsatellites have been developed for a growing number of wild vertebrate taxa, including fish (e.g., Goff et al., 1992), amphibians (e.g., Scribner et al., 1994), reptiles (e.g., N. Zucker, S. Loew, and R. C. Fleischer, *unpublished data*), mammals (e.g., Schlötterer et al., 1991; Paetkau and Strobeck, 1994; Taylor et al.,

1994), and birds (Ellegren, 1992; Hanotte et al., 1994; McDonald and Potts, 1994). Based on these studies, avian microsatellites appear to be as variable as those of mammals, having a range of 1–15 alleles and 40–92% heterozygosity per locus, but birds seem to have fewer microsatellites in their genome than other vertebrates (*personal observation;* T. Glenn and C. Hughes, *personal communications*), making it more difficult to isolate them from genomic libraries without enrichment (e.g., Fleischer and Loew.2).

The primary advantage to microsatellites is that each locus and allele can be typed exactly and put into a database: thus individuals run on different gels at different times can be directly compared. In addition, microsatellites can be amplified from very small amounts of DNA, including partly degraded DNA from shed hair (Morin and Woodruff, 1992), museum specimens (Ellegren, 1991; Taylor et al., 1994), and feces (Constable et al., 1995). The only real disadvantages to microsatellite amplification are (1) they are not as mutable (Nürnberg et al., 1989; Dallas, 1992) hence as variable as minisatellites and (2) primers can be difficult and time-consuming to develop; they are not universal, and they usually will cross-anneal only with species that are fairly closely related (same family or sometimes order; Schlötterer et al., 1991; Moore et al., 1991; Stallings et al., 1991; Ellegren, 1992; Hanotte et al., 1994; Garza et al., 1995; but see FitzSimmons et al., 1995). Therefore, until microsatellites are developed for a broad spectrum of taxa (5+ years from now?), they are likely to be most useful for long-term, intensive studies of single species or related groups of species, for which the development time and effort required will be repaid. A couple of other minor problems recently noted include nonamplifying or "null" alleles (Pemberton et al., 1995), which may predominate when primers are used beyond the taxon for which they were developed, and hidden variability (Gertsch et al., 1995). Although amplified microsatellites would be useful for any study of genetic mating systems, they are essential for studies of mating systems in taxa such as brood parasites or aquatic spawners, for which assignment of offspring to parents would be extremely difficult with multilocus fingerprinting.

Thus far, only a few studies have been published that use amplified VNTRs (microsatellites) to assess genetic mating systems in either nonhuman vertebrates (primates: Takasaki and Takenaka, 1991; Sugiyama et al., 1993; Morin et al., 1994; pinnipeds: Amos et al., 1995; birds: Primmer et al., 1995; bears: Craighead et al., 1995; and cetaceans: Amos et al., 1993) or invertebrates (see Strassmann et al., Chapter 8 in this volume). Most studies to date that utilize amplified microsatellites in vertebrates deal with forensics and linkage mapping in humans and domesticated species (e.g., Love et al., 1989; Fries et al., 1990; Edwards et al., 1992; Dietrich et al., 1992; Goff et al., 1992; Ostrander et al., 1993) or analyses of genetic structure and relatedness in natural populations (e.g., Bowcock et al., 1994; Gottelli et al., 1994; McDonald and Potts, 1994; Taylor et al., 1994). As noted above, however, amplified microsatellites are superior in many ways to other methods of assessing parentage in natural populations, and we should see a rapid increase in their use and in the rate of publication of results.

Randomly Amplified Polymorphic DNA (RAPD)

RAPD analysis (sometimes called AP-PCR, or arbitrarily primed PCR; Welsh and McClelland, 1990) uses random sequence priming to identify polymorphism in putatively random, anonymous DNA sequences. In its most common form (Williams et al., 1990), a single, randomly constructed 10-base oligonucleotide primer is used in a PCR reaction. The amplification requires that complementary, inverted primer sites occur in two locations along a continuous DNA sequence, and that they flank a region which is short enough (<2–3 kb) to be amplified. The resultant products (from 0 to 10 or more different-sized fragments) are electrophoresed in an agarose or polyacrylamide gel, stained with ethidium bromide or silver, and photographed. Some modified protocols have used isotope labels, denaturing polyacrylamide gels, and autoradiography to increase resolution and repeatability (e.g., Welsh and McClelland, 1990; McClelland et al., 1994).

The phenotype of the polymorphism is usually dominant and seen as the presence or absence of a fragment: presumably one variant amplifies and the other cannot because of deletions or substitutions in one or both priming sites, or large insertions or deletions between the priming sites (Williams et al., 1990). Most reports to date indicate that fragment inheritance is mendelian (e.g., Williams et al., 1990; Levitan and Grosberg, 1993), but some researchers have found artifactual variation seen as noninherited fragments (Riedy et al., 1992; Ellsworth et al., 1993; Ayliffe et al., 1994). In one study (Ayliffe et al., 1994), the extra, nonparental fragment was identified as a heteroduplex product of two variants that differ by 38 bp. Hadrys et al. (1993) dealt with this potential problem by amplifying a "synthetic offspring" that included equal amounts of DNA from both parents, and presumably produced any heteroduplex or chimerical products for comparison to actual offspring profiles. Other researchers (e.g., Jones et al., 1994) control artifacts by rigorous standardization of conditions such as DNA quantity and quality, buffer components, and amplification parameters. DNA degradation, in particular, can be a problematic cause for missing bands.

The dominance of RAPD markers is perhaps the greatest drawback to their use for parentage analysis because considerably more loci are required to identify parents than with codominant markers (Lewis and Snow, 1992; Milligan and McMurry, 1993). Interestingly, the probability of assignment increases with increasing recessive allele frequency of RAPDs (Lewis and Snow, 1992). Extra-pair parentage can be identified only if both parents lack a fragment (i.e., are recessive homozygotes) and an offspring has the fragment. Alternative methods of analysis include calculation of band-sharing coefficients (Hadrys et al., 1993) and cluster analysis (Levitan and Grosberg, 1993). The latter allowed clear identification of parents while minimizing the effects of artifactual fragments. To date, nearly all published studies of parentage using RAPD markers have dealt with plants or invertebrates. Unless methods develop to increase replicability and codominance of alleles, RAPD analysis will not likely be the usual method of choice for studies of genetic mating systems.

APPLICATIONS OF METHODS TO MATING SYSTEMS

Monogamy

Monogamy is the most common behavioral mating system in birds (>90%; Lack, 1968) and perhaps the rarest in mammals (<5%; Kleiman, 1977; Boness et al., 1993) and lower vertebrates (Birkhead and Møller, 1992). Thus most applications of genetic markers to assess whether behavioral monogamy equates with genetic monogamy have dealt with birds (but see Ribble, 1991; Dixon et al., 1994), and mostly with birds of temperate rather than tropical regions.

All parentage studies to date of a wide variety of behaviorally monogamous, *non*passerine birds have revealed relatively low EPF rates: the mean percentage of offspring resulting from EPF for 14 species is 4.0 ± 6.3% (range of 0–18%; snow goose, *Anser caerulescens;* blue duck, *Hymenolaimus malachorhynchos;* mallard, *Anas platyrhynchos;* swift, *Apus apus;* fulmar, *Fulmarus glacialis;* black vulture, *Coragyps atratus;* shag, *Phalacrocorax aristotelis;* sparrowhawk, *Accipiter nisus;* and oystercatcher, *Haematopus ostralegus,* from Appendix of Møller and Birkhead, 1994; short-tailed shearwater, *Puffinus tenuirostris,* from Austin et al., 1993; Cory's shearwater, *Calonectris diomedea,* from Swatschek et al., 1994; merlin, *Falco columbarius* from Warkentin et al., 1994; unhelped bee-eaters, *Merops apiaster,* and red-cockaded woodpeckers, *Picoides borealis,* from Haig et al., 1994). Thus genetic mating systems for nonpasserine birds mostly match behavioral mating systems.

Behaviorally monogamous passerine birds, on the other hand, exhibit a much higher variance in EPF rate: ranging from 0% to a high of 58% of offspring. The proportion of EPF may be related to sedentary versus migratory status, and perhaps also to living in the tropics. The few tropical species assessed thus far show very low rates of EPF per nestling (e.g., no evidence for EPF in the common myna, *Acridotheres tristis,* Telecky, 1989; the palila, *Loxioides balleui,* Fleischer et al., 1994; and the dusky antbird, *Cercomacra tyrannina,* C. Tarr et al., *unpublished data*). Sedentary monogamous species (including the tropical ones above) have relatively low levels of EPF (6.0 ± 6.2% of offspring, range of 0–17%, n = 17 studies of 14 species; values for jackdaws, *Corvus monedula;* Siberian jay, *Perisoreus infaustas;* zebra finch, *Taeniopygia guttata;* house finch, *Carpodacus mexicanus;* chaffinch, *Fringilla coelebs;* and dunnock, *Prunella modularis,* obtained from Appendix of Møller and Birkhead, 1994; great tit, *Parus major,* from Blakey, 1994; blue tit, *Parus caeruleus,* from Kempenaers et al., 1992; and both tit species from Gullberg et al., 1992; monogamous corn bunting, *Milaria calandra,* from Hartley et al., 1993; house sparrow, *Passer domesticus,* from Wetton et al., 1992, and Burke et al., 1991; and bull-headed shrike, *Lanius bucephalus,* from Yamagishi et al., 1992). Migratory monogamous species show, on average, a higher rate of EPF (23.2 ± 15.1% of offspring, range of 0–58%, n = 17 species; wheatear, *Oenanthe oenanthe;* eastern bluebird, *Sialia sialis;* reed warbler, *Acrocephalus schoenobaenus;* willow warbler, *Phylloscopus sibilatrix;* tree swallow, *Tachycineta bicolor;* purple martin, *Progne subis;* barn swallow, *Hirundo rustica;* cliff swallow,

Hirundo pyrrhonota; white-crowned sparrow, *Zonotrichia leucophrys;* dark-eyed junco, *Junco hyemalis;* indigo bunting, *Passerina cyanea;* and hooded warbler, *Wilsonia citrina,* from Appendix of Møller and Birkhead, 1994; house wren, *Troglodytes aedon,* and field sparrow, *Spizella pusilla, recalculated* from Price et al., 1989, and Petter et al., 1990, respectively; Wilson's warbler, *Wilsonia pusilla,* from Bereson et al., in press; Kentucky warbler, *Oporornus formosus,* from M. V. McDonald et al., *unpublished data;* and monogamous reed bunting, *Emberiza schoeniclus,* from Dixon et al., 1994). In summary, a remarkable number of these behaviorally monogamous taxa have turned out to have genetically promiscuous mating systems.

There have been a number of hypotheses proposed to account for the variation in rates of EPF in behaviorally monogamous species, and they generally include characteristics such as nesting density or mate guarding ability (reviewed in Westneat et al., 1990), sexual dimorphism (Møller and Birkhead, 1994), nesting synchrony (Westneat et al., 1990; Birkhead and Møller, 1992; Stutchbury and Morton, 1995), variation in male fertility (Sheldon, 1994), and habitat occludedness (Bereson et al., 1995). Almost certainly a multivariate approach will be necessary when assessing the contributions of various factors to EPF rate, and phylogenetic nonindependence should not be ignored. Only a small number of socially monogamous mammal species have been tested with molecular methods. Data from allozymes (Foltz, 1981) and DNA fingerprinting (Ribble, 1991) revealed no EPFs in *Peromyscus polionotus* and *P. californicus,* respectively, while an allozyme study on a third species, *P. leucopus,* showed direct evidence for multiple paternity (Xia and Millar, 1991).

Polygyny/Promiscuity

Most fish, amphibians, reptiles, mammals, and some birds fall into the categories wherein males typically mate with more than one female, while females do not (polygyny) or do (promiscuity or polygynandry) mate with more than one male. These mating systems are somewhat broadly defined and can include variants such as simultaneous and serial polygyny, harem polygyny, multi-male groups/communal breeders, scramble mating, aquatic spawning, and lek mating. In addition, promiscuous mating systems, in particular, can be more difficult to assess with molecular methods than monogamous or polygynous systems, primarily because there can be many more individuals to survey to exclude as parents. Thus "inclusion" of parentage is usually the priority for studies involving multiple males, unlike monogamy, in which parentage exclusion is the first priority.

Examples of classical polygyny that have been assessed by molecular methods include studies of red deer (*Cervus elaphus;* Pemberton et al., 1992), threespine sticklebacks (*Gasterosteus aculeatus;* Rico et al., 1992), red-winged blackbird (*Agelaius phoenecius;* Gibbs et al., 1990; Westneat, 1993), bobolink (*Dolichonyx oryzivorous;* Bollinger and Gavin, 1991), corn bunting (*Miliaria calandra;* Hartley et al., 1993), and harem-polygynous fur seals (*Arctocephalus* spp.; S. Goldsworthy et al., *unpublished data*). Both monogamous and polygynous corn buntings showed no evidence of EPF, whereas the other taxa all had significant rates of EPF. Remarkably, both red-winged blackbird studies revealed nearly identical rates of extra-pair fertilization (28% and 24%), but Gibbs et al. (1990) found that apparent reproductive success did not correlate with realized (or actual) reproductive success, while Westneat (1993) did. In the red deer and both red-winged blackbird studies total reproductive success of males was determined, and overall variance in reproductive success was greater than that predicted from behavioral data alone.

Another common assessment involving polygyny or promiscuity is of paternity in multi-male groups. Mammalian examples include primates (e.g., Pope, 1990; De Ruiter and van Hooff, 1993; Morin et al., 1994; S. Loew et al., *unpublished data*), lions (Packer et al., 1991), mongooses (Keane et al., 1994), and whales (Amos et al., 1991, 1993). Questions subject to DNA analysis included: Do all males in a group mate, or only dominant ones? Does relatedness of males affect their likelihood of paternity? In the primate, lion, and mongoose studies, all or nearly all paternity could be ascribed to males within the group, and multilocus DNA fingerprinting or allozyme (Pope, 1990) methods were generally adequate to resolve parentage. In all but one of the above studies the dominant (Pope, 1990) or high-ranking males obtained nearly all reproductive success. In lions and mongooses dominant males shared paternity primarily with *unrelated* high-ranking subordinates, suggesting match to models of power-sharing. However, in pilot whales (*Globicephala melas*), Amos et al. (1991, 1993) surprisingly found, from both multilocus fingerprinting and microsatellite amplification analyses, that virtually no offspring fetuses analyzed could have been fathered by males present in the pod at the time of capture. Amos et al. inferred that males remain in their natal pods with their families, mate only with unrelated females in other pods, but do not mate with the related females in their own pod.

Avian examples generally involve cooperatively breeding species. Researchers have used molecular methods to assess mating systems within natural populations of a variety of species, including woodpeckers, bee-eaters, fairy wrens, stripe-backed wrens (see Table 1 of Haig et al., 1994; Mulder et al., 1994), rallids (e.g., Jamieson et al., 1994; Lambert et al., 1994), and cuckoos (e.g., Quinn et al., 1994). Brood parasites such as cowbirds and cuckoos often represent a special case among birds in that both parents need to be identified with molecular methods. Hahn and Fleischer (1995) found evidence with multilocus DNA fingerprinting that some female brown-headed cowbirds associate with their own juvenile offspring at feeding sites.

Polyandry

Strict polyandry is a relatively rare social mating system and most applications of molecular methods have been conducted on polyandrous birds such as the dunnock (Burke et al., 1989), the Galapagos hawk (Faaborg et al., 1995), and the spotted sandpiper (*Actitis macularia*; Oring et al., 1992). In Galapagos hawks, Faaborg et al. (1995) used multilocus fingerprinting to confirm that offspring of a female were fathered in an egalitarian manner by most or all of the males in a group. Oring et al. (1992) provided some of the most compelling data for long-term sperm storage in a natural population. Female spotted sandpipers in Minnesota lay clutches serially for males to incubate. Males compete to be the incubator for a female's first clutch of the season, in spite of no observed fitness benefits accruing from being first. Eggs from first clutches only rarely have evidence of EPF, but eggs from later clutches show as high as 14% EPF, nearly all of which cannot be excluded from a female's prior mates. Many of these eggs could only have been fertilized by stored sperm because some previous mates disappeared from the study site long before the eggs were laid. This discovery, like many other examples presented above, could not have been made without the use of highly variable and reliable genetic markers.

ACKNOWLEDGMENTS

I would like to thank S. Loew, L. Wooninck, C. McIntosh, W. Piper, B. Reig, P. Escobar-Paramo, C. Tarr, and N. Zucker for laboratory assistance. I appreciate discussions of the topic with D. Boness, S. Loew, E. Morton, E. Perry, W. Piper, M. Sorenson, C. Tarr, and L. Wooninck, and comments and taxonomic equity edification by S. Loew and E. Perry. The manuscript benefitted from additional review by P. Parker, E. Morton, and J. Ferraris. My research in this area has been conducted with financial support from the National Science Foundation, U.S. Fish and Wildlife Service, Smithsonian Institution, and the Director's Circle and Friends of the National Zoo (FONZ).

REFERENCES

Alatalo RV, Gustafsson L, Lundberg A (1984): High frequency of cuckoldry in pied and collared flycatchers. *Oikos* 42:41–47.

Ali S, Müller CR, Epplen JT (1986): DNA fingerprinting by oligonucleotide probes specific for simple repeats. *Hum Genet* 74:239–243.

Amos B, Pemberton J (1992): DNA fingerprinting in non-human populations. *Curr Opinion Genet Dev* 2:857–860.

Amos B, Barrett J, Dover GA (1991): Breeding behaviour of pilot whales revealed by DNA fingerprinting. *Heredity* 67:49–55.

Amos B, Barrett JA, Pemberton JM (1992): DNA fingerprinting: parentage studies in natural populations and the importance of linkage analyses. *Proc R Soc Lond B* 249:157–162.

Amos B, Schlötterer C, Tautz D (1993): Social structure of pilot whales revealed by analytical DNA profiling. *Science* 260:670–672.

Amos B, Twiss S, Pomeroy P, Anderson S (1995): Evidence for mate fidelity in the gray seal. *Science* 268:1897–1899.

Armour JAL, Povey S, Jeremiah S, Jeffreys AJ (1990): Systematic cloning of human minisatellites from ordered array charomid libraries. *Genomics* 8:501–512.

Armour JAL, Crosier M, Jeffreys AJ (1992): Human minisatellite alleles detectable only after PCR amplification. *Genomics* 12:116–124.

Armour JAL, Neumann R, Gobert S, Jeffreys AJ (1994): Isolation of human simple repeat loci by hybridization selection. *Hum Mol Genet* 3:599–605.

Austin JJ, Carter RE, Parkin DT (1993): Genetic evidence for extra-pair fertilisations in socially monogamous short-tailed shearwaters, *Puffinus tenuirostris* (Procellariiformes: Procellariidae), using DNA fingerprinting. *Aust J Zool* 41:1–12.

Avise JC (1994): *Molecular markers, natural history and evolution.* New York: Chapman and Hall.

Avise JC, Bowen BW, Lamb T (1989): DNA fingerprints from hypervariable mitochondrial genotypes. *Mol Biol Evol* 6:258–269.

Ayliffe MA, Lawrence GJ, Ellis JG, Pryor AJ (1994): Heteroduplex molecules formed between allelic sequences cause nonparental RAPD bands. *Nucleic Acids Res* 22:1632–1636.

Barrowclough GR, Johnson NK, Zink RM (1985): On the nature of genic variation in birds. *Curr Ornithol* 2:135–154.

Bereson R, Rhymer J, Fleischer RC (in press): Extra-pair fertilizations in Wilson's warblers and correlates of cuckoldry. *Behav Ecol Sociobiol (in press).*

Birkhead TR, Møller AP (1992): *Sperm competition in birds: evolutionary causes and consequences.* London: Academic Press.

Birkhead TR, Burke T, Zann R, Hunter FM, Krupa AP (1990): Extra-pair paternity and intraspecific brood parasitism in wild zebra finches *Taeniopygia guttata,* revealed by DNA fingerprinting. *Behav Ecol Sociobiol* 27:315–324.

Blakey JK (1994): Genetic evidence for extra-pair fertilizations in a monogamous passerine, the great tit *Parus major.* Ibis 136:457–462.

Bollinger EK, Gavin TA (1991): Patterns of extra-pair fertilizations in bobolinks. *Behav Ecol Sociobiol* 29:1–7.

Boness DJ, Bowen WD, Francis JM (1993): Implications of DNA fingerprinting for mating systems and reproductive strategies of pinnipeds. *Symp Zool Soc Lond* 66:61–93.

Bray OE, Kenelly JJ, Guarino JL (1975): Fertility of eggs produced on territories of vasectomized red-winged blackbirds. *Wilson Bull* 87:187–195.

Brock MK, White BN (1991): Multifragment alleles in DNA fingerprints of the parrot *Amazona ventralis. J Hered* 82:209–212.

Bruford MW, Hanotte O, Brookfield JFY, Burke T (1992): Single-locus and multilocus DNA fingerprinting. In: Hoelzel AR (ed). *Molecular genetic analysis of populations.* New York: IRL Press/Oxford University Press.

Brown CR, Brown MB (1988): Genetic evidence of multiple parentage in broods of cliff swallows. *Behav Ecol Sociobiol* 23:379–387.

Bowcock AM, Ruiz-Linares A, Tomfohrde J, Minch E, Kidd JR, Cavalli-Sforza LL (1994): High resolution of human evolutionary trees with polymorphic microsatellites. *Nature* 368:455–457.

Burke, T (1989): DNA fingerprinting and other methods for the study of mating success. *Trends Ecol Evol* 4:139–144.

Burke T, Davies NB, Bruford MW, Hatchwell BJ (1989): Parental care and mating behavior of polyandrous dunnocks *Prunella modularis* related to paternity by DNA fingerprinting. *Nature* 338:249–251.

Burke T, Hanotte O, Bruford MW, Cairns E (1991): Multilocus and single locus minisatellite analysis in population biological studies. In: Burke T, Dolf G, Jeffreys AJ, Wolff R (eds). *DNA fingerprinting: approaches and applications.* Basel: Birkhäuser Verlag, pp 154–168.

Burns JT, Cheng KM, McKinney F (1980): Forced copulation in captive mallards. I. Fertilization of eggs. *Auk* 97:875–879.

Constable JJ, Packer C, Collins DA, Pusey AE (1995): Nuclear DNA from primate dung. *Nature* 373:393.

Craighead L, Paetkau P, Reynolds HV, Vyse ER, Stropeck C (1995): Microsatellite analysis of paternity & reproduction in Arctic grizzly bears. *J Hered* 86:255–261.

Dallas JF (1992): Estimation of microsatellite mutation rates in recombinant inbred strains of mouse. *Mamm Genome* 5:32–38.

Darwin C (1871): *The descent of man, and selection in relation to sex.* London: John Murray.

Decorte R, Cassiman J-J (1991): Detection of amplified VNTR alleles by direct chemiluminescence: application to the genetic identification of biological samples in forensic cases. In: Burke T, Dolf G, Jeffreys AJ, Wolff R (eds). *DNA fingerprinting: approaches and applications.* Basel: Birkhäuser Verlag, pp 371–390.

Dessauer HC, Cole CJ, Hafner MS (1990): Collection and storage of tissues. In: Hillis DM, Moritz C (eds). *Molecular systematics.* Sunderland, MA: Sinauer.

De Ruiter JR, van Hooff JARAM (1993): Male dominance rank and reproductive success in primate groups. *Primates* 34:513–523.

Dietrich W, Katz H, Lincoln SE, Shin H-S, Friedman J, Dracopoli N, Lander ES (1992): A genetic map of the mouse suitable for typing intraspecific crosses. *Genetics* 131:423–447.

Dixon A, Ross D, O'Malley SLC, Burke T (1994): Paternal investment inversely related to degree of extra-pair paternity in the reed bunting. *Nature* 371:698–700.

Dunn PO, Lifjeld JT (1994): Can extra-pair copulations be used to predict extra-pair paternity in birds? *Anim Behav* 47:983–985.

Edwards A, Hammond HA, Jin L, Caskey CT, Chakraborty R (1992): Genetic variation at five trimeric and tetrameric tandem repeat loci in four human groups. *Genomics* 12:241–253.

Ellegren H (1991): DNA typing of museum specimens of birds. *Nature* 354:113.

Ellegren H (1992): Polymerase-chain-reaction (PCR) analysis of microsatellites—a new approach to studies of genetic relationships in birds. *Auk* 109:886–895.

Ellegren H, Johansson M, Sandberg K, Andersson L (1992): Cloning of highly polymorphic microsatellites in the horse. *Anim Genet* 23:133–142.

Ellegren H, Johansson M, Hartman G, Andersson L (1994): DNA fingerprinting with the human 33.6 minisatellite probe identifies sex in beavers *Castor fiber. Mol Ecol* 3:273–274.

Ellsworth DL, Rittenhouse KD, Honeycutt RL (1993): Artifactual variation in randomly amplified polymorphic DNA banding patterns. *BioTech* 14:214–217.

Emlen ST, Oring LW (1977): Ecology, sexual selection and the evolution of mating systems. *Science* 197:215–223.

Epplen JT (1988): On simple repeated GA{T/C}A sequences in animal genomes: a critical reappraisal. *J Hered* 79:409–417.

Evans PGH (1987): Electrophoretic variability of gene products. In: Cooke F, Buckley PA (eds). *Avian genetics.* New York: Academic Press, pp 105–162.

Faaborg J, Parker PG, DeLay L, de Vries TJ, Bednarz JC, Maria Paz S, Naranjo J, Waite TA (1995): Confirmation of cooperative polyandry in the Galapagos hawk (*Buteo galapagoensis*). *Behav Ecol Sociobiol*, in press.

FitzSimmons NN, Moritz C, Moore SS (1995): Conservation and dynamics of microsatellite loci over 300 million years of marine turtle evolution. *Mol Biol Evol* 12:432–440.

Fleischer RC (1985): A new technique to identify and assess the dispersion of eggs of individual brood parasites. *Behav Ecol Sociobiol* 17:91–99.

Fleischer RC, Tarr CL, Pratt TK (1994): Genetic structure and mating system in the palila, an endangered Hawaiian honeycreeper, as assessed by DNA fingerprinting. *Mol Ecol* 3:383–392.

Foltz DW (1981): Genetic evidence for long-term monogamy in a small rodent, *Peromyscus polionotus*. *Amer Natur* 117:665–675.

Fries R, Eggen A, Stranzinger G (1990): The bovine genome contains polymorphic microsatellites. *Genomics* 8:403–406.

Galbraith DA, Boag PT, Gibbs HL, White BN (1991): Sizing bands on autoradiograms: a study of precision for scoring DNA fingerprints. *Electrophoresis* 12:210–220.

Garza JC, Slatkin M, Freimer NB (1995): Microsatellite allele frequencies in humans and chimpanzees, with implications for constraints on allele size. *Mol Biol Evol* 12:594–603.

Gertsch P, Pamilo P, Varvio S (1995): Microsatellites reveal high genetic diversity within colonies of *Camponotus* ants. *Mol Ecol* 4:257–260.

Gibbs HL, Weatherhead PJ, Boag PT, White BN, Tabak LM, Hoysak DJ (1990): Realized reproductive success of polygynous red-winged blackbirds revealed by DNA markers. *Science* 250:1394–1397.

Glass GE, Childs JE, LeDuc JW, Cassard SD, Donnenberg AD (1990): Determining matrilines by antibody response to exotic antigens. *J Mamm* 71:129–138.

Goff DJ, Galvin K, Katz H, Westerfield M, Lander ES, Tabin CJ (1992): Identification of polymorphic simple sequence repeats in the genome of the zebrafish. *Genomics* 14:200–202.

Gottelli D, Sillero-Zubiri C, Applebaum GD, Roy MS, Girman DJ, Garcia-Moreno JC, Ostrander EA, Wayne RK (1994): Molecular genetics of the most endangered canid: the Ethiopian wolf *Canis simensis*. *Mol Ecol* 3:301–312.

Gowaty PA, Bridges WC (1991): Behavioral, demographic, and environmental correlates of extra-pair fertilizations in eastern bluebirds *Sialia sialis*. *Behav Ecol* 2:339–350.

Gullberg A, Tegelström H, Gelter HP (1992): DNA fingerprinting reveals multiple paternity in families of great and blue tits (*Parus major* and *P. caeruleus*). *Hereditas* 117:103–108.

Hadrys H, Schierwater B, Dellaporta SL, DeSalle R, Buss LW (1993): Determination of paternity in dragonflies by random amplified polymorphic DNA fingerprinting. *Mol Ecol* 2:79–87.

Hahn DC, Fleischer RC (1995): DNA fingerprint similarity between female and juvenile brown-headed cowbirds trapped together. *Anim Behav* 49:1577–1580.

Haig SM, Walters JR, Plissner JD (1994): Genetic evidence for monogamy in the red-cockaded woodpecker, a cooperative breeder. *Behav Ecol Sociobiol* 34:295–303.

Hanken J, Sherman PW (1981): Multiple paternity in Belding's ground squirrel litters. *Science* 212:351–353.

Hanotte O, Burke T, Armour JAL, Jeffreys AJ (1991): Hypervariable minisatellite DNA sequences in the Indian peafowl *Pavo cristatus*. *Genomics* 9:587–597.

Hanotte O, Zanon C, Pugh A, Greig C, Dixon A, Burke T (1994): Isolation and characterization of microsatellite loci in a passerine bird: the reed bunting *Emberiza schoeniclus*. *Mol Ecol* 3:529–530.

Harris H, Hopkinson DA (1976): *Handbook of enzyme electrophoresis in human genetics.* New York: American Elsevier.

Hartley IR, Sheperd M, Robson T, Burke T (1993): Reproductive success of polygynous male corn buntings (*Milaria calandra*) as confirmed by DNA fingerprinting. *Behav Ecol* 4:310–317.

Hillis DM, Moritz C (1990): *Molecular systematics.* Sunderland, MA: Sinauer.

Hoelzel AR (1992): *Molecular genetic analysis of populations.* Oxford: IRL Press/Oxford University Press.

Hoelzel AR, Green A (1992): Analysis of population level variation by sequencing PCR-amplified DNA. In: Hoelzel AR (ed). *Molecular genetic analysis of populations.* Oxford: IRL Press/Oxford University Press.

Horn GT, Richards B, Klinger KW (1989): Amplification of a highly polymorphic VNTR segment by the polymerase chain reaction. *Nucleic Acids Res* 17:2140.

Hughes CR, Queller DC (1993): Detection of highly polymorphic microsatellite loci in a species with little allozyme polymorphism. *Mol Ecol* 2:131–137.

Hunter FM, Burke T, Watts SE (1992): Frequent copulation as a method of paternity assurance in the northern fulmar. *Anim Behav* 44:149–156.

Jamieson IG, Quinn JS, Rose PA, White BN (1994): Shared paternity among non-relatives is a result of an egalitarian mating system in a communally breeding bird, the pukeko. *Proc Roy Soc B* 255:271–277.

Jarman AP, Wells RA (1989): Hypervariable minisatellites: recombinators or innocent bystanders? *Trends Genet* 5:367–371.

Jeffreys AJ, Morton DB (1987): DNA fingerprints of dogs and cats. *Anim Gen* 18:1–15.

Jeffreys AJ, Wilson V, Thein SL (1985): Individual specific "fingerprints" of human DNA. *Nature* 316:76–79.

Jeffreys AJ, Wilson V, Thein SL, Weatherall DJ, Ponder BAJ (1986): DNA "fingerprints" and segregation analysis of multiple markers in human pedigrees. *Am J Hum Genet* 39:11–24.

Jeffreys AJ, Wilson V, Kelly R, Taylor BA, Bulfield G (1987): Mouse DNA "fingerprints": analysis of chromosome location and germ-line stability of hypervariable loci in recombinant inbred strains. *Nucleic Acids Res* 15:2823–2836.

Jeffreys AJ, Royle NJ, Wilson V, Wong Z (1988a): Spontaneous mutation rates to new length alleles at tandem-repetitive hypervariable loci in human DNA. *Nature* 332:278–281.

Jeffreys AJ, Wilson V, Neumann R, Keyte J (1988b): Amplification of human minisatellites by the polymerase chain reaction: towards DNA fingerprinting of single cells. *Nucleic Acids Res* 16:10953–10971.

Jeffreys AJ, Royle NJ, Patel I, Armour JAL, MacLeod A, Collick A, Gray IC, Neumann R, Gibbs M, Crosier M, Hill M, Signer E, Monckton D (1991): Principles and recent advances in human DNA fingerprinting. In: Burke T, Dolf G, Jeffreys AJ, Wolff R (eds). *DNA fingerprinting: approaches and applications.* Basel: Birkhäuser Verlag, pp 1–19.

Jones CS, Okamura B, Noble LR (1994): Parent and larval RAPD fingerprints reveal outcrossing in freshwater bryozoans. *Mol Ecol* 3:193–199.

Karl SA, Avise JC (1993): PCR-based assays of Mendelian polymorphisms from anonymous single-copy nuclear DNA: techniques and applications for population genetics. *Mol Biol Evol* 10:342–361.

Keane B, Waser PM, Creel SR, Creel NM, Elliott LF, Minchella DJ (1994): Subordinate reproduction in dwarf mongooses. *Anim Behav* 47:65–75.

Kempenaers B, Verheyen GR, Van den Broeck M, Burke T, Van Broeckhoven C, Dhondt AA (1992): Extra-pair paternity results from female preference for high-quality males in the blue tit. *Nature* 357:494–496.

Kleiman DG (1977): Monogamy in mammals. *Q Rev Biol* 52:39–69.

Kuhnlein U, Dawe Y, Zadworny D, Gavora JS (1989): DNA fingerprinting: a tool for determining genetic distances between strains of poultry. *Theor Appl Genet* 77:669–672.

Kuhnlein U, Zadworny D, Dawe Y, Fairfull RW, Gavora JS (1990): Assessment of inbreeding by DNA fingerprinting: development of a calibration curve using defined strains of chickens. *Genetics* 125:161–165.

Kwiatkowski DJ, Henske EP, Weimer K, Ozelius L, Gusella JF, Haines J (1992): Construction of a GT polymorphism map of human 9q. *Genomics* 12:229–240.

Lack D (1968): *Ecological adaptations for breeding in birds.* London: Chapman and Hall.

Lambert DM, Millar CD, Jack K, Anderson S, Craig JL (1994): Single- and multilocus DNA fingerprinting of communally breeding pukeko: do copulations or dominance ensure reproductive success? *Proc Natl Acad Sci* 91:9641–9645.

Lang JW, Aggarwal RK, Majumdar KC, Singh L (1993): Individualization and estimation of relatedness in crocodilians by DNA fingerprinting with a Bkm-derived probe. *Mol Gen Genet* 238:49–58.

Lessa EP, Applebaum G (1993): Screening techniques for detecting allelic variation in DNA sequences. *Mol Ecol* 2:119–129.

Levinson G, Gutman GA (1987): Slipped-strand mispairing: a major mechanism for DNA sequence evolution. *Mol Biol Evol* 4:203–221.

Levitan DR, Grosberg RK (1993): The analysis of paternity and maternity in the marine hydrozoan *Hydractinia symbiolongicarpus* using randomly amplified polymorphic DNA (RAPD) markers. *Mol Ecol* 2:315–326.

Lewis PO, Snow AA (1992): Deterministic paternity exclusion using RAPD markers. *Mol Ecol* 1:155–160.

Litt M, Luty JA (1989): A hypervariable microsatellite revealed by in vitro amplification of a dinucleotide repeat with the cardiac muscle actin gene. *Am J Hum Genet* 44:397–401.

Longmire JL, Lewis AK, Brown NC, Buckingham JM, Clark LM, Jones MD, Meincke LJ, Meyne J, Ratliff RL, Ray FA, Wagner RP, Moyzis RK (1988): Isolation and molecular characterization of a highly polymorphic centromeric tandem repeat in the family Falconidae. *Genomics* 2:14–24.

Love JM, Knight AM, McAleer MA, Todd J (1989): Towards construction of a high resolution map of the mouse genome using PCR-analyzed microsatellites. *Nucleic Acids Res* 18:4123–4130.

Lynch M (1990): The similarity index and DNA fingerprinting. *Mol Biol Evol* 7:478–484.

Marsden JE, May B (1984): Feather pulp: a non-destructive sampling technique for electrophoretic studies of birds. *Auk* 101:173–175.

McClelland M, Arensdorf H, Cheng R, Welsh J (1994): Arbitrarily primed PCR fingerprints resolved on SSCP gels. *Nucleic Acids Res* 22:1770–1771.

McDonald DB, Potts WK (1994): Cooperative display and relatedness among males in a lek-mating bird. *Science* 266:1030–1032.

Milligan BG, McMurry CK (1993): Dominant vs codominant genetic markers in the estimation of male mating success. *Mol Ecol* 2:275–283.

Mineau P, Cooke F (1979): Rape in the lesser snow goose. *Behaviour* 70:280–291.

Møller AP, Birkhead TR (1992): Validation of the heritability method to estimate extra-pair paternity in birds. *Oikos* 64:485–488.

Møller AP, Birkhead TR (1994): The evolution of plumage brightness in birds in relation to extrapair paternity. *Evolution* 48:1089–1100.

Moore SS, Sargeant LL, King TJ, Mattick JS, Georges M, Hetzel DJS (1991): The conservation of dinucleotide microsatellites among mammalian genomes allows the use of heterologous PCR primer pairs in closely related species. *Genomics* 10:654–660.

Morin PA, Ryder OA (1991): Founder contribution and pedigree inference in a captive breeding colony of lion-tailed macaques, using mitochondrial DNA and DNA fingerprint analyses. *Zoo Biol* 10:341–352.

Morin PA, Woodruff ES (1992): Paternity exclusion using multiple hypervariable microsatellite loci amplified from nuclear DNA of hair cells. In: Martin RD, Dixson AF, Wickings EJ (eds). *Paternity in primates: genetic tests and theories.* Basel: Karger, pp 63–81.

Morin PA, Wallis J, Moore JJ, Woodruff DS (1994): Paternity exclusion in a community of wild chimpanzees using hypervariable simple sequence repeats. *Mol Ecol* 3:469–478.

Morton ES, Forman L, Braun M (1990): Extra-pair fertilizations and the evolution of colonial breeding in purple martins. *Auk* 107:275–283.

Mulder RA, Dunn PO, Cockburn A, Lazenby-Cohen KA, Howell MJ (1994): Helpers liberate female fairy-wrens from constraints on extra-pair mate choice. *Proc Roy Soc B* 255:223–229.

Murphy RW, Sites JW Jr, But DG, Haufler CH (1990): Proteins I: isozyme electrophoresis. In: Hillis D, Moritz C (eds). *Molecular systematics.* Sunderland, MA: Sinauer.

Nakamura Y, Leppert M, O'Connell P, Wolff R, Holm T, Culver M, Martin C, Fujimoto E, Hoff M, Kumlin E, White R (1987): Variable number of tandem repeat (VNTR) markers for human gene mapping. *Science* 235:1616–1622.

Nei M (1987): *Molecular evolutionary genetics.* New York: Columbia University Press.

Norris KJ, Blakey JK (1989): Evidence for cuckoldry in the great tit *Parus major. Ibis* 131:436–442.

Nürnberg P, Rower L, Neitzel H, Sperling K, Pöpperl A, Hundreiser J, Pöche H, Epplen C, Zischler H, Epplen JT (1989): DNA fingerprinting with the oligonucleotide probe $(CAC)_5/(GTG)_5$: somatic stability and germline mutations. *Hum Genet* 84:75–78.

Oring LW, Fleischer RC, Reed JM, Marsden K (1992): Cuckoldry via sperm storage in the polyandrous spotted sandpiper. *Nature* 359:631–633.

Ostrander EA, Jong PM, Rine J, Duyk G (1992): Construction of small-insert genomic DNA libraries highly enriched for microsatellite repeat sequences. *Proc Natl Acad Sci USA* 89:3419–3423.

Ostrander EA, Sprague GF, Rine J (1993): Identification and characterization of dinucleotide repeat $(CA)_n$ markers for genetic mapping in dog. *Genomics* 16:207–213.

Packer C, Gilbert DA, Pusey AE, O'Brien SJ (1991): A molecular genetic analysis of kinship and cooperation in African lions. *Nature* 351:562–565.

Paetkau D, Strobeck C (1994): Microsatellite analysis of genetic variation in black bear populations. *Mol Ecol* 3:489–496.

Payne RB (1979): Sexual selection and intersexual differences in variation of mating success. *Am Nat* 114:447–452.

Payne RB, Payne LL (1989): Heritability estimates and behavior observations: extra-pair matings in indigo buntings. *Anim Behav* 38:457–467.

Pemberton JM, Albon SD, Guinness FE, Clutton-Brock TH, Dover GA (1992): Behavioral estimates of male mating success tested by DNA fingerprinting in a polygynous mammal. *Behav Ecol* 3:66–75.

Pemberton JM, Slate J, Bancroft DR, Barrett JA (1995): Nonamplifying alleles at microsatellite loci: a caution for parentage and population studies. *Mol Ecol* 4:249–252.

Petter SC, Miles DB, White MM (1990): Genetic evidence of a mixed reproductive strategy in a monogamous bird. *Condor* 92:702–708.

Price DK, Collier GE, Thompson CF (1989): Multiple parentage in broods of house wrens: genetic evidence. *J Hered* 80:1–5.

Primmer CR, Møller AP, Ellegren H (1995): Resolving genetic relationships with microsatellite markers: a parentage testing system for the swallow *Hirundo rustica*. *Mol Ecol* 4:493–498.

Pope TR (1990): The reproductive consequences of male cooperation in the red howler monkey: paternity exclusion in multi-male and single-male troops using genetic markers. *Behav Ecol Sociobiol* 27:439–446.

Quay WB (1988): Marking of insemination encounters with cloacal microspheres. *North Am Bird Bander* 13:36–40.

Queller DC, Strassmann JE, Hughes CR (1993): Microsatellites and kinship. *Trends Ecol Evol* 8:285–288.

Quinn TW, Quinn JS, Cooke F, White BN (1987): DNA marker analysis detects multiple maternity and paternity in single broods of the lesser snow goose. *Nature* 326:392–394.

Quinn JS, Macedo R, White BN (1994): Genetic relatedness of communally breeding guira cuckoos. *Anim Behav* 47:515–529.

Rabenold PP, Rabenold KN, Piper WH, Haydock J, Zack SW (1990): Shared paternity revealed by genetic analysis in cooperatively breeding tropical wrens. *Nature* 348:538–540.

Rabenold PP, Piper WH, Decker MD, Minchella DJ (1991): Polymorphic minisatellite amplified on avian W chromosome. *Genome* 34:489–493.

Rassmann K, Schlötterer C, Tautz D (1991): Isolation of simple-sequence loci for use in polymerase chain reaction-based DNA fingerprinting. *Electrophoresis* 12:113–118.

Rave EH, Fleischer RC, Duvall F, Black J (1994): Genetic analyses through DNA fingerprinting of captive populations of Hawaiian geese. *Conserv Biol* 8:744–751.

Ribble DO (1991): The monogamous mating system of *Peromyscus californicus* as revealed by DNA fingerprinting. *Behav Ecol Sociobiol* 29:161–166.

Rico C, Kuhnlein U, Fitzgerald GJ (1992): Male reproductive tactics in the threespine stickleback—an evaluation by DNA fingerprinting. *Mol Ecol* 1:79–87.

Riedy MF, Hamilton WJ III, Aquadro CF (1992): Excess of non-parental bands in offspring from known primate pedigrees assayed using RAPD-PCR. *Nucleic Acids Res* 20:918.

Sambrook J, Fritsch EF, Maniatis T (1989): *Molecular cloning: a laboratory manual* 2nd ed. Cold Spring Harbor, NY: Cold Spring Harbor Laboratory Press.

Sanger F, Nicklen S, Coulson AR (1977): DNA sequencing with chain terminating inhibitors. *Proc Natl Acad Sci USA* 74:5463–5467.

Schlötterer C, Tautz D (1992): Slippage synthesis of simple sequence DNA. *Nucleic Acids Res* 20:211–215.

Schlötterer C, Amos B, Tautz D (1991): Conservation of polymorphic simple sequences in cetacean species. *Nature* 354:63–65.

Scribner KT, Arntzen JW, Burke T (1994): Comparative analysis of intra- and interpopulation genetic diversity in *Bufo bufo,* using allozyme, single-locus microsatellite, minisatellite, and multilocus minisatellite data. *Mol Biol Evol* 11:737–748.

Selander RK (1976): Genic variation in natural populations. In: Ayala FJ (ed). *Molecular evolution.* Sunderland, MA: Sinauer, pp 21–45.

Selander RK, Smith MH, Yang SY, Johnson WE, Gentry JB (1971): Biochemical polymorphism and systematics in the genus *Peromyscus:* I. Variation in the old-field mouse (*Peromyscus polionotus*). Studies in Genetics VI, University of Texas Publication 7103:49–90.

Seutin G, White BN, Boag PT (1991): Preservation of avian blood and tissue samples for DNA analyses. *Can J Zool* 69:82–90.

Shin HS, Bargiello TA, Clark BT, Jackson FR, Young MW (1985): An unusual coding sequence from a *Drosophila* clock gene is conserved in vertebrates. *Nature* 317:445–448.

Slade RW, Moritz C, Heideman A, Hale PT (1994): Rapid assessment of single copy nuclear DNA variation in diverse species. *Mol Ecol* 3:359–373.

Smith GP (1976): Evolution of repeated DNA sequences by unequal crossover. *Science* 191:528–535.

Smith RL (1984): *Sperm competition and the evolution of animal mating systems.* New York: Academic Press.

Smith HG, Montgomerie R, Poldman T, White BN, Boag PT (1991): DNA fingerprinting reveals variation between tail ornaments and cuckoldry in barn swallows *Hirundo rustica. Behav Ecol* 2:90–98.

Smyth AP, Orr B, Fleischer RC (1993): Electrophoretic variants of egg white transferrin indicate a low rate of intraspecific brood parasitism in colonial cliff swallows in the Sierra Nevada, California. *Behav Ecol Sociobiol* 32:79–84.

Stallings RL, Ford AF, Nelson D, Torney DC, Hildebrand CE, Moyzis RK (1991): Evolution and distribution of (GT) repetitive sequences in mammalian genomes. *Genomics* 10:807–815.

Stephan W (1989): Tandem-repetitive noncoding DNA: forms and forces. *Mol Biol Evol* 6:198–212.

Stutchbury BJ, Rhymer JM, Morton ES (1994): Extrapair paternity in hooded warblers. *Behav Ecol* 5:384–392.

Stutchbury BJ, Morton ES (1995): The effect of breeding synchrony on extra-pair fertilization. *Behaviour* 132:675–690.

Sugiyama Y, Kawamoto S, Takenaka O, Kumazaki K, Miwa N (1993): Paternity discrimination and inter-group relationships of chimpanzees at Bossou. *Primates* 34:545–552.

Swatschek I, Ristow D, Wink M (1994): Mate fidelity and parentage in Cory's shearwater *Calonectris diomedea*—field studies and DNA fingerprinting. *Mol Ecol* 3:259–262.

Takasaki H, Takenaka O (1991): Paternity testing in chimpanzees with DNA amplification from hairs and buccal cells in wadges: a preliminary note. In: Ehara A, Kimura T, Takenaka O, Iwamoto M (eds). *Primatology today.* Amsterdam: Elsevier, pp 613–616.

Tamarin RH, Sheridan M, Levy CK (1983): Determining matrilineal kinship in natural populations of rodents using radionuclides. *Can J Zool* 61:271–274.

Tautz D (1989): Hypervariability of simple sequences as a general source for polymorphic DNA markers. *Nucleic Acids Res* 17:6463–6471.

Taylor AC, Sherwin WB, Wayne RK (1994): Genetic variation of microsatellite loci in a bottlenecked species: the northern hairy-nosed wombat *Lasiorhinus krefftii*. *Mol Ecol* 3:277–290.

Telecky T (1989): *The breeding biology and mating system of the common myna (Acridotheres tristis).* PhD thesis, University of Hawaii, Honolulu.

Tilley SG, Hansman JS (1976): Allozymic variation and occurrence of multiple inseminations in populations of the salamander *Desmognathus ochrophaeus*. *Copeia* 1976:734–741.

Trivers RL (1985): *Social evolution.* Menlo Park, CA: Benjamin/Cummings.

Vassart G, Georges M, Monsieur R, Brocas H, Lequarre AS, Christophe D (1987): A sequence in M13 phage detects hypervariable minisatellites in human and animal DNA. *Science* 235:683–684.

Verheyen GR, Kempenaers B, Burke T, Van Den Broeck M, Van Broeckhoven C, Dhondt A (1994): Identification of hypervariable single locus minisatellite DNA probes in the blue tit *Parus caeruleus*. *Mol Ecol* 3:137–143.

Warkentin IG, Curzon AD, Carter RE, Wetton JH, James PC, Oliphant LW, Parkin DT (1994): No evidence for extrapair fertilizations in the merlin revealed by DNA fingerprinting. *Mol Ecol* 3:229–234.

Weber JL, May PE (1989): Abundant class of human DNA polymorphisms which can be typed using the polymerase chain reaction. *Am J Hum Genet* 44:388–396.

Welsh J, McClelland M (1990): Fingerprinting genomes using PCR with arbitrary primers. *Nucleic Acids Res* 18:7213–7218.

Westneat DF (1987): Extra-pair fertilizations in a predominantly monogamous bird: genetic evidence. *Anim Behav* 35:877–886.

Westneat DF, Noon WA, Reeve HK, Aquadro CF (1988): Improved hybridization conditions for DNA fingerprints probed with M13. *Nucl Acids Res* 16:4161.

Westneat DF (1990): Genetic parentage in the indigo bunting: a study using DNA fingerprinting. *Behav Ecol Sociobiol* 27:67–76.

Westneat DF (1993): Polygyny and extrapair fertilizations in eastern red-winged blackbirds. *Behav Ecol* 4:49–60.

Westneat DF, Frederick PC, Wiley RH (1987): The use of genetic markers to estimate the frequency of successful alternative reproductive tactics. *Behav Ecol Sociobiol* 21:35–45.

Westneat DF, Sherman PW, Morton ML (1990): The ecology and evolution of extra-pair copulation in birds. *Curr Ornithol* 7:330–369.

Wetton JH, Parkin DT, Carter RE (1992): The use of genetic markers for parentage analysis in *Passer domesticus* (house sparrows). *Heredity* 69:243–254.

Williams JGK, Kubelik AR, Livak KJ, Rafalski JA, Tingey SV (1990): DNA polymorphisms amplified by arbitrary primers are useful as genetic markers. *Nucleic Acids Res* 18:6531–6535.

Wolff JO, Holleman DF (1978): Use of radioisotope labels to establish genetic relationships in free-ranging small mammals. *J Mamm* 59:859–860.

Wolff R, Nakamura Y, Odelberg S, Shiang R, White R (1991): Generation of variability at VNTR loci in human DNA. In: Burke T, Dolf G, Jeffreys AJ, Wolff R (eds). *DNA fingerprinting: approaches and applications*. Basel: Birkhäuser Verlag, pp 20–38.

Wong Z, Wilson V, Jeffreys AJ, Thein SL (1986): Cloning a selected fragment from a human DNA "fingerprint": isolation of an extremely polymorphic minisatellite. *Nucleic Acids Res* 14:4605–4616.

Wright JM (1994): Mutation at VNTR's: are minisatellites the evolutionary progeny of microsatellites? *Genome* 37:345–347.

Xia X, Millar JS (1991): Genetic evidence of promiscuity in *Peromyscus leucopus*. *Behav Ecol Sociobiol* 28:171–178.

Yamagishi S, Nishiumi I, Shimoda C (1992): Extrapair fertilization in monogamous bull-headed shrikes revealed by DNA fingerprinting. *Auk* 109:711–721.

CHAPTER 8

Strategies for Finding and Using Highly Polymorphic DNA Microsatellite Loci for Studies of Genetic Relatedness and Pedigrees

JOAN E. STRASSMANN, CARLOS R. SOLíS, JOHN M. PETERS,*
AND DAVID C. QUELLER

Department of Ecology and Evolutionary Biology, Rice University,
Houston, Texas 77005–1892

CONTENTS

*Present address: Department of Neurobiology and Behavior, Cornell University, Ithaca, New York 14853.

Molecular Zoology: Advances, Strategies, and Protocols, Edited by Joan D. Ferraris and Stephen R. Palumbi.
ISBN 0-471-14461-4 © 1996 Wiley-Liss, Inc.

SYNOPSIS

Detailed knowledge of the genealogical relationships between individuals is often essential to developing an understanding of the evolutionary significance of social interactions. Related individuals may cooperate in order to obtain indirect fitness benefits in situations in which unrelated individuals would not. Genetic information can also be critical for understanding the dynamics of social groups since it can indicate which individuals are reproducing, distinguish progeny of different parents, and reveal surreptitious reproduction. A codominant, neutral, single-locus class of genetic markers that are sufficiently polymorphic for relatedness estimation and parentage assignment has recently been identified and promises to revolutionize the field. These markers, known as DNA microsatellites, can be compared reliably across gels, which allows for comparisons among large numbers of individuals. They are short tandem repeats of simple motifs, such as AAT. The polymerase chain reaction can be used to amplify DNA regions that contain such repeats. These products are electrophoresed and then visualized, allowing precise determination of length polymorphisms. Polymorphic microsatellites have been used in gene mapping, population structure analyses, and most recently in assessment of relatedness and parentage. Microsatellites typically are found in rapidly evolving, noncoding DNA. Therefore flanking sequences used to prime the amplification of a microsatellite in one species are likely to work only in fairly closely related species. Although identifying microsatellites in a species of interest can be time-consuming, the techniques employed are well established and reliable. They involve constructing a partial genomic DNA library, identifying and sequencing clones containing microsatellites, designing polymerase chain reaction primers that flank the repeat region, amplifying the microsatellite in a population sample of individuals, and using those microsatellites that are polymorphic. Here we present straightforward protocols that should be applicable to all species.

INTRODUCTION

Evolutionary studies of social behavior benefit greatly from the availability of genetic markers capable of revealing parentage and other genetic relationships among individuals. Such markers are important because it is often neither feasible nor efficient to obtain relatedness information by observing individuals: females may mate with several males and copulations may be concealed. In social insects, the problem is particularly severe since even the egg layers cannot always be identified, and mating usually occurs outside the colony. In social species, there may also exist conflicts of interest among relatives, which can be understood with precise genetic data. Related individuals may cooperate in order to obtain indirect fitness benefits in situations in which unrelated individuals would not.

To date, a number of different techniques have been applied to the task of establishing genetic relatedness (Queller et al., 1993). Pedigrees can be based on inferring parentage from behavioral observations. This is the oldest technique. It is very useful in some cases but may not detect surreptitious matings or identify mothers in groups where many females are reproducing. Genetic techniques have indicated that surreptitious matings are much more common than was previously assumed (e.g., Birkhead and Moeller 1992; Westneat et al., 1987). Allozymes have been used extensively in studies of genetic relatedness (e.g., Gadagkar, 1985; Queller et al., 1993; Strassmann et al., 1989). They have been very useful for establishing relatedness among classes of individuals across groups. However, low genetic variability and a paucity of loci make this technique unsuitable for parentage assignment or within-group relatedness determinations. DNA fingerprinting has proved to be a reliable technique for parentage exclusion within family groups, but it is not suitable for examining larger groups or across groups because of the high variability of minisatellite loci and the difficulties associated with making comparisons between gels. In addition, fingerprinting requires large amounts of high-quality DNA. Other techniques discussed elsewhere in this book, such as RAPDs and SSCP, have advantages and disadvantages for various applications. For parentage assignment and relatedness estimation, an ideal marker would be mendelian, codominant, selectively neutral, and highly polymorphic. Scores would be unambiguous enough to be evaluated against a standard across gels. DNA microsatellites meet these criteria (Litt and Luty, 1989; Smeets et al., 1989; Tautz, 1989; Weber and May, 1989).

In this chapter, we discuss how to find and use microsatellites. We describe in detail techniques that have been used successfully in our laboratory on wasps, ants, and fish (Hughes and Queller, 1993; Queller et al., 1993). Since many investigators will find microsatellites from the literature or from GenBank, we first discuss using microsatellites that have previously been identified, and then go on to techniques for locating microsatellites in a new organism.

USING MICROSATELLITES

Finding Microsatellites

The simplest means of identifying microsatellites is by using previously described sequences. Partial sequence information for a growing number of species is available through GenBank or in the literature (Stallings et al., 1991). This is particularly true for species related to those popular with geneticists, such as *Drosophila*, mice, and humans, as well as agriculturally important species such as cattle and chickens (Cornall et al., 1991; Beckmann and Weber, 1992; Edwards et al., 1992). For reasons to be explained later, a disadvantage to this approach is that most previously identified microsatellites are dinucleotide repeats.

It is often possible to use microsatellite loci that have been identified for another species. Microsatellite primer pairs frequently amplify sequences common to most closely related species in any one genus and have been found occasionally to work across families. Currently, the success with which specific microsatellite loci can be used on other species must be empirically determined on a case-by-case basis. One study found that the relationship between repeat length and heterozygosity (Weber, 1990) present in the original species did not hold for new species (Moore et al., 1991). This suggests that all possible microsatellites should be explored in new species, not just those that were most polymorphic in the original species. Of 48 specific and polymorphic primers identified in cattle, 20 were variable in sheep, and none were variable in horses (though three amplified but were monomorphic) or humans (Moore et al., 1991). Arévalo et al. (1994) and Bhebhe et al. (1994) tried ten primers identified in goats (*Capra hircus*) on eleven species from eight other genera across the Bovidae. They found that four of the loci did not amplify product in any of the other species. Of the others, two were polymorphic in four new species, two were polymorphic in five species, one was polymorphic in six, and one was polymorphic in seven of the eleven species. Another locus that was mono-morphic in goats proved to be polymorphic in cows (E. Arévalo, *personal communication*). One or more of the six dinucleotide microsatellites identified from the reed bunting (*Emberiza schoeniclus*) worked on species from other families, including Muscicapidae, Sturnidae, Paridae, Hirundinidae, Sylviidae, and Passeridae, although they were not evaluated for polymorphisms in the new species (Hanotte et al., 1994). One of two loci from barn swallows were polymorphic in both bank swallows and house martins (Ellegren, 1992). Three dinucleotide repeats isolated from brown trout (*Salmo trutta*) were also present and polymorphic (exhibiting new alleles) in the rainbow trout (*Oncorhynchus mykiss;* Estoup et al., 1993). Primers identified in humans are useful and polymorphic in both species of chimpanzees (Morin and Woodruff, 1992; Morin et al., 1994). We have found that the primers for 40 or so trinucleotide microsatellites that we have identified in three species of polistine wasps in two genera generally work well on species in the same subgenus. We have evaluated a sample of all the genera in the Vespidae and have found at least two of our 40 microsatellite primers produce a PCR product of the right size on all genera assayed (V. O. Ezenwa et al., *in preparation;* J. Strassmann et al., *unpublished data*). To what extent these will prove polymorphic in the new species, and therefore useful, is still unknown. Thus, once a large number of microsatellites are available for a given group, a much wider range of taxa may be accessible to study.

DNA Extraction

One of the primary advantages of PCR-based microsatellite analysis is that nano-gram quantities of low-quality, partially degraded DNA can be used effectively, since fragments of only 100–300 base pairs are needed. By contrast, microgram quantities of high-quality DNA are necessary for traditional DNA fingerprinting. The minute quantities of DNA required for microsatellite analysis represent a signif-icant advantage for researchers studying small organisms or aiming to carry out nondestructive genetic sampling of moderate-sized organisms. Microsatellite stud-ies of genetic relatedness and parentage often require that many samples be an-alyzed. It is therefore important that the DNA extraction technique be easy and yield sufficient stable DNA to provide an adequate time window for the analysis. Conventional DNA isolation techniques designed to give high molecular weight DNA thus may be unnecessarily expensive and time-consuming for this application.

Contaminants, including EDTA and possibly insect cuticle pigments, inhibit the PCR reaction, so techniques that completely remove or avoid the use of these compounds are best (Strassmann.1). Our preferred protocol (Strassmann.1) works quickly and easily for whole insects, tissues, and eggs. A second protocol used extensively in our laboratory may be more successful if removal of pigment contaminants and lipids proves to be problematic (Strassmann.2). Some questions require that DNA be amplified from ancient, tiny, or degraded samples, samples stored in ethanol, sperm or eggs, blood, feathers, hair, scales, and other unconventional samples. Such samples may require different extraction techniques. We extract DNA from tissues where we can obtain a few micrograms (Strassmann.1, Strassmann.2). When only a tiny amount of tissue is available, as is the case for hymenopteran sperm from female spermathecae, we simply lyse the cells to make the DNA available to the PCR primers and avoid precipitating the DNA (Strassmann.3).

Carry-over of DNA between samples is a potentially serious problem since PCR can geometrically amplify even single DNA molecules. We use a fresh grinding surface for each sample (Strassmann.1). For both large, hard wasps and softer larvae we use a bead beater (Biospec Products), which pulverizes the tissues in a grinding buffer with two sizes of glass beads, which are then discarded. For wasp eggs, we use individual, disposable pestles made by melting a pipetteman tip in an eppendorf tube (Strassmann.1). Of course, many tissues can be broken down chemically without grinding of any kind.

We store genomic DNA at 4°C unless we do not plan to use it for 2 months or more, in which case, it is stored at −70°C. We find that this storage strategy minimizes DNA breakage, which could be caused by repeatedly freezing and thawing DNA samples.

Conditions for PCR

We run all radioactively labeled (with ^{35}S) PCR reactions in 10-μl volumes with an oil overlay under fairly standard conditions (Strassmann.4). We have found 10 μl to be the optimal volume since 5-μl volumes result in a higher number of samples failing to amplify on the first run. Volumes greater than 10 μl increase cost (more *Taq* polymerase and ^{35}S label must be used) and are unnecessary, since only 4 μl of the reaction, including loading buffer, are typically run out on the gel in the next step.

Visualizing Microsatellite Products and Size Standards

Before microsatellite polymorphisms can be scored, they must be visualized. This is typically done by running the product out on a gel (Strassmann.5). Some laboratories run products out on native, nondenaturing, thin (0.6 mm), acrylamide gels and then stain them with ethidium bromide or use silver staining protocols. Another means of visualizing length polymorphisms is through end-labeling one primer with ^{32}P. Although this method gives cleaner results for dinucleotides (because amplifying both strands increases stutter band numbers), it does, of course, use hazardous ^{32}P. If you plan to run thousands of samples, automated sequencing machines may be advantageous, though the use of fluorescent dyes makes this expensive. The most useful automated sequencing machines for microsatellite applications are those that read five different fluorescent labels per lane. These fluorescent labels are usually used to identify the four nucleotides and one size standard, but they can be used to label as many as nine different microsatellite products, if the microsatellites are of two nonoverlapping size classes. The microsatellite primers are end-labeled with the fluorescent labels (a different one for each same-sized microsatellite product to be run in that lane), which leaves one as a size standard. The resulting data can be imported directly into a program for analysis.

In our laboratory, we run denaturing gels exactly like those used for sequencing and visualize the microsatellites by incorporating ^{35}S into the PCR reaction (Figs. 1 and 2; Strassmann.5). Our technique has a number of advantages. A sequencing reaction can be run as a size standard on the denaturing gel. ^{35}S is a very low-energy beta emitter and, as such, is much less hazardous than ^{32}P. Another advantage is that the label is incorporated during the PCR reaction, so no extra steps are required to label primers. There has been some concern about volatile decomposition products from $^{35}SdATP$ in PCR (Trentmann et al., 1995), which make it preferable to keep the thermocycler in a vented hood. In our own laboratory frequent wipe tests inside and outside the lid of the thermocycler have not indicated elevated levels of radiation, perhaps because our volumes are so small.

It is possible to run more than one locus in a lane, if the fragments are either of nonoverlapping size classes or are loaded at different times. In the former case, samples are simply mixed and then loaded together. In the latter case, sample sets are loaded about an hour apart depending on the size, and separate size markers are needed for each set. We do not run more than one locus in the actual PCR reaction because interactions occasionally occur between the primers in some samples.

We score alleles for microsatellite loci by assigning each allele a size based on the M13mp18 sequence run in each gel (Strassmann.5). Samples may exhibit a complex banding pattern in which an allele may appear as several bands, particularly if they are dinucleotide repeat motifs. Care must be taken to always select the same band for scoring.

For behavioral ecologists, trinucleotide microsatellites are typically superior to dinucleotides. While not necessarily as variable or abundant as dinucleotides, in most species (even those in the Hymenoptera, which are frequently deficient in allozyme variation), trinucleotides have more than ample variability for precise relatedness estimation and for parentage assignment. Once identified, trinucleotide microsatellites are much more easily and accurately scored than shorter repeat motifs. Trinucleotides can be internally labeled with ^{35}S, whereas dinucleotides must often be end-labeled with ^{32}P or other relatively high-energy isotopes to reduce shadow bands. This adds an extra enzymatic step to the DNA analysis, as well as presenting a greater safety hazard. However, for population geneticists interested

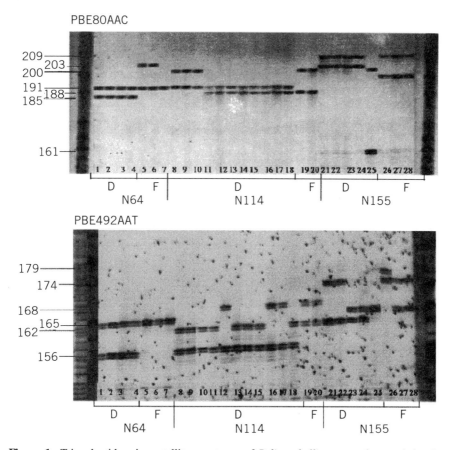

Figure 1. Trinucleotide microsatellite genotypes of *Polistes bellicosus* mothers and daughters on three colonies and two loci. Note the virtual absence of stutter bands from these trinucleotide repeats. The outside lanes are size markers from the C lane of an M13 sequence. For each colony, D indicates daughters and F indicates foundresses, or mothers. Lacking paternal genotypes, daughters cannot always be assigned unambiguously to one mother. However, it is clear from her score at PBE80AAC that daughter 25 is not the offspring of any of the foundresses on the colony. Daughter 25's score from PBE492AAT is consistent with this result, but this locus alone does not rule out any of the queens as mothers. Taken together, however, these data indicate that daughter 25 probably wandered over from another nest. (From Strassmann et al., 1995, by permission. http://www.rice.edu/wasps)

primarily in genome mapping, the relatively high frequency of dinucleotide repeats may be essential.

Error Checking

Eliminating all sources of error is essential when using microsatellites for parentage assignment. Thus all aspects of sample preparation, gel loading, and scoring must

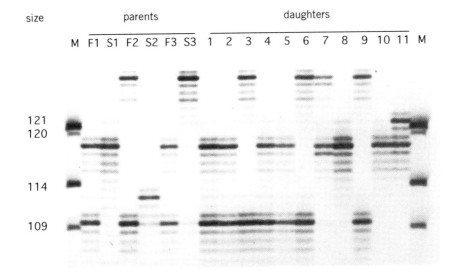

Figure 2. Microsatellite genotypes of *Polistes annularis* mothers, sperm from their spermathecae and daughters. The two outside lanes (M) are size markers (the C lane from an M13 sequence) labeled in nucleotides. The interior lanes contain amplified genotypes for locus PAN111bAAT from one colony. Three parent sets are shown, both foundress (F) and her stored sperm (S). Each sperm lane shows a single allele (each allele shows one strong band plus several weak stutter bands characteristic of microsatellites). The remaining 11 lanes are daughters. Lanes 1, 2, 4, 5, 8, and 10 show daughters of the first foundress and lanes 3, 6, 7, and 9 show daughters of the third foundress; note that these are distinguishable only because of the sperm. Lane 11 matches none of the collected foundresses and is the daughter of an uncollected, presumably dead, foundress. (From Peters et al., 1995, by permission.)

be done in ways that minimize error. Incorrect genotyping of an individual could result either in false exclusion of the true parent or in false acceptance of a nonparent. While both types of error are equally serious, to illustrate the importance of minimizing errors, we will focus here on errors leading to the false exclusion of the true parent. The analysis that follows could apply to errors in tube labeling, in adding the wrong DNA to the tube, in loading, in scoring, or in entering data.

Suppose we score n loci and the error rate per locus per individual is e. Suppose further that a fraction k of these errors leads to false exclusion of the true parent (some errors will not result in false exclusion; for example, true parent is ab, offspring is bb, but we score offspring as aa). When we score n loci, there are $2n$ chances to make an error for every parent–offspring pair. The probability of false exclusion, F, for a single pair is one minus the probability that no false exclusion error will happen $(1 - ke)$ in $2n$ tries or

$$F = 1 - (1 - ke)^{2n} \qquad (1)$$

The goal is clearly to make e as low as possible, but how low does it need to be? Suppose $k = 0.5$ and suppose we are scoring $n = 8$ loci (a fairly typical number). If our error rate, e, is 0.05, then 33% of the true parents will be excluded, which is clearly unacceptable. If we lower the error rate to 0.01, we exclude 8% of the true parents, and this is still too high. Perhaps we might consider a 1% rate acceptable (although even this is probably too high for reasons given in the next paragraph). Solving for e, we get 0.0012. In other words, we need to strive for an error rate of less than 1 in 1000.

The impact of errors on the accuracy of results is even more serious in the case of errors affecting many individuals. For example, if one skips a sample in loading a gel, all the samples loaded thereafter will be assigned to the wrong individual. A similar cascade effect can result from an error in entering data. Also, any error in a parent's genotype may affect all of its offspring, so a single error could produce incorrect results for an entire family.

Of course, F would be reduced by having lower k or n in equation 1. Obviously, however, this is not an option since high values of k and n are the very things that we need for parentage studies to succeed. If we are to succeed in our goal of correctly excluding individuals as parents, we must therefore screen many variable loci, each of which provides us with a good chance of being able to discriminate among individuals.

A number of procedures will reduce or eliminate error. The best check would be to do everything twice independently and then compare the results. But it is impractical to do all the molecular work twice. Tubes must be labeled carefully and ordered in ways to minimize mistakes. Checks can be made by running subsamples twice. We reduce mixups between people by supplying everyone with her/his own working box of primers and other reagents.

PCR reaction setup is a stage at which errors are both possible and serious. If a tube is skipped, the outcome may be errors in all subsequent tubes. The setup that we find minimizes errors is one that puts the genomic DNA and the PCR tubes in the same column in the same rack, so genomic DNA is only added to PCR tubes directly in front of it.

Gel loading is another point where mixups are possible and serious. Samples should be processed and loaded onto gels in numerical or alphabetical sequence, to prevent errors when recording loading patterns and make "skipped" samples immediately evident. Use a rectangular heat block to denature DNA (not round floaters). Each tube should be moved to a separate rack as loaded, to prevent double loading. Indicate skipped lanes, and locations of markers, by leaving gaps in the separate rack and mark them in pen on the front glass of the sequencing rig. After loading is complete, tube order should be rechecked.

Once an autoradiograph is produced, double scoring is inexpensive and feasible. In our laboratory, at least two people independently score each autorad and enter the data into computer files. The data sets are then compared in an Excel spreadsheet, using a function that detects nonmatching cell entries and enters a 0 or a 1 in a new matrix, depending on whether the cell contents match. Discrepancies are checked, and rerun if necessary. For unbiased scoring, all ink marks from previous scoring are removed from autorads with alcohol and each individual independently labels the ladder, sample IDs, and allele sizes. Since each locus has its own particular morphology, we provide scorers with a band summary sheet for each locus, drawn up after discussion of band characteristics at that locus (which band is to be scored, location of faint shadow bands, etc.).

As a check on the success of the above measures to eliminate error, we choose a random sample of individuals and primers for repeat runs. We also rerun cases where one locus gives results that are not consistent with the other loci for that individual. For example, if scoring of ten loci consistently assigns parentage to one individual, while scoring of an eleventh locus excludes that individual as a parent, we rerun that locus.

Analyzing Microsatellite Data

Microsatellites can provide a great deal of information on genetic relatedness and parentage assignment. Queller and Goodnight (1989) discuss estimation of Grafen's relatedness statistic (Grafen, 1985). This technique has been implemented in a user-friendly Macintosh program, Relatedness 4.2b (Goodnight and Queller 1994; currently obtainable, with its instruction manual, from our website, http:/www.rice.edu/wasps). This program allows the calculation of genetic relatedness in groups, between specific individuals, and jackknifes for error estimation. It also allows for population structure and inbreeding estimates.

If there are few candidate parents and progeny, genotypes can be lined up and compared by hand to assign the progeny to the appropriate parents. It is also desirable to estimate the power of the analysis by determining the frequency of a match due to chance alone. For more complex situations, maximum likelihood techniques that assign individuals to the most likely parents may be appropriate. Goodnight and Queller (*in preparation*) are working on such programs.

FINDING YOUR OWN MICROSATELLITES

Though somewhat tedious and time-consuming, there are no difficult steps involved in identifying microsatellites in a new species (Queller et al., 1993). Here, we follow the general approach of Hughes and Queller (1993) with some modifications. Though the protocols are very standard to molecular biologists, we nevertheless provide many of them because they are unlikely to be so familiar to behavioral ecologists. For the behavioral ecologist, inexperienced in molecular biology, the steps necessary for generating new microsatellites will involve learning quite a number of unfamiliar techniques. It is unlikely that all the steps will proceed smoothly at the first attempt, so it is important to not simply follow the protocols but to understand exactly what is going on at each step and know what is most likely to fail. The biochemical background need not be acquired before embarking on the project, but can be assimilated as you go by reading reference texts and by comparing protocols and browsing frequently through the newsgroup bionet.molbio.methds-reagnts. As you do this, you will see that there is a strong historical component to many protocols and newer ones may omit many steps or ingredients. Some of these shortcuts may be worth the time they save, others not. You should also watch the advertisements in the front of publications like *Nature* for new proprietary products that may greatly facilitate cloning and probing, though I would wait for assurance that they work from newsgroup members.

The general strategy for identifying microsatellites involves plating out on nylon membranes a partial genomic library of DNA fragments 300–600 base pairs long, inserted into plasmids that have been introduced into *Escherichia coli* cells. Replica membranes are made and probed with oligonucleotides such as $(AAT)_{10}$. Positive clones are picked up and grown in culture. Then the plasmid DNA is digested with a restriction enzyme to cut out the insert. This digest is run on an agarose gel. A Southern transfer of that gel is probed with the oligonucleotides to verify the positives and to estimate insert size. Positive clones are sequenced. Those with at least eight uninterrupted, identical, trinucleotide repeats are frequently variable in Hymenoptera (Hughes and Queller 1993; J. Strassmann et al., *unpublished data*). For these, PCR primer pairs are designed that amplify the repeat region. PCR conditions are optimized on nonradioactive samples run out on agarose gels and stained with ethidium bromide. A sample of 30 or 40 unrelated individuals is screened using radiolabeled PCR products run out on denaturing polyacrylamide gels to determine the degree of heterozygosity for the locus. The best loci can have 20 or more alleles, with the most common one occurring at a frequency of less than 0.20.

Much of the critical work is done by enzymes that act in very specific ways only under specific conditions, so any steps involving enzymes should be carried out precisely. If a procedure fails completely, bad enzyme or poor conditions for the enzyme are likely suspects. Although it is a rare occurrence, inactive or underactive enzymes have been supplied by all the companies with which we have dealt. The enzyme has usually been replaced at no cost. It is therefore often worth the extra effort to do controls that ascertain enzyme activity and prudent to remain with a supplier with whom you have had good experience.

Another general practice that can facilitate molecular work is to keep track of how much DNA you have by consistently running lambda standards of several known amounts next to your experimental DNA on agarose gels. We do this almost to the exclusion of other means of DNA quantification because it is easy, cheap, and also gives size information.

Extra steps in the overall process that provide additional information are often worth the trouble. For example, we do not begin sequencing immediately after picking up positives from a library. First, we cut out the inserts from the plasmid, run them on a gel, photograph it, and then probe a Southern transfer of that gel. This procedure verifies the presence of a microsatellite (if the Southern from that clone also yields a positive), gives an indication of the size of the insert, and reveals whether or not there were multiple inserts in one clone (they will be cut apart by the restriction enzyme and so show up as multiple bands in the lane). This information allows us to prioritize the clone candidates for sequencing.

Since many of the early steps in microsatellite development can be carried out on a large number of library membranes almost as quickly as on a few, we favor making and simultaneously probing very large libraries. The positives can then be picked up and frozen for eventual sequencing. Excess ligation reactions can also be frozen away to save time in making a larger library, if this proves necessary later. The libraries that generated most of our microsatellites came from two species of wasps probed simultaneously on a total of 30 large (132 mm) membranes averaging 1500 (range 1189–1772) colonies per membrane from one library (12 membranes total), and 656 (range 415–876) colonies per membrane from the other library (18 membranes total) (J. Strassmann et al., *unpublished data*). Probing these nearly 30,000 colonies required only an hour or two per day more than it would have taken to probe just one or two membranes. If too few colonies have been probed to generate enough useful microsatellites, the entire procedure must be repeated. This is time-consuming and frustrating. It is better to have more positives than you are likely to need frozen away for use. This is essential because many positives will have repeats that are stuck to the vector (so a flanking primer cannot be designed), have too many errors, or are too short.

In our laboratory, approximately six months of full-time effort was required to generate a minimum of 10 highly polymorphic trinucleotide microsatellite loci for an entirely new species in which microsatellites are not particularly rare. Though this represents a substantial time investment, the questions of parentage and kinship that can be addressed with microsatellites make it worthwhile.

Isolating High Molecular Weight DNA for Library Construction

When constructing a library, it is necessary to isolate high molecular weight DNA to ensure that the fragments for insertion have appropriate overhanging ends. There are many possible ways of obtaining high molecular weight DNA. One of the most important steps is the initial breakdown of tissues: overly vigorous grinding shears DNA. Grinding in liquid nitrogen is a simple and efficient means of isolating moderate amounts of high-quality genomic DNA from a variety of tissues (Strassmann.1, Strassmann.6).

Restriction Digests and Size Selection

The size of DNA to be inserted into the plasmid should not be longer than can be sequenced using forward and reverse primers that complement sites in the vector. However, if inserted DNA is too short, the chance is increased that repeats will fall so close to the edge of the inserted DNA that flanking PCR primers cannot be designed. Genomic DNA inserts between 300 and 600 base pairs are a good compromise. Although it may be difficult to obtain a readable sequence of 600 bases without the use of automatic sequencing, the use of forward and reverse primers will easily reveal any microsatellites in an insert of this size. If necessary, an internal primer can be designed to accurately obtain the rest of the sequence on the opposite side of the microsatellite. This primer can then be used as one of the two PCR primers, if a compatible primer can be designed on the other side of the repeat.

To cut up the genomic DNA for insertion into plasmids, we typically use SAU3A, a restriction enzyme with a four base pair recognition sequence that yields a large number of small fragments in the size range of interest. This is a very good choice of enzyme since it produces fragments with ends compatible with the sticky ends left by BamHI, a six cutter commonly chosen to cut open the plasmid. After digesting the high molecular weight DNA to completion, select the appropriate size fragments by running the DNA out on an agarose gel (Strassmann.7). If microsatellites are likely to be rare in your species, you may enrich the genomic DNA for microsatellites using techniques discussed by Fleischer in this volume. Bear in mind, however, that the extra steps entailed rely on the success of several enzymes (Kandpal et al., 1994).

Plasmid Selection, Cutting, and Dephosphorylation

The plasmid is processed at the same time that the genomic DNA is digested (Strassmann.7). We have used both pUC19 and pBluescript SK+ plasmids with success. It is important to identify three restriction sites within the polylinker region of the chosen plasmid. The restriction sites on either side of the insertion site are required when the insert is cut out in order to verify the presence of a microsatellite and determine the insert's length. Since these latter two restriction enzymes must work simultaneously, it is critical that they function well under the same buffer and temperature conditions. In pUC19 we used HindIII and EcoRI. In SURE cells we used HindIII and SACI. Details of restriction enzyme conditions and compatibility can be found in suppliers' catalogues. The middle site must create sticky ends in the plasmid polylinker complementary to the genomic DNA overhangs so that the genomic DNA and the plasmid are capable of base pairing during the ligation reaction. We use BamHI to cut the plasmid and produce sticky ends matching those of the genomic DNA cut with SAU3A. The cut end of the plasmid must then be dephosphorylated so that when ligase is added in the next step, the plasmid will not reanneal before receiving the insert. Since dephosphorylation sometimes fails, we use two different phosphatases in succession, and ensure that they are fresh. The success of the dephosphorylation can be checked by adding ligase in the absence of any insert before proceeding. Alternatively, this check can be carried out as a control to the simultaneous checking of insert to vector ratios.

After cutting and dephosphorylation, the plasmids are run through an agarose gel and the linear plasmid band is cut out of the gel and purified. If the digestion of the plasmid was complete, further purification by running the plasmid through a gel and selecting the linear band is probably unnecessary.

Ligation of Insert into the Plasmid

This is a critical step that depends on the success of the restriction enzyme digestion, on the dephosphorylation of the plasmid, and on the quantification of vector and insert DNA (Strassmann.8). Since the plasmid is about 2900 bases long and the insert averages 400 bases, a 1:1 sticky end ratio (i.e., one insert for each plasmid) would require a DNA concentration with approximately seven times as much vector as insert. However, it is not always possible to quantify vector and insert DNA very accurately, so we usually try a number of different ligation ratios and have generally had the most success with two to four times as many inserts as plasmids (Strassmann.8). Another check on the ligase may be done by setting up a tube of plasmid that has been cut but not dephosphorylated. The success of the ligation cannot easily be determined until transformation takes place.

Transformation and Plating Out of Cells

Competent cells can either be purchased or made by the researcher (Hanahan, 1983; Sambrook et al., 1989:1.76–1.81) and should be frozen in 100- or 200-μl aliquots at −70°C in 1.5-ml microfuge tubes. We usually make our own competent cells, but it is probably not worth the effort for someone who does not do this routinely. Catching the growing cells at the optimum point for competency can be a little tricky. We recommend a healthy, simple strain such as DH5alpha (from GibcoBRL) or XL1Blue (from Stratagene Inc., La Jolla, CA). We have also used SURE (from Stratagene) cells that the manufacturers claim reduce recombination. However, these cells were harder to grow, making many subsequent steps difficult. In any case, we have not found recombination to be a problem for microsatellites. Transformation itself is quite simple (Strassmann.9).

To avoid duplication, transformed cells should be grown for no more than 45 minutes and should then be plated out. Since this is a test of the initial ligation reaction, plating can be done on small (100 mm) agar plates to which antibiotic has been added. The antibiotic used must, of course, be one to which the chosen plasmid is resistant. We usually plate out two or three replicates of each ligation reaction, using 50 μl of transformed cells (Strassmann.10). The number of colonies on each plate should then be counted. There should be very few on the no-insert control plate, and many on one or more of the test ligations.

Next, plate out the transformations that yielded the most colonies on large (132 mm) nylon membranes (Hybond, Amersham) for the library (Strassmann.11). This should be done as soon as possible after selecting the successful ligation reactions and determining the quantity needed to result in about 1000 evenly distributed colonies per membrane (Strassmann.10). Plating out on large nylon membranes rather than directly on agar facilitates producing clear replicate membranes for probing, as well as making it easier to identify and pick up specific colonies.

After growing the initial plates for 10–14 hours, the colonies should appear as small, discrete raised bumps. At this point, make replicate filters for probing (Strassmann.12). It is important to make holes through the replicate membrane and the original so that the two can be lined up later. After the copy is made, allow colonies on both membranes to grow another 3 or 4 hours until they again appear as raised bumps. At this time, a reserve copy of the library can be made and stored at −70°C using the technique of Dreyer et al. (1991) (Strassmann.12).

Probing, Selection, and Sequencing of Positives

Once the library is made and a replicate is denatured, we probe it with a trinucleotide repeat containing 10 repeats that has been labeled with ^{32}P using the enzyme terminal deoxynucleotidyl transferase (TdT) (Strassmann.13; Rosenberg et al., 1990). Membranes can be probed with more than one repeat at a time, if their melting temperatures are similar.

At the point of picking up positives from plates, the researcher has usually accumulated a number of autoradiographs probed with different repeat oligonucleotides (Strassmann.14). If it is not possible to pick up all positives on the same day, it is best to pick all positives from a subset of filters. We pick up positives by touching a toothpick to the clone and dropping it into a culture tube containing about 3 ml of LB/ampicillin medium. Keep track of what possible repeats are on each autorad so that you probe the Southern with the appropriate oligonucleotides.

Unfortunately, we have not found any correlation between intensity of the positive signal and length of the microsatellite repeat. A very dark positive on both library membrane and Southern is no more likely to yield a long repeat than is a less dark positive. Variability in signal intensity probably results from factors other than repeat length influencing binding of the probe to the clone.

Sequence inserts of the best positives either by hand, using the protocols provided with Sequenase, or with an automated sequencer. Much time in this step can be saved by contracting the sequencing out to a reliable university sequencing facility.

Designing PCR Primers, Optimizing, Checking for Polymorphisms

Once you have identified repeat regions that have eight or more uninterrupted, identical trinucleotide repeats, or somewhat more not quite perfect repeats, it is time to design primers that amplify across the repeat. Though it can be done by hand, this is probably best done by entering the entire sequence into a program like Oligo or MacVector and using the program to help design the primers. The primers should have very similar lengths and melting temperatures and should be noncomplementary so they do not anneal to one another or form hairpin loops. They should produce products of 100–200 base pairs in length. Every effort should be made to find primer pairs that produce products in the shorter end of this range, since they will run faster on a gel and will be easier to score, though having nonoverlapping sizes is advantageous for multiplexing on the same gel. Longer products can be run on a 4% acrylamide gel, though it is somewhat more difficult to handle than the standard 6% gel.

The synthesized primers are then used in a polymerase chain reaction at the predicted annealing temperature and run out on agarose gels to test for the presence of correctly sized product. If the run fails to yield product at this temperature, the PCR should be repeated at higher and lower temperatures in 2.5 degree steps until a product of the correct size is obtained. In attempts to obtain product, it is also sometimes helpful to vary magnesium chloride concentrations. To assess the variability (i.e., the number of alleles) at the locus in question, the primers should then be tested on a sample of 30 unrelated individuals in a radiolabeled PCR reaction that is run out on a denaturing gel. We have found that over 80% of the loci chosen following the above guidelines have four or more alleles and are useful for studies of kinship.

Development of a good set of polymorphic trinucleotide microsatellite loci for a species provides a powerful tool for addressing many exciting questions involving kinship and parentage. The answers to such questions are then easily obtained, since the processes involved in genotyping individuals using microsatellite primers are simple enough that new students can often produce usable data within a week of entering the laboratory (Strassmann.1 to .5).

ACKNOWLEDGMENTS

We thank Madhusudan Choudhary and Colin Hughes for helping us work out microsatellite methods. We thank Elisabeth Arévalo for many insightful discussions of specific protocol steps and Jeanne Zeh for greatly improving the clarity of the manuscript. We also thank David Zeh and Jeremy Field for comments on the manuscript. We thank the W. M. Keck Center for Computational Biology for the use of shared equipment. Our own work on microsatellites has been supported by the National Science Foundation (NSF IBN-9210051, NSF BSR-9021514).

REFERENCES

Arévalo E, Holder D, Derr J, Bhebhe E, Linn R, Ruvuna F, Davis S, Taylor J (1994): Caprine microsatellite dinucleotide repeat polymorphisms at the SR-CRSP-1, SR-CRSP-2, SR-CRSP-3, SR-CRSP-4, and SR-CRSP-5 loci. *Anim Genet* 25:645.

Ausubel F, Brent R, Kingston R, Moore D, Seidman J, Smith J, Struhl K (1990): *Current protocols in molecular biology.* New York: Wiley.

Beckmann J, Weber J (1992): Survey of human and rat microsatellites. *Genomics* 12:627–631.

Bender WP, Spierer D, Hogness S (1983): Chromosomal walking and jumping to isolate DNA from the Ace and rosy loci and the bithorax complex in *Drosophila melanogaster.* *J Mol Biol* 168:17–33.

Bhebhe E, Kogi J, Holder D, Arévalo E, Derr J, Linn R, Ruvuna F, Davis S, Taylor J (1994): Caprine microsatellite dinucleotide repeat polymorphisms at the SR-CRSP-6, SR-CRSP-7, SR-CRSP-8, SR-CRSP-9, and SR-CRSP-10 loci. *Anim Genet* 25:646.

Birkhead T, Møller AP (1992): *Sperm competition in birds.* London: Academic Press.

Bruford MW, Hanotte O, Brookfield JF, Burke T (1992): Single-locus and multilocus DNA fingerprinting. In: Hoelzel, A.R. (ed). *Molecular genetic analysis of populations.* Oxford: Oxford University Press, pp 225–269.

Cornall R, Aitman T, Hearne C, Todd J (1991): The generation of a library of PCR-analyzed microsatellite variants for genetic mapping of the mouse genome. *Genomics* 10:874–881.

Dreyer K, Opalka B, Schulte-Holthausen H (1991): An improved method for long-term storage of colony filters. *Trends Genet* 7:178.

Edwards A, Hammond H, Jin L, Caskey C, Chakraborty R (1992): Genetic variation at five trimeric and tetrameric tandem repeat loci in four human population groups. *Genomics* 12:241–253.

Ellegren H (1992): Polymerase-chain-reaction (PCR) analysis of microsatellites—a new approach to studies of genetic relationships in birds. *Auk* 109:886–895.

Estoup A, Presa P, Kreig F, Vaiman D, Guyomard R (1993): $(CT)_n$ and $(GT)_n$ microsatellites: a new class of genetic markers for *Salmo trutta* L. (brown trout). *Heredity* 71:488–496.

Gadagkar R (1985): Evolution of insect sociality—a review of some attempts to test modern theories. *Proc Indian Acad Sci* 94:309–324.

Goodnight KF, Queller DC (1994): *Relatedness* 4.2b. Pascal Program for Macintosh computers. Available at http://www.rice.edu/wasps.

Grafen A (1985): A geometric view of relatedness. *Oxford Surveys of Evolutionary Biology* 2:28–89.

Gustincich S, Manfioletti G, Del Sal G, Schneider C, Carninci P (1991): A fast method for high-quality genomic DNA extraction from whole human blood. *Biotechniques* 11:298–301.

Hanahan D (1983): Studies on transformation of *Escherichia coli* with plasmids. *J Mol Biol* 166:557–580.

Hannotte O, Bruford MW, Burke T (1992): Multilocus DNA fingerprints in gallinaceous birds: general approach and problems. *Heredity* 68:481–494.

Hanotte O, Zanon C, Pugh A, Greig C, Dixon A, Burk T (1994): Isolation and characterization of microsatellite loci in a passerine bird: the reed bunting *Emberiza schoeniclus*. *Mol Ecol* 3:529–530.

Hughes C, Queller DC (1993): Detection of highly polymorphic microsatellite loci in a species with little allozyme polymorphism. *Mol Ecol* 2:131–137.

Hughes CR, Queller DC, Strassmann JE, Davis SK (1993): Relatedness and altruism in *Polistes* wasps. *Behav Ecol* 4:128–137.

Kandpal R, Kandpal G, Weissman S (1994): Construction of libraries enriched for sequence repeats and jumping clones, and hybridization selection for region specific markers. *Proc Natl Acad Sci USA* 91:88–92.

Litt M, Luty J (1989): A hypervariable microsatellite revealed by in vitro amplification of a dinucleotide repeat within the cardiac muscle actin gene. *Am J Hum Genet* 44:397–401.

Moore S, Sargeant L, King T, Mattick J, Georges M, Hetzel D (1991): The conservation of dinucleotide microsatellites among mammalian genomes allows the use of heterologous PCR primer pairs in closely related species. *Genomics* 10:654–660.

Morin P, Wallis J, Moore J, Woodruff D (1994): Paternity exclusion in a community of wild chimpanzees using hypervariable simple sequence repeats. *Mol Ecol* 3:469–478.

Morin P, Woodruff D (1992): Paternity exclusion using multiple hypervariable microsatellite loci amplified from nuclear DNA of hair cells. In: Martin R, Dixson A, Wickings E (eds). *Paternity in primates: genetic tests and theories*. Basel: Karger, pp 63–81.

Paetkau D, Strobeck C (1994): Microsatellite analysis of genetic variation in black bear populations. *Mol Ecol* 3:489–495.

Peters JM, Queller DC, Strassmann JE, Solís CR (1995): Maternity assignment and queen replacement in a social wasp. *Proc R Soc Lond B* 260:7–12.

Queller DC, Goodnight KF (1989): Estimating relatedness using genetic markers. *Evolution* 43:258–275.

Queller DC, Strassmann JE, Hughes CR (1988): Genetic relatedness in colonies of tropical wasps with multiple queens. *Science* 242:1155–1157.

Queller DC, Strassmann JE, Hughes CR (1993): Microsatellites and kinship. *Trends Ecol Evol* 8:285–288.

Rosenberg HF, Ackerman SJ, Tenen DG (1990): An alternative method for labeling oligonucleotide probes for screening cDNA libraries. *Biotechniques* 8:384.

Sambrook J, Fritsch EF, Maniatis T (1989): *Molecular cloning: a laboratory manual*, 2nd ed. Cold Spring Harbor, NY: Cold Spring Harbor Laboratory Press.

Smeets HB, Ropers H, Wieringa B (1989): Use of variable simple sequence motifs as genetic markers: application to the study of myotomic dystrophy. *Hum Genet* 83:245–251.

Stallings R, Ford A, Nelson D, Torney D, Hildebrand C, Moyzis R (1991): Evolution and distribution of $(GT)_n$ repetitive sequences in mammalian genomes. *Genomics* 10:807–815.

Strassmann JE, Hughes CR, Queller DC, Turillazzi S, Cervo R, Davis SK, Goodnight KF (1989): Genetic relatedness in primitively eusocial wasps. *Nature* 342:268–270.

Tautz D (1989): Hypervariability of simple sequences as a general source for polymorphic DNA markers. *Nucleic Acids Res* 17:6463–6471.

Thomas SM, Moreno RF, Tilzer LL (1989): DNA extraction with organic solvents in gel barrier tubes. *Nucleic Acids Res* 13:5411.

Trentmann SM, Knaap E van der, Kende H (1995): Alternatives to ^{35}S as a label for the differential display of eukaryotic messenger RNA. *Science* 267:1186–1187.

Weber J (1990): Informativeness of human $(dC-dA)_n$. $(dG-dT)_n$ polymorphisms. *Genomics* 7:524–530.

Weber J, May P (1989): Abundant class of human DNA polymorphisms which can be typed using the polymerase chain reaction. *Am J Hum Genet* 44:388–396.

Westneat DF (1987): Extra-pair fertilizations in a predominantly monogamous bird: genetic evidence. *Anim Behav* 35:877–886.

CHAPTER 9

PCR-Based Cloning Across Large Taxonomic Distances and Polymorphism Detection: MHC as a Case Study

WAYNE K. POTTS

Center for Mammalian Genetics and Department of Pathology and Laboratory Medicine, University of Florida, Gainesville, Florida 32610-0275

CONTENTS

SYNOPSIS

This chapter addresses two general molecular problems that will be increasingly encountered by zoologists. First, for genes characterized in one or more species,

Molecular Zoology: Advances, Strategies, and Protocols, Edited by Joan D. Ferraris and Stephen R. Palumbi.
ISBN 0-471-14461-4 © 1996 Wiley-Liss, Inc.

how does one acquire species-specific genetic information in other species? Second, how does one efficiently detect genetic variation so that populations can be screened effectively for either known or unknown variants? As an example I will use recent advances in studies on the major histocompatibility complex (MHC) to illustrate possible approaches. I will highlight some of the genetic tools that have allowed us to go from the discovery of these genes in *Homo* and *Mus* to cloning them in a variety of vertebrates, to the detection and analysis of genetic variability in natural populations. I discuss PCR-based strategies for cloning genes of interest in species that are distantly related to species with existing genetic information. Methods are provided for efficiently screening clones prior to sequencing. Then the following three methods for detecting genetic variation are discussed and methods provided: denaturing gradient gel electrophoresis (DGGE), single-strand conformation polymorphism (SSCP), and heteroduplex analysis. I also provide methods for ethidium bromide staining of large acrylamide gels, which can be used in these three methods as well as many other acrylamide-based methods such as microsatellite analysis.

INTRODUCTION

Much of what is interesting in biology is defined by phenotypic differences and many of those differences are based on genetic variation. Our knowledge of the genetic basis of phenotypes has just entered an explosive phase and the ability to study newly discovered genes in appropriate species will allow the power of the comparative approach to be used on interesting genetic systems. Genetic systems can only fully be understood as we characterize their evolutionary histories and compare their functions in species with different ecologies. So the ability to efficiently characterize genetic systems in new species will have ever increasing applicability to zoologists. Understanding the extreme genetic diversity found at major histocompatibility complex (MHC) genes is becoming a classic example of such a genetic system (e.g., Figueroa et al., 1988; Gyllensten and Erlich, 1989; Alberts and Ober, 1993; Klein et al., 1993; Potts and Wakeland, 1993; Brown and Eklund, 1994).

The MHC is one of the best characterized genetic systems in *Mus* and *Homo* and studies in these species suggest that MHC genes influence the following: resistance to infectious disease through their role in immune recognition (Pazderka et al., 1987; Hill et al., 1991), susceptibility to autoimmune disease (Tiwari and Terasaki, 1985; Klein, 1986), reproduction through maternal–fetal interactions (Alberts and Ober, 1993), and a variety of behaviors including mating preferences (Yamazaki et al., 1976, 1978, 1988; Egid and Brown, 1989; Potts et al., 1991) and cooperative behavior (Manning et al., 1992), via their role in controlling individual odors (Yamaguchi et al., 1981; Yamazaki et al., 1983; Singh et al., 1987). This is a large, intriguing, and often controversial list of important biological traits to be influenced by the same few genes. One important approach to understanding the evolution of such a complex system is to use the comparative approach and study how these genes function in other vertebrates. This requires cloning MHC genes and developing techniques for detecting genetic variation in these new species so that hypotheses about the functional significance of these variants can be tested. This basic approach will be applicable to many loci and genetic systems. Techniques involved in such an approach are the subject of the following sections. In the final section I comment on the artificial divisions that divide biology into molecular/cellular and organismal camps. These artificial divisions inhibit the flow of ideas and techniques and ultimately the progress of science.

PCR-BASED CLONING ACROSS LARGE TAXONOMIC DISTANCES

Here I use MHC as an example to address many of the problems faced by an investigator wanting to study an interesting gene or genetic system in a new species where the effects of genetic variation are expected to be important. There are two general approaches to these problems, based on either DNA–DNA hybridization techniques or PCR (polymerase chain reaction) techniques. Hybridization approaches take advantage of the fact that DNA with sequence identity of over 60–75% will hybridize: two single strands of DNA will anneal to form a stable double-strand molecule of DNA (under appropriate conditions). One can take advantage of this principle by labeling the cloned gene in the characterized species and using it to probe DNA from the new species. If the taxonomic distance, and therefore sequence divergence, is not too great between the two species, Southern blots of the new species can be probed and genetic variation can be detected directly via restriction fragment length polymorphism (RFLP) analysis, without cloning the gene. Hybridization techniques can also be used to clone the gene of interest by screening (probing) random clones of DNA in the new species (DNA libraries) with labeled DNA from the cloned gene of interest. Both cross-species RFLP analysis and library screening techniques are commonly used (Sambrook et al., 1989) but the former has limited ability to detect genetic variation and the later can be a relatively difficult technique compared to the PCR-based cloning approach outlined below.

The major limitations to the PCR-based approach are that the regions flanking the gene of interest may be too variable for successful amplification with degenerate primers designed from other species or the gene of interest may be too large for PCR amplification. In either of these two situations screening cDNA or genomic libraries should be considered.

Designing Degenerate PCR Primers

PCR primers will often work on congeneric species and sometimes at even greater taxonomic distances. When such primers do not work, we take the following basic approach to designing degenerate PCR primers. For the gene of interest, we compare the amino acid sequence of two or more species to identify relatively conserved regions. In our case we wanted to clone the polymorphic region of MHC class II genes (exon 2) in the avian order passeriformes (songbirds). The only avian sequences available were for domestic chickens (*Gallus gallus domesticus*), which

a)

```
         1         11        21        31        41        51        61        71        81
         |         |         |         |         |         |         |         |         |
     DQ  DFVYQFKGMC YFTNGTERVR LVSRSIYNRE EVVRFDSDVG EFRAVTLLGL PAAEYWNSQK DILERKRAAV DRVCRHNYQL ELRTTLQRR
     DR  R.LE.V.HE. H.F....... FLD.YF.HQ. .Y........ .Y....E..R .D........ .L..Q..... .TY.....GV GESF.V...
  B-Lβ81  F.F.GTIAE. HYL....... .LD.YF...Q .YTH...... K.V.DSP..E .Q......NA EF..SRMN.. .TY.....GV GESF.V..S
  B-Lβ41  F.QWT..AE. HYL....... FLE.H....Q QFMH...... KYV.D.P..E RQ..I...NA E...DEMN.. .TF.....GV GESF.V..S
                   ‾‾‾‾‾‾‾‾                                                          ‾‾‾‾‾‾‾‾
                   primer 1                                                          primer 2
```

b)

Primer 1: TG(C/T)CA(C/T)TA(C/T)(C/T)TNAA(C/T)GGNACNGA(A/G)(A/C)G

Primer 2: CC(A/G)TA(A/G)TT(A/G)TGNC(G/T)(A/G)CA(A/G)TANGT(A/G)TC

Figure 1. Development of degenerate primers for amplifying a portion of exon 2 of the beta strand of MHC class II genes. (a) An alignment of the amino acid sequence of two human (DQ and DR) and two chicken (B-Lβ81 and B-Lβ41) sequences. A period indicates amino acid identity between sequences. Primers 1 and 2 were chosen to coincide with two relatively conserved regions as indicated. (b) Primer sequences are given. Degenerate sites (nucleotides within parentheses) correspond to all possible DNA sequences that would yield either the conserved or the avian amino acids. Primers 1 and 2 are 8192- and 2048-fold degenerate, respectively. (*Note:* PrimerGen is a DOS-based program that determines degenerate PCR primers for protein sequences and is distributed by: The National Research Council of Canada, Institute for Biological Sciences, Attn: Dr. John Nash, Ottawa, Ontario, Canada.)

were in a different avian order (galliformes); these two orders diverged an estimated 80 million years ago. Two amino acid sequences from domestic chickens were compared to two human sequences (Fig. 1a). For the two outer conserved regions (amino acid positions 10–18 and 71–79) degenerate primers were developed that represented all possible DNA sequences coding for each conserved amino acid sequence (Fig. 1b). (Alignments and primer sequences were provided by Dr. James Kaufman.) Although these primers were 2048- and 8192-fold degenerate, they have successfully amplified a 209 base pair fragment, showing sequence identity to MHC class II genes from every avian species attempted (over ten species) and the American alligator. Details of this work are provided in Edwards et al. (1995a).

Cloning PCR Products and Colony Screening

If the PCR product using the degenerate primers is the correct size, the next steps are to clone these products, screen the colonies for clones bearing the correct size insert, and sequence these clones to check for sequence identity with the gene of interest. There are numerous commercially available kits for cloning PCR products into bacterial plasmids, which has made this technique relatively routine. We use the *TA Cloning*™ kit from *Invitrogen*.

Even if the PCR product to be cloned is a relatively pure fragment there will often be many clones bearing fragments of the wrong size. If the PCR has bands other than the one of interest, then this clone may actually be in the minority among the clones. For these reasons it is wise to screen the clones for fragments of the correct size before sequencing. This formerly required a plasmid preparation for each clone but it is now possible to use the PCR to amplify from a colony with only minimal pretreatment. This technique is provided in Potts.1. Due to the ease of this technique a single person can easily screen tens of colonies in a single experiment.

Colonies bearing inserts of the correct size are then sequenced to determine if the gene of interest has been cloned. If so, many research avenues are open. (1) Additional sequences can be obtained from different individuals, allowing the design of species-specific primers. (2) The clones can be used as species-specific probes for screening libraries or conducting RFLP studies. (3) PCR-based approaches are available for obtaining flanking sequence information (Troutt et al., 1992; Edwards et al., 1995b). (4) Genetic variation can be detected and characterized using approaches described in the next section.

DETECTING GENETIC VARIATION

Due to the extreme genetic diversity of MHC genes, detecting these polymorphisms is a particularly important aspect of studying these genes in natural populations. However, detecting genetic variants in other less polymorphic systems is often more difficult because the site of the polymorphism may not be known and the percent sequence difference between alleles is usually much lower than in the case of MHC genes (Potts and Wakeland, 1993).

There are numerous methods for detecting genetic variants and many of these methods have been reviewed recently (Cotton, 1993; Dianzani et al., 1993; Prosser, 1994). We have chosen to focus on three of these methods due to their simplicity and efficiency in detecting variants. These three methods are denaturing gradient gel electrophoresis (DGGE), single-strand conformation polymorphism (SSCP), and heteroduplex analysis. These techniques are discussed in the following four sections.

Ethidium Bromide Staining of Large Acrylamide Gels

The three methods outlined below for detecting genetic variation utilize acrylamide gel electrophoresis. DNA can be visualized in acrylamide gels by radiolabeling DNA, silver staining, or ethidium bromide staining. Due to the low cost and the relative simplicity of ethidium bromide staining, we use it when possible. Although ethidium bromide staining is the least sensitive of the three DNA visualization methods, it is usually adequate for PCR-based approaches. Ethidium bromide staining is extremely simple with small gels but becomes more difficult with large acrylamide gels such as the size normally used in sequencing (45 cm by 45–60 cm). Although it is often considered by many to be too difficult to ethidium bromide stain this size of acrylamide gel, we routinely use this technique for both microsatellite analysis and SSCP; with a little practice it becomes easy. We provide this technique

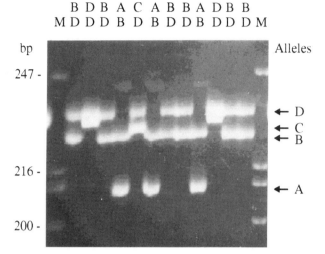

Figure 2. An example of an ethidium bromide stained large sequencing gel. A PCR amplification of a *Mus* microsatellite (dinucleotide) locus was run on a 45-cm by 60-cm acrylamide gel (86 lanes) of which we show 14 lanes here. Base pair size standards are provided on the left vertical axis as determined by the *PstI* cut *lambda* marker lanes (1 and 14). The four alleles (A–D) are designated on the right vertical axis. Lanes 2–13 are 12 *Mus* individuals and their genotypes are provided at the top of each lane. (M = marker; bp = base pairs.)

in Potts.2 and Figure 2 provides an example of a microsatellite locus run on a 45-cm by 60-cm acrylamide gel and visualized by ethidium bromide staining.

Denaturing Gradient Gel Electrophoresis (DGGE)

DGGE takes advantage of the principle that the point at which double-strand DNA denatures (melts), due either to thermal or chemical denaturant, is dependent on the nucleic acid sequence. Consequently, as two DNA molecules that are the same size but differ in sequence are electrophoresed through either a thermal or chemical denaturing gradient, they will often denature at different times during electrophoresis and will acquire new and often different electrophoretic mobilities. As double-strand DNA denatures its electrophoretic mobility is retarded. Consequently, the differential timing of denaturation and the subsequent differential mobilities lead to separation of these two populations of molecules during electrophoresis (Myers et al., 1987). DGGE is the most difficult of the three mutation detection techniques offered here and is the only technique requiring special equipment (beyond basic acrylamide electrophoresis) both for gel pouring and electrophoresis. DGGE works best for fragments between 200 and 1000 bp (methods provided in Potts.3).

It is estimated that approximately 50% of all single base substitutions can be resolved by DGGE and, by combining a GC clamp and heteroduplex analysis with DGGE, resolution approaches 100% (Myers et al., 1990). Double-strand DNA denatures sequentially in what are called melting domains. The sequence of each melting domain determines when denaturation occurs as the molecule moves through the denaturation gradient. Mutations in the last domain to melt are often undetectable because the mobility of the now single-strand DNA is sequence independent. This problem is solved by adding a GC clamp to the molecule. A GC clamp is approximately 40 bases, consisting of only guanines and cytosines. Due to the high melting temperature of the G–C bond, this GC clamp region becomes the last domain to melt, allowing mutation detection in all domains of the original molecule. GC clamps are usually added by incorporating them into one of the PCR primers (Sheffield et al., 1989) but they can also be added enzymatically (Abrams et al., 1990).

Combining the heteroduplex approach with DGGE also increases detection efficiency of mutations and is discussed in the section on heteroduplex analysis below.

The optimal range for the denaturing gradient may be estimated based on theoretical calculations but our experience suggests empirical determination is usually required to find the best conditions. Two empirical approaches may be employed. First, a perpendicular DGGE can be run, where the denaturation gradient is perpendicular to the direction of electrophoresis (Lessa, 1992). The approximate denaturation conditions at which the molecule of interest melts can be estimated directly from the gel, as shown in Figure 3, where the same sample [a PCR amplification of exon 2 of a class II MHC gene from an individual red jungle fowl (*Gallus gallus*)] has been loaded in all lanes and electrophoresed perpendicular to the denaturing gradient. At the point(s) along the gradient where denaturation occurs a sharp retardation in mobility occurs, allowing an estimate of the appropriate range of the gradient for the parallel DGGE. Figure 3 shows a single denaturation point around 50% and when the same locus is run on a parallel DGGE with a 30% to 70% gradient bracketing this 50% estimate, good resolution of alleles results (Fig. 4).

An alternative approach is to run a series of parallel DGGE gels that encompass different ranges of denaturing gradients. The largest range normally used is 0% to 80%. For example, one might start by running three gels with gradients of 0% to 40%, 20% to 60%, and 40% to 80%. Say 20% to 60% gave the best resolution; then the next step would be to determine if narrower ranges such as 20% to 40%, 30% to 50%, and 40% to 60% increases resolution. Optimizing denaturing conditions is best done using all possible alleles in the parallel or perpendicular DGGE trials, but this is often impossible when dealing with natural populations where many alleles may be unknown.

Figure 4 shows a DGGE gel of a PCR amplification of exon 2 of a class II MHC gene for seven red jungle fowl individuals from an outbred population. Each lane shows two bands representing each of two alleles in these heterozygous individuals. The additional two bands in lanes 6 and 7 are heteroduplex fragments and are discussed below. If the allelic forms require sequencing, the resolved bands can be excised from DGGE gels and used as template for sequencing (Lessa, 1992).

80% ◀— DENATURING GRADIENT◀— 0%

Figure 3. An example of perpendicular denaturing gradient gel electrophoresis (DGGE) using a PCR amplification of a class II MHC gene from red jungle fowl. A 0% to 80% denaturing gradient gel was poured and a single sample was loaded in all lanes and electrophoresed perpendicular to the gradient.

Single-Strand Conformation Polymorphism (SSCP)

If double-strand DNA is denatured by heating and placed directly in ice water, many of the single strands will not reanneal with their complement but, rather, will form their own secondary structure (or conformation) as short complementary stretches anneal within the molecule. The conformation of these single-strand molecules is sequence dependent and different conformations often have differential electrophoretic mobilities in native acrylamide gels. This is the basis of single-strand

1 2 3 4 5 6 7

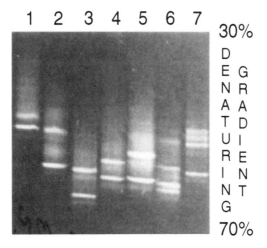

30%

D
E G
N R
A A
T D
U I
R E
I N
N T
G
70%

Figure 4. An example of parallel denaturing gradient gel electrophoresis (DGGE) using a PCR amplification of a class II MHC gene from seven red jungle fowl individuals. A 30% to 70% denaturing gradient gel was poured and the samples were electrophoresed parallel to the gradient.

conformation polymorphism (SSCP) analyses (Orita et al., 1989) and the methods for this technique are provided in Potts.4. SSCP works on molecules between 100 and 1000 bp but the size for optimal resolution is usually around 200 bp (Orita et al., 1989; Sheffield et al., 1993).

Figure 5 shows an SSCP gel of PCR amplifications of a class II MHC gene from seven inbred strains of domestic chickens known to differ at their MHC. All lanes show a common band, which is the double-strand DNA. In most other lanes two other slower migrating bands are seen, which are the two single-strand molecules.

1 2 3 4 5 6 7 8

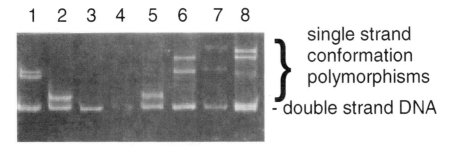

single strand
conformation
polymorphisms

- double strand DNA

Figure 5. An example of a single-strand conformation polymorphism (SSCP) gel using a PCR amplification of a class II MHC gene from eight inbred strains of domestic chickens known to have MHC serological differences. The region of the gel containing single-strand DNA fragments as well as the double-strand DNA fragment are marked. Lane 4 is blank.

Heteroduplex Analysis

When a PCR amplification containing two alleles is heated beyond denaturation temperature and allowed to cool slowly, the single-strand DNA molecules will reanneal both to their perfectly matched complement strands (homoduplexes) and to their imperfectly matched (allelic) complement strands (heteroduplexes). The mismatched regions of these heteroduplexes retard their electrophoretic mobilities relative to homoduplexes and thus provide a means for detecting the presence of allelic pairs that differ only by nucleic acid sequence (Clay et al., 1991; Keen et al., 1991; White et al., 1992). Methods for heteroduplex analysis are provided in Potts.5.

Figure 6 shows heteroduplex fragments that form due to sequence differences in PCR products of a house mouse (*Mus domesticus*) MHC gene. There are three alleles (b, d, and q) that each differ in sequence but are of the same length. No heteroduplex bands form in lanes with homozygote samples (first three sample lanes—b/b, d/d, and q/q). But in the lanes from heterozygotes (b/d, b/q, and d/q), three different heteroduplex patterns are observed in the region of the gel marked heteroduplex. The genotypes of each individual are denoted at the top of each lane.

Heteroduplex analysis can also be conducted on DGGE gels. The mismatch regions of the heteroduplex cause the double-strand DNA to denature prior to homoduplexes, retarding their electrophoretic mobilities both sooner and often differentially. An example of this can be seen in lanes 6 and 7 of Figure 4. None of the

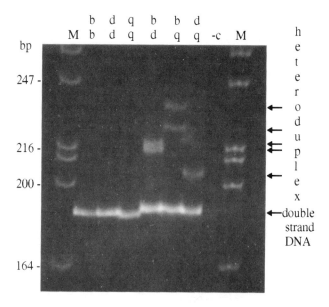

Figure 6. An example of a heteroduplex gel using a PCR amplification of a 190-bp intronic fragment in the house mouse (*Mus*) MHC class II A_β gene. The running positions of the five heteroduplex bands and the double-strand DNA band are given on the right and size standards are given on the left. (M = marker; bp = base pairs; −c = negative control.)

seven samples were treated to form heteroduplexes (as in Potts.5), but hetero-duplexes often form because the last cycle of the PCR may be inefficient. An inefficient cycle means that the denaturation step is followed by poor or nonexistent polymerase activity, thus allowing the nonreplicated single-strand DNA molecules to reanneal into both homoduplexes and heteroduplexes (when there are two or more alleles as templates). The conditions in an inefficient cycle are similar to the hetero-duplex forming conditions created in Potts.5. Lanes 1–5 probably had an efficient last cycle of PCR while lanes 6 and 7 did not, allowing heteroduplexes to form. The resolution of heteroduplexes can be different on DGGE gels relative to native acrylamide gels. Therefore, for maximal resolution, we conduct heteroduplex anal-ysis on both native acrylamide and DGGE gels.

Strategy for Detecting Polymorphisms Using DGGE, SSCP, and Heteroduplex

Because these three methodologies have different strengths and weaknesses, we employ the following five phases for detecting MHC polymorphisms in an un-characterized population. DGGE is the simplest method to interpret (each band corresponds to an allele) so we analyze the entire population with DGGE. The main weakness of DGGE is simply that it may not resolve all allelic differences. So we use SSCP and heteroduplex to resolve allelic difference not revealed by DGGE. In the second phase, all samples with either identical genotypes or containing identical alleles (as determined by DGGE) are run in adjacent lanes in an SSCP gel, which usually results in further resolution of allelic fragments. The third and fourth phases involve heteroduplex analysis on native and DGGE gels, respectively. Samples that appear to be identical homozygotes (as determined by DGGE and SSCP) are mixed in all pairwise combinations and both types of heteroduplex analyses are run. Only nonidentical pairs will show heteroduplex bands. Samples that appear to be "identi-cal heterozygotes" (as determined by DGGE and SSCP) are run in adjacent lanes in both types of heteroduplex analyses—DGGE and native acrylamide. Any differen-tial mobilities in the heteroduplex bands among "identical heterozygotes" indicates new alleles within these groups. The final phase is to sequence the alleles resolved in the first four phases. DGGE, SSCP, and heteroduplex fragments can be excised from acrylamide gels and used as templates for sequencing (Sambrook et al., 1989). Also, sequencing of "identical" bands can be used as a final check for identity. Sequencing of alleles may not be necessary if one simply needs to resolve most allelic differences, which the first four phases should accomplish.

EPILOGUE: BIOLOGY'S ARTIFICIAL DIVISIONS WILL SOON FALL

The molecular revolution started about three decades ago. Many of the new molecu-lar tools that were spawned by this revolution were equivalent to the discovery of the microscope; components of biology that had previously been invisible were seen for the first time. Such new tools allow major discoveries by description of new phenomena. This led to over two decades of major discoveries at the molecular level, often achieved by simple description of phenomena that had previously been invisible. This molecular revolution continues unabated today. But what should have been a marvelous flowering of all biology, through the laying of its own molecular foundations, has all too often become a house divided; molecular biologists and organismal biologists tended not to integrate and synthesize but rather to specialize, creating the artificial divisions we see today. Consequently, training in the biological sciences is often dichotomized with a focus on either molecular/cellular problems or organismal problems. But nature is continuous and great discoveries will come to those willing to train themselves broadly along this continuum of living systems. The major histocompatibility complex (MHC) has become a prime example of a genetic system that only began to fully yield its secrets when approached from multiple levels of biological inquiry (discussed above).

We are entering an era where the discovery of genetic systems underlying biological traits will outpace our ability to fully evaluate the functional significance of these genes. Biologists who can move easily between molecular and organismal worlds will play an important role in the synthesis of this information. In this climate I cannot see how biology's current artificial divisions can stand.

ACKNOWLEDGMENTS

I thank James Kaufman for the degenerate primers, Anthony Baker and Karl Nguyen for assistance with the microsatellite figure, Mats Grahn and Karen Achey for assistance with the DGGE figures, Amy Milton for assistance with the SSCP figure, Anthony Baker for assistance with the heteroduplex figure, and Scott Edwards and Patricia Slev for general technical and moral support. I thank Patricia Slev for comments on the manuscript. This work was conducted while W.K.P. was supported by grants from NSF and NIH.

REFERENCES

Abrams ES, Murdaugh SE, Lerman LS (1990): Comprehensive detection of single base changes in human genomic DNA using denaturing gradient gel electrophoresis and a GC clamp. *Genomics* 7:463–475.

Alberts SC, Ober C (1993): Genetic variability in the major histocompatibility complex: a review of non-pathogen-mediated selective mechanisms. *Phys Anthropol* 36:71–89.

Brown JL, Eklund A (1994): Kin recognition and the major histocompatibility complex: an integrative review. *Am Nat* 143:435–461.

Clay TM, Bidwell JL, Howard MR, Bradley BA (1991): PCR-fingerprinting for selection of HLA matched unrelated marrow. *Lancet* 337:1049–1052.

Cotton RGH (1993): Current methods of mutation detection. *Mutat Res* 285:125–144.

Dianzani I, Camaschelia C, Ponzone A, Cotton RGH (1993): Dilemmas and progress in mutation detection. *Trends Genet* 9:403–405.

Edwards SV, Grahn M, Potts WK (1995a): Dynamics of MHC evolution in birds and crocodilians: amplification of class II genes with degenerate primers. *Mol Ecol (in press).*

Edwards SV, Wakeland EK, Potts WK (1995b): Contrasting histories of avian and mammalian MHC genes revealed by class II B sequences from songbirds. *Proc Natl Acad Sci USA (in press).*

Egid K, Brown JL (1989): The major histocompatibility complex and female mating preferences in mice. *Anim Behav* 38:548–549.

Figueroa F, Gunther E, Klein J (1988): MHC polymorphism pre-dating speciation. *Nature* 335:265–267.

Gyllensten UB, Erlich HA (1989): Ancient roots for polymorphism at the HLA DQα locus in primates. *Proc Natl Acad Sci USA* 86:9986–9990.

Hill AVS, Allsop CEM, Kwiatkowski D, Anstey NM, Twumasi P, Rowe PA, Bennett S, Brewster D, McMichael AJ, Greenwood BM (1991): Common West African HLA antigens are associated with protection from severe malaria. *Nature* 352:595–600.

Keen J, Lester D, Inglehearn C, Curtis A, Bhattacharya S (1991): Rapid detection of single base mismatches as heteroduplexes on hydrolink gels. *Trends Genet* 7:5.

Klein D, Ono H, O'Huigin C, Vincek V, Goldschmidt T, Klein J (1993): Extensive MHC variability in cichlid fishes of Lake Malawi. *Nature* 364:330–334.

Klein J (1986): *Natural history of the major histocompatibility complex.* New York: Wiley.

Lessa E (1992): Rapid surveying of DNA sequence variation in natural populations. *Mol Biol Evol* 9:323–330.

Manning CJ, Wakeland EK, Potts WK (1992): Communal nesting patterns in mice implicate MHC genes in kin recognition. *Nature* 360:581–583.

Myers RM, Maniatis T, Lerman L (1987): Detection and localization of single base changes by denaturing gradient gel electrophoresis. *Methods Enzymol* 155:501–527.

Myers RM, Sheffield VC, Cox DR (1990): Mutation detection by PCR, GC-clamps, and denaturing gradient gel electrophoresis. In: Erlich HA (ed). *PCR technology: principles and applications for DNA amplification.* New York: Stockton Press, pp 71–88.

Orita M, Iwahana H, Kanazawa H, Sekiya T (1989): Detection of polymorphism of human DNA by gel electrophoresis as single-strand conformation polymorphisms. *Proc Natl Acad Sci USA* 86:2766–2770.

Pazderka F, Longenecker BM, Law GRJ, Stone HA, Ruth RF (1987): Histocompatibility of chicken populations selected for resistance to Marek's disease. *Hum Immunol* 19:155–162.

Potts WK, Manning CJ, Wakeland EK (1991): MHC genotype influences mating patterns in semi-natural populations of *Mus. Nature* 352:619–621.

Potts WK, Wakeland EK (1993): The evolution of MHC genetic diversity: a tale of incest, pestilence and sexual preference. *Trends Genet* 9:408–412.

Prosser J (1994): Detecting single-base mutations. *TIBTECH* 11:238–246.

Sambrook J, Fritsch EF, Maniatis T (1989): *Molecular cloning.* Cold Spring Harbor, NY: Cold Spring Harbor Laboratory Press.

Sheffield VC, Cox DR, Lerman L, Myers RM (1989): Attachment of a 40-base-pair G+C-rich sequence (GC-clamp) to genomic DNA fragments by the polymerase chain reaction results in improved detection of single-base changes. *Proc Natl Acad Sci USA* 86:232–236.

Sheffield VC, Beck JS, Kwitek AE, Sandstrom DW, Stone EM (1993): The sensitivity of single-strand conformation polymorphism analysis for the detection of single base substitutions. *Genomics* 16:325–332.

Singh PB, Brown RE, Roser B (1987): MHC antigens in urine as olfactory recognition cues. *Nature* 327:161–164.

Tiwari JL, Terasaki PI (1985): *HLA and disease associations.* New York: Springer-Verlag.

Troutt AB, McHeyzer-Williams MG, Pulendran B, Nossal GJ (1992): Ligation-anchored PCR. *Proc Natl Acad Sci USA* 89:9823–9825.

White MB, Carvalho M, Derse D, O'Brien SJ, Dean M (1992): Detecting single base substitutions as heteroduplex polymorphisms. *Genomics* 12:301–306.

Yamaguchi M, Yamazaki K, Beauchamp GK, Bard J, Thomas L, Boyse EA (1981): Distinctive urinary odors governed by the major histocompatibility locus of the mouse. *Proc Natl Acad Sci USA* 78:5817.

Yamazaki K, Boyse EA, Mike V, Thaler HT, Mathieson BJ, Abbott J, Boyse J, Zayas ZA, Thomas L (1976): Control of mating preferences in mice by genes in the major histocompatibility complex. *J Exp Med* 144:1324–1335.

Yamazaki K, Yamaguchi M, Andrews PW, Peake B, Boyse EA (1978): Mating preferences of F^2 segregants of crosses between MHC-congenic mouse strains. *Immunogenetics* 6:253–259.

Yamazaki K, Beauchamp GK, Egorov IK, Bard J, Thomas L, Boyse EA (1983): Sensory distinction between H-2^b and H-2^{bm1} mutant mice. *Proc Natl Acad Sci USA* 80:5685–5688.

Yamazaki K, Beauchamp GK, Kupniewski D, Bard J, Thomas L, Boyse EA (1988): Familial imprinting determines H-2 selective mating preferences. *Science* 240:1331–1332.

PART III

CLASSIC PROBLEMS IN THE EVOLUTION OF GROWTH AND DEVELOPMENT

CHAPTER 10

Strategies for Cloning Developmental Genes Using Closely Related Species

BILLIE J. SWALLA

Biology Department, Vanderbilt University, Nashville, Tennessee 37235

CONTENTS

SYNOPSIS

Ascidians are urochordates, chordates because the larvae have a notochord, dorsal hollow nerve tube, and muscle cells in their tails. Yet ascidians are also invertebrates because they never develop a vertebrate backbone either in the chordate larval stages or in the adults. Ascidian larvae exhibit the simplest chordate body plan and gastrulate early in development, after the 64-cell stage. Cleavage is determinate and invariant, and the cell lineage is completely known. There has been recent renewed interest in understanding ascidian embryonic development in the search for genes governing the evolution of the chordate body plan. Because of their unique and varied life histories, ascidians are also useful to study the evolution of larval morphology and life history. Two species that are closely related but have completely different larvae were used for the studies described here: *Molgula oculata*, which has a tailed larva exhibiting all of the chordate features, and *Molgula occulta*, which develops into a tailless (or anural) larva, lacking chordate features.

Molecular Zoology: Advances, Strategies, and Protocols, Edited by Joan D. Ferraris and Stephen R. Palumbi.
ISBN 0-471-14461-4   © 1996 Wiley-Liss, Inc.

This chapter describes how subtractive hybridization can be used between closely related species to isolate genes having species-specific expression. This can be successful if the genetic distance between the two species is small and comparable stages of development are used for the subtraction. In the system described here, the species-specific transcripts are correlated with a dramatic change in ascidian larval morphology and life history. These maternal genes are likely to play an important role in embryonic development, and in the evolution of ascidian species that have tailless larvae.

Analysis of the gene structure between the same gene differentially expressed in the two species may elucidate how or why the gene is not expressed in one species. These results allow examination of how molecular evolution in single-copy nuclear genes may result in rapid changes in morphology by affecting developmental processes during specification of the body plan.

INTRODUCTION

An ascidian embryo develops into a chordate larva, which swims for a period of time, then attaches to a suitable substrate and metamorphoses into a sessile adult that filter feeds by moving water through its two siphons. The adult ascidian makes a tough, leathery tunic, hence the historical common name, tunicate. Because the adult is immobile, the nonfeeding larva functions as the dispersal stage in the ascidian life history. The ascidian larva has all of the chordate features, including a tail with a central notochord flanked by muscle cells and a dorsal nerve chord (Figs. 1 and 2). The larval tail is formed by convergence and extension of the prospective notochord (Cloney, 1964; Miyamoto and Crowther, 1985), muscle, and epidermal cells in the posterior region of the embryo (reviewed in Swalla, 1993). Larval tail muscle cells can develop autonomously due to maternal determinants segregated from the egg into the muscle cell lineages (reviewed in Satoh et al., 1990; Swalla, 1992).

The larval head contains the brain, a few prominent pigmented sensory cells, and all the precursor tissues for the adult (Fig. 1). In contrast to the larval tail muscle, the sensory cells in the larval brain are specified by inductive signals (Nishida and Satoh, 1989), from presumptive notochord cells during gastrulation (reviewed in Venuti and Jeffery, 1989). The notochord cells may be specified by an earlier

Figure 1. A photomicrograph of a typical chordate ascidian larva. Note the pigment spot in the brain, located in the center of the head. There are 40 notochord cells visible in the center of the tail, and the muscle cells surround the notochord (see schematic cross-section in Fig. 2).

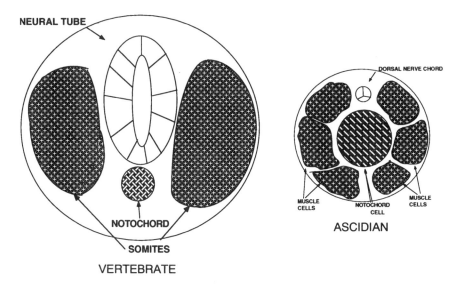

Figure 2. Drawings of cross sections through the trunk of a vertebrate embryo, such as a chicken embryo, and the tail of an ascidian larva. Drawings are not to scale. The ascidian larva would be much smaller, especially the species *Molgula oculata*. Note the tissues shown in the vertebrate embryo are composed of hundreds of cells, while the tail of the ascidian is made up of single cells. The dorsal nerve chord in ascidians is normally made up of 1–3 cells and is quite small relative to the other cell types, while the neural tube in vertebrates is a large, complex multicellular structure.

induction during cleavage (Nishida, 1992). Most ascidian species form a tailed (urodele) larva, which is the dispersal phase in the life cycle. However, there is one family of ascidians, the Molgulidae, in which some of the species have evolved completely different life histories (Fig. 3). Some species have totally eliminated the larval stage, while other species produce a "larva" that hatches but lacks the chordate larval morphology (Fig. 3; reviewed in Jeffery and Swalla, 1990, 1992a).

There are 15–20 species described in the family Molgulidae that do not develop into a chordate tadpole larva (Berrill, 1931; reviewed in Jeffery and Swalla, 1990), in contrast to about 3000 described ascidian species with a chordate larva, suggesting the tailed morphology is the ancestral one. Phylogenetic analysis suggests that tailless larvae have evolved at least four times independently, and that tailless larvae can evolve rapidly from tailed larvae in evolutionary time (Hadfield et al., 1995). I have extensively studied a clade of tailed and tailless ascidians called the Roscovite clade (Hadfield et al., 1995) in collaboration with Dr. William R. Jeffery. This group of ascidians is presently found off the northwest coast of Roscoff, France. The group contains multiple developmental modes existing sympatrically. This group of Molgulid ascidians allows us to examine natural variation on the chordate body plan, in which all of the larval variations are part of the life history of a viable species.

LIFE HISTORY

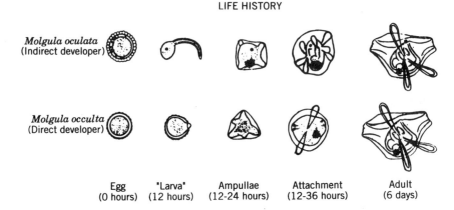

| Egg | "Larva" | Ampullae | Attachment | Adult |
| (0 hours) | (12 hours) | (12-24 hours) | (12-36 hours) | (6 days) |

Figure 3. Drawings of the life history of a tailed ascidian *Molgula oculata* (*top*) and an anural species, *Molgula occulta* (*bottom*). The anural larva lacks the neural sensory cell (melanin pigment spot) and the tail of the chordate larva and so cannot disperse by swimming. In both species, the larvae metamorphose into adults. (Redrawn with permission from Figures 1 and 8 in Berrill, 1931.)

The existence of sympatric, closely related species allowed subtractive hybridization between two species to isolate genes with species-specific expression (Swalla et al., 1993). The two species, *Molgula oculata* and *Molgula occulta,* may have diverged less than 5 million years ago. The two species retain the ability to cross-hybridize (Swalla and Jeffery, 1990) and show little genetic divergence in rRNA sequences (Hadfield et al., 1995). Subtractive hybridization has been a powerful means of isolating rare, single-copy genes by isolating mRNA from identical cultured cells treated differently, converting the RNA to cDNA, then allowing exhaustive hybridization of the DNA from the two different cell types. After removal of the double-strand DNA, the single-strand cDNA from the cell treatment of interest can be cloned and characterized. This procedure works best when there is an excess of mRNA from the control treatment. We used this strategy but hybridized mRNA between two different, closely related species. Subtractive hybridization between species is more likely to be successful when the genetic distance between the two species is small, and comparable stages of development are used for the subtraction. Subtractive hybridization was carried out between gonads of two different species to look for maternal transcripts involved in determining the chordate body plan. Here, I will summarize the findings of an original subtractive screen. Three candidate genes were isolated in that screen, *Uro-1, Uro-2,* and *Uro-11.* These three species-specific transcripts of the *Uro* (urodele-specific) genes are correlated with a dramatic change in ascidian larval morphology and life history. The three isolated genes are likely to play an important role in embryonic development (Swalla et al., 1993) and, subsequently, in development of the chordate body plan. We now have experimental evidence that one of the isolated genes, *manx* (*Uro-11*), is integral to the chordate body plan (Swalla and Jeffery, 1994).

MATERIALS AND METHODS

Molgula oculata (the tailed species) and the anural ascidian *Molgula occulta* were dredged from sand flats near Point de Bloscon, Roscoff, France. Gonads were dissected and mature eggs were inseminated and cultured as described previously (Swalla and Jeffery, 1990). The first subtractive hybridization was to isolate maternal messages (Swalla et al., 1993), and so entire ovotestes from the hermaphroditic animals were removed and homogenized in guanidine isothiocyanate (March et al., 1985). Poly(A)+ RNA was isolated with Oligotex-dT30 beads (Roche, Tokyo, Japan).

We first constructed libraries in lambda Zap II and Uni-Zap phage (Stratagene, La Jolla, CA), which allows RNA synthesis by use of T3 or T7 promoter sites in the flanking sequences (Swalla et al., 1993). The *M. oculata* cDNA was directionally cloned into Uni-Zap vector, so that the orientation of the clones would be apparent before sequencing. Inserts are cloned so the poly(A)+ tail is ligated into the XhoI site and the 5' end of the insert is ligated into the EcoRI site. One could start with total or poly(A)+ RNA, rather than synthesizing RNA from a cDNA library for subtraction. There are many subtractive procedures in which cDNA is subtracted with another control RNA preparation, then the library is constructed from single-strand cDNA, which may be more practical if the material is unlimited, as tissue culture cells. The construction of cDNA libraries before subtraction is the preferred method if RNA from a specific stage or species is difficult or costly to obtain, or if the availability of material varies due to location and/or seasonality. For ascidians, which must be collected by dredging or diving during the gravid season, it was essential to construct the libraries before attempting the subtractive hybridization. Also, the cDNA libraries can then be screened by more conventional methods with probes from other species. Obviously, construction of a good cDNA library is critical because it must contain representative amounts of the mRNAs in the tissue. This is a complex procedure and best attempted only by experienced molecular biologists. Since the initial construction of the cDNA libraries first used for subtraction (Swalla et al., 1993), we routinely have custom libraries prepared by companies for use in the laboratory. Though there are always rumors of researchers being dissatisfied with commercial libraries, we have had good luck with them. Considering the time and expense involved in constructing libraries, I would recommend the commercial route if you have little experience with molecular techniques.

The subtractive procedure has been published elsewhere (Swalla et al., 1993), but will be reviewed here, and a detailed protocol (Swalla.1) can be found in the back of this book. An aliquot of *M. oculata* and *M. occulta* cDNA gonad library was amplified by PCR with T3 and T7 primers to obtain double-strand DNA corresponding to the inserts (Timblin et al., 1990). Libraries were amplified separately, for less than 20 cycles. Increasing the number of cycles favored short inserts over long inserts. The amplified DNA from each library was then phenol extracted, and ethanol precipitated. DNA was resuspended in an RNA synthesis buffer as outlined in Van Gelder et al. (1990). Antisense RNA was synthesized from either *M. oculata* or *M. occulta* cDNA library with T7 RNA polymerase (Bethesda Research Laboratories, Bethesda, MD) and RQ1 DNase (Promega, Madison, WI) was added to digest the DNA template at the end of the reaction. The *M. oculata* RNA was then converted into single-strand cDNA using the T3 primer, and RNA was subsequently hydrolyzed in 1N NaOH, 0.5% SDS. The *M. occulta* cDNA library was not directionally cloned, so RNA synthesized from the T7 promoter consisted of both sense and antisense messages. This RNA was labeled with photoactivatable biotin (Clontech Laboratories, Inc., Palo Alto, CA) by use of a PAB sunlamp (Clontech). Subtractive hybridization was then performed with the *M. oculata* single-strand cDNA hybridized with tenfold excess of biotin-labeled *M. occulta* library RNA (Sive and St. John, 1988) for three rounds of hybridization. Following subtractive hybridization, biotin-labeled transcripts (from the *M. occulta* library) were removed by addition of streptavidin and phenol extraction. Subtracted cDNA was then random primer labeled (BMB; Boehringer-Mannheim Biochemicals, Indianapolis, IN) with radioactivity and used to screen the *M. oculata* cDNA library at high stringency for positive clones.

A total of 5×10^4 phages were screened with the subtracted probe at low plaque density (5×10^3 per 10×14 cm plate) in 60 ml of hybridization solution (50% formamide, 6X SSPE, 5X Denhardt's, 0.5% SDS, 100 µg/ml salmon sperm DNA) with 2×10^6 cpm/ml probe. The 24 positive *M. oculata* clones obtained were then hybridized with a cDNA probe prepared from RNA synthesized *in vitro* from the *M. occulta* cDNA library, according to the method described above, except that washes were at low stringency (5 washes at $2\times$ SSC, 0.1% SDS for 30 min at 55°C; 4 washes at $0.1\times$ SSC, 0.1% SDS for 25 min at 50°C), and 8 negative clones were selected. (Most of the clones were positive for both species.) One of the 16 clones positive for both species served as a control for RNA slot blot hybridization.

Clones positive for *M. oculata* cDNA but negative with *M. occulta* cDNA were further characterized by RNA slot blot hybridization. RNA synthesized *in vitro* from the *M. oculata* and *M. occulta* cDNA libraries was blotted onto Hybond N+ filter paper (Amersham) using the formamide/formaldehyde method (Sambrook et al., 1989). The filters were hybridized separately in Rapid Hybridization Buffer (Amersham) with 1×10^6 cpm of random-primed (BMB) ^{32}P-labeled (3000 Ci/mmol, Amersham) cDNA. Probes were prepared from templates obtained by PCR amplification of the individual inserts and a control clone positive for both species. Three cDNA clones, *Uro-1, Uro-2,* and *Uro-11* (of the original 24), hybridized preferentially to *M. oculata* RNA; the remaining clones hybridized to the same extent as the control clone to RNA of both species. Sequencing and expression of the clones were accomplished by previously described procedures (Swalla et al., 1993).

RESULTS AND DISCUSSION

Comparative studies of closely related species with varying developmental modes may further our understanding about how developmental changes may underlie evolutionary differences in body plans and life histories (Raff, 1987; Jeffery and

TABLE 1. *Uro* **Genes Isolated by Subtractive Screen of Gonad Libraries**

Gene[a]	Protein Motifs	Gonads	Embryos	Hybrids
Uro-1	Tryosine kinase ankyrin repeats	*M. oculata,* but not *M. occulta*	Not expressed	Not expressed
Uro-2[b] (#L19340)	Leucine zipper	*M. oculata,* but not *M. occulta*	Not expressed	Not expressed
Uro-11[b] (#L19339)	Zinc cluster; nuclear signal	*M. oculata,* but not *M. occulta*	*M. oculata* only at gastrulation	Expressed at gastrulation

[a]GenBank accession # is listed under the gene.
[b]For details of *Uro-2* and *Uro-11* expression, see Swalla et al. (1993). *Uro-1* results are in preparation (Makabe et al., 1995). *Molgula oculata* eggs develop into tailed larvae, while *M. occulta* develop into tailless larvae.

Swalla, 1992a). Previous studies have shown that changes in both maternal and zygotic factors are involved in the evolution of anural development in ascidians (Swalla and Jeffery, 1990; Jeffery and Swalla, 1992b). We used subtractive hybridization to identify maternal genes expressed differentially in two ascidian species in an attempt to isolate regulatory genes involved in the evolutionary transition from chordate to anural development. We identified three cDNAs by subtractive procedures encoding *Uro* (urodele-specific) genes. Each of these genes are expressed during oogenesis in the tailed ascidian *M. oculata* but are inactive or down-regulated in the anural ascidian *M. occulta* (Swalla et al., 1993; Table 1).

It is possible to isolate both novel developmental genes from ascidians and to isolate homologous genes first described in other species. *Uro-1* belongs to the shark family of tyrosine kinases. These are unusual tyrosine kinases with 5 ankyrin repeats and 2 SH_2 domains at the 5' end (Ferrante et al., 1995; Makabe et al., 1995; Table 1). The first gene isolated from this family was described in hydra (Chan et al., 1994) and later also found in *Drosophila* (Ferrante et al., 1995). In *Drosophila* the protein is expressed on the apical surface of ectodermal cells, and genetic evidence suggests that it may be involved in signaling by *crumbs,* but the exact function of it is unknown (Ferrante et al., 1995). Ankyrin repeats have been identified in a number of diverse proteins and have been implicated in protein–protein interactions (Michaely and Bennett, 1992). It is intriguing that a specific tyrosine kinase might be anchored in the egg cortex of the tailed species in order for the initial events of axis formation to occur during the first cell cycle that will eventually lead to the proper movements during gastrulation (Swalla, 1993).

Consistent with this possibility, the myoplasm, an egg cytoplasmic region that is segregated to muscle lineages in ascidian embryos (Swalla, 1992), is modified in eggs of anural species (Swalla et al., 1991; Jeffery and Swalla, 1992b). The modification appears to occur during oogenesis, after vitellogenesis begins (Swalla et al., 1991). In the tailed species *M. oculata,* and in every tailed ascidian species examined, a specific protein, p58, became localized in the egg cortex during vitellogenesis and remained cortical, even after egg maturation (Swalla et al., 1991). Later, this cortical cytoplasm is inherited by tail muscle cells, which then undergo myogenesis (Swalla, 1992). In contrast, *M. occulta* and other anural species' eggs contained p58 but failed to initiate and maintain the cortical localization of protein (Swalla et al., 1991), and the embryos also later lack muscle cells (Swalla and Jeffery, 1990; Jeffery and Swalla, 1992a). The mechanism(s) of the changes in oogenesis are not fully understood, but it has also been shown that all species in the family Molgulidae lack an ankyrin-like protein in eggs and embryos (Jeffery and Swalla, 1993). This raises the interesting possibility that the Molgulid ancestor(s) lost the ankyrin-like protein, which may be a preadaptation to the further changes during oogenesis that are seen in all tailless Molgulid species. Further research will focus on how changes in oogenesis and gene expression result in the complete morphological changes seen in the tadpole larvae in closely related *Molgula* species.

The other two novel genes that were isolated by subtractive hybridization may be transcription factors and are likely candidates for regulatory factors involved in changing the mode of larval development (Swalla et al., 1993). Both of these genes showed the characteristic expression pattern selected in the subtractive screen, that is, a high level of transcripts in ovaries of the tailed species (*M. oculata*), but little or no expression in the tailless species (*M. occulta*) (Swalla et al., 1993 and summarized in Table 1). *Uro-2* has a leucine zipper motif (Landschultz et al., 1988; Buckland and Wild, 1989; McCormack et al., 1989; White and Weber, 1989), suggesting that it may be able to dimerize with other proteins, though further studies of *Uro-2* have required generation of polyclonal antibodies to the protein encoded by the gene. *Uro-2* mRNA expression is restricted to oogenesis (Swalla et al., 1993; Table 1), making gene manipulation difficult. It is likely that *Uro-2* is activated during oogenesis and that the protein is translated and localized in the egg, later affecting development of the larva. This possibility will be investigated with specific antibodies, which are currently being prepared and characterized (W. R. Jeffery, *personal communication*).

The third gene isolated in an initial subtractive screen, *Uro-11* or *manx*, has a nuclear localization signal (Kalderon et al., 1984) plus a zinc finger similar to other known transcription factors (Vallee et al., 1991; Swalla et al., 1993; Table 1). This suggests that *manx* is a nuclear protein and may be a transcription factor (Reddy et al., 1992). *Manx* expression is high during oogenesis in the tailed species, *M. oculata*, but is low during oogenesis of the anural species, *M. occulta* (Swalla et al., 1993; Table 1). In addition, *manx* also has an early embryonic expression period, increasing just before gastrulation, high in the ectoderm and caudal (tail) mesoderm during gastrulation, then decreasing at neurulation until transcripts are no longer detected at the tailbud stage (Swalla et al., 1993). Expression is seen primarily in the neural tube during neurulation and is not seen during brain pigment cell formation, but the expression period coincides with the time when inductive interactions important in tissue specification are known to occur (Nishida and Satoh, 1989). The timing and location of *manx* expression in the presumptive muscle and notochord cells of the future tail also suggest a role for this gene during inductive processes occurring in notochord (Nakatani and Nishida, 1994) and secondary muscle cells (Venuti and Jeffery, 1989) during gastrulation. The tissues expressing *manx* are those that will be undergoing convergence and extension to form the larval tail, and the neural tube, which develops into the brain as well as the dorsal nerve chord of the tail (see Swalla, 1993, for review).

Further evidence suggesting a causal role for *manx* in the formation of these structures are experiments inhibiting zygotic expression of *manx* with antisense oligodeoxynucleotides, which also prevented the formation of otoliths and tails in hybrid embryos (Swalla and Jeffery, 1994, 1995). *Molgula oculata* and *M. occulta* are able to develop into hybrid embryos if cross-fertilized in the laboratory (Swalla and Jeffery, 1990). The phenotype of the larvae that develop is highly dependent on the egg. Tailed species' eggs develop into normal chordate larvae, whether the sperm is from the tailed or anural species. However, in the reciprocal cross (anural *M. occulta* eggs fertilized with tailed *M. oculata* sperm), in some cases the hybrid larvae show chordate phenotypes, including the formation of a short tail and a brain pigment cell (Swalla and Jeffery, 1990). These hybrid larvae develop from eggs with slight p58 localizations and are dependent on an active *M. oculata* genome to develop the chordate features (Jeffery and Swalla, 1992b). When the expression of the *Uro* genes was examined in hybrid embryos with small tails and pigment spots, only *Uro-11* (*manx*) showed high levels of expression (Swalla and Jeffery, 1994, 1995). The expression of *manx* in hybrid embryos was similar to the zygotic expression of *manx* in *M. oculata* embryos (Swalla et al., 1993); transcripts were found during gastrulation in the ectoderm and presumptive tail mesoderm, transiently in the neural tube, and were not detectable by the time of tail formation. Treatment of hybrid cultures with antisense oligodeoxynucleotides to *manx* RNA resulted in greatly reduced amounts of *manx* RNA detected by *in situ* hybridization (Swalla and Jeffery, 1994). The antisense-treated hybrid cultures also contained very few hybrid embryos that developed short tails and brain pigment spots. Cultures treated with sense oligodeoxynucleotides to *manx* RNA contained high levels of *manx* RNA at gastrulation, and later a high percentage developed into hybrid larvae with short tails and brain pigment spots (Swalla and Jeffery, 1994). Fluorescent oligodeoxynucleotides were used to show that the embryos readily absorbed the s-modified oligos (both sense and antisense) from the seawater, and that the oligos persisted for up to 5 hours in the cytoplasm of the embryos.

Because the *manx* gene has been shown to have a dramatic effect on the phenotype of ascidian larvae, it is important to examine the gene structure in the two species used in the experiments above. When the *manx* gene was isolated and sequenced in both *Molgula oculata* and *M. occulta,* it was found to contain a second gene within one of the *manx* introns, an RNA helicase (Pederson et al., 1994). Surprisingly, even though the coding sequence for *manx* was divergent between the two species, there were no stops found in the coding sequence in *M. occulta,* which does not normally express *manx*. Further analysis of the *manx* gene will be necessary to determine why the gene is not expressed in *M. occulta* eggs and embryos, and whether *manx* is also silenced in the other anural Molgulid species. We are also currently investigating the importance of the cortical cytoplasm in the localization and maintenance of a cortical complex that may normally include one or more of the Uro proteins made during oogenesis.

ACKNOWLEDGMENTS

Many thanks are due to Dr. Joan D. Ferraris and Dr. Stephen R. Palumbi for organizing the ASZ symposium in St. Louis and completing this volume. I also would like to thank Dr. William R. Jeffery, who collaborated on all the studies done on this system, for his help and support. Many of the experiments described were done in his laboratory at Bodega Marine Laboratory, the University of California at Davis, or the Station Biologique in Roscoff, France. The collection staff at Station Biologique, Roscoff, is thanked for their assistance in obtaining *M. occulta* and *M. oculata* and Dr. Laurent Meijer is thanked for the generous use of his laboratory at Roscoff. Many thanks are also due to the Director of the Station Biologique, Dr. Andre Toulmond, and Nicole Sanseau for their help in obtaining living and laboratory space at the Station Biologique. Tony Carroll designed and executed the graphic artwork used in Figure 2. This paper is dedicated to my advisor and mentor, Dr. Michael Solursh. This research was supported by NSF grant IBN-9304958.

REFERENCES

Berrill NJ (1931): Studies in tunicate development. Part II. Abbreviation of development in the Molgulidae. *Philos Trans R Soc Lond B Biol Sci* 219:281–346.

Buckland R, Wild F (1989): Leucine zipper motif extends. *Nature* 338:547.

Chan TA, Chu CA, Rauen KA, Kroiher M, Tatarewicz SM, Steele RE (1994): Identification of a gene encoding a novel protein-tyrosine kinase containing SH$_2$ domains and ankyrin-like repeats. *Oncogene* 9:1253–1259.

Cloney RA (1964): Development of the ascidian notochord. *Acta Embryol Morphol Exp* 7:111–130.

Ferrante AW Jr, Reinke R, Stanley ER (1995): Shark, a Src homology 2, ankyrin repeat, tyrosine kinase, is expressed on the apical surfaces of ectodermal epithelia. *Proc Natl Acad Sci* 92:1911–1915.

Hadfield KA, Swalla BJ, Jeffery WR (1995): Multiple origins of anural development in ascidians inferred from rDNA sequences. *J Mol Evol* 40:413–427.

Jeffery WR, Swalla BJ (1990): Anural development in ascidians: evolutionary modification and elimination of the tadpole larva. *Semin Dev Biol* 1:253–261.

Jeffery WR, Swalla BJ (1992a): Evolution of alternate modes of development in ascidians. *Bioessays* 14:219–226.

Jeffery WR, Swalla BJ (1992b): Factors necessary for restoring an evolutionary change in an anural ascidian embryo. *Dev Biol* 153:194–205.

Jeffery WR, Swalla BJ (1993): An ankyrin-like protein in ascidian eggs and its role in the evolution of direct development. *Zygote* 1:197–208.

Kalderon D, Roberts BL, Richardson WD, Smith AE (1984): A short amino acid sequence able to specify nuclear localization. *Cell* 39:499–509.

Landschultz WH, Johnson P, McKnight SL (1988): The leucine zipper: a hypothetical structure common to a new class of DNA binding proteins. *Science* 240:1759–1764.

Makabe KW, Swalla BJ, Reardon JA, Satoh N, Jeffery WR (1996): Characterization of *cymric,* a new member of the *shark* family of non-receptor tyrosine kinases, which is expressed differentially in ascidians with alternate modes of development. (*in preparation*).

March CJ, Mosley B, Larsen A, Cerretti DP, Braedt G, Price V, Gillis S, Henney CS, Kronheim SR, Grabstein K, Conlon PJ, Hopp TP, Cosman D (1985): Cloning, sequence and expression of two distinct human interleukin-1 complementary DNAs. *Nature* 315:641–647.

McCormack K, Campanelli JT, Ramaswami M, Mathew MK, Tanouye MA, Iverson LE, Rudy B (1989): Leucine-zipper motif update. *Nature* 340:103.

Michaely P, Bennett V (1992): The ANK repeat: a ubiquitous motif involved in macromolecular recognition. *Trends Cell Biol* 2:127–129.

Miyamoto DM, Crowther RJ (1985): Formation of the notochord in living ascidian embryos. *J Embryol Exp Morphol* 86:1–17.

Nakatani Y, Nishida H (1994): Induction of notochord during ascidian embryogenesis. *Dev Biol* 166:289–299.

Nishida H (1992): Developmental potential for tissue differentiation of fully dissociated cells of the ascidian embryo. *Roux's Arch Dev Biol* 201:81–87.

Nishida H, Satoh N (1989): Determination and regulation in the pigment cell lineage of the ascidian embryo. *Dev Biol* 132:355–367.

Pederson EL, Swalla BJ, Just ML, Jeffery WR (1994): An expressed RNA helicase gene is embedded in *manx*, a gene required for development of the ascidian tailed larva. *Mol Biol Cell Suppl* 5:225a(#1312).

Raff RA (1987): Constraint, flexibility, and phylogenetic history in the evolution of direct development in sea urchins. *Dev Biol* 119:6–19.

Reddy BA, Etkin LD, Freemont PS (1992): A novel zinc finger coiled-coil domain in a family of nuclear proteins. *Trends Biochem Sci* 17:344–345.

Sambrook J, Fritsch EF, Maniatis T (1989): *Molecular cloning: a laboratory manual*, 2nd ed., Cold Spring Harbor, NY: Cold Spring Harbor Laboratory Press.

Satoh N, Deno T, Nishida H, Nishikata T, Makabe KW (1990): Cellular and molecular mechanisms of muscle cell differentiation in ascidian embryos. *Int Rev Cytol* 122:221–258.

Sive HL, St John T (1988): A simple subtractive hybridization technique employing photoactivatable biotin and phenol extraction. *Nucleic Acids Res* 16:10937.

Swalla BJ (1992): The role of maternal factors in ascidian muscle development. *Semin Dev Biol* 3:287–295.

Swalla BJ (1993): Mechanisms of gastrulation and tail formation in ascidians. *Microsc Res Tech* 26:274–284.

Swalla BJ, Jeffery WR (1990): Interspecific hybridization between an anural and urodele ascidian: differential expression of urodele features suggests multiple mechanisms control anural development. *Dev Biol* 142:319–334.

Swalla BJ, Jeffery WR (1994): Expression of the *manx* gene is required for specifying the larval body plan in ascidians. *Mol Biol Cell Suppl* 5:225a(#1313).

Swalla BJ, Jeffery WR (1995): The *manx* gene is required for the restoration of chordate features in ascidian larvae. *Nature* (*submitted*).

Swalla BJ, Badgett MR, Jeffery WR (1991): Identification of a cytoskeletal protein localized in the myoplasm of ascidian eggs: localization is modified during anural development. *Development* 111:425–436.

Swalla BJ, Makabe KW, Satoh N, Jeffery WR (1993): Novel genes expressed differentially in ascidians with alternate modes of development. *Development* 119:307–318.

Timblin C, Battey J, Kuehl WM (1990): Amplification for PCR technology to subtractive cDNA cloning: identification of genes expressed specifically in murine plasmacytoma cells. *Nucleic Acids Res* 18:1587–1593.

Vallee BL, Coleman JE, Auld DS (1991): Zinc fingers, zinc clusters, and zinc twists in DNA binding proteins. *Proc Natl Acad Sci USA* 88:999–1003.

Van Gelder RN, von Zastrow ME, Yool A, Denent WC, Barchas JD, Eberwine JH (1990): Amplified RNA synthesized from limited quantities of heterogeneous cDNA. *Proc Natl Acad Sci USA* 87:1663–1667.

Venuti JM, Jeffery WR (1989): Cell lineage and determination of cell fate in ascidian embryos. *Int J Dev Biol* 33:197–212.

White MK, Weber MJ (1989): Leucine-zipper motif update. *Nature* 340:103–104.

CHAPTER 11

Using Nematode Vulva Development to Model the Evolution of Developmental Systems

RALF J. SOMMER* AND PAUL W. STERNBERG

Howard Hughes Medical Institute and Division of Biology 156-29, California Institute of Technology, Pasadena, California 91125

CONTENTS

SYNOPSIS

In the nematode model organism *Caenorhabditis elegans* developmental processes can be analyzed at various levels. The invariance of cell lineage allows a high-resolution morphological description of development and an experimental approach

*Present address: Max-Planck Institut für Entwicklungsbiologie, Spemannstrasse 35, 72076 Tübingen, Germany.

Molecular Zoology: Advances, Strategies, and Protocols, Edited by Joan D. Ferraris and Stephen R. Palumbi.
ISBN 0-471-14461-4 © 1996 Wiley-Liss, Inc.

by ablation of individual cells. Isolation and characterization of genetic mutations reveal the basis of the genetic program underlying particular developmental processes. DNA-mediated transformation can be used to analyze the function of cloned genes and thus finally generate also a molecular understanding of the process under investigation. Evolutionary approaches to development are rare and so far consist only of a subset of techniques used in the reference model system. By using the complete set of techniques from the model organism, including genetics, in related but morphologically distinct species, one can get a detailed comparison of a developmental process. Here we describe our attempt to establish the techniques known in the model organism *Caenorhabditis* of the Rhabditidae in other free-living nematodes, including *Pristionchus pacificus* of the Neodiplogastridae.

INTRODUCTION

All morphological structures are the end product of developmental processes; thus morphological changes during evolution arise by the modification of ontogeny. In spite of this obvious connection between evolution and development, the "synthetic theory of evolution" that has dominated evolutionary thinking over the last 50 years largely ignores developmental input on evolution (for more details see Raff and Kauffman, 1983; Buss, 1987; Thomson, 1988). Thus, during most of the twentieth century, the impact of the analysis of developmental processes on evolutionary biology has been minor.

Based on the genetic approach to development in a small number of genetic model organisms, especially *Drosophila melanogaster* and *Caenorhabditis elegans*, this picture has started to change over the last few years. A variety of developmental processes have been studied at the genetic and molecular levels, giving major insights into molecular interactions driving development. Having an understanding of the genetic program controlling ontogeny in special model organisms allows us to examine the evolution of these organisms from a new perspective. Increasingly, studies on this topic in recent years show surprising similarities between major groups of organisms (see Akam et al., 1994). Most developmentally important genes are conserved during evolution, and thus counterparts with sequence similarity can be found in many organisms. In addition, for some regulatory pathways functional similarities also are beginning to be uncovered (see Akam et al., 1994). Thus, at the molecular level, not only the basic biochemical machinery but also regulatory aspects of development are at least related among major organisms. These results reflect the relatedness of all eukaryotic organisms and establish a framework for the still unanswered major questions of developmental evolution: What causes the reorganization of phenotypes over evolutionary times? To what extent can a "universal pathway" be modified to underlie the diversity of development?

The known similarities at the molecular level cannot explain the generation of different morphologies. This paradox—astonishing similarity at the molecular level and broad diversity at the organismic level—is simply due to our limited genetic understanding of development in most organisms. Only in some model systems can we currently combine genetic and molecular analysis in the study of development. In contrast, evolutionary comparisons rely mostly on molecular data only. Thus a more profound analysis of the evolution of developmental processes and mechanisms is necessary to explain the known complexity at the morphological level. In principle, we can gain insight into this problem only by studying the differences among related organisms at the same taxonomic level and with the same techniques as to study the model organisms themselves.

Here we describe our attempts to establish such a situation by using free-living nematodes related to *Caenorhabditis elegans*. In the rhabditid nematode *C. elegans,* developmental processes can be studied at the cellular, genetic, and molecular levels (Brenner, 1974; Wood, 1988). A useful evolutionary analysis of a particular developmental process in other nematode species relies on the use of the same set of techniques as in *C. elegans*. We describe our current effort in some nematode species to establish all the methodological circumstances to fulfill these requirements.

RESULTS AND DISCUSSION

The Nematode Life Cycle

Nematodes can easily be collected from soil samples. In the laboratory, many of these species are cultured on agar petri plates with a lawn of *Escherichia coli* bacteria as food source, as it is used for *Caenorhabditis* (Sommer.1). Although morphologically very different from each other, free-living nematodes all have a very similar life cycle. In general, the life cycle can be divided into two phases, embryonic and postembryonic development (Chitwood and Chitwood, 1950). The number of larval stages is limited to four in most species (Chitwood and Chitwood, 1950). All major organs are generated during embryogenesis. The first stage (L1) larvae look like miniature adults, missing only the reproductive organs. An enormous increase in size occurs during the third and fourth larval stages (L3 and L4) and reflects the growth of the reproductive system.

Caenorhabditis elegans and *Pristionchus pacificus,* a species of the family of the Neodiplogastridae, which we will focus on during the paper, are protoandric self-fertilizing hermaphrodites. The adult hermaphrodites first produce a limited number of sperm and switch later to the development of oocytes. Males occur in normal populations only at a very low frequency. Crosses of males and hermaphrodites result in about 50% males.

One of the structures whose development has been studied intensively in *Caenorhabditis* at the experimental, genetic, and molecular levels is the vulva, the egg-laying system of nematode hermaphrodites or females. The vulva develops during the third larval stage (L3) and is seen as an invagination in the ventral epidermis of worms in the L3 and the early L4 stages.

Vulva Development in *C. elegans*

In *C. elegans,* as in all nematodes, the vulva is formed as part of the ventral epidermis and its development is easily accessible by cell lineage analysis (Sommer.2) (Sulston and Horvitz, 1977). Twelve epidermal precursor cells (P1.p to P12.p from anterior to posterior) generate the ventral hypodermis in a position-specific manner (Fig. 1A). Six of the central cells, P(3–8).p, are tripotent vulva precursor cells (VPCs) and form an equivalence group. In intact worms, P(5–7).p respond to an inductive signal from the gonadal anchor cell (AC) and divide to generate the 22 nuclei that finally form the vulva (Figs. 1B and 2A) (Kimble, 1981). By cell lineage and cell ablation experiments we can distinguish three different cell fates among the VPCs. P6.p has the primary cell fate and generates eight progeny; P5.p and P7.p have the secondary cell fate and form only seven progeny (Fig. 2A) (Sternberg and Horvitz, 1986). The three other cells have a nonvulval fate, named tertiary, and generate epidermis (Fig. 2A).

Genetic studies helped in understanding basic features of these developmental processes. In *C. elegans,* mutations can easily be obtained for every screenable phenotype. Concerning the vulva, two opposite phenotypes that are easily recognizable under the dissecting microscope have been described: in some mutations a normal vulva cannot be formed. These "vulvaless" animals cannot lay eggs so that the larvae hatch inside their mother. In "multivulva" animals, excess vulva differentiation occurs because not only three but all six potential VPCs divide and generate vulval tissue.

Genetic studies characterized at least three intercellular signals involved in vulva formation. In addition to the inductive signal from the gonadal anchor cell (AC), a lateral signal between neighboring VPCs and a negative signal from the surrounding hypodermis have been identified (see Sternberg, 1993, for a review). DNA-mediated transformation was used to clone several genes involved in vulva development. The genes in the inductive signaling process are very similar to genes involved in oncogenesis in mammals or a variety of developmental processes in *Drosophila* (Fig. 1C) (see Sternberg, 1993, for a review). The inductive signal is encoded by the *lin-3* gene and is an epidermal growth factor (EGF)-like molecule (Hill and Sternberg, 1992). This positional information provided by the localized inductive signal is transmitted via a signal transduction pathway comprising an EGF-receptor like tyrosine kinase, a Ras protein, and a Raf-like serine–threonine kinase (Fig. 1C; for details, see below) (see Sternberg, 1993, for a review).

The Evolution of Vulval Cell Lineages in Different Nematode Species

The basis for an evolutionary analysis of development processes is a detailed description of morphology. Since free-living nematodes have a small and typically invariant number of cells, the anatomy of organs can be determined relatively easily and with cellular resolution. The development can be analyzed readily at the level of individual cell divisions and cell fates. We have analyzed the cell lineages and pattern formation in the vulva equivalence group among several species belonging

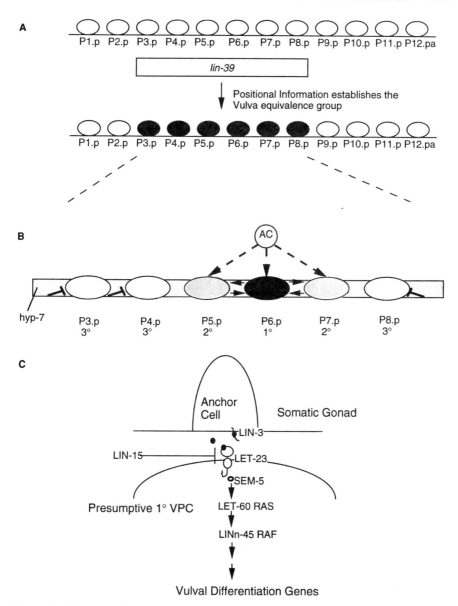

Figure 1. Schematic of the different specification events during vulva development. (A) *lin-39*, one of the genes of the *Hom-C* complex, contributes positional information to the specification of the vulva equivalence group (black ovals). (B) An inductive signal (dashed arrows) from the gonadal anchor cell initiates cell fate specification within the vulva equivalence group. Lateral signaling (plain arrows) between neighboring VPCs is involved in fate specification as well. The hypodermal syncytium is indicated in the grey box and is involved in negative signaling (T-bars). (C) Molecular description of the inductive signaling event via a signal transduction pathway. See text for further details.

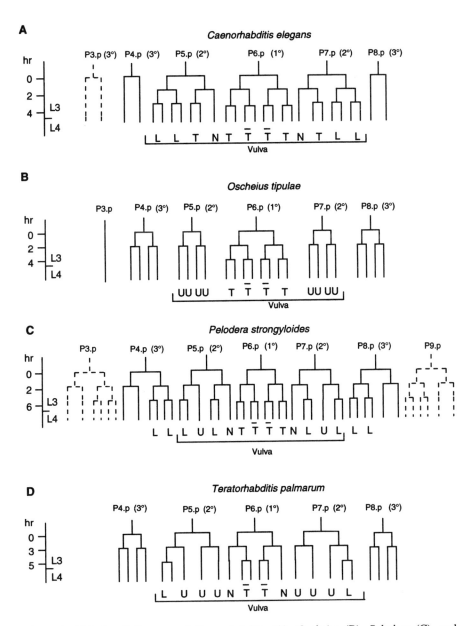

Figure 2. Vulval cell lineages of *Caenorhabditis* (A), *Oscheius* (B), *Pelodera* (C), and
Teratorhabditis (D). The vulva is formed by P(5–7).p in all these species. P6.p has the 1° cell
fate, P5.p and P7.p have the 2° cell fate, and P4.p and P8.p have the 3° cell fate. In addition,
P3.p and/or P9.p can adopt the 3° fate in a variable, but species-specific, manner. Dashed
lines indicate variable cell divisions. The exact lineage of all three cell fates is species-
specific. L, U, T, and N are descriptors that characterize the third round of cell division more
precisely. L, longitudinal division; T, transverse division; U, undividing cell; N, no division,
but compact nucleolus. Terms are used according to the definition of Sternberg and Horvitz
(1986) and Sommer and Sternberg (1995). Overlines indicate cells that attach to the anchor
cell during morphogenesis of the egg-laying system.

to different families of nematodes (Sternberg and Horvitz, 1982; Sommer and Sternberg, 1994, 1995). Concerning the ventral epidermis, homologous epidermal precursor cells can be found at the end of embryogenesis in all these species. The position-dependent development of these cells can then be studied during larval development.

Most species form the vulva in the central body region, like *Caenorhabditis*. Among these species we find three major evolutionary alterations in cell lineage. First, the number of cells constituting the vulval equivalence group differs; that is, different numbers of cells have the potential to generate vulval cells. Furthermore, the actual cell lineage of a particular vulval precursor cell fate differs among most species. Figure 2 shows the vulval cell lineages of *Caenorhabditis* (Fig. 2A) in comparison to a less complex lineage observed in the species *Oscheius* (Fig. 2B), and to a more complex lineage of the species *Pelodera* (Fig. 2C).

Second, different types of asymmetric vulva cell lineages have been observed (i.e., 2° lineage of *Caenorhabditis* and 2° and 3° lineage of *Pelodera*, Fig. 2A,C). Cell ablation experiments suggest that these asymmetries are oriented by the gonad in interactions different from vulva induction (Sommer and Sternberg, 1995; W. Katz and P. Sternberg, *in preparation*). Finally, variability among individuals of the same species has been observed in two different circumstances. In *Pelodera*, the number of cells having the tertiary cell fate is variable among individuals (Fig. 2B). In *Protorhabditis*, variability of the position of the AC is correlated with an anterior shift of cells generating the vulva (Fig. 3B). Although the canonical lineage in *Protorhabditis* is similar to the vulva lineage in *Caenorhabditis*, a frequent number of *Protorhabditis* animals show the AC dorsal to P5.p instead of P6.p. These animals have a vulva formed by the progeny of P(4–6).p, thus an anterior shift of cells generating the vulva. In rare animals, more than the usual three VPCs formed vulval tissue.

This variability of AC position and the correlated anterior shift of the cells generating vulval tissue are phenotypically similar to animals described for the *lon-1* mutation in *Caenorhabditis* (Fig. 3B) (Sternberg and Horvitz, 1986). The *lon-1* hermaphrodites are up to 50% longer than wild-type (Brenner, 1974) and have occasionally vulvae generated by VPCs other than P(5–7).p. The majority of these animals also have the AC dorsal to P5.p and a vulva formed by P(4–6).p.

A different AC position is also known in *Panagrellus redivivus* of the family Panagrolaimidae (Sternberg and Horvitz, 1982). In *Panagrellus* the vulva is formed by four instead of three VPCs, and these are P(5–8).p (Fig. 3C). Two VPCs have the primary cell fate, P6.p and P7.p, whereas P5.p and P8.p have the secondary fate. The AC is located between P6.p and P7.p. We speculate that this, presumably derived, character evolved from a situation with a variable AC position, like the ones observed in *Protorhabditis* or the *lon-1* mutation in *Caenorhabditis*.

The vulva in *Panagrellus* is located at 60% body length, slightly more posterior than the usual 50% body length of species with a central vulva. Other species, such as *Mesorhabditis* of the family Rhabditidae, form the vulva far more posterior at 80% body length (Fig. 3D). In *Mesorhabditis*, the vulva is formed by VPCs originally located in the central body region (Sommer and Sternberg, 1994). During the L2 stage P(4–8).p migrate toward the posterior and in an intact animal the vulva is

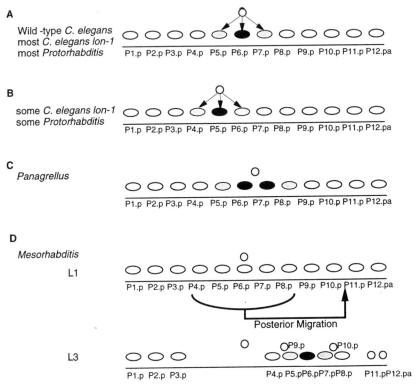

Figure 3. Schematic of various examples of differences in AC position. (A) *Caenorhabditis* wild-type, most *Protorhabditis* animals, most *Caenorhabditis lon-1* mutant animals. The AC is located dorsal to P6.p. (B) Some *Protorhabditis* animals, some *Caenorhabditis lon-1* animals. The AC is located dorsal to P5.p. This variability of AC position correlates with an anterior shift of cells generating the vulva. (C) *Panagrellus* has a vulva generated by P(5–8).p. The AC is located at a position between P6.p and P7.p, which both have the 1° cell fate. (D) *Mesorhabditis* has a posterior vulva that is formed by P(5–7).p after migration toward the posterior during the second larval stage. The AC is not involved in vulva induction in this species.

formed by P(5–7).p, like in species with a central vulva. A dramatic evolutionary change occurs in this species, because the vulva is no longer induced by the gonadal AC, and furthermore, the VPCs are not equivalent to each other in their potentials to generate vulval cells. Thus, in some species with a posterior vulva, changes of induction and competence occurred during evolution (Sommer and Sternberg, 1994).

Molecular Biology of Vulva Development Reveals Conservation of Developmental Genes

The molecular understanding of vulva development in *C. elegans* increased intensively over the last few years. In principle, several different specification events can be distinguished.

The *Hom-C* genes provide the key information establishing the position-specific differences among the P-ectoblasts along the anterior–posterior axis (see Kenyon, 1994, for a review). One of the genes in the *Hom-C* cluster, *lin-39*, determines the vulva equivalence group (Fig. 1A) (Clark et al., 1993; Wang et al., 1993). Loss-of-function mutations of the *lin-39* gene confer a vulvaless phenotype because the vulval equivalence group cannot be established. Specifically, P(3–8).p adopt the fate of the more anterior and posterior Pn.p cells; which differentiate into non-specialized epidermis without division.

Once the equivalence group is established, the gonadal AC induces the formation of the vulva (Fig. 1B). As mentioned above, the signaling pathway involved in transmitting this signal comprises an EGF-like molecule, a growth factor (LIN-3), a transmembrane receptor with presumed tyrosine kinase activity (LET-23), an adaptor protein with src-homology-2 and src-homology-3 domains (SEM-5), a ras protein (LET-60), a raf-like serine–threonine kinase (LIN-45), and several downstream kinases (see Sternberg, 1993, for a review). Lateral signaling between the VPCs involves LIN-12, a Notch-like cell surface receptor (see Greenwald, 1994, for a review). A negative signal from the surrounding hypodermis limits vulva development and involves the activity of LIN-15 (Herman and Hedgecock, 1990; Huang et al., 1994).

We have started to analyze the conservation of some of the major molecules involved in vulva development using PCR and genomic or cDNA cloning with the fragments originally obtained by PCR (Sommer.3). The homeobox fragments of some of the *Hom-C* genes have been cloned from other nematode species by PCR (R. J. Sommer and P. W. Sternberg, *unpublished observations*). These fragments are highly conserved, but, surprisingly, new introns were found within the homebox domain in species-specific positions (R. J. Sommer and P. W. Sternberg, *unpublished observations*).

Within the vulva signaling pathway the *let-60* ras gene plays an important role as a molecular switch. Gain-of-function mutations in *let-60* ras cause multivulva phenotypes in *Caenorhabditis,* whereas loss-of-function alleles cause vulvaless phenotype (Beitel et al., 1990; Han and Sternberg, 1990). Furthermore, overexpression of *let-60* ras also results in multivulva animals. We have cloned a ras-homologue from *Pristionchus* by PCR, again finding an intron in a position that does not contain an intron in *Caenorhabditis.* The amino acid sequence comparison of this PCR fragment in comparison to the corresponding region of the *Caenorhabditis, Drosophila,* and human ras gene indicates high sequence similarities (Fig. 4). We interpret the extent of amino acid divergence between *Caenorhabditis* and *Pristionchus* to represent roughly 100 million years. Furthermore, the *Pristionchus* ras gene revealed additional changes of the exon–intron structure of the gene (R. J. Sommer and P. W. Sternberg, *unpublished observations*).

Genetic Analysis

The application of genetics in the study of development was very fruitful over the last 30 years (Gilbert, 1994). Isolation and characterization of mutations followed by molecular cloning gave major insights into the genetic program controlling

```
Caenorhabditis    115   N K C D L S S R S V D F R Y V S E T A K G   136
Pristionchus            - - - - - A G - A - - S - V - Q D - - R A
Human                   - - - - - P T - T - - T K Q A H - L - - S
Drosophila              - - - - - A - W N - N N E Q A R - V - - Q
Yeast                   - - S - - E N E K Q V S Y QDG L N M - - Q


Caenorhabditis    137   Y G I P N V D T S A K T R M G V D E A F   156
Pristionchus            - - - - E - - - - - - - - - - - - D - -
Human                   - - - - F I E - - - - - - Q - - E D - -
Drosophila              - - - - Y I E - - - - - - - - - - D - -
Yeast                   M N A - F L E - - - - Q A I N - E - - -
```

Figure 4. Amino acid sequence of a *let-60* ras fragment of *Pristionchus*, cloned by PCR. The fragment spans the region from amino acids 115 to 156 in Han and Sternberg (1990). The comparison contains the corresponding sequence of the *Caenorhabditis*, *Drosophila*, human, and yeast ras gene. See Han and Sternberg (1990) for references.

development. A comparative evolutionary–developmental approach is best if the same set of techniques can be applied in all species under investigation. Thus genetics, as a key technique in the model systems, has also to be used in the other species under investigation.

Nematodes other than *Caenorhabditis* are an attractive system for genetic studies because many species are bacterial feeders and thus easily culturable (Sommer et al., 1994). Furthermore, a number of species have a fast life cycle and a small number of chromosomes, which are both requirements for genetic analysis. As an example, *Pristionchus* has a life cycle of 3 days and contains only six chromosomes, as does *Caenorhabditis;* thus it fulfills requirements for a genetic analysis (R. J. Sommer, L. Carta, and P. W. Sternberg, *in preparation*). In a pilot screen of just several thousand genomes, we were able to isolate mutations with phenotypes similar to mutations isolated from *Caenorhabditis*. By screening 3000 genomes we isolated three dumpy (Dpy) mutants, three uncoordinated (Unc) mutations, and nearly twenty egg-laying defective (Egl) mutations (R. J. Sommer and P. W. Sternberg, *unpublished observations*). Dpy mutants are smaller in size than wild-type worms, Unc mutants show various movement defects, and Egl mutants can be defective in various aspects of egg laying including defects in the development of the vulva. Two of the Egl mutants isolated in the pilot screen have a vulvaless phenotype similar to the corresponding mutants described for *Caenorhabditis* (R. J. Sommer and P. W. Sternberg, *unpublished observation*). Thus, in nematode species belonging to families different from the Rhabditidae, the isolation of mutants is easily accessible and can be used in the future to address developmental questions, like the generation of vulva development.

Transgenic Technology

One additional technique important in *Caenorhabditis* developmental biology is DNA-mediated transformation (Sommer.4). Rescuing a mutation by injection of genomic DNA containing the wild-type copy of the corresponding gene is the first step toward cloning a gene in *Caenorhabditis* (Sommer.5). Furthermore, once a gene is cloned, DNA-mediated transformation experiments can be used to overexpress or misexpress a gene under the control of a heat-shock promoter (Sommer.5). With these types of experiment genes can be activated at different times during development or in body regions, where the gene is not expressed normally.

We are currently trying to establish transgenic technology in nematode species other than *Caenorhabditis* using the technology described by Mello et al. (1991). A first experiment in *Pristionchus* using a construct containing a human heat shock promoter fused to the *lacZ* gene of *E. coli* was successful, indicating that transgenic technology can be transferred to other nematode species (R. J. Sommer and P. W. Sternberg, *unpublished observations*). We are currently searching for the best markers to use for cross-species transformation, or endogenous markers.

ACKNOWLEDGMENTS

We thank our many colleagues for discussion on the evolution of nematode development and L. Carta and M. Felix for comments on the manuscript. This research was supported by an NSF Presidential Young Investigator Award to P.W.S., an investigator of the Howard Hughes Medical Institute; and by an EMBO long-term fellowship to R.J.S.

REFERENCES

Akam M, Holland P, Wray G (eds) (1994): Evolution of developmental mechanisms. *Dev Suppl*

Avery L, Horvitz HR (1987): A cell that dies during wild-type *C. elegans* development can function as a neuron in a ced-3 mutant. *Cell* 51:1071–1078.

Beitel GJ, Clark SG, Horvitz HR (1990): *Caenorhabditis elegans ras* gene *let-60* acts as a switch in the pathway of vulval induction. *Nature* 348:503–509.

Brenner S (1974): The genetics of *Caenorhabditis elegans*. *Genetics* 77:71–94.

Buss L (1987): *Evolution of individuality.* Princeton NJ: Princeton University Press.

Chitwood BG, Chitwood MB (1950): *Introduction to nematology.* Baltimore: University Park Press.

Clark SG, Chisholm AD, Horvitz HR (1993): Control of cell fates in the central body region of *C. elegans* by the homeobox gene *lin-39*. *Cell* 74:43–55.

Fire A (1992): *GATA* 9:151–158.

Gilbert SF (1994): *Developmental biology.* Sunderland, MA: Sinauer Associates.

Greenwald I (1994): Structure/function studies of *lin-12/Notch* proteins. *Curr Opin Genet Dev* 4:556–562.

Han M, Sternberg PW (1990): *let-60*, a gene that specifies cell fates during *C. elegans* vulval induction encodes a *ras* protein. *Cell* 63:921–931.

Herman RK, Hedgecock EM (1990): The size of the *C. elegans* vulval primordium is limited by *lin-15* expression in surrounding hypodermis. *Nature* 348:169–171.

Hill RJ, Sternberg PW (1992): The *lin-3* gene encodes an inductive signal for vulval development in *C. elegans. Nature* 358:470–476.

Huang LS, Tzou P, Sternberg PW (1994): The *lin-15* locus encodes two negative regulators of *C. elegans* vulval development. *Mol Biol Cell* 5:395–412.

Kenyon C (1994): If birds can fly, why can't we? Homeotic genes and evolution. *Cell* 78:178–180.

Kimble J (1981): Lineage alterations after ablation of cells in the somatic gonad of *Caenorhabditis elegans. Dev Biol* 87:286–300.

Mello CC, Fire A (1995): In Skates A, Epstein H (eds). *Methods in cell biology: C. elegans.* San Diego, CA: Academic Press.

Mello CC, Kramer JM, Stinchcomb D, Ambros V (1991): Efficient gene transfer in *C. elegans:* extrachromosomal maintenance and integration of transforming sequences. *EMBO* 10:3959–3970.

Raff RA, Kaufman TC (1983): *Embryos, genes, and evolution.* Bloomington: Indiana University Press.

Sambrook J, Fritsch EF, Maniatis T (1989): *Molecular cloning: a laboratory manual,* 2nd ed. Cold Spring Harbor, NY: Cold Spring Harbor Laboratory Press.

Sommer RJ, Sternberg PW (1994): Changes of induction and competence during the evolution of vulva development in nematodes. *Science* 265:114–118.

Sommer RJ, Sternberg PW (1995): Evolution of cell lineage and pattern formation in the vulval equivalence group of Rhabditid nematodes. *Dev Biol* 167:61–74.

Sommer RJ, Carta LK, Sternberg PW (1994): The evolution of cell lineage in nematodes. *Dev Suppl* 85–95.

Sternberg PW (1993): Intercellular signaling and signal transduction in *C. elegans. Annu Rev Genet* 27:551–574.

Sternberg PW, Horvitz HR (1982): Postembryonic nongonadal cell lineages of the nematode *Panagrellus redivivus:* description and comparison with those of *Caenorhabditis elegans. Dev Biol* 93:181–205.

Sternberg PW, Horvitz HR (1986): Pattern formation during vulval development in *C. elegans. Cell* 44:761–772.

Stringham EG, et al (1982): *Mol Biol Cell* 3:221–233.

Sulston JE, Horvitz HR (1977): Postembryonic cell lineage of the nematode *Caenorhabditis elegans. Dev Biol* 56:110–156.

Thomson KS (1988): *Morphogenesis and evolution.* New York: Oxford University Press.

Wang BB, Mueller-Immergluck MM, Austin J, Robinson NT, Chisholm A, Kenyon C (1993): A homeotic gene cluster patterns the anteroposterior body axis of *C. elegans. Cell* 74:29–42.

Wood WB (ed) (1988): *The nematode Caenorhabditis elegans.* Cold Spring Harbor, NY: Cold Spring Harbor Laboratory Press.

CHAPTER 12

The Analysis of Lineage-Specific Gene Activity During Sea Urchin Development

R. ANDREW CAMERON, ROBERT W. ZELLER,* JAMES A. COFFMAN, and ERIC H. DAVIDSON

Division of Biology 156-29, California Institute of Technology, Pasadena, California 91125

CONTENTS

SYNOPSIS

The experimental approach presented here aims to understand the mechanisms by which differential genomic expression is first established in the various spatial

*Present address: Department of Biology, University of California–San Diego, 9500 Gilman Drive, La Jolla, California 92093.

Molecular Zoology: Advances, Strategies, and Protocols, Edited by Joan D. Ferraris and Stephen R. Palumbi.

domains of the sea urchin embryo. In terms of regulatory organization, the sea urchin and most invertebrate taxa share a general form of embryogenesis in which cell lineage plays an important role. Regions upstream of genes whose expression patterns identify particular embryonic territories are used to construct *in vitro* reporter genes, which are easily injected into sea urchin embryos. Various reporter gene constructs are utilized to identify regulatory domains of the territorial marker genes. By experimentally altering the sequences in the regulatory domains, we can ascertain the functional relationship among the components. Because sea urchin embryos are easily obtained in large quantity, a straightforward strategy for the biochemical isolation of their nuclear proteins has proved successful. Techniques including affinity chromatography, gel mobility assays, and protein gel blots probed with DNA fragments permit the direct isolation and identification of proteins that specifically bind to segments of the regulatory domains. When interpreted against a background of cell lineage and fate, the mechanisms of transcriptional regulation thus revealed produce a vertically integrated explanation of blastomere specification in the early sea urchin embryo.

INTRODUCTION

We are interested in the mechanisms by which differential genomic expression is first established in the various spatial domains of the early sea urchin embryo. The rapidly dividing cells of the embryo manifest divergent phenotypes both through regional localization or activation of maternal gene products stored in the egg and through spatially and temporally regulated zygotic gene expression. Furthermore, development is often regulated by interactions between differentially specified, phenotypically divergent cell types. Thus small changes in the regulation of genes involved in cell specification can have a profound effect on phenotype, and understanding the function and regulation of such genes has obvious implications for the study not only of development but of evolution as well.

Gene expression is, in general, a function of transcriptional regulation, which results from the combinatorial binding and interaction of sequence-specific DNA binding proteins with the regulatory region of a gene. A number of reviews have been written (e.g., Ptashne, 1988; Mitchell and Tjian, 1989; Clark and Doherty, 1993; Kingston and Green, 1994) that describe in detail how these proteins bind to DNA and interact to either activate, modulate, or repress transcription. In an embryo, such combinatorial mechanisms come into play in both the spatial and temporal regulation of gene expression. Thus one route to understanding the establishment of territory-specific patterns of gene expression at a molecular level lies in the analysis of interactions between specific DNA-binding proteins and regulatory sequences controlling genes expressed in the early embryo.

In terms of the regulatory organization, there are three major types of embryogenesis: *Type 1,* a general form in which lineage plays an important role in the spatial organization of early embryos, and cell specification occurs *in situ,* by both autonomous and conditional mechanisms; *Type 2*, the vertebrate form of embryo-genesis, which proceeds by mechanisms that are essentially independent of cell lineage, in which diffusible morphogens and extensive early cell migration are particularly important; *Type 3*, the form exemplified by long germ band insects in which several different regulatory mechanisms are used to generate precise patterns of nuclear gene expression prior to cellularization (reviewed in Davidson, 1991). The sea urchin and most of the invertebrate taxa undergo *Type 1* embryogenesis. The distinctive suite of key developmental features, which are shared by these disparate forms, is as follows. (1) The cleavage of each species is invariant so that the cell lineages descended from specific blastomeres display predictable fates. Furthermore, the early cell divisions segregate unique groups of founder cells, which become the primordia for the exclusive territories of gene expression. The implication is that the founder cells constitute spatial domains of egg cytoplasm within which specific sets of genes will be active. (2) One axis of the egg is preformed while the second is set up after fertilization. (3) Where studied with molecular markers, specification of cell types and the appearance of an initial set of differentiated cell types are found to occur before gastrulation and to precede any large-scale cell migrations. (4) Some founder cells for the lineages found in the early embryo are specified autonomously, while others are specified conditionally (see Davidson, 1989, 1990, 1991).

Comparative considerations of development can be immensely illuminating and were often the subject of embryological research over the past century. As the superb molecular technology of today carries us toward a more thorough under-standing of transcriptional regulation in developmental systems such as the sea urchin, opportunities for comparison among the invertebrate groups emerge at a rapid rate. Here we describe the approach we use to understand spatially restricted gene expression in the sea urchin embryo.

SEA URCHIN EMBRYOS HAVE SPATIALLY RESTRICTED GENE EXPRESSION

In the sea urchin embryo the invariant cleavage pattern leads to the establishment of five territories—the aboral ectoderm, the oral ectoderm, the vegetal plate, the skeletogenic mesenchyme, and the small micromeres (Cameron et al., 1987; Cam-eron and Davidson, 1992; Fig. 1). Each territory is defined by its unique cell lineage, an individual pattern of gene expression, and the presence of one or more distinct cell types (Davidson, 1989, 1990). As cleavage proceeds, the fate of the blastomeres becomes progressively restricted until a given blastomere will only produce cells of a single territory. At this point, the blastomere is said to become a "founder" cell for that territory. The descendants of a group of founder cells there-fore compose a territory within which the molecular interactions that regulate spe-cific gene expression are established (Cameron and Davidson, 1992). Only the skeletogenic mesenchyme territory is autonomously specified; that is, the cells can

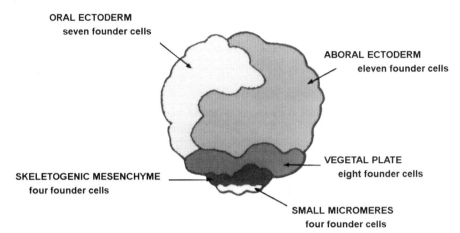

ORAL ECTODERM
seven founder cells

ABORAL ECTODERM
eleven founder cells

VEGETAL PLATE
eight founder cells

SKELETOGENIC MESENCHYME
four founder cells

SMALL MICROMERES
four founder cells

Figure 1. A schematic representation of the sea urchin embryonic territories at the seventh cleavage displaying the founder cell populations. Each of the territories is composed of cells with a unique lineage history, an individual pattern of gene expression, and the presence of one or more distinct cell types. Because exclusive segregation of a blastomere to a territory may occur between the third and seventh cleavage, the number of founder cells does not equal the number of cells at the canonical 128-cell stage (Cameron and Davidson, 1992).

be removed to culture and they will complete their developmental program. All the other territories are conditionally specified; that is, they require some form of cellular interaction to develop correctly.

The spatial pattern of gene expression in sea urchin embryos has been detected through the use of *in situ* hybridization to mRNA. Initially, radioactive hybridization probes were employed (Angerer and Davidson, 1984) but more recently whole mount techniques using digoxigenin-labeled probes have become widely used (e.g., see Harkey et al., 1992; Ransick et al., 1993; Cameron.1). The genes identified in this manner are listed in Table 1 (Coffman and Davidson, 1992). By examining the regulation of genes expressed in various spatial and temporal patterns, we can begin to elucidate how regions of differential gene expression are initiated in the early embryo. In the sea urchin, *Strongylocentrotus purpuratus,* genes expressed exclusively in the aboral ectoderm include the cytoskeletal actin genes *CyIIIa* and *CyIIIb* (Cox et al., 1986) as well as the genes encoding the calcium binding proteins, *Spec1* and *Spec2* (Lynn et al., 1983; Hardin et al., 1988). Two other cloned genes are expressed in this territory—the gene encoding arylsulfatase (Yang et al., 1989) and a metallothionein gene (Nemer et al., 1991). Several genes have been characterized from the skeletogenic mesenchyme territory. Among these are the genes encoding the spicule matrix proteins *SM30* (George et al., 1991) and *SM50* (Benson et al., 1987), and the gene encoding the cell surface glycoprotein *MSP130* (Parr et al., 1989; Harkey et al., 1992). The spicule matrix proteins form the organic matrix of the embryonic skeleton, and MSP130 is involved in calcium uptake (Farach et al., 1987). The *Endo16* gene encodes an extracellular protein initially expressed in the

TABLE 1. Spatially Regulated Genes in the Sea Urchin Embryo

Gene	Protein Product	Territory[a]	Species[b]	References
Spec1	Ca^{2+} binding protein	AE	S.p.	Lynn et al. (1983)
Spec2 (a–d)	Ca^{2+} binding protein		S.p.	Hardin et al. (1988)
LvS1	Ca^{2+} binding protein		L.v.	Wessel et al. (1989b)
LpS2 (a,b)	Ca^{2+} binding protein		L.p.	Xiang et al. (1988, 1991)
SpMTa	Metallothionein		S.p.	Harlow et al. (1989), Wilkinson and Nemer (1987), Nemer et al. (1991)
CyIIIa	Cytoskeletal actin		S.p.	Cox et al. (1986)
CyIIIb	Cytoskeletal actin		S.p.	Cox et al. (1986)
SpARS	Arylsulfatase		S.p.	Yang et al. (1989)
ARS	Arylsulfatase		H.p.	Akasaka et al. (1990)
Hbox1	Transcription factor		S.p., T.g.	Angerer et al. (1989)
SpEGFI	ECM protein	AE + OE	S.p.	Grimwald et al. (1991), Bisgrove et al. (1991)
SpEGFII	ECM protein		S.p.	Grimwald et al. (1991)
SpHe	Metalloendoprotease		S.p.	Reynolds et al. (1992)
SpAN	Metalloendoprotease		S.p.	Reynolds et al. (1992)
Spec3	Ciliary protein		S.p.	Eldon et al. (1987)

(*continued*)

TABLE 1. (*Continued*)

Gene	Protein Product	Territory[a]	Species[b]	References
SpMTb1	Metallothionein	Global	S.p.	Nemer et al. (1991)
CyI	Cytoskeletal actin	Early	S.p.	Cox et al. (1986)
CyIIb	Cytoskeletal actin	OE + V Late	S.p.	Cox et al. (1986)
Endo16	Cell surface protein	V	S.p.	Nocente-McGrath et al. (1989)
LvN1.2	Cell surface protein		L.v.	Wessel et al. (1989a)
CyIIa	Cytoskeletal actin	V + SM	S.p.	Cox et al. (1986)
Collagen	ECM protein		S.p.	Angerer et al. (1988)
COLL1a	ECM protein		P.l.	D'Alessio et al. (1989)
COLL2a	ECM protein		P.l.	D'Alessio et al. (1990)
SM50	Spicule matrix protein	SM	S.p.	Benson et al. (1987), Katoh-Fukui et al. (1991), Sucov et al. (1987)
SM30	Spicule matrix protein		S.p.	George et al. (1991)
MSP130	Cell-surface glycoprotein		S.p.	Parr et al. (1989), Harkey et al. (1992)

[a]Territory abbreviations: AE, aboral ectoderm; OE, oral ectoderm; SM, skeletogenic mesenchyme; V, vegetal plate.

[b]Species abbreviations: H.p., *Hemicentrotus pulcherrimus*; L.p., *Lytechinus pictus*; L.v., *Lytechinus variegatus*; P.l., *Paracentrotus lividus*; S.p., *Strongylocentrotus purpuratus*; T.g., *Tripneustes gratilla*.

Source: Coffmann and Davidson (1992).

vegetal plate; later its expression is restricted to the midgut (Nocente-McGrath et al., 1989; Ransick et al., 1993). Some genes are expressed in multiple territories such as the cytoskeletal actins CyI and CyIIb, which are initially expressed throughout the embryo and are later restricted to the oral ectoderm and vegetal territories (Cox et al., 1986).

The identification of *cis*-regulatory regions for sea urchin embryonic genes has been greatly facilitated through the use of exogenous reporter genes. McMahon et al. (1985) and Flytzanis et al. (1985) developed a technique for microinjecting solutions containing reporter genes into sea urchin eggs or zygotes (Cameron.2). Reporter genes are composed of fragments of DNA containing *cis*-regulatory elements linked to DNA sequences encoding an easily detected protein that is not normally expressed in the sea urchin embryo. The occupation of *cis*-regulatory sites by *trans*-acting factors (DNA-binding proteins) activates the reporter gene, resulting in the production of the foreign protein. Quantitation of enzyme activity from reporter genes expressing chloramphenicol acetyl transferase (CAT) or luciferase is used to measure the level of reporter gene expression in the embryo. *In situ* hybridization to reporter gene mRNA using probes derived from cloned DNA or the enzymatic detection of the protein product encoded by β-galactosidase reporter genes reveals the spatial patterns of reporter gene expression. By comparing reporter gene expression to endogenous gene expression, it is possible to determine the promoter sequences that are necessary and sufficient for proper temporal and spatial regulatory control.

The regulatory regions of genes expressed in three different embryonic territories have now been characterized using the gene transfer methods described above. Zygotic injections of *CyIIIa-CAT* or *CyIIIb-CAT* have identified the regulatory domains for two different cytoskeletal actin genes, *CyIIIa* and *CyIIIb,* expressed exclusively in the aboral ectoderm territory. The *CyIIIa* regulatory domain occupies 2.3 kb of sequence upstream from the start of transcription and contains regulatory elements necessary and sufficient for proper temporal and spatial gene expression (Flytzanis et al., 1987; Hough-Evans et al., 1987, 1988; Zeller et al., 1992, 1995a). A 2.2-kb region is sufficient for correct temporal expression of the *CyIIIb-CAT* gene (Niemeyer and Flytzanis, 1993). Spec reporter genes employing a CAT reporter, containing ~5.6 kb, 1.5 kb, and 5.6 kb of upstream region from the *Spec1, Spec2a,* and *Spec2c* genes, respectively, were found to have proper temporal expression after microinjection into eggs (Gan et al., 1990b). These same promoter regions were fused to a β-galactosidase coding region, instead, and injected into eggs. Then the proper spatial expression was only observed for the *Spec2a-lacZ* construct (Gan et al., 1990a), suggesting that *Spec1* and *Spec2c* reporter genes are missing *cis*-regulatory elements, which control proper aboral ectoderm expression.

In the early embryo, *Endo16* is first expressed in the vegetal plate, then in the gut, and later restricted to only the midgut (Ransick et al., 1993). A 2.3-kb region of the *Endo16* promoter was found to confer proper temporal and spatial expression when fused to a CAT reporter (Yuh et al., 1994). The regulatory domain of *Endo16* contains a large number of DNA-binding protein target sites that most likely control the complicated temporal/spatial expression of this gene. The regulatory domains of two spicule matrix protein genes have also been identified. An *SM50-CAT* reporter gene containing about 450 bp of regulatory region is properly expressed in the sea urchin embryo (Sucov et al., 1988). A 2.6-kb of *SM30* upstream region directs proper mesenchyme-specific expression of a CAT reporter (Akasaka et al., 1994).

THE IDENTIFICATION OF SPECIFIC DNA-BINDING SITES IN CIS-REGULATORY REGIONS

Once a regulatory domain is delimited, the identification of sequences that are specifically recognized by DNA-binding proteins is relatively straightforward. Here two different territory-specific genes serve as examples, *CyIIIa* and *Endo16*. The procedure used to localize DNA-binding protein target sites is diagrammed in Figure 2 (Thézé et al., 1990; Yuh et al., 1994; Cameron.3 and .4). About 20 sites of specific DNA–protein interactions are found in the *CyIIIa* regulatory domain (Thézé et al., 1990; Fig. 2), and 38 sites in the *Endo16* promoter (Yuh et al., 1994).

CyIIIa Regulatory Domain

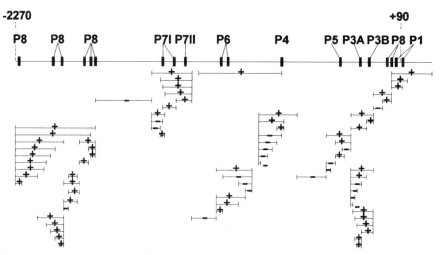

Figure 2. Strategy for the rapid identification of DNA-binding protein target sites and a map of the *CyIIIa* regulatory domain. The 2.3-kb *CyIIIa* regulatory domain has been mapped by subdividing the promoter into small fragments, which are examined for the presence of *cis*-target sites by gel shift assays as described in the text. Fragments that were bound specifically by DNA-binding proteins are labeled "+"; those that did not show evidence of DNA–protein interactions are labeled "−." Comparing the positions of the various fragments allows sites of DNA–protein interaction to be assigned; these are shown as filled boxes. There are around 20 DNA-binding protein target sites serviced by about 10 different transcription factors in the *CyIIIa* promoter (Calzone et al., 1988; Thézé et al., 1990; Coffman et al., 1992). This diagram is redrawn from Thézé et al. (1990).

The basic strategy is to use restriction enzymes to subdivide the regulatory domain into small fragments of 300–400 bp in length, which are then used as probes in gel mobility shift assays (Fried and Crothers, 1981; Calzone et al., 1988; Thézé et al., 1990; Yuh et al., 1994). When fragments containing regulatory target sites are bound by a DNA-binding protein, the migration of the DNA–protein complex is retarded relative to unbound probe. An autoradiograph of the gel shift pattern resulting from the binding of SpGCF1 protein to the P8 site is shown in Figure 3a. Positive fragments are then further subdivided and tested in the same fashion; eventually regulatory target sites are narrowed down to a 20–40 bp region. Further examination of target sites with oligonucleotide probes defines the DNA sequence best suited for the isolation of the DNA-binding protein.

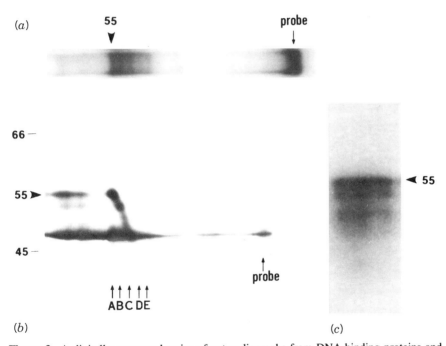

Figure 3. A digitally composed series of autoradiographs from DNA-binding proteins and the radioactive DNA probes to which they bind. The three preparations are arranged to illustrate the correspondence between: (a) one-dimensional gel shift of the radiolabeled P8 site of the *CyIIIa* regulatory region with affinity-purified nuclear extract. Because the protein that binds P8, SpGCF1, exists in five forms, five separate bands are apparent. The largest complex that contains the 55-kDa form of the protein (55) and the unbound probe (probe) are indicated. (b) A two-dimensional gel shift of SpGCF1 and the radiolabeled P8 DNA fragment. The gel shift dimension is left to right [as in (a)] and the SDS polyacrylamide–protein gel dimension is from top to bottom [as in (c)]. The five complexes of DNA and protein are resolved into a diagonal array of spots (marked A–E). (c) A protein gel blot of nuclear extract to which a radiolabeled P8 DNA fragment has been allowed to bind. The five forms of SpGCF1 are detected with the largest at 55 kDa.

This method is very rapid and detects proteins with high affinities for DNA target sites. Specific target site affinities are described by the dimensionless term K_r, which is simply the ratio of the equilibrium constant for the interaction of the protein with its specific target site, divided by the equilibrium constant for the interaction of the protein with nonspecific DNA (Calzone et al., 1988). Measurements of K_r are made by challenging DNA–protein complexes with increasing amounts of unlabeled, specific target sites ("cold probe"). As the concentration of target sites increases, more protein begins to interact with the cold probe and the amount of labeled complex will decrease. Large K_r values are thus indicative of high-affinity DNA–protein interactions; the *CyIIIa* factors have K_r values of $\geq 10^4$. Since transcriptional control is mediated by high-affinity protein interactions at *cis*-regulatory sites, the rapid mapping procedure described above identifies those target sites that are likely to have significant regulatory functions. All of the *CyIIIa* target sites mapped in this manner have subsequently been shown to have regulatory functions (Franks et al., 1990; Hough-Evans et al., 1990; C. Kirchhamer and E. Davidson, *in press*) (see below).

THE FUNCTION OF SITES IN THE REGULATORY DOMAIN THAT BIND TO PROTEIN

Reporter genes present the opportunity to examine the function of those sequences within the control region that have been identified as sites where proteins bind. Discrete regions can be changed or competed with and the subsequent effect on the expression of the reporter gene observed. The most extensive investigation of this sort has been performed on the *S. purpuratus* cytoskeletal actin gene *CyIIIa* (Franks et al., 1990; Hough-Evans et al., 1990; C. Kirchhamer and E. Davidson, *in press*). *In vivo* competition studies in which an excess of a particular DNA fragment was injected into the egg along with the reporter gene carrying the entire regulatory domain first indicated the subregions that function as either positive or negative regulators (Franks et al., 1990; Hough-Evans et al., 1990). For example, a competing fragment that causes a decrease in expression must bind a positive regulatory protein. Two subregions that did not compete in this way subsequently produced ectopic expression, however. These elements exercise negative spatial regulatory functions (Hough-Evans et al., 1990). An extensive series of deletion and mutation studies have dissected the function of the *CyIIIa* control region in detail (Wang et al., 1995; C. Kirchhamer and E. Davidson, *in press*). In synopsis these functions are illustrated in Figure 4. Coarsely, the control region can be divided into three modular regions based on function, (1) a proximal region responsible for early specification; (2) a midregion responsible for late spatial control, and (3) a distal enhancer region. The proximal region has correct spatial expression but it generates only 5% of the normal expression level of the entire region, and it shows a different temporal profile. Within the proximal module are two sites, the "P3A" sites, which function as negative spatial controllers, while the P4, P5, P3B, and P1 sites all bind proteins to confer positive regulation (see Fig. 2). In the midregion late spatial module, the

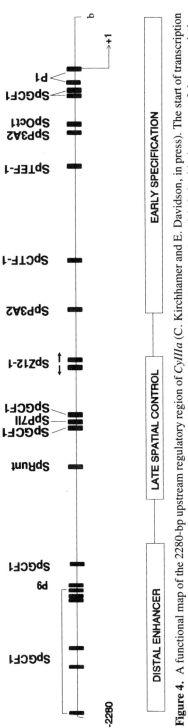

Figure 4. A functional map of the 2280-bp upstream regulatory region of *CyIIIa* (C. Kirchhamer and E. Davidson, in press). The start of transcription is indicated with a right-pointing arrow. The various binding sites are depicted as black boxes and are labeled with the names of the transcription factors that bind there, where they are known; otherwise with the name of the site (e.g., P1). Below the map is a series of open boxes that indicate the position of the three functional modules of this regulatory domain.

"P7II" site acts as a negative spatial regulator that prevents expression in the oral ectoderm and skeletogenic mesenchyme; the "P6" site functions as a negative spatial regulator for skeletogenic mesenchyme (Wang et al., 1995) and the "P7I" site acts as a positive regulatory site that probably also confers temporal regulation. The P8 sites positioned in all three modules are thought to be intermodule communicators, which facilitate the interaction between the various modules (Zeller et al., 1995a). This interpretation of the function of the various sites implies a two-level regulatory organization: first, activated proteins bind and interact within a module and, second, the modules interact with each other and the site of polymerase binding to control transcription.

THE ISOLATION OF SITE-SPECIFIC DNA-BINDING PROTEINS

A common method, often referred to as "ligand-based expression screening," depends on the functional expression of recombinant DNA-binding proteins and their specific recognition of DNA target sites (reviewed by Singh, 1993). In practice, a library is constructed in which proteins encoded by cDNA inserts are expressed in an active form that will bind DNA. Duplicate filters of protein-expressing clones are then reacted with a labeled DNA target site probe. Positive clones contain cDNA inserts encoding DNA-binding proteins, which recognize the DNA target site used as a probe. A major limitation of this technique is that there are often many false-positive signals. In addition, many proteins fail to renature properly and will thus go undetected by this method. Since first described (Vinson et al., 1988; Singh et al., 1988), this method has been used to identify a number of different DNA-binding proteins. However, we have been successful in recovering only 2 out of 12 different *CyIIIa* clones for which we searched: that is, *SpZ12-1* (Wang et al., 1994) and *SpP3A1* (Höög et al., 1991). The major advantage of this method over the biochemical method described below is that it allows direct cloning of transcription factors in cases where the large quantities of nuclear extract required for biochemical purification are not available.

Affinity chromatography is an efficient method for isolating proteins that bind to DNA (reviewed by Kadonaga, 1991; Jarrett, 1993; Cameron.3 and .5). It is the method we use for the isolation of sea urchin transcription factors (Calzone et al., 1991; Coffman et al., 1992). In general, if DNA-binding activity for a particular target site probe is present in nuclear extract, affinity chromatography efficiently purifies the protein or proteins responsible. Proteins are purified from a nuclear extract when the extract is passed over a column bearing multimerized oligonucleotides that contain the DNA target sites for those proteins (Kadonaga and Tjian, 1986). The ability of the affinity column to retain transcription factors is due to the relatively high affinity of these proteins for their specific DNA target sites. Nonspecific proteins are removed from the column during subsequent washes and specific DNA-binding proteins are eluted from the column with a salt gradient. Eluted proteins are used to generate amino acid sequence, which is used to derive nucleic acid probes for the isolation of the gene encoding the DNA-binding protein.

Typical proteins can be purified 100- to 1000-fold with $\geq$30% yields (Calzone et al., 1991; Kadonaga, 1991; Coffman et al., 1992).

Affinity column fractions often contain a substantial number of proteins that may hinder identification of the DNA-binding protein of interest. Thus it is usually necessary to further identify which protein in the mixture is responsible for specific DNA binding. The first step is to identify the column fractions that contain the previously observed DNA-binding activity using gel retardation assays. Once the DNA-binding activity has been localized to a set of column fractions, several different procedures may be used for the identification of the protein responsible for specific complex formation. A procedure often used to identify sea urchin transcription factors is the "Southwestern" blot (Calzone et al., 1991; Coffman et al., 1992; Yuh et al., 1994; Cameron.6). In this procedure, samples of column fractions containing DNA-binding activity are separated by standard SDS–protein gels. The separated proteins are transferred to nitrocellulose, allowed to renature, and incubated with a probe containing the DNA target site of interest. Proteins that bind the probe are likely to be the proteins responsible for complex formation. An example of this procedure is illustrated in Figure 3c, where the five different forms of the protein SpGCF1 are bound by probe.

The two-dimensional gel shift, another method to detect DNA–protein complexes, can also be used to determine the molecular weight of affinity purified DNA-binding proteins (Coffman et al., 1992; Yuh et al., 1994; Zeller et al., 1995b; Cameron.7). In this assay, complexes formed in a standard gel shift reaction are separated by nondenaturing electrophoresis (the first dimension, generally performed in a tube gel), then the gel containing the DNA–protein complexes is denatured in SDS and placed across the top of an SDS–protein gel. Electrophoresis in the second dimension separates the components of the DNA–protein complexes and resolves the individual polypeptides into discrete bands. Analysis of the resulting pattern of polypeptides can elucidate the stoichiometries of the proteins binding to their DNA target site. The two-dimensional gel shift pattern for the multiple forms of the SpGCF1 protein is shown in Figure 3b. Between the two methods described here, we have been able to identify, by molecular weight, all the proteins that bind to *cis*-regulatory sites in the *CyIIIa* regulatory domain.

As mentioned previously, affinity chromatography requires large amounts of nuclear material for the purification of sufficient quantities of transcription factors to permit amino acid sequencing. An advantage of the sea urchin system is the feasibility of rearing very large quantities of synchronously growing embryos, to a desired developmental stage, which can then be processed for nuclear proteins (Calzone et al., 1991; Coffman et al., 1992). To date, we have grown and processed over 10^{11} embryos, which represent about 4×10^{13} nuclei. We have developed an automated affinity chromatography system that allows the simultaneous isolation of up to a dozen different DNA-binding proteins (Coffman et al., 1992). The recoveries of DNA-binding protein activity from several factor's binding target sites in the *CyIIIa* promoter range between 10% and 50% yield, with an enrichment of 50- to 500-fold (Coffman et al., 1992; Calzone et al., 1991). Factors that are present in very low abundance, such as the factor binding to the P4 site of the *CyIIIa* regulatory region, may be purified from several different rounds of affinity chromatography until enough protein has accumulated for protein sequence analysis. Factors that bind to specific target sites in the *Endo16* and *SM50* genes are now being isolated by automated affinity chromatography (C.-H. Yuh, K. Makabe, J. Coffman, and E. Davidson, *unpublished data*).

Once a protein has been accumulated in sufficient quantities, partial amino acid sequence is obtained and used to design oligonucleotide probes for the identification of the corresponding gene. In our experience, N-terminal protein sequencing is not possible as the proteins are most likely modified at the N-terminal residue. We therefore generate a series of peptides using a lysylendopeptidase (*Achromobacter* protease I, often called "Lys-C"). This protease specifically cleaves the carboxy terminal of lysine residues and generally produces peptides of sufficient length for subsequent gene cloning procedures. The peptides are separated by reverse phase HPLC (Char et al., 1993; Coffman and Davidson, 1994; Zeller et al., 1995b) and individual peaks, containing single peptide species, are collected. Amino acid sequence is obtained from automated Edman degradation sequencing as described (Char et al., 1993), and several different peptide sequences are usually obtained from each purified protein. The procedures for the actual identification of the gene encoding the transcription factor are described in the next section.

IDENTIFICATION OF TRANSCRIPTION FACTOR GENES

The mRNAs encoding several different sea urchin transcription factors belong to the "rare" class of messenger RNAs (Cutting et al., 1990; Wang et al., 1994; Zeller et al., 1995b). Hence the frequency of transcription factor mRNAs in a population of embryo mRNAs is expected to be relatively low, and the detection of cDNA clones encoding transcription factors is problematic. As the amino acid code is degenerate, one must account for all possible codon possibilities when designing nucleic acid probes. The degeneracy of such a probe refers to the number of different sequences present in the population of probe molecules. For example, an oligonucleotide encoding a pentapeptide of methionine residues has a degeneracy of one since methionine is encoded by a single codon. An oligonucleotide encoding a pentapeptide of prolines accounting for all possible codons is 1024-fold degenerate (four codons, five positions $= 4^5$). Peptides composed of amino acids encoded by only a few different codons should therefore be selected to design nucleic acid probes.

We have used three different screening strategies for isolating the cDNA clones encoding sea urchin transcription factors. The first method is the direct screening of a phage cDNA library with a pool of degenerate oligonucleotide probes. After hybridization, probe that is not specifically hybridized to complementary target sequences is removed with washes of a quaternary alkylammonium salt such as tetramethyl ammonium chloride (Wood et al., 1985; Calzone et al., 1991; Anderson et al., 1994). Unlike normal stringency washes using sodium salts, the stringency of quaternary ammonium salt washes does not depend on the GC content of the probe, but only on the probe length (DiLella and Woo, 1987), which is advantageous when using pools of degenerate probes of variable GC content. Using this method, Anderson et al. (1994) screened a genomic library for DNA-binding protein target sites and were able to wash to a stringency of 4–5 base mismatches out of a total probe length of 29 nucleotides. The quaternary alkylammonium salt method was also used to isolate the cDNA clone encoding the P3A2 transcription factor (Calzone et al., 1991).

There are a number of disadvantages to the direct screening of a cDNA library with a pool of degenerate oligonucleotides, the main disadvantage being the low concentration of specific probe in the pool of degenerate oligonucleotides that actually detects the mRNA. This results in a probe of low specific activity, which makes the identification of positive clones extremely difficult. If we suppose that a plaque contains 10^7 phage, of which 50% are exposed for probe binding, and a single sequence oligonucleotide probe is labeled with high specific activity ^{32}P (e.g., 4×10^8 cpm/50 ng of a 30-mer oligonucleotide), then one might expect about 500 cpm of probe binding per positive plaque. If a degenerate probe is used directly in the library screen, the number of labeled molecules that contain the sequence complementary to the actual clone will decrease as the degeneracy of the probe pool increases. This can reach a point in which insufficient numbers of probe molecules are bound to target sites to be readily detectable.

As a way to increase the relative concentration of target sequences to probe sequences we and others have utilized a plasmid-based cDNA library screening method. This provides a minimum of an order of magnitude increase in the number of cDNA target sites available for hybridizing to the degenerate probe pool. The plasmid cDNA library is subdivided into aliquots of 1000–2000 clones. The plasmids of each aliquot are recovered and digested to release the cDNA inserts, which are then separated by gel electrophoresis and transferred to an appropriate membrane. Using this method, the number of specific target sequences present in a single band is $\geq 10^8$ (given a plasmid MW of 3.3×10^6, digesting 1 µg of plasmid from a 2500-clone pool and assuming that the pool includes only one copy of the plasmid; more may of course be present if the transcript is modestly prevalent). The membrane containing the digested clones is then hybridized and washed with the quaternary ammonium salts described above. Aliquots containing positive clones are further subdivided and rescreened until a small number of possible clones remain. The desired clone is then recovered by the direct hybridization of the probe to the bacterial colonies. This method has been used to isolate clones for the SpTEF-1 homologue (J. Xian and E. Davidson, *unpublished data*) and clones for at least one homeodomain protein (P. Martinez and E. Davidson, *unpublished data*).

Nondegenerate probes may be obtained from partial amino acid sequences through the use of the polymerase chain reaction (PCR). The PCR process amplifies DNA sequence that is flanked by two oligonucleotide primers from minute amounts of nucleic acid template. If two degenerate primers are used for PCR, the amplified nucleic acid sequence will serve as an excellent, nondegenerate probe for subsequent library screening. Two different strategies may be employed. Degenerate primers derived from the terminal 6–7 amino acids of a particularly long peptide may be used with PCR to amplify the nucleic acid sequence encoding the internal amino acids (Lee et al., 1988). This method was used to identify the cDNAs encoding proteins binding to the *CyIIIa* P7I and P7II sites (J. Coffman, *unpublished data;* Coffman and Davidson, 1994) and the SpGCF1 protein that binds to the *CyIIIa* P8 site (Zeller et al., 1995b). If two short peptide sequences are available, degenerate primers may be designed to amplify the nucleic acid sequence encoding the protein between the two primers.

The PCR method of identifying nondegenerate transcription factor probes has a number of advantages over the direct use of degenerate oligonucleotide probes. First, since the probes derived from PCR are nondegenerate, higher stringency conditions may be used to isolate an appropriate clone. Second, a variety of different templates, such as cDNAs from many different tissues or developmental stages, may be examined simultaneously. This eliminates the need to screen many different libraries to identify the clone of interest. Oligonucleotides that are too degenerate in sequence for direct library screening may be used successfully in PCR amplification. In addition, oligonucleotides derived from two short peptide sequences, which are of insufficient length to be used for direct library screening, may work perfectly well in PCR amplification.

CONCLUSIONS

The number of transcription factors that regulate early sea urchin gene expression has been estimated by two-dimensional gel electrophoresis (Harrington et al., 1992). DNA-binding proteins usually interact with DNA sequences through a region of basic residues and this feature allows the enrichment of transcription factors in nuclear extract with cation-exchange chromatography. When nuclear extracts from 24 hr sea urchin embryos were chromatographically fractionated and analyzed by two-dimensional electrophoresis, about 265 polypeptides, representing about 100 different transcription factors, were observed (Harrington et al., 1992). This suggests that genes expressed in the early sea urchin embryo are regulated by a tractable number of DNA-binding proteins (Coffman and Davidson, 1994). With the availability of large quantities of nuclear extract, the identification of *cis*-regulatory target sites from differentially expressed genes, and the rapid development of techniques to sequence smaller amounts of proteins, it is conceivable that we could isolate a significant fraction of the relevant transcription factors present in the sea urchin embryo. This will provide a library of molecular components that will be necessary to understand the transcriptional regulatory network of the sea urchin embryo.

Although the emphasis here has been on the biochemical steps in the dissection of transcriptional regulation of spatially restricted gene expression in early sea urchin development, the overall approach is a vertically integrated one, which also includes cell lineage, cell fate, and cell interaction studies as well as gene transfer and biochemical methodologies. Blastomere recombination experiments (reviewed in Hörstadius, 1939; Ransick et al., 1993; Wilt, 1987) and cell lineage studies of axis specification (Cameron et al., 1987) imply that specification of early embryo blastomeres results from blastomere interaction during cleavage. Thus the activation of territorially expressed genes such as *CyIIIa* and *Endo16* must depend in part on the downstream consequences of these interactions. These must ultimately govern the activation of particular transcription factors in particular domains of the embryo. Now that we are determining the proximal components of this system (i.e., the transcription factors), the mechanisms by which cell interactions influence gene expression can be studied.

ACKNOWLEDGMENTS

The original research from the Davidson laboratory cited in this chapter was supported by grants from the NIH and the NSF. R.W.Z. was supported by the ONR AASERT program (N00014-93-1-1400).

REFERENCES

Akasaka K, Ueda T, Higashiakagawa T, Yamada K, Shimada H (1990): Spatial pattern of arylsulfatase mRNA expression in sea urchin embryos. *Dev Growth Differ* 32:9–13.

Akasaka K, Frudakis T, Killian CE, George NC, Yamasu K, Khaner O, Wilt FH (1994): Genomic organization of a gene encoding the spicule matrix protein SM30 in the sea urchin *Strongylocentrotus purpuratus*. *J Biol Chem* 269:20592–20958.

Anderson R, Britten RJ, Davidson EH (1994): Repeated sequence target sites for maternal DNA-binding proteins in genes activated in early sea urchin development. *Dev Biol* 163:11–18.

Angerer RC, Davidson EH (1984): Molecular indices of cell lineage specification in sea urchin embryos. *Science* 226:1153–1160.

Angerer LM, Chambers SA, Yang Q, Venkatesan M, Angerer RC, Simpson RT (1988): Expression of a collagen gene in mesenchyme lineages of the *Strongylocentrotus purpuratus* embryo. *Genes Dev* 2:239–246.

Angerer LM, Dolecki GJ, Gagnon M, Lum R, Wang G, Yang Q, Humphries T, Angerer RC (1989): Progressively restricted expression of a homeobox gene within the aboral ectoderm of developing sea urchin embryos. *Genes Dev* 3:370–383.

Benson S, Sucov H, Stephens L, Davidson E, Wilt F (1987): A lineage-specific gene encoding a major spicule matrix protein of the sea urchin spicule. I. Authentification of the cloned gene and its developmental expression. *Dev Biol* 120:499–506.

Bisgrove BW, Andrews ME, Raff RA (1991): Fibropellins, products of an EGF repeat-containing gene, form a unique extracellular matrix structure that surrounds the sea urchin embryo. *Dev Biol* 146:89–99.

Calzone FJ, Thézé N, Thiebaud P, Hill RL, Britten RJ, Davidson EH (1988): The developmental appearance of factors that bind specifically to *cis*-regulatory sequences of a gene expressed in the sea urchin embryo. *Genes Dev* 2:1074–1088.

Calzone FJ, Höög CH, Teplow DB, Cutting AE, Zeller RW, Britten RJ, Davidson EH (1991): Gene regulatory factors of the sea urchin embryo. I. Purification by affinity chromatography and cloning of P3A2, a novel DNA-binding protein. *Development* 112:335–350.

Cameron RA, Davidson EH (1992): Cell type specification during sea urchin development. *Trends Genet* 7:212–218.

Cameron RA, Hough-Evans BR, Britten RJ, Davidson EH (1987): Lineage and fate of each blastomere of the eight-cell sea urchin embryo. *Genes Dev* 1:75–84.

Char BR, Bell JR, Dovala J, Coffman JA, Harrington MG, Becerra JC, Davidson EH, Calzone FJ, Maxson R (1993): SpOct, a gene encoding the major octamer-binding protein in sea urchin embryos: expression profile, evolutionary relationships, and DNA binding of expressed protein. *Dev Biol* 158:350–363.

Clark AR, Doherty K (1993): Negative regulation of transcription in eukaryotes. *Biochem J* 296:521–541.

Coffman JA, Davidson EH (1992): Expression of spatially regulated genes in the sea urchin embryo. *Curr Opin Genet Dev* 2:260–268.

Coffman JA, Davidson EH (1994): Regulation of gene expression in the sea urchin embryo. *J Mar Biol Assoc UK* 74:17–26.

Coffman JA, Moore JG, Calzone FJ, Britten RJ, Hood LE, Davidson EH (1992): Automated sequential affinity chromatography of sea urchin DNA binding proteins. *Mol Mar Biol Biotechnol* 1:136–146.

Colin AM (1986): Rapid repetitive microinjection. In: Schroeder TE (ed). *Methods in cell biology, Volume 27, Echinoderm eggs and embryos*. Orlando, FL: Academic Press, pp. 395–405.

Cox KH, Angerer LM, Lee JJ, Davidson EH, Angerer RC (1986): Cell lineage-specific programs of expression of multiple actin genes during sea urchin embryogenesis. *J Mol Biol* 188:159–172.

Cutting AE, Hoog CH, Calzone FJ, Britten RJ, Davidson EH (1990): Rare maternal mRNAs code for regulatory proteins that control lineage specific gene expression in the sea urchin embryo. *Proc Natl Acad Sci USA*. 87:7953–7959.

D'Alessio M, Ramirez F, Suzuki H, Solursh M, Gambino R (1989): Structure and developmental expression of a sea urchin fibrillar collagen gene. *Proc Natl Acad Sci USA* 86:9303–9307.

D'Alessio M, Ramirez F, Suzuki H, Solursh M, Gambino R (1990): Cloning of a fibrillar collagen gene expressed in the mesenchymal cells of the developing sea urchin embryo. *J Biol Chem* 265:7050–7054.

Davidson EH (1989): Lineage-specific gene expression and the regulative capacities of the sea urchin embryo: a proposed mechanism. *Development* 105:421–445.

Davidson EH (1990): How embryos work: a comparative view of diverse modes of cell fate specification. *Development* 108:365–389.

Davidson EH (1991): Spatial mechanisms of gene regulation in metazoan embryos. *Development* 113:1–26.

DiLella AG, Woo SLC (1987): Hybridization of genomic DNA to oligonucleotide probes in the presence of tetramethylammonium chloride. *Methods Enzymol* 152:447–451.

Eldon ED, Angerer LM, Angerer RC, Klein WH (1987): Spec3: embryonic expression of a sea urchin gene whose product is involved in ectoderm ciliogenesis. *Genes Dev* 1:1280–1292.

Farach MC, Valdizan M, Park HR, Decker GL, Lennarz WJ (1987): Developmental expression of a cell-surface protein involved in calcium uptake and skeleton formation in sea urchin embryos. *Dev Biol* 122:320–331.

Flytzanis CN, McMahon AP, Hough-Evans BR, Katula KS, Britten RJ, Davidson EH (1985): Persistence and integration of cloned DNA in postembryonic sea urchins. *Dev Biol* 108:431–442.

Flytzanis CN, Britten RJ, Davidson EH (1987): Ontogenic activation of a fusion gene introduced into sea urchin eggs. *Proc Natl Acad Sci USA* 84:151–155.

Franks RR, Hough-Evans BR, Anderson RF, Britten RJ, Davidson EH (1990): Competitive titration in living sea urchin embryos of regulatory factors required for expression of the CyIIIa actin gene. *Development* 110:31–40.

Fried MG, Crothers DM (1981): Equilibria and kinetics of lac repressor–operator interactions by polyacrylamide gel electrophoresis. *Nucleic Acids Res* 9:6505–6524.

Gan L, Wessel GM, Klein WH (1990a): Regulatory elements from the related Spec genes of *Strongylocentrotus purpuratus* yield different spatial patterns with a *lacZ* reporter gene. *Dev Biol* 142:346–359.

Gan L, Zhang W, Klein WH (1990b): Repetitive DNA sequences linked to the sea urchin Spec genes contain transcriptional enhancer-like elements. *Dev Biol* 139:186–196.

George NC, Killian CE, Wilt FH (1991): Characterization and expression of a gene encoding a 30.6 kD *Strongylocentrotus purpuratus* spicule matrix protein. *Dev Biol* 147:334–342.

Grimwald JE, Gagnon ML, Yang Q, Angerer RC, Angerer LM (1991): Expression of two mRNAs encoding EGF-related proteins identifies subregions of sea urchin embryonic ectoderm. *Dev Biol* 143:44–57.

Hardin PE, Angerer LM, Hardin SH, Angerer RC, Klein WH (1988): Spec2 genes of *Strongylocentrotus purpuratus:* structure and differential expression in embryonic aboral ectoderm cells. *J Mol Biol* 202:417–431.

Harkey MA, Whiteley HR, Whiteley AH (1992): Differential expression of the MSP130 gene among skeletal lineage cells in the sea urchin embryo—a 3-dimensional in situ hybridization analysis. *Mech Dev* 37:173–184.

Harlow P, Watkins E, Thornton RD, Nemer M (1989): Structure of an ectodermally expressed sea urchin metallothionein gene and characterization of its metal-responsive region. *Mol Cell Biol* 9:5445–5455.

Harrington MG, Coffman JA, Calzone FJ, Hood LE, Britten RJ, Davidson EH (1992): Complexity of sea urchin embryo nuclear proteins that contain basic domains. *Proc Natl Acad Sci USA* 89:6252–6256.

Höög CH, Calzone FJ, Cutting AE, Britten RJ, Davidson EH (1991): Gene regulatory factors of the sea urchin embryo II. Two dissimilar proteins, P3A1 and P3A2, bind to the same target sites that are required for early territorial gene expression. *Development* 112:351–364.

Hörstadius S (1939): The mechanisms of sea urchin development studied by operative methods. *Biol Rev* 14:132–179.

Hough-Evans BR, Franks RR, Cameron RA, Britten RJ, Davidson EH (1987): Correct cell type-specific expression of a fusion gene injected into sea urchin eggs. *Dev Biol* 121:576–579.

Hough-Evans BR, Britten RJ, Davidson EH (1988): Mosaic incorporation of an exogenous fusion gene expressed exclusively in aboral ectoderm cells of the sea urchin embryo. *Dev Biol* 129:198–208.

Hough-Evans BR, Franks RR, Zeller RW, Britten RJ, Davidson EH (1990): Negative spatial regulation of the lineage specific CyIIIa actin gene in the sea urchin embryo. *Development* 110:41–50.

Hu Y-F, Lüscher B, Admon A, Mermod N, Tjian R (1990): Transcription factor AP-4 contains multiple dimerization domains that regulate dimer specificity. *Genes Dev* 4:1741–1752.

Hwang J-J, Chambon P, Davidson I (1993): Characterization of the transcription activation function and the DNA binding domain of transcriptional enhancer factor-1. *EMBO J* 12:2337–2348.

Jarrett HW (1993): Affinity chromatography with nucleic acid polymers. *J Chromatogr* 618:315–339.

Johnson PF, McKnight SL (1989): Eukaryotic transcriptional regulatory proteins. *Annu Rev Biochem* 58:799–839.

Kadonaga JT (1991): Purification of sequence-specific binding proteins by DNA affinity chromatography. *Methods Enzymol* 208:10–23.

Kadonaga JT, Tjian R (1986): Affinity purification of sequence-specific DNA binding proteins. *Proc Natl Acad Sci USA* 83:5889–5893.

Katoh-Fukui Y, Noce T, Ueda T, Fujiwara Y, Hashimoto N, Higashinakagawa T, Killian CE, Livingston BT, Wilt FH, Benson SC, Sucov HM, Davidson EH (1991): The corrected structure of the SM50 spicule matrix protein of *Strongylocentrotus purpuratus*. *Dev Biol* 145:201–202.

Kieran M, Blank V, Logeat F, Vandekerckhove J, Lottspeich F, Le Bail O, Urban MB, Kourilsky P, Baeuerle PA, Isreal A (1990): The DNA binding subunit of NF-kB is identical to factor KBF1 and homologous to the *rel* oncogene product. *Cell* 62:1007–1018.

Kingston RE, Green MR (1994): Modeling eukaryotic transcriptional activation. *Curr Biol* 4:325–332.

Kirchhamer CV, Davidson EH (1996): Spatial and temporal information processing in the sea urchin embryo: modular and intramodular organization of the *CyIIIa* gene *cis*-regulatory system. *Development* 122:*in press*.

Laemmli UK (1970): Cleavage of structural proteins during the assembly of the head of the bacteriophage, T4. *Nature* 227:680–685.

Lee CC, Wu X, Gibbs RA, Cook RG, Muzny DM, Caskey CT (1988): Generation of cDNA probes directed by amino acid sequence: cloning of urate oxidase. *Science* 239:1288–1291.

Li L, Olson EN (1992): Regulation of muscle cell growth and differentiation by the MyoD family of helix–loop–helix proteins. *Adv Cancer Res* 58:95–119.

Lynn DA, Angerer LM, Bruskin AM, Klein WH, Angerer RC (1983): Localization of a family of mRNAs in a single cell type and its precursors in sea urchin embryos. *Proc Natl Acad Sci USA* 80:2656–2660.

McMahon AP, Flytzanis CN, Hough-Evans BR, Katula KS, Britten RJ, Davidson EH (1985): Introduction of cloned DNA into sea urchin egg cytoplasm: replication and persistence during embryogenesis. *Dev Biol* 108:420–430.

Mitchell PJ, Tjian R (1989): Transcriptional regulation in mammalian cells by sequence-specific DNA binding proteins. *Science* 245:371–378.

Nemer M, Thornton RD, Stuebing EW, Harlow P (1991): Structure, spatial, and temporal expression of two sea urchin metallothionein genes, $SpMTB_1$ and SpMTA. *J Biol Chem* 266:6586–6593.

Niemeyer CC, Flytzanis CN (1993): Upstream elements involved in the embryonic regulation of the sea urchin CyIIIb actin gene: temporal and spatial specific interactions at a single *cis*-acting element. *Dev Biol* 156:293–302.

Nocente-McGrath C, Brenner CA, Ernst SG (1989): Endo16, a lineage-specific protein of the sea urchin embryo, is first expressed just prior to gastrulation. *Dev Biol* 136:264–272.

Norman C, Runswick M, Pollock R, Treisman R (1988): Isolation and properties of cDNA clones encoding SRF, a transcription factor that binds to the *c-fos* serum response element. *Cell* 55:989–1003.

Ogawa E, Inuzuka M, Maruyama M, Satake M, Naito-Fujimoto M, Ito Y, Shigesada K (1993a): Molecular cloning and characterization of PEBP2b, the heterodimeric partner of a novel *Drosophila runt*-related DNA binding protein PEBP2a. *Virology* 194:314–331.

Ogawa E, Maruyama M, Kagoshima H, Inuzuka M, Lu J, Satake M, Shigesada K, Ito Y (1993b): PEBP2/PEA2 represents a family of transcription factors homologous to the products of the *Drosophila* runt gene and the human *AML1* gene. *Proc Natl Acad Sci USA* 90:6859–6863.

Pabo CO, Sauer RT (1992): Transcription factors: structural families and principles of DNA recognition. *Annu Rev Biochem* 61:1053–1095.

Parr BA, Parks AL, Raff RA (1989): Promoter structure and protein sequence of Msp130, a sea urchin primary mesenchyme cell glycoprotein. *J Biol Chem* 265:1408–1413.

Perkins KK, Admon A, Patel N, Tjian R (1990): The *Drosophila* fos-related AP-1 protein is a developmentally regulated transcription factor. *Genes Dev* 4:822–834.

Ptashne M (1988): How eukaryotic transcriptional activators work. *Nature* 335:683–689.

Ransick A, Ernst S, Britten RJ, Davidson EH (1993): Whole mount *in situ* hybridization shows Endo16 to be a marker for the vegetal plate territory in sea urchin embryos. *Mech Dev* 42:117–124.

Reynolds SD, Angerer LM, Palis J, Nasir A, Angerer RC (1992): Early mRNAs, spatially restricted along the animal-vegetal axis of sea urchin embryos, include one encoding a protein related to tolloid and BMP-1. *Development* 114:769–786.

Singh H (1993): Specific recognition site probes for isolating genes encoding DNA-binding proteins. *Methods Enzymol* 218:551–567.

Singh H, LeBowitz H, Baldwin AS, Sharp PA (1988): Molecular cloning of an enhancer binding protein: isolation by screening of an expression library with a recognition site DNA. *Cell* 52:415–423.

Soeller WC, Oh CE, Kornberg TB (1993): Isolation of cDNAs encoding the *Drosophila* GAGA transcription factor. *Mol Cell Biol* 13:7961–7970.

Sucov HM, Benson S, Robinson JJ, Britten RJ, Wilt FH, Davidson EH (1987): A lineage-specific gene encoding a major spicule matrix protein of the sea urchin embryo. II. Structure of the gene and derived sequence of the protein. *Dev Biol* 120:507–519.

Sucov HM, Hough-Evans BR, Franks RR, Britten RJ, Davidson EH (1988): A regulatory domain that directs lineage-specific expression of a skeletal matrix protein gene in the sea urchin embryo. *Genes Dev* 2:1238–1250.

Summers RG, Stricker SA, Cameron RA (1993): Applications of confocal microscopy to studies of sea urchin embryogenesis. In: Matsumoto B (ed). *Methods in cell biology,* vol 38. Orlando, FL: Academic Press, Chap 10.

Thézé N, Calzone FJ, Thiebaud P, Hill RL, Britten RJ, Davidson EH (1990): Sequences of the CyIIIa actin gene regulatory domain bound specifically by sea urchin embryo nuclear proteins. *Mol Reprod Dev* 25:110–122.

Vinson CR, LaMarco KL, Johnson PF, Landschultz WH, McKnight SL (1988): In situ detection of sequence-specific DNA binding activity specified by a recombinant bacteriophage. *Genes Dev* 2:801–806.

Virbasius C-MA, Virbasius JV, Scarpulla RC (1993): NRF-1, an activator involved in nuclear–mitochondrial interactions, utilizes a new DNA-binding domain conserved in a family of developmental regulators. *Genes Dev* 7:2431–2445.

Wakao H, Gouilleux F, Groner B (1994): Mammary gland factor (MGF) is a novel member of the cytokine transcription factor gene family and confers the prolactin response. *EMBO J* 13:2182–2191.

Wang L-H, Tsai SY, Cook RG, Beattie WG, Tsai M-J, O'Malley BW (1989): COUP transcription factor is a member of the steroid receptor superfamily. *Nature* 340:163–166.

Wang DG-W, Kirchhamer CV, Britten RJ, Davidson EH (1995): SpZ12-1, a negative regulator required for spatial control of the territory-specific CyIIIa gene in the sea urchin embryo. *Development* 121:1111–1122.

Wessel GM, Goldberg L, Lennarz WJ, Klein WH (1989a): Gastrulation in the sea urchin embryo is accompanied by the accumulation of an endoderm-specific mRNA. *Dev Biol* 136:526–536.

Wessel GM, Zhang W, Tomlinson CR, Lennarz WJ, Klein WH (1989b): Transcription of the Spec1-like gene of *Lytechinus* is selectively inhibited in response to disruption of the extracellular matrix. *Development* 106:355–365.

Wiederrecht G, Seto D, Parker CS (1988): Isolation of the gene encoding the *S. cerevisiae* heat shock transcription factor. *Cell* 54:841–853.

Wilkinson DG, Nemer M (1987): Metallothionein genes MTa and MTb expressed under distinct quantitative and tissue-specific regulation in sea urchin embryos. *Mol Cell Biol* 7:48–58.

Williams T, Admon A, Lüscher B, Tjian R (1988): Cloning and expression of AP-2, a cell-type-specific transcription factor that activates inducible enhancer elements. *Genes Dev* 2:1557–1569.

Wilt F (1987): Determination and morphogenesis in the sea urchin embryo. *Development* 100:559–575.

Wood WI, Gitschier J, Laskey LA, Lawn RM (1985): Base composition-independent hybridization in tetramethyl ammonium chloride—a method for oligonucleotide screening of highly complex gene libraries. *Proc Natl Acad Sci USA* 82:1585–1588.

Xiang M, Bedard P-A, Wessel GM, Filion M, Brandhorst BP, Klein WH (1988): Tandem duplication and divergence of a sea urchin protein belonging to the troponin C superfamily. *J Biol Chem* 263:17173–17180.

Xiang M, Ge T, Tomlinson CR, Klein WH (1991): Structure and promoter activity of the LpS1 genes of *Lytechinus pictus:* duplicated exons account for LpS1 proteins with eight different calcium-binding domains. *J Biol Chem* 266:10524–10533.

Xiao JH, Davidson I, Matthes H, Garnier J-M, Chambon P (1991): Cloning, expression, and transcriptional properties of the human enhancer factor TEF-1. *Cell* 65:551–568.

Yang Q, Angerer LM, Angerer RC (1989): Structure and tissue-specific developmental expression of a sea urchin arylsulfatase gene. *Dev Biol* 135:53–65.

Yuh C-H, Ransick A, Martinez P, Britten RJ, Davidson EH (1994): Complexity and organization of DNA–protein interactions in the 5′-regulatory region of an endoderm-specific marker gene in the sea urchin embryo. *Mech Dev* 47:165–186.

Zeller RW, Cameron RA, Franks RR, Britten RJ, Davidson EH (1992): Territorial expression of three different *trans*-genes in early sea urchin embryos detected by a whole-mount fluorescence procedure. *Dev Biol* 151:382–390.

Zeller RW, Griffith JD, Moore JD, Kirchhamer CV, Britten RJ, Davidson EH (1995a): A multimerizing transcription factor of sea urchin embryos capable of looping DNA. *Proc Natl Acad Sci USA* 92:2989–2993.

Zeller RW, Coffman JA, Harrington MG, Britten RJ, Davidson EH (1995b): SpGCF1, a sea urchin embryo DNA binding protein, exists as five nested variants encoded by a single mRNA. *Dev Biol* 169:713–727.

CHAPTER 13

Evolutionary Approaches to Analyzing Development

RUDOLF A. RAFF and ELLEN M. POPODI

Indiana Institute for Molecular and Cellular Biology and Department of Biology, Indiana University, Bloomington, Indiana 47405

CONTENTS

SYNOPSIS

We discuss the complementary issues of evolutionary approaches to developmental biology and the use of developmental/genetic information to understand constraints on evolutionary change. Data derived from experimental embryology and developmental genetics can now be considered from an evolutionary perspective. The

Molecular Zoology: Advances, Strategies, and Protocols, Edited by Joan D. Ferraris and Stephen R. Palumbi.
ISBN 0-471-14461-4   © 1996 Wiley-Liss, Inc.

discovery of conserved regulatory genes in model systems can be extended to nonmodel organisms. The knowledge of these genes and the techniques of molecular biology make possible comparative studies that allow the evaluation of alternate forms of homologous developmental features or processes. Placement of comparative data in a phylogenetic context allows the direction of evolutionary change, and in some cases its rate, to be inferred. We present two examples of studies in evolutionary developmental biology from our work on the evolution of development in sea urchins. The first explores the mechanisms that allow radical evolutionary changes in mode of early development. The second uses conserved patterns of Hox gene expression to understand evolutionary changes in body plan in the transition from a bilaterally symmetric deuterostome ancestor to a pentameral echinoderm.

INTRODUCTION

Although textbooks feature the standard mode of development for the most intensively studied animals, there is a wide diversity in developmental modes even among closely related species. As developmental biologists, we seek the mechanisms that transform genomic information into an individual organism. The majority of this work is performed on a few model organisms. The major advantage of most of the currently fashionable model organisms is their amenability to study by genetic approaches and their ease of maintenance as standardized, genetically homogeneous laboratory stocks. However, as Bolker (1995) has pointed out, species used as model systems may be very untypical members of their groups. Nonetheless, model systems have allowed immense progress to be made in developmental biology and have provided the tools that allow phylogenetically broader investigations, such as those we will describe.

Developmental biologists use methods such as mutational disruption of genes, ablation or transplantation of embryonic cells, tracing of cell lineages, and surgical manipulation of embryos. These approaches focus on developmental processes and do not require that we consider development in an evolutionary context. We argue here that evolutionary concepts and approaches provide both powerful means of analyzing development and ways to assess the role of developmental regulatory genes and processes in the evolution of form.

HOW TO STUDY THE EVOLUTION OF DEVELOPMENT

We can explore these processes by means of evolutionary comparisons between model systems and species that have divergent developmental patterns. The use of several taxa allows us to seek commonalities of mechanism. In some cases, similar evolutionary changes in development may have evolved independently in disparate taxa. Convergent evolution may indicate powerful constraints on the architecture of developing systems or on the processes available (Wake, 1991).

There are two distinct approaches to connecting developmental and evolutionary biology. The most common is to compare phylogenetically distant model systems. These comparisons show that major developmental regulatory genes are generally highly conserved in sequence and may play similar or divergent roles in development. The homeotic genes of the Hox cluster that mediate axial polarity in flies also do so in vertebrates, although in importantly different ways. The gene *hedgehog,* which produces a signaling protein involved in specification of segment polarity in *Drosophila,* also generates polarities in the development of the vertebrate central nervous system and limbs. These and other analogous observations have stunned developmental biologists [see discussions of the role of *hedgehog* in *Drosophila* by Ingham and Martinez-Arias (1992); in vertebrates by Echelard et al. (1993), Krauss et al. (1993), and Riddle et al. (1993)]. Comparisons between gene sequences have also revealed conserved domains in regions upstream of the coding sequences, important in binding of conserved transcription factors. Evolutionary conservation has thus become a major tool for developmental biologists.

Because we want to know what makes ontogenies different, rather than what makes them the same, we find a second approach more useful. This involves seeking what mechanisms underlie evolutionary changes in development among closely related species that are not necessarily model system animals. The approach is analogous to that used in developmental genetics, where mutations are used to disrupt the functions of developmentally important genes. The resulting phenotypes are compared with the wild type, and the defects are analyzed to define the role of the mutated gene. In evolutionary studies of development, the modified pattern is compared to the primitive mode of development, which serves the same role as the wild type in a genetic comparison. This approach has been effectively used by Warren et al. (1994), for example, in showing how regulatory changes in Hox gene expression and in downstream gene responses underlie morphological differences between flies, which lack larval abdominal legs and a second pair of wings, and butterflies, which possess these features.

To identify the primitive mode of development, a phylogeny, based on features other than developmental ones, should be inferred and developmental features mapped onto it. In many cases, gene sequences offer the best approach to finding informative characters. Molecular analyses also allow inference of minimum rates of change by allowing estimates of times of divergence of closely related lineages (Raff et al., 1994). Once a robust phylogeny is obtained, the direction or polarity of evolutionary change can be inferred from the distribution of character states in the tree.

Evolutionary comparisons also require the establishment of homologies. Using closely related species generally allows homologies between developmental features to be established more readily than they can be between distant forms. The general criteria for establishing homology are shared position within a complex of features, similarity in structure and composition, and historical transition (generally obtained from the fossil record) (Riedl, 1978). These criteria, with adjustment for the fact that embryonic rather than adult features are being used, can be applied in comparisons between cells and structures within embryos.

Establishing homologies among developing features requires detailed data from both primitive and derived embryos. Origins and fates of cells provide a particularly important source of information. Simple observation is insufficient for tracing where cells arise or end up, because cells and even whole sheets of cells can change position in the embryo without a dramatic change in overall shape. Mapping of cell lineages is done by labeling precursor cells early in development, and observing the fates of marked cells at some later stage. The resulting data yield both cell lineages and cell fate maps for the embryo. Cell lineage tracing can yield data that fulfill the positional criterion for homology.

Some caution is necessary because, as noted by deBeer (1971), Roth (1988), and Shubin and Alberch (1986), developmental features do evolve. Homology criteria based on developmental features can thus be misleading. Elements of cell behavior or gene expression in development can be lost, truncated, or shifted in timing, which may seriously affect inference of cellular homologies. However, without at least a provisional inference of homology, evolutionary comparisons cannot be made. The use of closely related species makes the inference of homology more certain.

MODES OF DEVELOPMENT IN SEA URCHINS

The requirements listed above for investigating evolutionary changes in development are met in only a few organisms, but the list is growing. Closely related sea urchins that differ in early development meet the requirements and are accessible for comparative studies. Most sea urchins develop in a stereotypic manner, as diagrammed in Figure 1a. The embryo cleaves radially to form a three-tiered 16-cell embryo (Hörstadius, 1973). The mesomeres give rise to larval ectoderm; macromeres to larval ectoderm, gut, secondary mesenchyme cells, and coelomic cells; and the micromeres to the primary mesenchyme cells that secrete the skeleton and to coelomic cells.

Once the hollow blastula is reached, primary mesenchyme cells ingress into the blastocoel, and invagination of the archenteron or larval gut begins (Ettensohn, 1984; Burke et al., 1991; Hardin, 1989; Hardin and Cheng, 1986). Filopodia produced by secondary mesenchyme cells at the tip of the archenteron guide it to the roof of the blastocoel and help establish the site of the larval mouth (Hardin and McClay, 1991). The skeleton of calcareous spicular rods is secreted by the primary mesenchyme cells. A ciliary band arrayed along the surfaces of the pluteus arms is used for swimming and food capture. A larval nervous system appears at the base of the arms. The tip of the archenteron fuses with the oral ectoderm to form the larval mouth, and the archenteron differentiates to form a functioning gut.

The pluteus ultimately develops eight arms, and the precursor of the definitive juvenile (the echinus rudiment) develops from the ectodermal vestibule that forms on the left side of the larva, and the hydrocoel that arises from the left coelomic cavity. The pluteus has a history extending 250 million years or more (Wray, 1992). However, about 20% of living species of sea urchins lack a feeding pluteus and develop more or less directly from gastrula into juvenile (Fig. 1b). The evolution

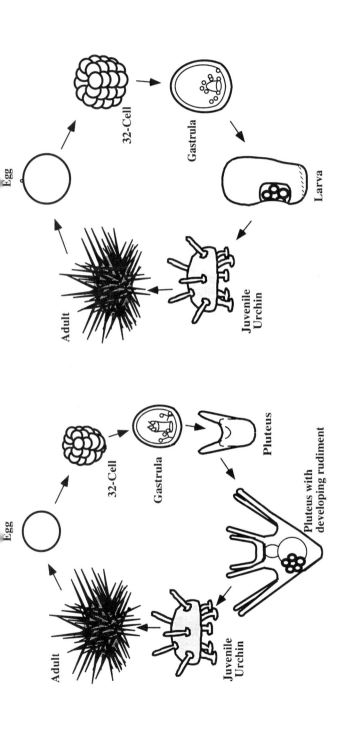

(a) **INDIRECT DEVELOPER**　　　(b) **HELIOCIDARIS ERYTHROGRAMMA**

Figure 1. (a) Life cycle of a typical indirect-developing sea urchin. Embryos are oriented with the animal pole up. The pluteus is drawn looking down on its animal pole. Its oral face and mouth face up. Development proceeds from egg through a feeding pluteus larva. The prospective adult forms as an echinus rudiment on the left side of the pluteus. Upon metamorphoses, the juvenile everts, and some parts of the pluteus, such as the larval arms, are lost. (b) Life cycle of *H. erythrogramma*, a direct-developing sea urchin. The embryos are oriented animal pole up. The ventral side is to the left and shows the vestibule with the tube feet of the developing juvenile sea urchin visible. The ciliated band is visible under the vestibule.

from feeding to nonfeeding larval forms is a common phenomenon in the evolution of larval stages of marine animals (Strathmann, 1978). In sea urchins, the indirect mode via a pluteus is inferred to be primitive.

Individual sea urchin species that possess nonfeeding larvae are embedded in the phylogenetic tree among species that possess plutei (Fig. 2), supporting the inference that the direct-developing larva evolved from a pluteus. Direct development has evolved independently among members of almost all orders of sea urchins (Emlet et al., 1987; Raff, 1987; Wray and Raff, 1991a, 1991b). Egg size correlates with developmental mode [as, surprisingly, do sperm head size and genome size (Raff et al., 1990)]. Indirect-developing sea urchins produce eggs ranging from 65 to 320 μm in diameter, with most around 100 μm, whereas direct developers produce eggs ranging from 300 to 2000 μm.

Larger-egged forms produce floating (lecithotrophic) larvae. As in the case of *Asthenosoma ijimai* (Amemiya and Emlet, 1992), they may secrete reduced larval skeletal elements in stubby arms. In *Heliocidaris erythrogramma,* which retains no

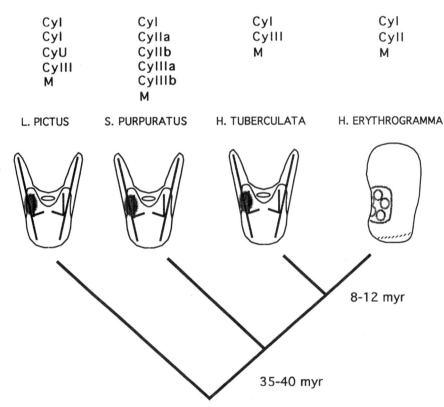

Cyl
Cyl
CyU
CyIII
M

Cyl
CyIIa
CyIIb
CyIIIa
CyIIIb
M

Cyl
CyIII
M

Cyl
CyII
M

L. PICTUS   S. PURPURATUS   H. TUBERCULATA   H. ERYTHROGRAMMA

8-12 myr

35-40 myr

Figure 2. Phylogenetic position of *H. erythrogramma* within related indirect developers. The distribution of cytoplasmic and muscle actin genes in sea urchin phylogeny is also indicated. The gene family shows both expansion and contraction of members as well as changes in patterns of gene expression in early development.

external trace of arms, they produce a tiny pair of reduced pluteus arm rods (Emlet, 1995). The largest known eggs, such as the 1.3-mm egg of the brooding Antarctic heart urchin *Abatus cordatus,* retain no trace of the pluteus (Schatt, 1985). Developmental modes can differ among closely related species. For example, *Phyllacanthus imperialis* has a partial nonfeeding four-armed pluteus, whereas *P. parvispinus* has a football-shaped larva without pluteus features (Parks et al., 1989; Olsen et al., 1993).

The most common southeastern Australian shallow-water sea urchins are *Heliocidaris tuberculata* and *H. erythrogramma*. The first has a 90-μm egg and produces a typical pluteus. In contrast, *H. erythrogramma* has a 430-μm egg, 100 times the volume of the *H. tuberculata* egg, and undergoes direct development from egg to juvenile in about 90 hours (Williams and Anderson 1975; Raff, 1987). Single-copy DNA hybridization data are consistent with a 10–13 million year divergence between the species, and the mitochondrial distances with a 5–8 million year divergence (Smith et al., 1990; McMillan et al., 1992). The transition to direct development probably took place relatively soon after the separation of the lineages, with additional modifications occurring thereafter.

THE REORGANIZATION OF DEVELOPMENT

The evolution of direct development in sea urchins was rapid, but it was not a simple event. Extensive modifications in both internal and external cell fates show that no unitary mechanism or single gene change underlies the switch to direct development. A first hypothesis about *H. erythrogramma* was that direct development arose by heterochrony (Raff, 1987, 1988). Development of the archenteron and larval skeleton is abbreviated, and development of coelomic cavities and adult skeleton begins early and is accelerated. However, these heterochronies are more accurately considered as the results of more fundamental underlying changes in morphogenesis of *H. erythrogramma* (Raff, 1992).

Our initial experimental approach was to determine cell fates in *H. erythrogramma* for comparison with those in *S. purpuratus* for which a complete fate map has been generated (Cameron et al., 1987, 1989, 1991). This was done by injection of fluorescently labeled dextran into single cells of 8- or 32-cell-stage embryos (Wray and Raff, 1989, 1990). By observing the patches of labeled cells in gastrula and later stages, it was possible to trace cell lineages and fates of cells derived from each of the embryonic blastomeres. The fate map of *H. erythrogramma* shows how profoundly cell lineages and fates have been remodeled (Wray and Raff, 1989, 1990).

The animal–vegetal axis of sea urchin eggs is established in the unfertilized egg. The dorsoventral axis of indirect developers is established during early embryonic development. We investigated the establishment of dorsoventral symmetry by separation and culturing of the first two cells formed by the fertilized egg of *H. erythrogramma* (Henry and Raff, 1990; Henry et al., 1990, 1992). Neither the site of sperm entry nor the plane of the first cleavage sets up the dorsoventral axis in *H.*

erythrogramma. Thus our studies indicated that the maternal organization of the egg cytoplasm has changed in *H. erythrogramma* such that the dorsoventral axis has become maternally determined. Although the dorsoventral pattern is determined maternally in *H. erythrogramma*, its expression requires subsequent embryonic gene expression and cell–cell communication (Henry and Raff, 1994). Separated dorsal and ventral halves show a progressive decline in the ability of dorsal halves to produce ventral structures, consistent with a repressive signal propagated along the dorsoventral axis. The use of maternal information to establish pattern, followed by increasingly complex patterns of cell–cell interaction and transcriptional specialization of groups of cells within the embryo, is a general strategy in early development. These mechanisms turn out to be strikingly flexible in evolution.

Morphogenesis of *H. erythrogramma* is distinct from that of indirect developers. As in other large-egged echinoderms, a wrinkled blastula is formed. The wrinkles disappear through a process of cell elongation and tighter packing, and a typical-looking blastula is formed (Henry et al., 1991). A large number of mesenchyme cells, about 2000, enter the blastocoel, and the archenteron invaginates (Williams and Anderson, 1975; Parks et al., 1988). Thus the processes that lead to gastrulation in *H. erythrogramma* have been substantially modified. Despite the expectation that gastrulation should be evolutionarily conservative, profound changes in gastrulation of *H. erythrogramma* have occurred (Wray and Raff, 1991b). In indirect development, archenteron elongation results from cell rearrangement, but in *H. erythrogramma* a sheet of cells involutes from the ventral side of the embryo. That change was very simply demonstrated by injection of fluorescent dye into the ventral blastomere at the two-cell stage of *H. erythrogramma*, and observing the movement of the labeled cell sheet derived from the ventral half. Because the *H. erythrogramma* embryo does not grow, this means of elongating the archenteron has evolved not to produce a larval gut, but to provide raw material for the early production of the coelom required for accelerated adult morphogenesis. The novel process cannot have evolved by modification of the ancestral cell rearrangement mechanism.

SEEKING GENE DIFFERENCES UNDERLYING DEVELOPMENTAL EVOLUTION

Differential gene expression is thought to underlie differentiation and morphogenesis. The different morphologies of indirect- and direct-developing sea urchin embryos imply that their patterns of gene expression should be distinct. Two hypotheses suggest how gene expression might underlie the evolutionary changes in morphology observed in *H. erythrogramma*. The first is that the direct-developing embryo has lost many pluteus features, with loss of expression of genes involved in constructing the pluteus. Alternatively, the direct developer might have evolved novel developmental features, with new genes as well as changes in timing of expression of genes shared by the two *Heliocidaris* species. *Heliocidaris erythrogramma* has a 30% larger genome than *H. tuberculata* (Raff et al., 1990). Much of that is single-copy DNA (Smith et al., 1990). Thus no obvious simplification is involved.

We are taking three approaches to studying evolutionary changes in gene regulation. These are to (1) examine mechanisms of change in expression of structural genes, (2) examine changes in expression of regulatory genes known from model systems, and (3) search for differentially expressed novel regulatory genes.

Structural Genes: A Mesenchyme-Specific Protein

One of the most pronounced changes in *H. erythrogramma* development is a heterochronic delay in secretion of the skeleton, and a corresponding delay in the expression of a mesenchyme-specific protein (msp130) involved in skeleton secretion relative to its expression in indirect development (Parks et al., 1988). We are studying transcriptional regulatory sites upstream of the gene by means of promoter constructs that contain the msp130 promoter (or a truncated version of that promoter) attached to a β-galactosidase gene as a reporter. The constructs are injected into eggs of an indirect-developing sea urchin, *Lytechinus pictus*, and site and timing of expression of the β-galactosidase are determined by microscopic examination. The reporter protein can readily be detected by testing for its activity with a detection system that produces a blue color if active β-galactosidase is present (Gan et al., 1990; Harkey et al., 1995). Promoters from *S. purpuratus, H. erythrogramma*, and *H. tuberculata* are very similar and are expressed like the host msp130. Thus the heterochrony in *msp130* expression is the result of changes in transcriptional behavior of the cell, not in the promoter of the structural gene (K. Klueg and R.A. Raff, *in preparation*).

Structural Genes: Actins

We initially cloned the actin gene family from the *Heliocidaris* species because we wanted evolutionarily static markers. The logic was that the actin genes of *S. purpuratus* were highly regulated and cell lineage-specific. We reasoned that determining the sites of expression of these conserved genes in *H. erythrogramma* should allow us to homologize cell lineages between embryos of different developmental modes. However, we found that the actin gene family evolves over relatively short intervals of geological time, and even more surprisingly, patterns of expression of orthologous members of the family change rapidly. Both *Heliocidaris* species (as shown by RNA blots from eggs and embryos) express only two cytoplasmic actin genes (J.C. Kissinger, J.H. Hahn, and R.A. Raff, *in preparation*). These are a subset of the *S. purpuratus* genes.

Like *S. purpuratus, H. tuberculata* develops a pluteus larva. Nevertheless, both the number of actin genes and the tissue-specific patterns of expression of these genes have changed (Hahn et al., 1995; Kissinger et al., 1996). *Strongylocentrotus purpuratus* embryos express CyI, CyIIa, CyIIb, CyIIIa, and CyIIIb cytoplasmic actin genes and a muscle actin gene (Lee et al., 1984). *Heliocidaris tuberculata* expresses *CyI, CyIII,* and a muscle actin. The *H. erythrogramma* embryos also express only two cytoplasmic actin genes (*CyI* and *CyII,* and muscle actin), a different subset of the *S. purpuratus* genes. One actin is shared with *H. tuberculata,* the other is not. *CyIII* is represented only by a pseudogene in *H. erythrogramma* (demonstrated by cloning the gene and showing that it has a four-base insertion in its coding sequence). Loss of the *CyIII* gene correlates with developmental mode, in that *H. erythrogramma* develops no ectoderm similar to the aboral ectoderm of plutei.

The evolutionary behavior of the actin gene family is illustrated in Figure 2. *Strongylocentrotus purpuratus* and *Lytechinus pictus* are closely related outgroups to the two *Heliocidaris* species. *Lytechinus pictus* has four cytoplasmic and one muscle actin gene (Fang and Brandhorst, 1994). The family has expanded and contracted over modest evolutionary time scales, 10–40 million years.

We (Kissinger et al., 1996) determined the sites of expression of the actin genes in sections of embryos of both *Heliocidaris* species by *in situ* hybridization, using ^{33}P-labeled nucleotides in the *in situ* protocol of Angerer and Angerer (1991). Use of this isotope gives high specific activity probes, combined with a relatively low energy β particle, and thus the ability to rapidly obtain *in situ* data with good resolution. Because it is needed for rare mRNAs, this is our standard procedure, even though it is not necessary for the prevalent actin mRNAs.

Some aspects of actin expression were conserved between the two *Heliocidaris* species and *S. purpuratus,* but other aspects of the patterns have changed. With the limited number of cytoplasmic actin genes used by the *Heliocidaris* species, cell type expression patterns that are divided among several actin genes in *S. purpuratus* are covered by the two actin genes expressed in *Heliocidaris* species. Genes and gene expression patterns have evolved in sea urchins in ways that suggest that both the passage of time (neutral evolution) and evolution of developmental mode (adaptive evolution) play important roles.

Regulatory Genes: Known Genes

Detection of regulatory genes discovered in model systems in other organisms is currently a favored research approach in developmental biology. It offers known genes and defined probes for cloning, or sequences from which PCR primers can be prepared. For evolutionary developmental biology, known regulators present the investigator who is using a nonmodel organism with genes whose expression and function can be compared with a better known system. Because this is such a prevalent approach, we will consider one example here.

A study of genes involved in patterning of the pentameral and radially symmetric juvenile of *H. erythrogramma* (as a convenient model for echinoderms in general) should reveal how the patterning of radial symmetry evolved from the patterning systems of a bilaterally symmetric deuterostome ancestor. One aspect of this problem, involving Hox genes, is discussed further in the last section of this chapter. As an example of the general application of this approach to seeking known patterning genes, Ferkowicz and Raff (*in progress*) are using PCR, cDNA, and genomic library screens to isolate clones of members of the *wnt* gene family, which play roles in cell–cell signaling in a number of developmental settings (Nusse and Varmus, 1992; Parr and McMahon, 1994). Several *wnt* genes (*wnt-3A, wnt-5A,* and *wnt-5B*) are expressed in mouse primitive streak (Takada et al., 1994). We isolated a homologue of the *wnt-5A* gene in our screen. It has a complex set of expression patterns in vertebrates and has been implicated in morphogenetic movements (Moon et al., 1993). *In situ* hybridization reveals that in early *H. erythrogramma* its major expression is in the coelomic pouch, which is destined to contribute the mesodermal component of the juvenile (M. Ferkowicz and R.A. Raff, *unpublished data*). In later embryos, the *wnt-5A* expression pattern resolves into a pentameral array. Other mesoderm-expressed genes involved in patterning of vertebrates (which are deuterostomes, and thus related to echinoderms) are being isolated and investigated as well. Expression patterns of these genes will help to establish homologies between axial elements of vertebrate and echinoderm development.

Regulatory Genes: Searching for New Genes

Individual gene differences cannot tell us how different or similar gene deployment is, nor can it reveal any novel gene associated with an evolutionary change. We are seeking genetic differences between *H. erythrogramma* and *H. tuberculata* by a global screen for gene expression differences. We have devised a differential screen that will reveal gene expression specific to the direct developer. Duplicate lifts of an *H. erythrogramma* late gastrula/early larval cDNA library are screened with radioactively labeled cDNA probes derived from *H. erythrogramma* (first set of lifts) and *H. tuberculata* (second set of lifts) (E. Haag and R.A. Raff, *in progress*). The lifts are put on x-ray film to reveal clones to which the labeled probes have hybridized. A clone labeled by the *H. erythrogramma* probe but not by the *H. tuberculata* probe is a candidate for differential expression and is isolated and characterized further. From the late gastrula through early larval stage, *H. tuberculata* is differentiating a skeleton and a functioning gut, whereas *H. erythrogramma* is not yet secreting a skeleton but is forming a large coelomic pouch. The coelomic pouches of *H. tuberculata* will form 2 or 3 days later. Nevertheless, the vast majority of mRNAs are shared, with differences represented by fewer than 0.1% of clones visualized by hybridization. This means that despite the nascent differences in morphogenesis, the transcript inventory is conserved overall. Although the transcript catalogs of *H. tuberculata* and *H. erythrogramma* are so similar, two kinds of differences have been detected in our screen. These are gene expression heterochronies and apparently *H. erythrogramma*-specific sequences. Important genes may also be expressed by both species, but in different locations within the embryo, or in different processed forms. The differences in the expression of these genes will not be detected by a "catalog" screen of mRNAs.

An analogous study of tailed and tailless development in congeneric ascidians revealed transcripts specific to the tailed species. Swalla et al. (1993) isolated three sequences that may encode regulatory genes. Their study differs from ours in two important ways. First, they focused on maternal transcripts. Second, they looked for losses of genes in the switch to an abbreviated mode of development, whereas we have focused on gains in the direct-developing sea urchin. Losses can readily be investigated in our study as well by screening a *H. tuberculata* library as outlined above. We are also conducting a screen of maternal templates (J. Bolker and R.A. Raff, *in progress*).

THE ECHINODERM RADIAL BODY PLAN

The origin of the radial body plan of echinoderms has posed a difficult problem not resolved by the fossil record. Recent discoveries in development and genetics have suggested a novel approach to this evolutionary problem. A number of genes involved in embryonic pattern formation have been identified in *Drosophila*. Many of these genes have been conserved through evolution and appear to have the same role in diverse animals (the subject of many recent review articles, e.g., Akam et al., 1994). Comparative studies of genes that specify pattern can provide insight into the means by which the radial body plan of echinoderms evolved from the bilateral deuterostome ancestor.

Living echinoderms share a body plan that includes a pentamerally symmetric adult, and a unique ring-shaped coelomic water vascular system that generally has extensions into each of the five ambulacra (an arm or radius and its associated tube feet). The central nervous system (with some complexities that we discuss briefly later) consists of a central nerve ring from which radial nerves project into the ambulacra (Brusca and Brusca, 1990). Echinoderms are the sister group of hemichordates and chordates (Raff et al., 1994; Turbeville et al., 1994). The common ancestor of echinoderms, hemichordates, and chordates was most likely bilaterally symmetrical, with a differentiated anterior end and a linear nervous system. Our question is how this linear body plan was transformed into a radially symmetric one.

Pentameral symmetry arose after the origin of the echinoderms. Many Cambrian echinoderms were nonpentameral, with asymmetric or nearly bilaterally symmetric body plans. The extant classes, including the sea urchins, appeared some 30–50 million years later, during the Ordovician. Older taxa have died out and ancestral features have been lost. Only pentameral echinoderms remain today (Paul and Smith, 1984).

The topology of the central nervous system provides a key to the transformation of the linear body plan shared by the ancestors of all deuterostome phyla. The central nervous system is deeply integrated into the body plan of animal groups and develops as part of the phylotypic stage. The developmental mechanisms responsible for nervous system formation appear to be conserved throughout animal groups. Nervous tissue is ectodermal in origin and morphologically similar among animals. It develops early in ontogeny, and many of the genes involved in its development are conserved and exhibit conserved expression patterns (e.g., Lo et al., 1991; Oliver et al., 1993; Simeone et al., 1994).

The echinoderm nervous system is not highly centralized or cephalized. Living echinoderms have three distinguishable nervous systems, elaborated to varying degrees among extant classes (Hyman, 1955). The oral (ectoneural) system consists of a pentagonal ring surrounding the esophagus and five radial nerves. The radial nerves radiate from the nerve ring and run down the center of each ambulacral tract. The deep oral (hyponeural) system is usually present as five radial tracts running along the coelomic surface of the radial nerves of the oral system. The aboral (entoneural) system resembles the oral system in its organization, but its ring is located at the aboral pole of the animal. In sea urchins, the aboral and hyponeural systems are poorly developed; the oral system innervates the animal.

A group of conserved genes whose pattern of expression may shed some light on the evolution of pentameral symmetry in echinoderms are represented by the homeotic genes of *Drosophila*. These were among the first of the genes affecting embryonic pattern formation to be described (Lewis, 1978). These genes are involved in specifying identity along the anterior–posterior axis and are clustered into two linkage groups in flies (ANTP and BX). As molecular studies progressed, these genes were discovered to share a common structural motif called the homeobox, a 180 nucleotide stretch that encodes a 60 amino acid DNA binding domain (McGinnis et al., 1984; Scott and Weiner, 1984). Although homeoboxes have subsequently been found in a large number of different genes in a variety of different organisms, those found in the homeotic gene clusters are closely related to one another. These are called HOM genes in insects and Hox genes in other animals. Hox/HOM genes are clustered and their order in the cluster has been conserved. In all species examined they exhibit an anterior to posterior pattern of expression that reflects their order in the genome: this is termed "collinearity." These genes are expressed in a number of different tissues at different times during development. In both protostomes and deuterostomes they are expressed in a collinear fashion in the central nervous system early in its development (reviewed in Duboule, 1994). Therefore it is likely that they are expressed in a patterned order during the development of the echinoderm central system.

To investigate the transformation from an ancestral linear nervous system to the radial one of echinoderms, we have exploited the collinearity and nervous system expression of the genes in the Hox cluster. We reasoned that the expression pattern of the Hox genes in the developing sea urchin nervous system should reveal the topological transformation. We present three models for this transformation. For the purpose of model building we assumed the deuterostome ancestor had a single nerve cord. Echinoderms, like the related hemichordates, possess a subepidermal nerve net as well as a centralized nervous system. Nerve nets are common to "primitive" taxa and may well have been present in the deuterostome ancestor. However, as all modern deuterostomes have at least one central nerve cord, it seems likely that the ancestor did also. There is more than one ring nervous system in echinoderms; again, to simplify the models, we are focusing on just one.

Our models are presented in Figure 3 as a set of hypothetical transformations, along with the predicted consequences for Hox gene expression patterns. In the first, the radial nerves are proposed to represent the ancestral nerve cord. The five radial nerves of the echinoderm would have arisen by replication of the original nerve cord, followed by evolution of a ring commissure to link the nerve cords

MODELS

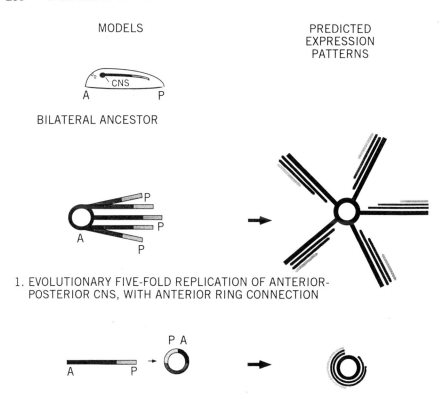

BILATERAL ANCESTOR

1. EVOLUTIONARY FIVE-FOLD REPLICATION OF ANTERIOR-POSTERIOR CNS, WITH ANTERIOR RING CONNECTION

2. LINEAR CNS CIRCULARIZED BY JOINING A TO P

3. BIFURCATION OF A LINEAR CNS

Figure 3. Hypothetical transformations of a linear nervous system into a radial nervous system. *Left side:* A cartoon of a proposed ancestral deuterostome and three models of the transformation of a linear nerve cord into a radially symmetric nervous system. (A = anterior; P = posterior.) *Right side:* Diagrams of the products of each transformation and the predicted patterns of expression of three Hox genes for each model. [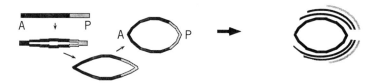, expression of a gene located toward the 3′ ("head") end of the cluster relative; ▮, expression of a gene in the center; ▮, expression of a gene located toward the 5′ ("tail") end of the cluster.]

together. This hypothesis predicts that the anterior to posterior pattern of Hox gene expression would be replicated in each radial nerve.

The other two hypotheses suggest that the ancestral nerve cord was itself transformed into a ring. The second postulates that the ring was formed by a head-to-tail fusion of the ancestral cord. The ancestral Hox gene expression pattern would be expected to be expressed in the ring. The third hypothesis suggests that the ancestral nerve cord consisted of bilaterally paired strands. A split of the center portion would have produced a ring, but one with a different Hox polarity than the head-to-tail ring. In both of these, the radial nerves would have arisen subsequent to the origin of the ring.

The hypothetical topological transformations of the echinoderm nervous system are testable by localizing the products of the Hox genes in the developing adult central nervous system. A stretch of at least 500 base pairs of sequence specific for a single gene is needed to detect the expression of that gene's message *in situ*. While the homeobox is highly conserved across phyla, very little conservation is found outside the homeobox region. In order to investigate the expression of the Hox genes in sea urchins several different sea urchin Hox genes needed to be isolated. Portions of a few Hox genes had been cloned from sea urchins and a little was known about expression in the pluteus larva and a few adult tissues (Dolecki et al., 1986, 1988; Wang et al., 1990; Angerer et al., 1989; Zhao, 1992). Nothing is known about the expression of these genes during the formation of the adult body. The direct developing sea urchin provides access to the developing rudiment.

We have cloned five Hox genes from a genomic library constructed from *H. erythrogramma* DNA. We chose to screen genomic DNA as opposed to cDNA because we knew so little of the expression patterns of the genes. In vertebrates the majority of Hox genes exhibit a conserved genomic structure of two exons, with the homeobox contained within the last (3' most) exon (e.g., Tournier-Lasserve et al., 1989; Ericson et al., 1993). Occasionally an additional exon upstream of the homeobox is observed (e.g., Cho et al., 1988; Benson et al., 1995). In sea urchins the homeobox is contained in the last exon of each of the genes cloned; we do not yet know the structure of the upstream portions of these genes. We sequenced the entire 3' exon of each gene and determined the class of homeobox. Two genes are of the *Abd-B* type (paralog groups 9–10) and three are of the *Antp* type (paralog groups 6–8).

These genes represent the middle and one end of the cluster. To demonstrate that the sea urchin genes are clustered, we tested the linkage of four genes on genomic DNA using Southern blots of very large fragments of DNA separated by pulse-field gel electrophoresis. All four genes hybridized to the same 300-kilobase band of DNA. This is the best evidence to date that these genes are clustered. Our data, as well as that of others (Ruddle et al., 1994), are consistent with the existence of only one cluster in sea urchins. This is similar to the situation in the cephalochordate amphioxus (Garcia-Fernandez and Holland, 1994).

Homeobox genes are under investigation in several other sea urchin species. There have been a total of eight different Hox-type genes cloned from sea urchins (Dolecki et al., 1986, 1988; Pfeffer and von Holt, 1991; Wang et al., 1990; Zhao, 1992; G. Spinelli, *personal communication;* E.M. Popodi, M.E. Andrews, and

R.A. Raff, *in preparation*). Of the five Hox genes we isolated, two had not been isolated from other sea urchins. Figure 4 summarizes the current state of knowledge of the Hox genes in sea urchins. The putative sea urchin cluster is drawn under a generalized chordate cluster. Sea urchin genes representing the middle, head, and tail ends of the cluster have been cloned. One gene is most similar to paralogs 2–3 at the "head" end. Three genes are most similar to paralogs 9–10 at the "tail" end, and four represent paralogs found in the middle of the Hox cluster. If the collinearity of expression of Hox genes is conserved in sea urchins, the five genes we have cloned will be sufficient to distinguish between the models we've proposed.

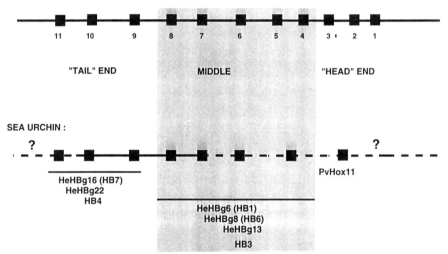

Figure 4. Summary of sea urchin Hox gene cluster. The top of the figure depicts a generalized chordate Hox cluster (Garcia-Fernandez and Holland, 1994). The number of the paralog group is underneath the box representing each gene. Genes of the middle group (paralogs 4–8 or *Scr-, Dfd-, Antp-, Ubx-,* and *Abd-A*-type genes) are in the shaded box. Eight Hox-type genes have been cloned from sea urchins. These are diagrammed in the bottom of the figure. The specific order of the sea urchin genes has not been established. Three are like paralogs 9–11 (*Abd-B*-type). Four are most similar to the middle group of genes (paralogs 4–8). One is similar to paralogs 2–3 (*pb*-type). We have shown at least four genes are linked; these are joined by a solid line. The dashed line indicates that the remaining genes are likely to be linked but this hasn't been demonstrated. The question marks (?) indicate that we don't know if there are additional Hox genes. Homeobox genes are under investigation in several other sea urchin species. Two of the genes we isolated, HeHBg*13* and HeHBg*22*, had not been isolated in other sea urchins. HeHBg16 is the same as SpHbox 7 isolated in *Strongylocentrotus purpuratus* (Zhao, 1992). HeHBg6 is the same as HB*1*, which had been isolated in *Tripnuestes gratilla* (Dolecki et al., 1986) and in *S. purpuratus* (Angerer et al., 1989). HeHBg8 is the same as HB6 originally isolated in *T. gratilla* (Wang et al., 1990). HB3 and HB4 were isolated in *T. gratilla* (Dolecki et al., 1988) and PvHox11 was isolated in *Paracentrotus lividus* (G. Spinelli, *personal communication*).

We are beginning our *in situ* analyses. Using a cDNA of a nervous system-specific message that we have isolated from a library of radial nerve cDNAs, we can easily detect the developing central nervous system in *H. erythrogramma* larvae and juveniles. Engrailed genes are another class of homeobox containing genes that have a conserved role in the development of the central nervous system (Patel et al., 1989). Preliminary in situ localization of engrailed mRNA in larvae and juveniles shows engrailed expression in the radial nerves, a pattern consistent with model 1. We are beginning to examine the expression pattern of the *HB1* gene, one of the middle group Hox genes. Characterizing these gene expression patterns represents a first step toward distinguishing among our proposed models for the topological transformation in the origin of the echinoderm body plan.

ACKNOWLEDGMENTS

This work was supported by USPHS Grants R01 Hd 21337 (to R.A.R.) and R01 NS 31661 (to R.A.R. and E.M.P.). We thank Dr. J. Bolker for critically reading the manuscript. We are deeply indebted to the University of Sydney and the Sydney Aquarium for their hospitality and assistance during our work in Australia.

REFERENCES

Akam M, Holland P, Ingham P, Wray G (eds) (1994): The evolution of developmental mechanisms. *Development 1994 Supplement.*

Amemiya S, Emlet RB (1992): The development and larval form of an echinothurioid echinoid, *Asthenosoma ijimai,* revisited. *Biol Bull* 182:15–30.

Angerer LM, Angerer RC (1991): Localization of mRNAs by in situ hybridization. *Methods Cell Biol* 35:37–71.

Angerer LM, Dolecki GJ, Gagnon ML, Lum R, Wang G, Yang Q, Humphreys T, Angerer TC (1989): Progressively restricted expression of a homeobox gene within the aboral ectoderm of developing sea urchin embryos. *Genes Dev* 3:370–383.

Benson GV, Nguyen T-HE, Maas RL (1995): The expression pattern of the murine Hoxa-10 gene and the sequence recognition of its homeodomain reveal specific properties of *Abdominal B*-like genes. *Mol Cell Biol* 15:1591–1601.

Bolker JA (1995): Model systems in developmental biology. *BioEssays* 17:451–455.

Brusca RC, Brusca GJ (1990): *Invertebrates.* Sunderland, MA: Sinauer.

Burke RD, Myers RL, Sexton TL, Jackson C (1991): Cell movements during the initial phase of gastrulation in the sea urchin embryo. *Dev Biol* 146:542–557.

Cameron RA, Hough-Evans BR, Britten RJ, Davidson EH (1987): Lineage and fate of each blastomere of the eight-cell sea urchin embryo. *Genes Dev* 1:75–85.

Cameron RA, Fraser SE, Britten RJ, Davidson EH (1989): The oral–aboral axis of a sea urchin is specified by first cleavage. *Development* 106:641–647.

Cameron RA, Fraser SF, Britten RJ, Davidson EH (1991): Macromere cell fates during sea urchin development. *Development* 113:1085–1091.

Cho KWY, Goetz J, Wright CVE, Fritz A, Hardwicke J, DeRobertis EM (1988): Differential utilization of the same reading frame in a *Xenopus* homeobox gene encodes two related proteins sharing the same DNA-binding specificity. *EMBO J* 7:2139–2149.

deBeer G (1971): Homology, an unsolved problem. In: Head JJ, Lowenstein OE (eds). *Oxford biology readers*. London: Oxford University Press, pp 3–16.

Dolecki GJ, Wannakrairoj S, Lum R, Wang G, Riley HD, Carlos R, Wang A, Humphreys T (1986): Stage-specific expression of a homeobox-containing gene in the non-segmented sea urchin embryo. *EMBO J* 5:925–930.

Dolecki GJ, Wang G, Humphreys T (1988): Stage- and tissue-specific expression of two homeobox genes in sea urchin embryos and adults. *Nucleic Acids Res* 16:11543–11558.

Duboule D (ed) (1994): *Guidebook to the homeobox genes*. Oxford: Oxford University Press.

Echelard Y, Epstein DJ, St-Jacques B, Shen L, Mohler J, McMahon JA, McMahon AP (1993): *Sonic hedgehog,* a member of a family of putative signaling molecules, is implicated in the regulation of CNS polarity. *Cell* 75:1417–1430.

Emlet RB (1995): Larval spicules, cilia, and symmetry as remnants of indirect development in the direct developing sea urchin *Heliocidaris erythrogramma*. *Dev Biol* 167:405–415.

Emlet RB, McEdward LR, Strathmann RR (1987): Echinoderm larval ecology viewed from the egg. In: Jangoux M, Lawrence JM (eds). *Echinoderm studies 2*. Rotterdam: Balkema, pp 55–136.

Ericson JU, Krauss S, Fjose A (1993): Genomic sequence and embryonic expression of the zebrafish homeobox gene *hox-3.4*. *Int J Dev Biol* 37:263–272.

Ettensohn CA (1984): Primary invagination of the vegetal plate during sea urchin gastrulation. *Am Zool* 24:571–588.

Fang H, Brandhorst BP (1994): Evolution of actin gene families of sea urchins. *J Mol Evol* 39:347–356.

Gan L, Wessel GM, Klein WH (1990): Regulatory elements from the related Spec genes of *Strongylocentrotus purpuratus* yield different spatial patterns with a *LacZ* reporter gene. *Dev Biol* 142:346–359.

Garcia-Fernandez J, Holland PWH (1994): Archetypal organization of the amphioxus Hox gene cluster. *Nature* 370:563–566.

Hahn J-H, Kissinger JC, Raff RA (1995): Structure and evolution of CyI cytoplasmic actin-encoding genes in the indirect- and direct-developing sea urchins *Heliocidaris tuberculata* and *Heliocidaris erythrogramma*. *Gene* 153:219–224.

Hardin J (1989): Local shifts in position and polarized motility drive cell rearrangements during sea urchin gastrulation. *Dev Biol* 136:430–445.

Hardin J, Cheng LY (1986): The mechanisms and mechanics of archenteron elongation during sea urchin gastrulation. *Dev Biol* 115:490–501.

Hardin J, McClay DR (1991): Target recognition by the archenteron during sea urchin gastrulation. *Dev Biol* 142:86–102.

Harkey MA, Klueg K, Sheppard P, Raff RA (1995): Structure, expression, and extracellular targeting of PM27, a skeletal protein associated specifically with growth of the sea urchin larval spicule. *Dev Biol* 168:549–566.

Henry JJ, Raff RA (1990): Evolutionary change in the process of dorsoventral axis determination in the direct developing sea urchin, *Heliocidaris erythrogramma*. *Dev Biol* 141:55–69.

Henry JJ, Raff RA (1994): Progressive determination of cell fates along the dorsoventral axis in the sea urchin *Heliocidaris erythrogramma*. *Roux's Arch Dev Biol* 204:62–69.

Henry JJ, Wray GA, Raff RA (1990): The dorsoventral axis is specified prior to first cleavage in the direct developing sea urchin *Heliocidaris erythrogramma*. *Development* 110:875–884.

Henry JJ, Wray GA, Raff RA (1991): Mechanism of an alternate type of echinoderm blastula formation: the wrinkled blastula of the sea urchin *Heliocidaris erythrogramma*. *Dev Growth Differ* 33:317–328.

Henry JJ, Klueg KM, Raff RA (1992): Evolutionary dissociation between cleavage, cell lineage and embryonic axes in sea urchin embryos. *Development* 108:107–119.

Hörstadius S (1973): *Experimental embryology of echinoderms*. Oxford: Clarendon Press.

Hyman LH (1955): *The invertebrates: Echinodermata. The coelomate bilateria,* vol IV. New York: McGraw-Hill.

Ingham PW, Martinez-Arias A (1992): Boundaries and fields in early embryos. *Cell* 68:221–235.

Kissinger JC, Hahn J-H, Raff RA (1996): Evolutionary changes in sites and timing of expression of actin genes in embryos of the direct- and indirect-developing sea urchins *Heliocidaris erythrogramma* and *H. tuberculata*. (*submitted*).

Krauss S, Concordet J-P, Ingham PW (1993): A functionally conserved homolog of the *Drosophila* segment polarity gene *hh* is expressed in tissues with polarizing activity in zebrafish embryos. *Cell* 75:1431–1444.

Lee JJ, Shott RJ, Rose SJ III, Thomas TL, Britten RJ, Davidson EH (1984): Sea urchin actin gene subtypes. Gene number, linkage and evolution. *J Mol Biol* 172:149–176.

Lewis EB (1978): A gene complex controlling segmentation in *Drosophila*. *Nature* 276:565–570.

Lo L-C, Johnson JE, Wuenschell CW, Saito T, Anderson DJ (1991): Mammalian *achaete-scute* homolog 1 is transiently expressed by spatially restricted subsets of early neuroepithelial and neural crest cells. *Genes Dev* 5:1524–1537.

McGinnis W, Levine MS, Hafen E, Kuroiwa A, Gehring WJ (1984): A conserved DNA sequence in homeotic genes of the *Drosophila* Antennapedia and bithorax complexes. *Nature* 308:428–433.

McMillan WO, Raff RA, Palumbi SR (1992): Population genetic consequences of developmental evolution in sea urchins (Genus *Heliocidaris*). *Evolution* 46:1299–1312.

Moon RT, Campbell RM, Christian JL, McGrew LL, Shih J, Fraser S (1993): *Xwnt-5A:* a maternal *wnt* that affects morphogenetic movements after overexpression in embryos of *Xenopus laevis*. *Development* 119:97–111.

Nusse R, Varmus H (1992): *Wnt* genes. *Cell* 69:1073–1087.

Oliver G, Sosa-Pineda B, Geisendorf S, Spana EP, Doe C, Gruss P (1993): Prox 1, a *prospero*-related homeobox gene expressed during mouse development. *Mech Dev* 44:3–16.

Olsen RR, Cameron JL, Young CM (1993): Larval development (with observations on spawning) of the pensil urchin *Phyllacanthus imperialis:* a new intermediate larval form? *Biol Bull* 185:77–85.

Parks AL, Parr BA, Chin JE, Leaf DS, Raff RA (1988): Molecular analysis of heterochrony in the evolution of direct development in sea urchins. *J Evol Biol* 1:27–44.

Parks AL, Bisgrove BW, Wray GA, Raff RA (1989): Direct development in the sea urchin *Phyllacanthus parvispinus* (Cidaroidea): phylogenetic history and functional modification. *Biol Bull* 177:96–109.

Parr BA, McMahon AP (1994): *Wnt* genes and vertebrate development. *Curr Opin Genet Dev* 4:523–528.

Patel NH, Martin-Blanco E, Coleman KG, Poole SJ, Ellis MC, Kornberg TB, Goodman CS (1989): Expression of engrailed proteins in Arthropods, Annelids, and Chordates. *Cell* 58:955–968.

Paul CRC, Smith AB (1984): The early radiation and phylogeny of echinoderms. *Biol Rev* 59:443–481.

Pfeffer PL, von Holt C (1991): Stage- and adult tissue-specific expression of a homeobox gene in embryo and adult *Parechinus angulosus* sea urchins. *Gene* 108:219–226.

Raff RA (1987): Constraint, flexibility, and phylogenetic change in the evolution of direct development in sea urchins. *Dev Biol* 119:6–19.

Raff RA (1988): Direct developing sea urchins: a system for the study of developmental processes in evolution. In: Burke RD, Mladenov PV, Lambert P, Parsley RL (eds). *Echinoderm biology*. Rotterdam: Balkema, pp 63–69.

Raff RA (1992): Direct-developing sea urchins and the evolutionary reorganization of early development. *BioEssays* 14:211–218.

Raff RA, Herlands L, Morris VB, Healy J (1990): Evolutionary modification of echinoid sperm correlates with developmental mode. *Dev Growth Differ* 32:283–291.

Raff RA, Marshall CR, Turbeville JM (1994): Using DNA sequences to unravel the Cambrian radiation of the animal phyla. *Annu Rev Ecol Syst* 25:351–375.

Riddle RD, Johnson RL, Laufer E, Tabin C (1993): *Sonic hedgehog* mediates the polarizing activity of the ZPA. *Cell* 75:1401–1416.

Riedl R (1978): *Order in living organisms*. Chichester: Wiley.

Roth VL (1988): The biological basis of homology. In: Humphries CJ (ed). *Ontogeny and systematics*. New York: Columbia University Press, pp 1–26.

Ruddle FH, Bentley KL, Murtha MT, Risch N (1994): Gene loss and gain in the evolution of the vertebrates. *Development 1994 Supplement*, pp 155–161.

Schatt P (1985): *Development et Croissance embryonaire de L'oursin incubant Abatus cordatus (Echinoidea: Spatangoida)*. Doctoral dissertation, Pierre and Marie Curie University, National Museum of Natural History, Paris.

Scott MP, Weiner AJ (1984): Structural relationships among genes that control development: sequence homology between the *Antennapedia, Ultrabithorax,* and *fushi tarazu* loci of *Drosophila*. *Proc Natl Acad Sci USA* 81:4115–4119.

Shubin NH, Alberch P (1986): A morphogenetic approach to the origin and basic organization of the tetrapod limb. *Evol Biol* 20:319–387.

Simeone A, D'Apice MR, Nigro V, Casonova J, Graziani F, Acampora D, Avantaggiato V (1994): *Orthopedia*, a novel homeobox-containing gene expressed in the developing CNS of both mouse and *Drosophila*. *Neuron* 13:83–101.

Smith MJ, Boom JD, Raff RA (1990): Single copy DNA distances between two congeneric sea urchin species exhibiting radically different modes of development. *Mol Biol Evol* 7:315–326.

Strathmann RR (1978): The evolution and loss of feeding larval stages of marine invertebrates. *Evolution* 32:894–906.

Swalla BJ, Makabe KW, Satoh N, Jeffery WR (1993): Novel genes expressed differentially in ascidians with alternate modes of development. *Development* 119:307–318.

Takada S, Stark KL, Shea MJ, Vassileva G, McMahon JA, McMahon AP (1994): *Wnt-3a* regulates somite and tailbud formation in the mouse embryo. *Genes Dev* 8:174–189.

Tournier-Lasserve E, Odenwald WF, Garbern J, Trojanowski J, Lazzarini RA (1989): Remarkable intron and exon sequence conservation in human and mouse homeobox Hox 1.3 genes. *Mol Cell Biol* 9:2273–2278.

Turbeville JM, Schulz JR, Raff RA (1994): Deuterostome phylogeny and the sister group of the chordates: evidence from molecules and morphology. *Mol Biol Evol* 11:648–655.

Wake DB (1991): Homoplasy: the result of natural selection, or evidence of design limitations. *Am Nat* 138:543–567.

Wang GVL, Dolecki GJ, Carlos R, Humphreys T (1990): Characterization and expression of two sea urchin homeobox gene sequences. *Dev Genet* 11:77–87.

Warren RW, Nagy L, Selegue J, Gates J, Carroll S (1994): Evolution of homeotic gene regulation and function in flies and butterflies. *Nature* 372:458–461.

Williams DHC, Anderson DT (1975): The reproductive system, embryonic development, larval development, and metamorphosis of the sea urchin *Heliocidaris erythrogramma* (Val.) (Echinoidea: Echinometridae). *Aust J Zool* 23:371–403.

Wray GA (1992): The evolution of larval morphology during the post-Paleozoic radiation of echinoids. *Paleobiology* 18:258–287.

Wray GA, Raff RA (1989): Evolutionary modification of cell lineage in the direct-developing sea urchin *Heliocidaris erythrogramma*. *Dev Biol* 132:458–470.

Wray GA, Raff RA (1990): Novel origins of lineage founder cells in the direct-developing sea urchin *Heliocidaris erythrogramma*. *Dev Biol* 141:41–54.

Wray GA, Raff RA (1991a): The evolution of developmental strategy in marine invertebrates. *Trends Ecol Evol* 6:45–50.

Wray GA, Raff RA (1991b): Rapid evolution of gastrulation mechanisms in a sea urchin with lecithotrophic larvae. *Evolution* 45:1741–1750.

Zhao AZ (1992): *Temporal regulation of the L1 late H2B histone gene during sea urchin embryonic development by an Antennapedia class homeoprotein.* Ph.D. Dissertation, University of Southern California, Department of Biology, Los Angeles, California.

CHAPTER 14

Revealing Homologies Between Body Parts of Distantly Related Animals by *in Situ* Hybridization to Developmental Genes: Amphioxus Versus Vertebrates

LINDA Z. HOLLAND

Scripps Institution of Oceanography, University of California–San Diego, La Jolla, California 92093-0202

PETER W.H. HOLLAND

School of Animal and Microbial Sciences, University of Reading, Whiteknights, Reading, RG6 2AJ United Kingdom

NICHOLAS D. HOLLAND

Scripps Institution of Oceanography, University of California–San Diego, La Jolla, California 92093-0202

CONTENTS

Molecular Zoology: Advances, Strategies, and Protocols, Edited by Joan D. Ferraris and Stephen R. Palumbi.
ISBN 0-471-14461-4   © 1996 Wiley-Liss, Inc.

SYNOPSIS

During the last decade, molecular geneticists have demonstrated that genes directing development are often remarkably conserved—structurally and, to some extent, functionally—between one animal phylum and the next. From an evolutionary perspective, comparative molecular genetics can be used to construct developmental gene trees, to reveal developmental gene duplications in evolutionary lineages, and to help establish homologies between body parts of distantly related animals. This last use, which is the focus here, employs expression patterns of homologous developmental genes as new phenotypic characters to infer homologies. The approach is illustrated by comparing amphioxus (lancelets) with vertebrates to address the problem of the origin of the vertebrates from the invertebrates. For the zoologist who is not well-versed in molecular biology, we provide an overview of the methodology involved in going all the way from live animals to determining the expression patterns of genes by *in situ* hybridization. The emphasis is on the pros and cons of alternative methods and on avoiding pitfalls. Some specific methods—including those for whole-mount *in situ* hybridization of riboprobes to small invertebrate larvae—that are not readily available in general molecular methods texts, are presented in Holland.1, Holland.2, and Holland.3.

INTRODUCTION

The object of this volume is to show how techniques of molecular biology can help resolve some outstanding difficulties in classical zoology. One such difficulty arises when one attempts to establish body part homologies between relatively distantly related animals. The 1830 debate between Cuvier and Geoffroy Saint-Hilaire centered on the question of over how wide a spectrum of the animal kingdom homologies can be established. Geoffroy had no reservations about comparing organisms with very different body plans such as molluscs and vertebrates, whereas Cuvier believed such comparisons were only valid for closely related organisms such as birds and mammals. For more than 150 years after this famous debate, this issue remained unresolved (Appel, 1991).

In the last few years molecular biology has started to provide insights into homologies over a wider range of taxa than Cuvier believed possible. The approach is to use the expression patterns (i.e., where and when during development the mRNA for a particular gene is present) of developmental genes as phenotypic characters to reveal previously cryptic homologies. It has been particularly effective in demonstrating homologies between animals as distant as vertebrates and amphioxus (the lancelet), generally believed to be the nearest living invertebrate relative of the vertebrates, and is providing new perspectives on how the vertebrates evolved from the invertebrates. Just how far across the animal kingdom this approach can be extended is currently under debate and will require a comparison of the expression patterns of many developmental genes among a wide spectrum of organisms.

Molecular biology is often said to be "cookbook." To some extent this is true, and, like many cookbooks, methods texts often assume that the scientist knows the order and combination of recipes to be followed. However, the beginner often does not know what procedures are most appropriate and struggles with the telegraphic and jargon-riddled methods published in research articles. Therefore the present chapter consists of two parts: first, a short discussion of homology and the rationale of our approach and, second, an overview of the experimental strategy intended for zoologists largely unfamiliar with the techniques of molecular biology. The emphasis is on the advantages and disadvantages of the approach and the pitfalls inherent in the methods. Useful information rarely, if ever, mentioned in published protocols is given, but specific protocols are included only for those not readily available in such general methods texts as Berger and Kimmel (1987), Sambrook et al. (1989), and Stern and Holland (1993). The amphioxus–vertebrate comparison is used as an example; however, there are notations where methods developed for amphioxus larvae may have to be modified for other organisms.

THE CONCEPT OF HOMOLOGY AND THE RATIONALE OF THE APPROACH

The concepts and criteria of homology are thoroughly discussed in the chapters in Hall (1994) and are summarized in N.D. Holland (1996). For our present purposes, it is sufficient to state that our concept of homology is historical (i.e., a homologous feature in two different organisms results from its inheritance from a common ancestor) and that our criterion of homology is that of special quality (i.e., the testes of all vertebrates have special features like seminiferous tubules). This homology criterion is used twice in succession: first, from sequence data we decide whether particular developmental genes (i.e., genes coding for proteins such as transcription factors that are turned on during development) are homologous; then, we use the expression patterns of these genes as phenotypic characters to help identify homologous body parts between different animals. Importantly, we are not proposing that gene expression patterns should necessarily take precedence over traditional (usually morphologic) characters. Instead, the most acceptable homologies result when all the available characters make sense when considered together.

Our approach works best when the organisms being compared are neither too similar nor too dissimilar. Comparisons of the expression domains of developmental genes suggest homologies that are both thought-provoking and convincing when one compares animals with moderately different body plans such as agnathan vertebrates versus gnathostomes or amphioxus and vertebrates. For example, the *engrailed* gene, which codes for a transcription factor and is named for its mutant phenotype in *Drosophila,* is expressed during development in certain jaw muscles of gnathostomes and in some muscles near the mouth in agnathans as well as in a stripe at the midbrain–hindbrain junction in both groups (N.D. Holland et al., 1993). Comparative anatomy alone points to the homology of the midbrain–hindbrain junction in gnathostomes and agnathans; the expression of *engrailed* at this junction is just one more piece of evidence. On the other hand, since agnathans lack jaws, even though the muscles in question are similarly innervated in the two groups, there is insufficient anatomy to infer homology. However, in this case *engrailed* expression suggests that the similarly innervated muscles are truly homologous.

An example of just how well the method can work is shown by a comparison of gene expression patterns within the amphioxus and vertebrate nerve cords. This example shows first of all that while conclusions can sometimes be made on the basis of a comparison of just one gene and its homolog, they can be considerably strengthened and extended by a comparison of a suite of genes within the same structure—in other words, three genes are better than one. Second, it shows that detailed anatomical studies can be very useful adjuncts for interpreting data obtained from gene expression patterns. Amphioxus and vertebrates have fairly similar overall body plans and it is usually supposed that their dorsal nerve cords are homologous. However, what part, if any, of the amphioxus nerve cord is homologous to the vertebrate brain has been contentious because, except for a slight anterior swelling, the cerebral vesicle, the amphioxus nerve cord has no obvious divisions, whereas the vertebrate nerve cord has an anterior brain divided into forebrain, midbrain, and hindbrain, which in development is transiently segmented into rhombomeres. The most common view has been that the cerebral vesicle is homologous to the entire vertebrate brain. However, a comparison of gene expression patterns in the amphioxus and vertebrate nerve cords shows that this view is incorrect.

Figure 1 shows the expression patterns of vertebrate *HoxB1, HoxB3,* and *engrailed* and their amphioxus homologs *AmphiHox1, AmphiHox3,* and *AmphiEn* in their respective embryonic dorsal nerve cords. In vertebrates, a major region of expression of *engrailed* is at the midbrain–hindbrain boundary, while *HoxB1* is expressed in rhombomere 4 in the hindbrain and *HoxB3* is expressed posterior to the rhombomere 4/5 boundary. In developing amphioxus *AmphiEn* is expressed in a few cells within the cerebral vesicle, *AmphiHox1* in a stripe at the level of somites 4 and 5 (P.W.H. Holland and Garcia-Fernàndez, 1996), and *AmphiHox3* posterior to the level of the somite 4/5 boundary (P.W.H. Holland et al., 1992). The conclusion is that in spite of anatomical differences, the amphioxus nerve cord does have an extensive homolog of the vertebrate hindbrain with its anterior limit within the cerebral vesicle. Taken together with neuroanatomy (Lacalli et al., 1994), which indicates that the *AmphiEn*-expressing cells have the morphology of infundibular cells (which are located in vertebrates in the posterior region of the forebrain), the

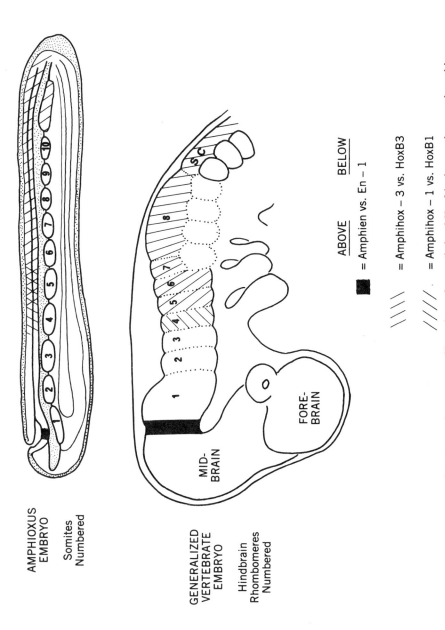

AMPHIOXUS EMBRYO

Somites Numbered

GENERALIZED VERTEBRATE EMBRYO

Hindbrain Rhombomeres Numbered

MID-BRAIN

FORE-BRAIN

ABOVE   BELOW

= Amphien vs. En – 1

= Amphihox – 3 vs. HoxB3

= Amphihox – 1 vs. HoxB1

Figure 1. A comparison of the expression patterns of homologs of *engrailed* and *Hox3* in the vertebrate and amphioxus.

tentative conclusion is that amphioxus may lack a homolog of the vertebrate mid-brain. Thus a midbrain may have evolved after the amphioxus and vertebrate lineages split.

How distantly related organisms can be before this approach ceases to give insights into homologies remains to be seen. It has been effective in comparing arthropods, annelids, and molluscs (Wedeen and Weisblat, 1991; Jacobs et al., 1994) and, as discussed above, in comparing amphioxus to vertebrates. It might work well in comparing hemichordates and vertebrates, although as yet there is no information on gene expression patterns in hemichordates. However, when comparing phyla with very different body plans—for example, nematodes versus vertebrates—apparent similarities in gene expression patterns may or may not reflect homologies between specific body parts. A notable example is the *Drosophila* gene *eyeless* and its vertebrate homolog *Pax-6*. Either *eyeless* or mouse *Pax-6* when expressed in *Drosophila* imaginal discs other than the eye disc induces the formation of ectopic eyes in adult flies (Halder et al., 1995). This suggests a common ancestry for photoreceptor cells of all animals, although the gene cascades downstream from *Pax-6/eyeless* must be elucidated before the course of evolution of image-forming eyes can be plotted. A second example is an attempt to homologize the ventral nerve cord of protostomes with the dorsal nerve cord of deuterostomes on the grounds, first of all, that the *Drosophila* gene *decapentaplegic,* which controls differentiation of dorsal structures, is expressed dorsally in *Drosophila* embryos, while its vertebrate homolog *BMP-4* is expressed ventrally and has ventralizing activity in vertebrate embryos, and, secondly, that *achaete-scute* in *Drosophila* and its vertebrate homologs are both expressed during development in their respective nerve cords, that is, ventral in *Drosophila* and dorsal in vertebrates (Arendt and Nübler-Jung, 1994). Similarly, the gene *short gastrulation (Sog)* is expressed in ventrolateral cells that are precursors of the *Drosophila* nerve cord, while its *Xenopus* homolog *chordin* is expressed in the dorsal blastopore lip and dorsal mesoderm in *Xenopus* embryos. Injection of either *chordin* or *Sog* mRNA into *Xenopus* embryos has dorsalizing effects, while injection of either *chordin* or *Sog* mRNA into *Drosophila* has ventralizing effects (Holley et al., 1995). Thus the idea is that there was an inversion of the dorsal–ventral body axis at the base of the chordates. Although this idea is tantalizing, it should be kept in mind that a particular cell type may be homologous (e.g., a nerve cell) without the structures that include that cell type (e.g., the nerve cords) being necessarily homologous. Furthermore, old genes may be used for new functions. What is needed to resolve these issues are the expression patterns in more animal phyla of connected suites of genes expressed in different spatiotemporal patterns within a given anatomical structure.

In using gene expression patterns as indicators of anatomical homologies, there are many genes to choose from. It is only necessary that the genes being compared in the two organisms are homologous and have restricted spatiotemporal expression patterns during development. A plethora of developmental genes has already been cloned from commonly studied organisms, for example, vertebrates, the fruitfly *Drosophila,* and the nematode *Caenorhabditis.* Many of these genes are transcription factors with evolutionarily conserved DNA-binding sequences. Examples of transcription factors that have restricted expression patterns are *engrailed* and the *Hox* and *Pax* families (Patel et al., 1989; De Robertis et al., 1990; Krauss et al., 1991). However, some useful genes have other functions; for example, *wingless* codes for a secreted protein (Schmidt-Ott and Technau, 1992). Other developmental genes are not useful in determining homologies since they are relatively ubiquitously expressed, for example, *Id,* a repressor of differentiation (Evans and O'Brien, 1993).

In the protein products of developmental genes, the conserved regions of amino acid sequence usually correspond to functional domains—for example, domains for DNA–protein or protein–protein interaction. Such motifs include the homeodomain (which codes for the homeobox), zinc finger, leucine zipper, basic helix–loop–helix, and paired domain (which codes for the paired box) (Suzuki, 1993; Nelson, 1995). While many genes have only one such motif—for example, *Hox* and *engrailed* each have one homeodomain—some have two or more, for example, most genes in the *Pax* family have both a homeodomain and a paired domain. Homologies are deduced by comparing, in order of importance, the amino acid sequences of these domains (amino acid sequences are more conserved than the base sequence), their position within the protein, and the intron–exon organization of the gene. For example, Table 1 shows that, within the homeobox, homologs of *engrailed* and *Hox3* from amphioxus, vertebrates, and *Drosophila* are about 90% identical. Convergence of protein or DNA sequence can occur, but only rarely: the most famous case being the digestive enzymes of lemurs and cows (Swanson et al., 1993).

Gene or protein sequences for comparison can be obtained from databanks by e-mail to the National Center of Biotechnology Information. Send e-mail messages with the single word Help as the text to blast@ncbi.nlm.nih.gov for instructions on comparing unknown sequences and to retrieve@ncbi.nlm.nih.gov for instructions on obtaining a given sequence. In addition, the computer program CLUSTAL V is useful for performing multiple sequence alignments and can be obtained from the EMBL file server at Netserv@EMBL-Heidelberg.DE by sending an e-mail message with the words Help and Help Software on two lines.

TABLE 1. Amino Acid Identities in the Homeobox Between Amphioxus and Other Organisms[a]

	Percentage Identity Between		
AmphiEn and:		AmphiHox 3 and:	
Mouse en-1	87%	Mouse HoxB3	97%
Mouse en-2	85%	Mouse HoxA3 and D3	87%
Other vertebrate en	80%	All other vertebrate	60–80%
Fruitfly en	80%	and fruitfly Horm/	
Vertebrate and fruitfly Horm/ Hox	45–50%	Hox	

See Lonai et al. (1987), Scott et al. (1989), and, PWH Holland et al. (1992).

EXPERIMENTAL STRATEGY

Many steps are involved in going from a live animal to the demonstration of a gene expression pattern in its larvae. DNA and/or RNA is isolated from the organism, a genomic DNA or cDNA library constructed, a probe for screening the library generated, and the library screened. Positive clones are then transferred into plasmids, sequenced, and linearized to make a template for synthesis of an antisense RNA probe (riboprobe). Larvae are raised and fixed, a riboprobe is synthesized, and the larvae are hybridized with the probe, photographed, embedded in plastic, sectioned, and photographed again. Fortunately, techniques are constantly being improved such that one person working full-time can now reasonably expect to complete such a project in about a year. Subsequent projects may only require 6–7 months, since DNA libraries can be amplified and remain viable for years when stored at 4°C or longer at −80°C, and fixed larvae stored in 70% ethanol will retain mRNA for at least a year.

The following is an overview of each of the steps involved. The emphasis is on alternative methods and their pros and cons. References to published protocols are given. Specific protocols not previously published for purification of genomic DNA, for purification of lambda phage DNA, and for *in situ* hybridization of riboprobes to amphioxus larvae (probably applicable to small larvae in general) are listed in Holland.1, Holland.2, and Holland.3. When trying a molecular biological technique for the first time it is often best to buy kits when they are available. Such kits, though invariably expensive, usually have the procedures optimized. One word of caution is that kits are sometimes rushed to market before the manufacturer has optimized the procedure and occasionally have an incorrectly formulated ingredient. All kits mentioned below have been shown to work well. There are many alternative kits on the market, and new ones are constantly appearing. With experience one learns when a kit is cost-effective. All companies have a technical services department, usually reached by a toll-free number, for help in trouble-shooting their products.

DNA Libraries

For riboprobe synthesis, cDNA clones, obtained from reverse transcription of mRNA, are preferable to genomic DNA clones, obtained from genomic DNA. Genomic clones have introns that sometimes break up the transcribed DNA into such small pieces that suitable riboprobes cannot be synthesized and are harder to work with than cDNA clones. However, cDNA libraries include only those genes that are being expressed at the stage and tissue from which the library was made, the representation of a particular gene being in proportion to the number of its mRNAs. Thus cDNA clones for transcription factors can be relatively rare. In contrast, genomic DNA clones represent all DNAs in proportion to the number of copies of the gene. It is therefore advisable first to screen a cDNA library made from larvae or embryos and, if that fails, to screen a genomic library. Other scientists with cDNA or genomic DNA libraries will typically send an aliquot on request if there is no conflict of interest. If no such library is available, it is necessary to construct one.

Different lambda phage vectors are usually used for cDNA and genomic DNA libraries, for example, Lambda Zap II for cDNA and Lambda Fix for genomic DNA (Stratagene Inc., La Jolla, CA). Specific methods for cDNA library construction are in Kakizuka et al. (1993). Only a brief summary with a few useful hints will be given here. For cDNA libraries, begin by isolating total RNA from the desired developmental stage. The guanidinium isothiocyanate (GuSCN) method of Chomczynski and Sacchi (1987) works well. For the carbohydrate-rich tissues of many invertebrates and plants use hexadecyltrimethyl ammonium bromide (CTAB) (Dellacorte, 1994). Some tissues and larvae such as those of sea urchins contain RNases and must be frozen and ground up in liquid N_2 before adding the GuSCN. Other larvae such as those of amphioxus can simply be centrifuged in a chilled centrifuge tube to prevent them from resuming swimming before the supernatant seawater can be removed and the homogenization solution added. For isolating mRNA from total RNA, OligodT Dynabeads (Dynal Inc., Great Neck, NY) provide a high yield. Theoretically, a library with 1×10^6 independent clones can be obtained from 5 µg or less of mRNA. If possible, however, begin with 10 µg mRNA to compensate for losses during the procedure. Sufficient RNA can be obtained from about 500 µl of packed, relatively nonyolky larvae like those of amphioxus.

The mRNA serves as a template for synthesis of double-stranded cDNA. Several kits are available that work well, for example, Amersham cDNA synthesis system Plus kit (USB-Amersham Life Science, Inc. Arlington Heights, IL). To measure incorporation of radioactivity into first and second strand cDNA use Whatman DE81 filters. Spot a small aliquot of the sample on the filter, count in a scintillation counter without adding scintillation cocktail. Wash the filter three times in 0.5 M phosphate buffer pH 7.0 and recount. Typically 2–10% of the starting mRNA is transcribed into first strand cDNA and >95% of the first strand cDNA into second strand cDNA.

Double-stranded cDNA is then cloned into a bacteriophage lambda vector. Commercially available vectors are engineered to contain a series of sites for restriction enzymes, called the multiple cloning site (MCS), into which fragments of double-stranded DNA terminating in compatible restriction sites can be ligated. Vectors can be purchased already cut at a restriction site within the MCS and dephosphorylated to prevent self-ligation. Lambda Zap II (Stratagene Inc., La Jolla, CA) allows nucleotide and antibody screening and also has the advantage of the MCS being within the pBluescript plasmid as part of the phage DNA. When positive clones are identified, this plasmid can be excised from the phage *in vivo* to facilitate further manipulations (see below). To clone double-stranded cDNA into a vector, phosphorylated linkers (short pieces of DNA with a restriction site compatible with a site in the multiple cloning site of the vector) are blunt-end ligated onto the double-stranded cDNA. (Do not purchase dephosphorylated linkers as they cannot be ligated to a dephosphorylated vector.) If cloning into the EcoRI site of Lambda Zap II, before blunt-end ligating EcoRI linkers to the cDNA, methylate the cDNA with EcoRI methylase [the substrate usually supplied with the enzyme adenosyl-L -methionine (SAM) does not keep well so it is best to buy it fresh].

After ligation of the linkers, size fractionate the cDNA on a Sepharose CL-4B column (pour the column in a 5-ml disposable pipette with a plug of autoclaved glass wool or polyester fluff at the tip) to eliminate excess linkers and smaller fragments of cDNA. Digest the cDNA with EcoRI (the prior methylation protects EcoRI sites within the cDNA) and ligate it into the vector. For all steps requiring precipitation of the DNA, carefully monitor the radioactivity in the pellet (put the microfuge tube into a scintillation vial and count the radioactivity without adding scintillation fluid) and increase centrifugation times if necessary. When working with very small amounts of DNA, a 20-min centrifugation in a microfuge frequently suffices to bring down only 50% of the DNA. Increasing centrifugation time to 1 hr will increase the yield to about 75%. Package into lambda phage coats using a high-efficiency commercial packaging extract consisting of lambda phage head and tail proteins. Purchase the highest-efficiency packaging extract available and follow the manufacturer's instructions. Do not try to save money by increasing the volume of cDNA packaged. When using Stratagene's Gigapak® Gold II packaging extracts, package no more than 1.5 μl of cDNA at a time; the efficiency of packaging drops precipitously as the volume of DNA increases. Libraries can be amplified once (follow the instructions supplied with the vector). Lambda phage suspensions will keep at 4°C for several years if a drop of chloroform is added to inhibit bacterial growth; alternatively, add 9% DMSO and freeze the suspension at −80°C.

Genomic libraries can be made in either lambda phage or cosmids. Lambda phage libraries have DNA inserts about 8–23 kb long and are more widely used for screening for individual genes. Cosmid libraries have larger inserts of 30–42 kb and are well-suited for genomic walking. However, they are more difficult to handle than libraries in lambda phage. Cosmid libraries will not be considered further here. For methods and a discussion of cosmid libraries see DiLella and Woo (1987) and Sambrook et al. (1989).

For genomic libraries in lambda phage, high molecular weight DNA is necessary. A simple method is found in Holland.1. It is often difficult to purify DNA from tissues rich in carbohydrate. In that case, if possible, use the testis from a ripe male or extract tissue with CTAB (Holland, 1993; Dellacorte, 1994). The DNA is fragmented into 8–20 kb pieces by partial digestion with a restriction enzyme such as Sau3a and ligated into a vector such as Lambda Fix II (Stratagene Inc., La Jolla, CA). Follow the manufacturer's instructions.

Screening Libraries

Methods for screening libraries in lambda phage in increasing order of usefulness are (1) antibodies, good only for screening cDNA libraries in an expression vector; (2) oligonucleotide probes 25–100 nucleotides long chemically synthesized on the basis of known protein sequence; (3) clones of the homologous gene from another species; and (4) short clones (200–400 bp) derived by the polymerase chain reaction (PCR) from the same species.

Antibodies. Antibodies have two major drawbacks: (1) only clones in the proper reading frame can be detected and (2) developmental genes often have limited expression, so that it is technically impossible to obtain sufficient protein for antibody generation. This limitation can be partially overcome by using antibodies raised against the homologous fusion protein, that is, overexpressed protein made by bacteria carrying a plasmid with an inserted DNA sequence of a closely related organism. However, the number of false positives, which is typically high at best, would increase.

Single-Stranded Oligonucleotide Probes. Single-stranded oligonucleotide probes, particularly those over 60 nucleotides long, generally have a lower percentage of false positives than antibodies and will detect clones regardless of the reading frame. Shorter probes often give unacceptably low signal/noise ratios. For probe synthesis it is best to have amino acid sequence of the protein from the same species. However, if stretches of amino acid sequence are identical across taxa, then an oligonucleotide probe incorporating redundancies in the genetic code can be synthesized. Redundancies should be avoided for the three most 3' bases, which usually must be a perfect match, and can be reduced based on the codon usage of the particular species. Probes are radioactively end-labeled with ^{32}P using the Klenow fragment of DNA polymerase (Sambrook et al., 1989). Labeled probes are purified away from unincorporated nucleotides by chromatography on a NucTrap™ push column (Stratagene Inc., La Jolla, CA).

Clones from Other Species. Clones of the homologous gene from another species are most useful when the species are fairly closely related and the genes highly conserved. It is advisable to subclone, if necessary, a 200–800 bp portion of the most conserved part of the gene such as the homeobox. Such probes from one vertebrate have been successful for screening libraries of other vertebrates, and probes from *Xenopus* and *Drosophila* have been successful in isolating genes from amphioxus libraries.

PCR Clones from the Same Species. The best probe is 150–800 bp amplified by the polymerase chain reaction (PCR) from DNA isolated from the organism of interest. Genomic DNA or DNA purified from a cDNA library can be used. Amplification is with degenerate primers based on the sequences of a conserved region (e.g., the homeobox) of the homologs from other organisms. If genomic DNA is used, introns must be avoided. If amplifying amphioxus DNA, primer design should be based on sequences from insect and fish or insect and mammal rather than on fish and mammal. This method is described in detail in P.W.H. Holland (1993), which includes methods for small-scale purification of genomic DNA. Methods for purification of DNA from a cDNA library are in Sambrook et al. (1989). Unless the library titer is very high, start with 3 ml of lambda phage suspension. When using cDNA, PCR can be performed with one gene-specific primer and one primer in the vector (anchored PCR). In theory this should yield the entire 3' or 5' end of the gene of interest. However, with only one gene-specific primer, especially if the gene of interest is rare, the wrong genes are often amplified. Nested PCR in which the first PCR product is reamplified with new primers internal to the first set increases the rate of success. The PCR product obtained is cloned into a plasmid and sequenced.

In performing PCR it is especially important to avoid contamination with foreign

DNA. Wash out the barrels of pipetters with 1 M HCl, followed by water and ethanol to dry, in order to destroy any DNA therein, wear gloves, and when pipetting DNA use aerosol filter pipette tips. After mixing together the components of the PCR reaction except for the DNA and DNA polymerase, irradiate the tube with short-wave UV for 3 min to break up contaminating DNA. If amplifying DNA from predatory animals either avoid extracting the gut or allow the animals to starve for a few days first. Cloning PCR products into a plasmid is facilitated with the PCR-Script™ cloning system (Stratagene Inc., La Jolla, CA).

Using Clones for Library Screening and Riboprobe Synthesis

Library Screening. For library screening the insert is excised from the plasmid with the appropriate restriction enzyme(s) and purified on a 5% acrylamide gel. Estimate the amount of DNA extracted from the gel with the ethidium bromide drop assay (Sambrook et al., 1989). Label the probe with ^{32}P by nick-translation (Sambrook et al., 1989) or random priming (Feinberg and Vogelstein, 1983). Random priming is the preferred method as it yields probes of a higher specific activity. Labeling 75 ng of DNA in a 30-μl reaction typically yields about 6×10^7 cpm with a specific activity of close to 1×10^9 cpm/μg if the isotope is no more than 3 days past the assay date. Do not use ^{32}P that is more than 1 week past the assay date since the specific activity of the probe will be low. After labeling, monitor the incorporation of ^{32}P with DE81 paper. Probes should be used at 1×10^6 cpm/ml. For library screening, if the incorporation of radioactivity into DNA is greater than 90% it is not necessary to remove the unincorporated nucleotides. Detailed methods for library screening are typically provided by the manufacturer of the vector. For either cDNA or genomic DNA libraries the *Escherichia coli* strain LE392 is useful since it grows rapidly and plaques form within 6–8 hr. To prepare bacteria of this strain for phage infection grow up an overnight culture in TB (5 g NaCl, 10 g bactotryptone/liter) supplemented with 10 mM MgSO$_4$ (required for infection) and 0.2% maltose. Centrifuge and resuspend to an OD$_{600}$ of 0.5 in 10 mM MgSO$_4$. Infect with lambda phage at a concentration to obtain about 5000–7000 phage particles/9 cm petri dish. Plate out on petri dishes with a medium containing 10 mM MgSO$_4$. (Instructions for plating out are usually included with the vector.) If the library is good, 40 plates should suffice. Once plaques are about 1–2 mm across, plates should be refrigerated several hours or overnight before transfer of plaques to nitrocellulose filters. If using a probe >200 bp long, a 24-hr exposure of filters to x-ray film should suffice. Oligonucleotide probes may require an exposure of several days. Positive plaques are cored with a 1-ml pipetter tip (cut the tip off with a razor blade to enlarge it), transferred to phage dilution buffer, and used to reinfect bacteria. This process is repeated until a 100% positive suspension of phage particles is obtained. It is important that the population be pure since clones with short or no insert can overgrow ones with long inserts.

Working with Positive Clones. Once a pure population of positive clones is obtained, it is transferred from lambda phage to a plasmid. With some cDNA vectors, for example, Lambda Zap II (Stratagene Inc., La Jolla, CA), this is easy, since a plasmid containing the insert can be excised from the lambda phage *in vivo* by coinfection of bacteria with the lambda phage and a filamentous phage. Follow the manufacturer's instructions. With genomic DNA vectors such as Lambda Fix (Stratagene Inc., La Jolla, CA) this is difficult. Because genomic DNA clones are typically 10–20 kb long, they must be cut into smaller pieces to be cloned into a plasmid. Therefore it is necessary to purify phage DNA (see Holland.2), digest the inserts with restriction enzymes, separate them on an agarose gel (include a lane of molecular weight standards, e.g., lambda phage DNA cut with HindIII), transfer the DNA to a nylon filter (Southern blot), and probe with the same probe used to screen the library. A second gel is run, and fragments hybridizing with the probe are isolated from the gel and cloned into a plasmid such as pBluescript (Stratagene Inc., La Jolla, CA) with T7 and T3 RNA polymerase sites, antibiotic resistance, and the *lacZ* gene for blue/white selection of recombinants. If the second gel is run in low gelling temperature agarose, visualize the bands with a long-wave UV lamp, excise appropriate bands with a razor blade, and purify the DNA according to the methods in Sambrook et al. (1989). However, the resolution of low gelling point agarose is typically not as good as that for high gelling point agarose. Therefore, if high gelling temperature agarose is used, insert a piece of Whatman #1 filter paper into a 1-ml syringe, add the excised gel band, and collect the liquid that comes through the filter when the plunger is depressed. Since the yield is typically low, the following steps are to maximize chances of cloning success. The recovered DNA should be concentrated by precipitation. Before precipitating add as a carrier the appropriate amount of vector DNA pretreated with alkaline phosphatase to prevent self-ligation. Wash the pellet with 70% ethanol, dry in a SpeedVac (Savant Instruments Inc., Farmingdale, NY), and take up in 9 μl of distilled water; add 1 μl 10× ligation buffer (supplied with the enzyme), 0.5 μl 10 mM rATP, and 1 μl T4 DNA ligase (4 U/μl). Incubate at 15°C overnight or longer. Transform into commercially available supercompetent bacteria with *lacZ* complementation, for example, Epicurean Coli™ supercompetent cells (Stratagene Inc., La Jolla, CA) following the manufacturer's instructions. Select white colonies (those in which the *lacZ* gene is interrupted by inserted DNA, prepare an overnight 2-ml culture with the appropriate antibiotic, and extract the plasmid DNA (Sambrook et al., 1989).

DNA from such small-scale preparations can be used for sequencing and riboprobe synthesis. An aliquot of the bacteria can be mixed with 9% DMSO, frozen on dry ice, and stored at −80°C. Analyze the clones by digesting 5 μl of plasmid DNA with restriction enzyme(s) present in the multiple cloning site, subject the digest to electrophoresis on a 1% agarose gel, and perform a Southern blot (Sambrook et al., 1989). Inserts should hybridize strongly to the DNA probe used to screen the library.

Sequencing. Before genomic DNA and cDNA clones can be used for riboprobe synthesis, they must be sequenced. Double-stranded sequencing of the small-scale plasmid preparations is possible using ^{32}P or ^{33}P end-labeled primers and the thermocycler (PCR machine). Numerous kits, for example, fmol™ DNA sequencing kit

(Promega Inc., Madison, WI), work well. Alternatively, if available, a DNA sequencing machine (e.g., Applied Biosystems Inc., Foster City, CA) can be used. While results with such machines can surpass those obtained with end-labeled primers, more DNA must be used (1 μg) and the DNA must be very clean. For details consult technical services for the sequencing machine.

Whole-Mount in Situ Hybridization. For *in situ* hybridization, an antisense riboprobe should be about 500–1000 bp long and complementary to a nonconserved portion of the mRNA. The template for riboprobe synthesis consists of all or part of the transcribed portion of the gene of interest cloned into a plasmid such as pBluescript (Stratagene Inc., La Jolla, CA) that contains RNA polymerase initiation sites at either end of the insert. Only if the organism has just one cognate of the gene of interest is it safe to include in the template a highly conserved portion of the gene such as the homeobox. Generally, templates are constructed to include the 3′ untranslated region (UTR) since the clones obtained from screening the cDNA library can be used directly without subcloning. However, in amphioxus and other marine invertebrates from wild populations, gene polymorphism is common and if the 3′ UTR contains large insertions and deletions from individual to individual, it may be desirable to construct the template from the 5′ UTR and adjacent coding regions. This may require subcloning a suitable fragment. Subcloning may also be necessary if genomic clones are used as templates to avoid transcription of introns or sequences 3′ of the 3′ UTR. To subclone, determine the restriction sites of the clone with a computer program such as DNASIS for Windows or MacDNASIS for MacIntosh available from National Biosciences Inc. (Plymouth, MN) or Hitachi Software Engineering America Ltd. (San Bruno, CA). If possible, choose restriction sites within the gene compatible with the multiple cloning site of the vector. If there are none, the overhangs left by an incompatible enzyme can be filled in to create blunt ends suitable for cloning. After cutting with restriction enzyme(s), the fragment is separated from the remainder of the plasmid on either a 5% acrylamide gel (fragments under 800 bp) or a 1% agarose gel (fragments over 800 bp) and ligated into the appropriate vector cut with appropriate restriction enzymes (specific methods are in Sambrook et al., 1989). To synthesize a riboprobe once suitable clones are in hand, the plasmid is cut with a restriction enzyme at the 5′ end of the portion to be transcribed. Avoid restriction enzymes generating 3′ overhangs. ATP, CTP, GTP, and digoxigenin-UTP in an appropriate buffer plus the appropriate RNA polymerase are added and the mixture is incubated at 37°C. The nonradioactive RNA synthesis kit from Boehringer-Mannheim (Indianapolis, IN) works well. Multiple copies of the RNA transcript are synthesized for each strand of starting DNA. After riboprobe synthesis the DNA template is digested with RNase-free DNase. The probe is incubated with the larvae and the digoxigenin detected with an anti-digoxigenin antibody coupled to alkaline phosphatase, which in turn catalyzes a reaction with an artificial substrate yielding a purple product. The dogma is that whole-mounts cannot be embedded in resin and sectioned because the purple product is soluble in organic solvents. However, we have developed a method for embedding the whole-mounts in Spurr's resin, which involves dehydration in alcohol in which the purple product is insoluble. A detailed method for *in situ* hybridization is given in Holland.3. This method is time-consuming, but it gives reliable results and is probably applicable to many other small (100–500 μm) larvae. For methods developed for larger larvae, see Wilkinson (1993).

ACKNOWLEDGMENTS

This work was supported in part by National Science Foundation research grant IBN 92-21622 to N.D.H. and an SERC/BBSRC grant to P.W.H.H. We thank Dr. J.M. Lawrence for providing laboratory facilities at the University of South Florida, Tampa, during the summer breeding season of amphioxus.

REFERENCES

Appel T (1991): *The Cuvier–Geoffroy debate*. Oxford: Oxford University Press.

Arendt D, Nübler-Jung K (1994): Inversion of dorsoventral axis. *Nature* 371:26.

Berger SL, Kimmel AR (eds) (1987): *Guide to molecular cloning techniques*. Meth Enzymol v. 152. Orlando, FL: Academic Press.

Chomczynski P, Sacchi N (1987): Single-step method of RNA isolation by acid guanidinium thiocyanate-phenol-chloroform extraction. *Anal Biochem* 162:156–159.

Dellacorte C (1994): Isolation of nucleic acids from the sea anemone *Condylactis gigantea* (Cnidaria: Anthozoa). *Tissue Cell* 126:613–619.

De Robertis EM, Oliver G, Wright CVE (1990): Homeobox genes and the vertebrate body plan. *Sci Am* 263:46–52.

DiLella AG, Woo SLC (1987): Cloning large segments of genomic DNA using cosmid vectors. In: Berger SL, Kimmel AR (eds). *Meth Enzymol* 152:199–212.

Evans SM, O'Brien TX (1993): Expression of the helix–loop–helix factor *Id* during mouse embryonic development. *Dev Biol* 159:485–499.

Feinberg AP, Vogelstein B (1983): A technique for radiolabeling DNA restriction endonuclease fragments to high specific activity. *Anal Biochem* 132:6–13.

Halder G, Callaerts P, Gehring WJ (1995): Induction of ectopic eyes by targeted expression of the *eyeless* gene in *Drosophila*. *Science* 267:1788–1792.

Hall BK (1994): Introduction. In: Hall BK (ed). *Homology: the hierarchical basis of comparative biology*. San Diego: Academic Press, pp 1–19.

Holland ND (1996): Homology, homeobox genes, and the early evolution of the vertebrates. Mem California Academy of Sciences, in press.

Holland ND, Holland LZ, Honma Y, Fujii T (1993): *Engrailed* expression during development of a lamprey, *Lampetra japonica:* a possible clue to homologies between agnathan and gnathostome muscles of the mandibular arch. *Dev Growth Differ* 35:153–160.

Holland PWH (1993): Cloning genes using the polymerase chain reaction. In: Stern CD, Holland PWH (eds). *Essential developmental biology. A practical approach*. Oxford, UK: IRL Press, pp 243–256.

Holland PWH, Holland LZ, Williams NA, Holland ND (1992): An amphioxus homeobox gene: sequence conservation, spatial expression during development and insights into vertebrate evolution. *Development* 116:653–661.

Holland PWH, Garcia-Fernàndez J (1996): *Hox* genes and chordate evolution. *Dev Biol* in press.

Holley SA, Jackson PD, Sasai Y, Lu B, De Robertis EM, Hoffman FM, Ferguson EL (1995): A conserved system for dorsal-ventral patterning in insects and vertebrates involving *sog* and *chordin*. *Nature* 376:249–253.

Jacobs DK, De Salle R, Wedeen C (1994): Engrailed: homology of metameric units, molluscan phylogeny and relationship to other homeodomains. *Dev Biol* 163:536.

Kakizuka A, Yu RT, Evans RM, Umesono K (1993): cDNA library construction. In: Stern CD, Holland PWH (eds). *Essential developmental biology. A practical approach.* Oxford, UK: IRL Press, pp 223–232.

Krauss S, Johansen T, Korzh V, Fjose A (1991): Expression of zebrafish *Pax* genes suggests a role in early brain regionalization. *Nature* 353:267–270.

Lacalli TC, Holland ND, West JE (1994): Landmarks in the anterior central nervous system of amphioxus larvae. *Philos Trans R Soc Lond B* 344:165–185.

Lonai P, Arman E, Czosnek H, Ruddle FH (1987): New murine homeoboxes: structure, chromosomal assignment, and differential expression in adult erythropoiesis. *DNA* 6:409–418.

Nelson HCM (1995): Structure and function of DNA-binding proteins. *Curr Opin Genes Dev* 5:180–189.

Patel NH, Martin-Blanco E, Coleman KG, Poole SJ, Ellis MC, Kornberg TB, Goodman CS (1989): Expression of *engrailed* proteins in arthropods, annelids, and chordates. *Cell* 58:955–968.

Sambrook J, Fritsch EF, Maniatis T (1989): *Molecular cloning. A laboratory manual,* 2nd ed. Cold Spring Harbor, NY: Cold Spring Harbor Laboratory Press.

Schmidt-Ott U, Technau GM (1992): Expression of *en* and *wg* in the embryonic head and brain of *Drosophila* indicates a refolded band of seven segment remnants. *Development* 116:111–125.

Scott MP, Tamkum JW, Harzell GW (1989): The structure and function of the homeodomain. *Biochim Biophys Acta* 989:25–48.

Stern CD, Holland PWH (eds) (1993): *Essential developmental biology. A practical approach.* Oxford, UK: IRL Press.

Suzuki M (1993): Common features in DNA recognition helices of eukaryotic transcription factors. *EMBO J* 12:3221–3226.

Swanson KW, Irwin DM, Wilson AC (1993): Stomach lysozyme gene of the langur monkey: tests for convergence and positive selection. *J Mol Evol* 33:418–425.

Wedeen CJ, Weisblat DA (1991): Segmental expression of an *engrailed*-class gene during early development and neurogenesis in an annelid. *Development* 113:805–814.

Wilkinson DG (1993): *In situ* hybridization. In: Stern CD, Holland PWH (eds). *Essential developmental biology. A practical approach.* Oxford, UK: IRL Press, pp 257–276.

CHAPTER 15

Evolutionary Transposition and the Vertebrate *Hox* Genes: Comparing Morphology to Gene Expression Boundaries with *in Situ* Hybridization

ANN CAMPBELL BURKE and CRAIG E. NELSON

Department of Genetics, Harvard Medical School, Boston, Massachusetts 02115

CONTENTS

SYNOPSIS

In situ hybridization allows the visualization of gene expression during ontogeny and has greatly augmented studies of development. However, interpretation of *in*

Molecular Zoology: Advances, Strategies, and Protocols, Edited by Joan D. Ferraris and Stephen R. Palumbi.
ISBN 0-471-14461-4 © 1996 Wiley-Liss, Inc.

situ hybridizations is subject to some of the same ambiguities inherent in classical methods, and molecular reagents introduce other complicating factors that will be discussed. The vertebrate *Hox* genes encode a class of DNA-binding proteins with putative roles as regulators of pattern formation during development. The *Hox* genes are expressed in a sequential pattern along the anterior–posterior axis of the embryo, with distinct expression boundaries at different levels in the paraxial mesoderm. Misexpression of *Hox* genes in the mouse results in homeotic transformations of vertebrae. The number of vertebrae contributing to different regions of the axial skeleton varies between different vertebrate taxa, in a phenomenon known as vertebral transposition. To examine the evolutionary role of *Hox* genes in transposition, the axial *Hox* code in the paraxial mesoderm of chickens and mice has been compared using whole-mount *in situ* hybridization. The gene expression boundaries examined are transposed together with anatomical boundaries, confirming an evolutionary role for *Hox* genes in transposition.

INTRODUCTION

The fields of zoology and evolution have long classical traditions that rely largely on observational principles and the comparative method to answer questions about the patterns and processes in nature. This volume provides examples of how modern molecular techniques can be used to address questions whose answers have been left incomplete by classical methods of analysis.

The classical problem of vertebral transposition, as described by E. S. Goodrich in 1913, is an excellent example of how an old question can be re-addressed with new data. The vertebral column is characterized by different regional specializations along the anterior–posterior (A-P) axis. Goodrich used the term "transposition" to describe the wide range of evolutionary variation in the number of vertebral segments that contribute to a given region of the axial skeleton and the variable positions of the paired appendages along the axis in different taxa. The "axial formula" of a given taxon describes the number of segments in each morphologically defined region. The axial formula of the mouse is: 5 occipital, 7 cervical, 13 thoracic, 6 lumbar, and 4 sacral somites or vertebrae. The axial formula of the chicken is: 5 occipital, 14 cervical, 7 thoracic, 9 lumbar, and 4 sacral somites or vertebrae. Though the total number of presacral vertebrae in these two animals differs only by one segment (34 versus 35), there is obvious transposition of regional boundaries between the two taxa (Fig. 1).

The vertebrate *Hox* genes encode transcription factors that are expressed in a colinear manner along the A-P axis in both the central nervous system and the paraxial mesoderm of the vertebrate embryo (Graham et al., 1989). Given the regulatory role of these genes and their wide phylogenetic conservation, it is likely that they play a significant role in the evolution of morphology. Perturbations of members of the *Hox* family have been shown to cause "homeotic" transformations of vertebral type, and hence changes in the axial formula of the mouse. It has been suggested that the specification of vertebral morphology is dependent on a particular

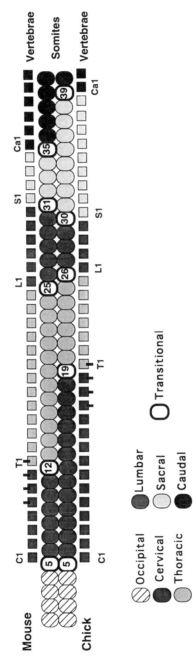

Figure 1. A schematic representation of the axial formulae of the mouse and the chick. The ovals represent somites, and the squares represent vertebrae (not to scale). The spinal nerves of the brachial plexus are represented by vertical black bars.

combination of *Hox* gene expression (Kessel and Gruss, 1990, 1991). To examine the evolutionary role of *Hox* genes in vertebral transposition, the axial *Hox* code in the paraxial mesoderm of chickens and mice has been compared using whole-mount *in situ* hybridization. The conclusions of this study are briefly summarized here and presented in detail elsewhere (Burke et al., 1995).

In Situ Hybridization

The method of *in situ* hybridization allows for the temporal and spatial localization of specific gene transcripts through the hybridization of a synthetic, labeled, antisense ribonucleic acid probe to native mRNA within the embryo during development. The signal carried by the probe can be visualized within the cells in which a particular gene is being transcribed. This is extremely important for studies of developmental mechanisms, gene function, and gene interaction and also can be used in a comparative context to reveal the phylogenetically conserved or variable aspects of the causal relationship between gene expression and morphology.

Initially, *in situ* hybridization techniques used radioactive probes on sectioned material. Later, a whole-mount protocol using nonradioactive probes was worked out for the *Drosophila* embryo (Berleth et al., 1988; Tautz and Pfeifle, 1989). This basic technique has been modified for many different organisms including a variety of vertebrates (cf. Harland, 1991; Rosen and Beddington, 1993). Nonradioactive methods utilize a probe labeled with an antigen such as digoxigenin or biotin and a secondary antibody conjugated to a histochemical tag. Many such conjugates are available, but we have had the greatest success with anti-digoxigenin conjugated to alkaline phosphatase. The following sections offer a brief, somewhat anecdotal description of the *in situ* hybridization procedure used successfully in our laboratory on chick and mouse embryos. A detailed protocol can be found in Burke.1.

Preparation of Embryos and Probe

First, the embryos must be prepared and a riboprobe synthesized from template DNA. Embryos are stripped of their membranes, and the brain ventricles and the heart are opened to prevent entrapment of reagents. Chick embryos ranging from Hamburger and Hamilton (H&H, 1951) stage 10 through about stage 27, and mice from embryonic day 7 to day 14 can generally be used for whole mounts. Older, larger embryos may need to be dissected prior to the procedure, in order to ensure access of the probe to relevant tissues. Individual organs can be hybridized independently. Younger embryos, chick H&H stage 4–10, and mice up to embryonic day 7, are small, fragile, and require additional care. The mild fixative, paraformaldehyde, is made up fresh to 4% in PBS (see Burke.1 for constituents) and filtered before use. Embryos are fixed overnight at 4°C, though fixation for a few hours at room temperature is also sufficient. The embryos are then rinsed in PBT (see Burke.1), and dehydrated into 100% methanol. They can be stored at −20°C in methanol for months.

We have found a great deal of variation in signal detection among different transcripts. Different regions of the same cDNA used as a template can affect the signal detected and it is difficult to determine if variable results are gene specific, probe-prep specific, or sequence specific. It is strongly recommended that several different probe constructs be tried to optimize the results for each gene. Templates ranging in size from cDNA clones as large as 3 kilobases, or as small as 400 base pairs, can give positive results with different genes. A template is best prepared from clean, plasmid prepared DNA. For an antisense probe, the template is linearized by restriction digestion at sites unique to the poly-linker on the 5' end of the insert. The template should be phenol extracted and alcohol precipitated before transcription of riboprobe.

In general, it is wise to use regions of the cDNA that are not highly conserved to prevent cross-reactivity with other genes. This is especially true of members of large gene families such as the *Hox* genes. In our experience with the chick *Hox* genes, the most commonly successful region of sequence included some or all of the 5' end of the gene, plus some or all of the homeobox region (Fig. 2). In certain cases, however, the 3' end of the gene was successful when probes from the 5' end gave negative results (see Fig. 3).

Due to the variability among transcripts it is very important to have suitable controls for each new probe attempted. The specificity of an antisense transcript is assessed by running a parallel hybridization with a sense transcript made from the same gene template linearized at the 3' end of the insert. A positive control, in the form of a well-characterized probe, is useful for testing the quality of the reagents in an individual *in situ* hybridization run and also controls for the signal-to-noise ratio. Taking an embryo through all the steps of the procedure but omitting the labeled probe is a good negative control for the background level of the detection system. If the expression patterns of some of the members of a gene family are known, they

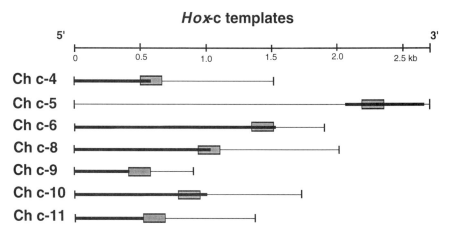

Figure 2. The position of the DNA templates for members of the *Hox*-c cluster from which successful riboprobes were transcribed for whole-mount *in situ* hybridization are shown here by heavy lines relative to the complete cDNA. Shaded boxes represent the homeobox.

Ch c-6 full lenth cDNA

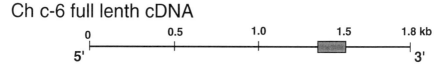

Subclones used as templates:

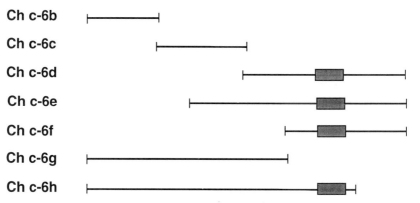

Figure 3. Seven different subclones of chicken *Hox* c-6 were used as templates for riboprobes. Subclones c-6b, c-6c, c-6e, and c-6f gave unsatisfactory labeling. Subclones c-6d and c-6h gave the same labeling pattern, though labeling with c-6h was far more intense. Shaded boxes represent the homeobox.

can be useful as internal controls. For example, the axial expression boundaries of the *Hox* genes are very similar within paralog groups, but the limb expression varies considerably between the *Hox* clusters. Limb expression can thus be used to control for cross-reactivity between paralog members.

After the embryos and the labeled transcripts have been prepared, the total whole-mount *in situ* hybridization procedure itself takes 4 days to complete. It may be possible to shorten this to 3 days, but shortcuts can result in increased background.

Day 1

The first day consists of a series of pretreatments that render the tissue accessible to probe. The embryos stored in methanol are rehydrated into PBT. This dehydration and rehydration step seems to increase the permeability of the embryos and reduces background. Going directly from fixative into PBT results in poor quality signal at the end of the whole-mount procedure. The next step is a hydrogen peroxide bleaching step that may be unnecessary. It can be left out completely; however, there is some evidence that it reduces background and increases the contrast between positive and negative areas. The next step involves treatment of embryos with proteinase-K, which permeabilizes the tissue and allows for probe penetration. The degree of enzymatic digestion required by different stages, different tissue types, and possibly different levels of mRNA signal vary considerably, and the timing and concentration of proteinase-K should be titrated for different conditions. For instance, *Bmp-2*, a member of the bone morphogenetic protein family, is expressed in posterior mesoderm in the limbs and also in the ectoderm of the apical ectodermal ridge (AER). Using the standard concentration and duration of proteinase-K treatment (10 μg/ml for 15 minutes), the mesodermal staining in a stage 23 chick limb is clear, but the AER staining is absent, as the ectoderm has been partially digested away under these conditions. At a lower concentration (2.5 μg/ml) the labeling is very intense in both the mesoderm and the ectoderm. Decreasing the concentration further, to 0.15 μg/ml, allows visualization of the signal, but it is much less intense, indicating an inadequate degree of permeabilization (Fig. 4).

In younger, more fragile embryos, proteinase-K digestion can be too harsh under

Proteinase-K treatment [ ]

10μg/ml

2.5μg/ml

0.15μg/ml

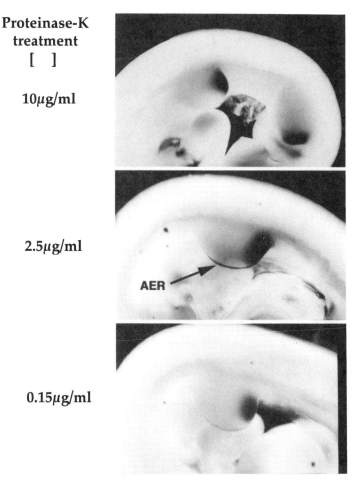

Figure 4. A titration series of proteinase-K treatment showing the variable effects on ectodermal (AER) and mesodermal labeling for expression of *Bmp-2* in the chick limb bud. In this case the medium concentration of proteinase-K gives the best result. See text for details. (From R. Johnson, with permission.)

any conditions. A treatment with a mixture of ionic and nonionic detergents (RIPA) can be used instead, replacing the hydrogen peroxide, the proteinase-K, and the glycine steps (the latter used to neutralize the proteinase-K). In embryos larger than a stage 28 chick or day 15 mouse, permeability can be a problem, especially in dense tissues like cartilage or liver. In this situation, proteinase-K can be increased up to 10 times the standard concentration. After the permeabilization step, it is important to stabilize the tissue by postfixation in 4% paraformaldehyde with 0.2% gluteraldehyde.

The embryos can be stored in prehybridization mix at −20°C for several weeks, though prolonged storage increases background levels. The hybridization step is performed overnight, at 70°C. This temperature ensures high stringency for hybridization. Stringency can be changed by changing the salt content or the formamide concentration of the hybridization buffer. The temperature can also be altered to effect stringency, but lower temperatures increase the likelihood of cross-reactivity.

For the prehybridization, hybridization, and antibody incubation steps, early protocols minimized the volume of solution by transferring embryos into 1.5 ml, screw-top tubes. This serves to minimize the amount of probe and antibody required, but the transfer is time-consuming and severely limits the size and number of embryos that can be processed. We have found that all steps can be performed in scintillation vials on as many as 15 embryos (at the size of a H&H stage 20 chick), increasing the solution volume to cover the embryos, but maintaining the concentration of probe or antibody in the larger volume.

Days 2 and 3

The second day of the protocol is designed to wash excess free probe out of the tissue and prepare it for incubation with antibody to the labeled probe. Originally, digestion with ribonuclease was performed to eliminate any single-strand RNA remaining in the tissue and reduce cross-reactivity and background. We have found that the RNase step can drastically decrease the level of hybridization signal, especially with probes that yield a weak signal. A trial run, with and without the RNase step, will determine if this step decreases background or signal.

The embryos are preincubated in 10% sheep serum to prevent nonspecific antibody interactions. Preabsorption of the antibody with species-specific embryo powder will also reduce nonspecific binding. The embryos are then incubated in antibody overnight at 4°C. The third day of the procedure is devoted to removing unbound antibody from the tissue with a buffer that includes levamisole to inhibit endogenous alkaline phosphatase, and 1% detergent (Tween-20). The levamisole has been left out without ill effect.

Day 4, Detection and Storage

The antibody is conjugated to a marker, usually alkaline phosphatase, and a histochemical reaction is used to visualize the bound antibody. The reaction of 5-bromo-4-chloro-3-indolylphosphate (BCIP or X-phos) with the alkaline phosphatase results in a blue precipitate, which can be enhanced to a darker purple with the addition of nitroblue tetrazolium (NBT). This color reaction is done at high pH (pH 9.5) and can be stopped with washes of lower pH (PBT pH 5.5). The time required for the detection reaction can vary from 5 minutes to 48 hours and must be determined empirically for each probe. Individual runs of the procedure can also vary in the amount of time it takes for the detection reaction, and it is important to monitor the reaction carefully each time. Reaction time can affect the signal-to-noise ratio and reactions that have been left too long, or stopped prematurely, lose details of labeling distribution. Stopping the reaction is always a judgment call, and for this reason *in situ* hybridization data cannot be used to make accurate quantitative statements about gene expression.

After the reaction is stopped, the embryos are postfixed again to stabilize the reaction product in the tissue. They can be stored at 4°C in PBT with sodium azide (0.03%) to prevent mold. If clearing is necessary, the embryos can be incubated in a graded series of glycerol:PBT and stored in 90% glycerol.

General Notes on Background and Trouble-shooting

Because of the length of the protocol and the number of different steps involved, it is often very difficult to determine where problems are arising. Trial-and-error is often the best way to trouble-shoot. The signal-to-noise ratio can be affected by many different factors. If the signal is strong, a high level of background can be tolerated, and if background is very low, weak signals still carry information. Weak signal can be due to low abundance of message, to low efficiency of hybridization, or to a problem of probe penetration. The larger the transcript, the more label it will carry. However, larger transcripts will suffer from a lower hybridization efficiency. Extremely weak signals can be enhanced by adding polyvinyl alcohol to the detection reaction, to increase the sensitivity of the X-Phos–NBT alkaline phosphatase reaction (De Block and Debrouwer, 1993; Barth and Ivarie, 1994).

If you are working with a well-characterized probe and encountering problems with background, fresh antibody should be used, and all of the solutions should be replaced, especially the hybridization solution. Ribonuclease contamination will only be an issue in preparing and storing the labeled transcripts, and in the steps of the protocol leading up to hybridization. RNase contamination in water can be eliminated by treatment with diethyl pyrocarbonate (DePC, Sigma). DePC-treated water should be used in all the solutions up to this point and the work area should be kept free of ribonuclease. While DePC-treated water is not required after the hybridization step, all solutions should be made with sterile distilled water and filtered prior to use. When changing solutions, the amount of time the embryos are exposed to air should be minimized to reduce background. Gentle agitation of the embryos is required. This can be done with almost any standard shaker, but the "hula-dancer" action of the Orbitron or Nutator seems particularly effective.

Double Labeling

Whole-mount *in situ* hybridization labeling for two genes simultaneously is possible using probes labeled with different antigens. A combination of probes labeled with digoxigenin–UTP and fluorescein–UTP can be hybridized at the same time and then detected sequentially. There are a number of different ways to detect this labeling (see Jowett and Lettice, 1994), and we have had the greatest success by first detecting the digoxigenin-labeled probe using an anti-digoxigenin antibody conjugated to alkaline phosphatase, exactly as in the standard protocol. After the reaction and postfixing, the embryos are heated to 65°C to destroy residual alkaline phosphatase activity. The protocol then begins again at the TBST washes on day 2 and embryos are reincubated with an alkaline phosphatase conjugated anti-fluorescein antibody and detected with magenta-phos (Biosynth International). This substrate gives a pinkish-purple precipitate that is distinguishable from the purple X-phos–NBT reaction product. The second reaction always gives a much weaker signal, probably due to the sensitivity of probe-bound RNA to high pH. The longer the length of the first reaction (at pH 9.5), the weaker the second signal becomes. Therefore, if you know that one of your probes will yield a robust signal quickly, that labeled transcript should be processed first.

Sectioning Whole Mounts

The whole-mount method is excellent for the visualization of gene expression in three dimensions on the surface of the embryo. However, the exact tissue distribution of label within the embryo, and the subtleties of distribution are only visible in sectioned material. Embryos that have been prepared as *in situ* hybridization whole mounts can easily be sectioned at 20 μm on a cryostat after being embedded in gelatin and OCT. In order to maximize the amount of signal in a section, the whole-mount detection reaction should be let go as long as possible. If penetration into older embryos or dense tissue types is a problem, the *in situ* hybridization protocol can be performed on sectioned material.

Interpretation of Results and Conclusions

It has been pointed out elsewhere (Kostic and Capecchi, 1994; Burke et al., 1995) that the technique of whole-mount *in situ* hybridization does not have the sensitivity to enable the determination of absolute boundaries. Like most morphological data therefore, the data produced are of a qualitative nature. In this study the difficulty in assigning an exact boundary between positive and negative areas (in this case somites), is conceded by designating axial boundary levels by two segment numbers. Despite this ±2 error, we found that the anterior boundary of expression of *Hox* genes in the paraxial mesoderm of the chick and the mouse embryo map consistently to particular morphological regions rather than to ordinal segment number (Burke et al., 1995; and Fig. 5). The correlation of gene expression pattern

---

Figure 5. A summary diagram of the paraxial mesoderm expression of 12 of the *Hox* genes in the mouse and chick, mapped against a schematic of their axial formulae and regional anatomy. The horizontal bars are placed to show the anterior expression boundary. The upper bar represents the mouse level, the lower bar the chick level. The bar lengths are arbitrary and do not correspond to a posterior expression boundary (data from Burke et al., 1995).

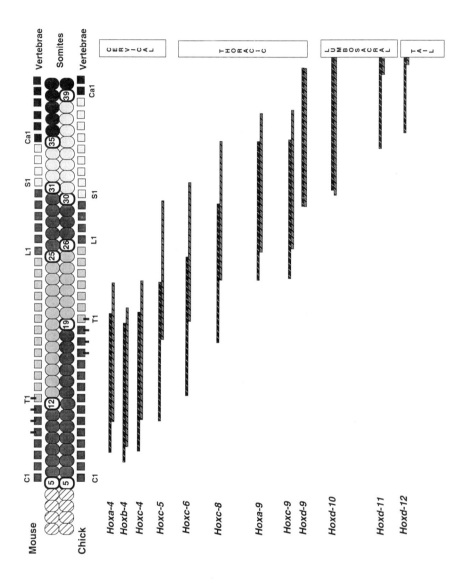

with axial anatomy in two morphologically distinct species indicates that changes in *Hox* gene expression patterns are a proximal cause in the morphological evolution represented by vertebral transposition.

The transposition of gene expression boundaries along the A-P axis is in some areas subtle, and in others more dramatic. For instance, all the members of the *Hox*-4 paralog group have anterior borders of expression in anterior cervical vertebrae of both the chick and the mouse. The borders differ in segment number only by 2 or 3 somites (e.g., *Hox* c-4 is expressed in somite 7–8 in the mouse, and in somite 10–11 in the chick, Fig. 5). Representative members of paralog groups 5 and 6 show a more dramatic transposition. The chick has 14 cervical vertebrae, the mouse has only seven. The cervical–thoracic transition is thus transposed by seven segments between the two species. In both animals, *Hox* c-5 is expressed at the very end of the cervical series, and *Hox* c-6 comes on at the very beginning of the thoracic series. The expression boundaries of these genes are also transposed by seven segments, in concert with anatomy (e.g., *Hox* c-6 is expressed at somite 12–13 in the mouse and at somite 19–20 in the chick). Members of the other paralog groups follow this trend, though the registration of anatomical boundaries is offset by only one or two segments in the lumbosacral region of these animals.

Molecular Evidence for Developmental Homologies

Goodrich, as an evolutionary morphologist, accepted that vertebrae are both serially homologous as individual segments, and historically or phylogenetically homologous in functional units defined by regional morphology. He was not completely satisfied with this explanation, however, and argued that "some theory of the redistribution of the formative substances to which morphological differentiation is due is necessary if we are to explain homology" (Goodrich, 1913, p. 237). The *Hox* genes are excellent candidates for the "formative substance." Homology long recognized on morphological grounds has been given a molecular aspect.

As pointed out by Holland et al. (Chapter 14 in this volume) the phylogenetic distance between two organisms affects the criteria for homology that are most appropriate for a comparative analysis. In comparing the dorsal nerve cords of amphioxus and vertebrates, gene expression can be used to assign homology in the absence of morphological characters. In comparing the correlation of gene expression patterns to morphology within the vertebrates, we are using classically determined homology to test the regulatory role of genes in the specification of morphology.

ACKNOWLEDGMENTS

The whole-mount *in situ* procedure discussed here originated from D. Wilkinson and was taught to A.C.B. by M. Hollyday, and further developed in the laboratory of Cliff Tabin. The following people added significant empirical data to the protocol: R. Johnson, E. Laufer, S. Bruhn, A Vorkamp, and D. Roberts. We thank E. Laufer, J. Golden, and S. Bruhn for critical reading of the manuscript.

REFERENCES

Barth J, Ivarie R (1994): Polyvinyl alcohol enhances detection of low abundance transcripts in early stage quail embryos in a nonradioactive whole mount *in situ* hybridization technique. *Biotechniques* 17:324–327.

Berleth T, Burri M, Thoma G, Bopp D, Richstein S, Frigerio G, Noll M, Nusslein-Volhard C (1988): The role of localization of bicoid RNA in organizing the anterior pattern of the *Drosophila* embryo. *EMBO J* 7:1749–1756.

Burke AC, Nelson CE, Morgan BA, Tabin C (1995): *Hox* genes and the evolution of vertebrate axial morphology. *Development* 212:333–346.

De Block M, Debrouwer D (1993): RNA–RNA *in situ* hybridization using digoxigenin-labeled probes: the use of high-molecular-weight polyvinyl alcohol in the alkaline phosphatase indoxyl-nitroblue tetrazolium reaction. *Anal Biochem* 215:86–89.

Goodrich ES (1913): Metameric segmentation and homology. *Q J Microsc Sci* 59:227.

Graham A, Papalopulu N, Krumlauf R (1989): The murine and *Drosophila* homeobox gene complexes have common features of organisation and expression. *Cell* 57:367–378.

Hamburger V, Hamilton HL (1951): A series of normal stages in the development of the chick embryo. *J Exp Zool* 88:49–92.

Harland RM (1991): *In situ* hybridization: an improved whole-mount method for *Xenopus* embryos. *Methods Cell Biol* 36:685–695.

Jowett T, Lettice L (1994): Whole-mount *in situ* hybridizations on zebrafish embryos using a mixture of digoxigenin and fluorescein-labelled probes. *Trends Genet* 10:73–74.

Kessel M, Gruss P (1990): Murine developmental control genes. *Science* 249:347–379.

Kessel M, Gruss P (1991): Homeotic transformations of murine vertebrae and concomitant alteration of *Hox* codes induced by retinoic acid. *Cell* 76:89–104.

Kostic D, Capecchi MR (1994): Targeted disruption of the murine *Hox*a-4 and *Hox*a-6 genes result in homeotic transformations of the vertebral column. *Mech Dev* 46:231–247.

Rosen B, Beddington RSP (1993): Whole-mount *in situ* hybridization in the mouse embryo: gene expression in three dimensions. *Trends Genet* 9:162–167.

Sambrook J, Fritsch EF, Maniatis T (1989): *Molecular cloning: a laboratory manual*, 2nd ed. Cold Spring Harbor, NY: Cold Spring Harbor Laboratory Press.

Tautz D, Pfeifle C (1989): A non-radioactive *in situ* hybridization method for the localization of specific RNAs in *Drosophila* embryos reveals translational control of the segmentation gene *hunchback*. *Chromosoma* 98:81–85.

Wilkinson DG (ed) (1992): *In situ hybridization: a practical approach*. Oxford: Oxford University Press, pp 75–83.

CHAPTER 16

Lineage Analysis Using Retroviral Vectors

JEFFREY A. GOLDEN and CONSTANCE L. CEPKO

Department of Genetics, Harvard Medical School, Boston, Massachusetts 02115

CONTENTS

SYNOPSIS

The mechanisms used to generate the diversity of cell types during development is an area of intense investigation in developmental biology. Lineage analysis is one method used to characterize the mode(s) of cell division of progenitor cells and allows one to follow the fate and distribution of clonally related progeny cells. Retroviruses provide a genetic tool for lineage mapping. We have developed a retroviral vector library for determining lineage that facilitates the identification of sibling relationships. A replication incompetent avian retrovirus encoding the human placental alkaline phosphatase gene was engineered to encode a library of

Molecular Zoology: Advances, Strategies, and Protocols, Edited by Joan D. Ferraris and Stephen R. Palumbi.
ISBN 0-471-14461-4   © 1996 Wiley-Liss, Inc.

molecular tags. The tags are derived from a degenerate oligonucleotide synthesized to have a complexity of 1.7×10^7. The oligonucleotide tags provide unique markers to identify progeny from single progenitors. We have used this library to analyze lineal relationships in the developing chick central nervous system. Our data have shown novel patterns of clonal expansion and diverse fate of progeny from a single progenitor cell. The avian retroviral library should prove valuable for studying lineal relationships in many systems.

INTRODUCTION

Lineage mapping via the introduction of a traceable marker is a classic technique in developmental biology. It has been used to generate models of genealogical relationships among cells and has provided insight into the timing of decisions concerning cell fates. The basic principle is to label a progenitor cell with an easily identified marker that is faithfully passed to all progeny but cannot be passed horizontally to neighboring cells. Although a variety of methods have been developed to trace lineage, three primary methods exist in vertebrates: (1) the production of chimeric animals, (2) single cell injection, and (3) retrovirally mediated lineage mapping. Each method carries inherent advantages and disadvantages.

The production of chick/quail hybrids (Le Douarin, 1973) provided the basis for many early studies investigating lineage. Le Douarin and colleagues exploited the unique appearance of quail cell nuclei to follow the fate of quail progenitors transplanted into embryonic chicks. In order to study neural crest cell migration, they performed transplantation of neural tube segments. Transplantation of small segments from a quail into a chick resulted in the rapid integration of the quail tissue. Cells migrating from the transplanted segment gave rise to the neural crest cells in the corresponding segment of the adult chimeric animal. The specific cells arising from the quail neural crest cells were easily recognized by their nuclear characteristics. This technique was instrumental in identifying the cell types generated from a population of progenitor cells as well as the migrational pathways used by neural crest cells (Le Douarin, 1982). The small fragments of neural tube transplanted in these experiments included many progenitors, and thus clonal analysis was not possible.

In the 1980s a technique was developed that allowed one to determine the fate of siblings from a single progenitor by filling the cytoplasm of a single cell with a fluorescent tag (Gimlick and Cook, 1983; Gimlick and Braun, 1985; also see Cameron et al., Chapter 12 in this volume). Using lysinated rhodamine dextran (LRD), a large molecule incapable of passing through gap or other cell junctions, cytoplasmic staining remains confined to a single cell. At each cell division, approximately 50% of the marker is passed to each daughter cell. Cells can then be tracked for many cell generations, allowing direct observation of the mode of replication, migrational pathways, and the cell types generated from a single progenitor. This technique has several advantages: the anatomic site of the progenitor cell at the time of injection is known, one can track live labeled cells (although extinction of the fluorescent label represents a potential limitation), and it is easy to evaluate sibling relationships. However, this technique is limited by the technical difficulties of precise injection, as well as the dilution of the dye over many generations, preventing characterization of clones after many cell divisions. The second limitation can preclude evaluation of the final cell fate of the progeny of early progenitor cells.

Retroviruses provide another method to study lineal relationships (Price, 1987; Cepko, 1988; Sanes, 1989). Retroviruses use RNA as their genome but produce a DNA copy of the genome upon entry into a cell using the viral encoded reverse transcriptase. The reverse transcribed DNA is integrated into the host cell genome of cycling cells in an M-phase dependent process (Roe et al., 1993). Vectors for lineage analysis have been rendered replication-incompetent by the removal of essential viral genes, and thus cannot spread to neighboring cells. The advantage of these vectors is that they integrate into a cycling cell and will be passed genetically to all daughter cells, but will not spread among neighboring cells. The introduction of a histochemically assayable gene into the replication-incompetent retroviruses creates a simple system for determining which cells harbor the provirus. The *E scherichia coli lacZ* gene encoding β-galactosidase and the human placental alkaline phosphatase (PLAP) gene have been successfully cloned into vectors and used for lineage analysis (Price et al., 1987; Fields-Berry et al., 1992). The promoter located within the viral long terminal repeat region (LTR) (Price et al., 1987; Fields-Berry et al., 1992) or a nonretroviral promoter, such as the SV40 early promoter (Sanes, et al., 1986), is used to drive the gene encoding the histochemical marker. Since there is usually more than one cell infected at the injection site, one must later sort out who is related to whom among the labeled cells. That is, how many clones are there and which cells belong to each clone? This can be a difficult problem when the progeny of clones intermingle and/or travel long distances. We have defined the two possible types of error in lineage mapping with retroviral vectors as splitting errors and lumping errors.

Lumping errors can be made when a conclusion is reached about a single progenitor giving rise to multiple cell types. This type of error is more likely to occur when injections are made into solid tissue or a region where the spread of the inoculum is limited. Multiple progenitor cells are infected by an injection of retroviruses because the inoculum used to infect each embryo is typically composed of more than 1 infectious virion. The possibility exists that two adjacent progenitors will be infected, giving rise to two clones of two different cell types, in an overlapping region. Since both clones express the same histochemical marker, cells from each clone appear indistinguishable. An erroneous interpretation might be that a single progenitor is multipotent, giving rise to both cell types identified with the histochemical marker. This would be considered a lumping error.

The possibility of a splitting error arises when a conclusion is drawn about clones being composed of a single cell type. If, for example, a retrovirally infected progenitor cell divides, resulting in two mitotically active daughter progenitors, the two daughters could migrate away from each other prior to the generation of further progeny. The subsequent cell divisions from these daughter cells could give rise to spatially distinct subclones. If each of the first two daughter cells produced different cell types, then the original infected cell was at least bipotential. This may not be apparent as each subclone may be thought to be the entire clone and the erroneous conclusion could be drawn that the original infected cell gave rise to only one cell type.

Previous studies using retroviruses addressed lumping errors by performing infections with a serially diluted stock and analyzing clone compositions that resulted from the various dilutions (Turner and Cepko, 1987; Turner et al., 1990) and with the use of viral stocks containing more than one histochemically distinguishable marker (Galileo et al., 1990; Hughes and Blau, 1990; Fields-Berry et al., 1992; Levison et al., 1993; Fekete et al., 1994). A more direct approach to address lumping and splitting errors was developed by Walsh and Cepko (1992), who constructed a library of viruses. Each virus of the library carried one member from a pool of approximately 80 DNA fragments from *Arabidopsis thalliana* DNA, in addition to the lacZ gene. Infected cells, recognized by their β-galactosidase activity, were mapped and the positive cells cut from cryostat sections. The *A. thalliana* DNA was amplified by PCR and characterized by size and restriction enzyme digestion patterns. If the size and restriction digestion pattern of the PCR product from two or more cells were the same, they were considered siblings with a probability calculated on the basis of the number of infections in that brain and the complexity of the library (Walsh and Cepko, 1992). Lineage analysis using this library revealed novel lineal relationships in the cerebral cortex. This study demonstrated that neurons and glia arise from a common progenitor and that neurons dispersed widely across functional boundaries in the cerebral cortex. The limited number of unique members in the library made from *A. thalliana* DNA restrained the analysis to tissues with low infection rates (Walsh and Cepko, 1992).

More data could be acquired with each experiment and additional questions could be addressed in the central nervous system and other tissues with a more complex library containing a greater number of DNA tags. We have therefore constructed a retroviral vector, CHAPOL (*c*hick *a*lkaline *p*hosphatase with *o*ligonucleotide *l*ibrary), that includes a DNA tag composed of a degenerate oligonucleotide that has a theoretical complexity of 1.7×10^7. Studies in the developing nervous system of the chick have demonstrated that progenitor cells give rise to both neurons and glia and that siblings are widely dispersed (Golden and Cepko, 1996). This retroviral library and general strategy should prove valuable for lineage analysis, not only in the nervous system but in any tissue or tissues in which lineal relationships are sought.

METHODS

Construction of CHAPOL

A detailed description of the construction of the retroviral vector CHAPOL can be found elsewhere (Golden et al., 1995). Briefly, a population of double-strand DNA molecules that included a short degenerate region, $[(G \text{ or } C)(A \text{ or } T)]_{12}$, was generated by PCR amplification of a chemically constructed single-strand oligonucleotide population of the same sequence.

The avian replication-incompetent virus CHAP, encoding the human placental alkaline phosphatase (PLAP) gene, was modified to accept inserts of an oligonucleotide pool (Ryder and Cepko, 1994). CHAP was linearized, purified, and mixed with the population of degenerate oligonucleotides in the presence of ligase and aliquots of the resulting ligation products were used to transform *E. coli* DH5α. Following transformation, all aliquots were pooled. One hundred microliters of the pool were plated at varying dilutions on ampicillin plates. The remainder of the pool was divided and added to eight 2-liter flasks containing 1 liter of LB medium with 50 μg/ml ampicillin. The cultures were shaken overnight at 37°C. Plasmid DNA was extracted from these cultures by the triton lysis procedure and purified on CsCl gradients (Sambrook et al., 1989).

Preparation of CHAPOL Virus Stocks

The pool of CHAPOL DNA was transfected into the avian virus packaging line Q_2bn (Stoker and Bissell, 1987), and the transiently produced virus was collected and concentrated (see Golden et al., 1995). Aliquots of $CaPO_4$ precipitates of 100 μg CHAPOL DNA were made in 10 ml of HBS (HEPES-buffered saline). The precipitate in each aliquot was then distributed equally on ten 10-cm plates of Q_2bn and glycerol shock was carried out for 90 seconds at room temperature 4 hours later. At 24 hours post glycerol shock, the supernatants were collected and pooled. This was repeated at 48 hours. The supernatants from the 24- and 48-hour harvests were pooled and the titer was calculated by infection of QT6 cell and assay of the PLAP activity as described (Cepko, 1992). The stock was filtered through a 0.45-μm filter and concentrated by centrifugation in an SW27 rotor at 4°C, 20K, for 2 hours. The concentrated stock was titered on QT6 and tested for helper virus, which proved negative. The titer of CHAPOL was determined to be 1.1×10^7 CFU/ml. The same stock was used for all experiments described herein.

In Vivo Infection

Fertilized virus-free White Leghorn chicken eggs were obtained from SPAFAS (Norwich, CT) and kept at 4°C until they were transferred to a high-humidity, rocking incubator (Petersime, Gettysburg, OH) at 38°C, which was designated time 0. At approximately 18–42 hours incubation (Hamburger and Hamilton stage 10–17) (Hamburger and Hamilton, 1951), the eggs were removed and 0.1–1.0 μl of CHAPOL was injected with a final concentration of 0.025% fast green dye to fill the ventricles (Fekete and Cepko, 1993). The infected eggs were incubated to embryonic day 8 (E8) or E18, at which time the brains were dissected out in PBS followed by overnight fixation in 4% paraformaldehyde made in PBS (pH 7.4) at 4°C. Brains recovered at E8 were fixed and processed for alkaline phosphatase (AP) activity with X-phos (5-bromo-4-chloro-3-indolylphosphate) and NBT (nitroblue tetrazolium) (Golden.1). E18 brains were washed overnight in three changes of PBS and then cryoprotected in 30% sucrose. After cryoprotection, the brains were embedded in OCT media and cut on a Reichart-Jung 3000 cryostat at 60 μm. Sections were histochemically reacted with X-phos and NBT (Golden.1). Cells infected with the retrovirus and expressing AP were easily identified by the purple precipitate. Coverslips were mounted with gelvatol.

The outline of each section was drawn by camera lucida and the locations and types of cells were labeled on each section. Once the location of all positive cells and cluster of cells was recorded, a single cell or cluster of cells with a small group of surrounding cells was removed using a heat pulled glass micropipette and transferred to a 96-well PCR plate for proteinase-K digestion (Golden.2). Following digestion, nested PCR was performed (Golden.3). The product of each PCR was run on a 1.5% agarose gel to determine if a product of the appropriate size had been amplified (Fig. 1). Sequencing of the oligonucleotide insert (Golden.4) was per-

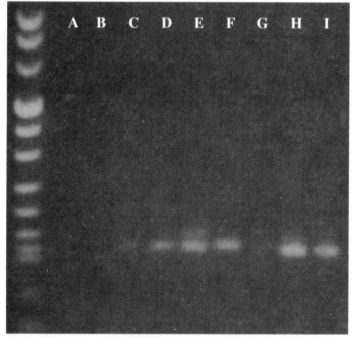

Figure 1. The products of nine nested PCR are analyzed on a 1.5% agarose gel. The first lane is a molecular weight marker (MWM VIII, Boehringer-Mannheim). Lane A is a control lane of the product of PCR without DNA. Lane B is a control lane with the PCR products from tissue picked from an infected brain in a region having no purple precipitate. Lanes C–I are all the PCR products from tissue fragments with 1 to 5 AP+ cells. A single band at the expected size of 121 base pairs is found in lanes C, D, F, H, and I. A suggestion of a faint band is present in lane G, but no definite band is present. Lane E shows the expected band and a slightly larger band. The lighter extra band represents some priming in the second PCR from primer carried over from the first reaction.

formed on all reactions that gave the expected product on the agarose gel analysis. All sequences were stored in the software program GCG (1991).

All common sequences were pulled out of the database created in GCG to identify cells that are clonally related. Sections were then aligned to determine the three-dimensional boundaries of clonal expansion. The cell types of sibling cells were also recorded to determine the variety of cells that can arise from a single progenitor.

RESULTS

The ability of CHAPOL to infect chick brains was assessed by performing AP histochemistry on infected brains and looking for purple cells in whole-mount preparations. Alkaline phosphatase positive (AP+) columns of cells were easily recognized in brains analyzed at E8 after injection at E1.5 to E2.5. Brains showed variable levels of infection (from 0 to >200 AP+ columns of cells), with some being heavily infected (Fig. 2). To determine the cell types and the full extent of clonal expansion, chicks infected at E1.5 to E2.5 were analyzed at E18. At this stage the chick diencephalon was developed and all the nuclei of the adult structure were recognizable. The purple NBT product from the AP reaction sufficiently outlined cells to identify between 30% and 40% of the cells as either neuronal or glial (Fig. 3). Once cells had been localized and identified as to cell type, they were picked using a heat pulled glass micropipette (Fig. 4). Following digestion with

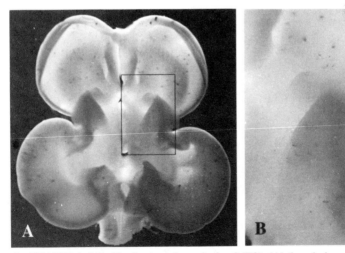

Figure 2. CHAPOL infected brains at embryonic day 8 (E8). (A) dorsal view of an E8 brain fixed and reacted with X-phos and NBT as a whole mount. The ventricles have been opened to expose >100 AP+ columns of cells. Each column is likely to represent an independent clone. The complexity of CHAPOL allows analysis of brains with high infection rates. The square in (A) defines the area, which is magnified in (B) to better appreciate the columnar organization of AP+ cells.

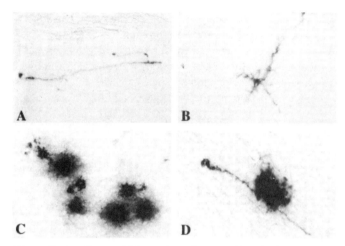

Figure 3. AP+ cells in 60-μm thick sections from an E18 brain. The morphology of multiple cell types can be appreciated. (A) A radial glial cell. (B) A neuron. (C) A cluster of glial cells. (D) A neuron and adjacent glial cell.

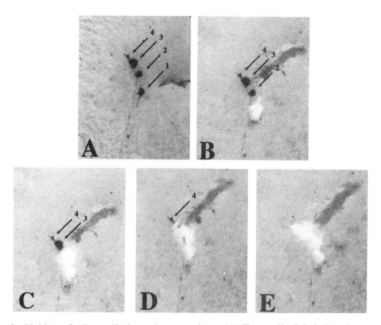

Figure 4. Picking of AP+ cells from tissue sections. (A) Four cells (labeled 1–4) are present in the section. Note the long processes projecting toward the bottom of the panel. These processes were joining a white matter tract and allowed the cells to be defined as neurons. (B) After removal of cell 1. (C) After removal of cell 2. (D) After removal of cell 3. (E) All cells have been removed.

proteinase-K and nested PCR, the PCR product was sequenced. Sequencing of the DNA insert gave an unambiguous identity to the viral insert in each infected cell or small cluster of infected cells (Fig. 5). Amplification and sequencing, including the amplification from single cells, was performed on 323 samples. DNA was amplified from 212 (74.3%) of the picks and sequence obtained on 134 of the 165 (81.2%) PCR products.

We have analysed lineal relationships in the mature chick diencephalon (Golden and Cepko, 1996). AP+ cells were found throughout the diencephalon. A representative clone from diencephalon is diagrammed in Figure 6. The clone in Figure 6 is from an E18 brain and contained 19 cells with the same insert. The probability that these cells arose from separate progenitors, infected with the vector carrying the same insert, is $<10^{-10}$ (Church, 1992, in Walsh and Cepko, 1992), given that the previously determined complexity of the CHAPOL library is likely to be greater than 6×10^4 (Golden et al., 1995). The value of using a complex library of vectors was illustrated in more heavily infected brains where closely aggregated AP+ cells, which would have been lumped into a single clone based on proximity, carried different inserts, indicating they arose from separate progenitors (Fig. 7). Analysis of other siblings revealed that both neurons and glia arise from a single progenitor.

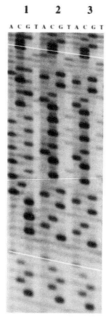

Figure 5. The sequencing reactions from the PCR products of three different samples are shown (1–3). The white lines delineate the beginning and end of the degenerate oligonucleotide sequence. PCR product #1 reveals a unique sequence. The sequence of PCR products #2 and #3 are the same, indicating that they are from siblings.

Figure 6. A composite of serial coronal sections through the diencephalon. The long vertical structure in the midline is the third ventricle. Emanating to the left from the midportion of the third ventricle are markings for the location of multiple AP+ cells. PCR and sequence analysis of each cell indicated they are all members of a single clone. The clone extends from the medial third ventricle to the lateral wall of the diencephalon. Cells from the clone were present on ³/₄ of the coronal sections, indicating extensive anterior to posterior spread.

DISCUSSION

Lineage analysis requires the ability to track the progeny from a single progenitor cell. Although many techniques have been used, retroviruses possess several inherent advantages for determining lineal relationships (also see Cepko et al., 1993). First, the histochemical marker is faithfully and indelibly transferred to all progeny. No problems due to label dilution are encountered as occurs when a dye is injected into a single cell. However, there can be variable expression of the gene or lack of expression in some cells. Since a single cell is usually infected by a single retroviral vector, and the vector infecting a cell cannot spread to neighboring cells, the progeny of single progenitors can be traced. The introduction of a DNA marker, such as the degenerate oligonucleotide used in CHAPOL, has also addressed the previous problems of retroviral studies regarding lumping and splitting errors. However, the retroviral method has a disadvantage in that one does not always know the precise location of the original infected cell.

   Retroviruses are engineered for lineage analysis by rendering them incapable of replication. They are able to infect dividing cells but cannot replicate to produce

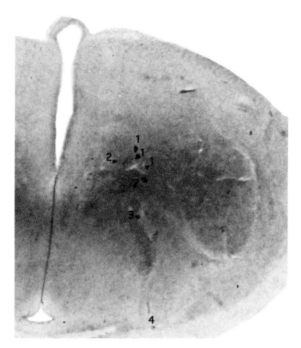

Figure 7. A representative 60-μM section of the right diencephalon from a brain infected with CHAPOL and analyzed at E18. The AP+ cells were removed and the inserts were sequenced following PCR. The AP+ cells labeled "1" were each found to carry the same insert. The two adjacent cells labeled "2" each carried the same insert, which was different from those in clone "1." Two other cells, "3" and "4," in the same section each carried unique inserts that were also different from either "1" or "2." The proximity of the cells labeled "1" and "2" would have resulted in a lumping error if clonal definitions were based on geometric boundaries.

infectious particles. The defective retrovirus will stably integrate into the host cell DNA and be passed to all progeny by simple mendelian inheritance along with host chromosomes. By cloning a histochemically assayable gene in place of the genes required for production of infectious viral particles, a method for detecting infected cells is provided (Sanes et al., 1986; Price et al., 1987; Fields-Berry et al., 1992). Since viral promoters are generally strong promoters in most tissues, the histo-chemically assayable gene is transcribed at relatively high levels in the cells carry-ing the provirus. Thus the progeny of an infected progenitor cell, regardless of the number of cells in the clone, can be identified.

Lineage analysis with retroviruses can be limited by lumping and splitting errors (as defined in the Introduction). CHAPOL has been engineered to directly address both of these possibilities. The utility of CHAPOL in addressing these potential problems is due to the DNA tags. Two issues are important for determining the value of this type of library of DNA markers: (1) the number of unique members in the library and (2) the distribution of the library members (Walsh et al., 1992). If only two members exist in the library, for example, then there is a one in two chance that the tag will be selected in two consecutive picks. If 100 members exist in the library, the chance that two picks come up with the same member is reduced to 10^{-2}. The second important variable determining the quality of the library is the distribution of the members within the library. This can be illustrated as follows. Consider a library composed of 10^6 members, with 50% of the library composed of one member. If two neighboring or distant cells are found to carry the over-represented insert, even though the library may contain 10^6 members, the probability that the two cells arose from separate clones is still 0.5. CHAPOL was previously shown to have an equal distribution in that each of the inserts picked to date ($n = 320$) has occurred independently only once (Golden et al., 1995).

To demonstrate the utility of CHAPOL, progenitor cells in the developing chick embryo nervous system were infected at stages 9–12. The results showed that infection of progenitor cells did occur and progenitor cells continued to divide, giving rise to multiple cells. The vector was found to express AP in sufficient levels to allow recognition of infected cells and in many cases determine the cell type (see Fig. 3). Success of the method also requires sequencing of the DNA tag integrated into the host cell's genome. PCR allows the amplification of very small amounts of starting material and recovery of the PCR product was successful in greater than 80% of the cell picks that were analyzed. Sequence was obtained from the majority of these products. This allowed accurate determination of siblings in the majority of clones. We have found in practice that this method of tag identification is in fact easier than our previous method based on the analysis of the size and restriction digestion pattern of *A. thalliana* DNA.

The retroviral library we have constructed and characterized is extremely powerful in delineating the progeny of a single progenitor. This system should be widely applicable to performing lineage analysis in many parts of the developing organism and, where appropriate, in the mature animal. Recent advances in virology suggest that retroviral vectors may be available to infect a wide range of species (Burns et al., 1993; Kasahara et al., 1994). Others have recently shown that retroviruses can even be used for infection and gene expression in germ line cells of previously uninfectable species, such as zebrafish (Allioli et al., 1994; Chen, 1995 Ch. 22 in this volume). Infection of non-neural tissue with CHAPOL has also been observed, as one would expect, indicating that CHAPOL can be used to investigate lineage in many tissue types. As with previous studies, analysis of lineage will likely lead to the development of models for how progenitor cells replicate, when cell fate decisions are made, and allow the definition of the migrational pathways used by siblings. Results from these types of studies can then lead to experiments to test the models generated.

REFERENCES

Allioli N, Thomas J-L, Chenbloune Y, Nigon V-M, Verdier G, Legras C (1994): Use of retroviral vectors to introduce and express the β-galactosidase marker gene in cultured chicken primordial germ cells. *Dev Biol* 165:30–37.

Burns J, Friedmann T, Driever W, Burrascano M, Yee J-K (1993): Vesicular stomatitis virus G glycoprotein pseudotyped retroviral vectors: concentration to very high titer and efficient gene transfer into mammalian and nonmammalian cells. *Proc Natl Acad Sci USA* 90:8033–8037.

Cepko C (1988): Retrovirus vectors and their applications in neurobiology [review]. *Neuron* 1:345–353.

Cepko C (1992): Transduction of genes using retroviral vectors. In: Ausubel F, et al (eds). *Current protocols in molecular biology.* New York: Greene Publishing Associates and Wiley-Interscience, Unit 9.11, vol 1.

Cepko CL, Ryder EF, Austin CP, Walsh C, Fekete DM (1993): Lineage analysis using retrovirus vectors. *Methods Enzymol* 225:933–960.

Church G (1992): MONTAG. Available through anonymous internet ftp from rascal.med.harvard.edu.

Fekete D, Cepko C (1993): Replication-competent retroviral vectors encoding alkaline phosphatase reveal spatial restriction of viral gene expression/transduction in the chick embryo. *Mol Cell Biol* 13:2604–2613.

Fekete D, Perez-Miguelsanz J, Ryder EF, Cepko CL (1994): Clonal analysis in the chicken retina reveals tangential dispersion of clonally related cells. *Dev Biol* 166:666–682.

Fields-Berry SC, Halliday AL, Cepko CL (1992): A recombinant retrovirus encoding alkaline phosphatase confirms clonal boundary assignment in lineage analysis of murine retina. *Pro Nat Acad Sci USA* 89:693–697.

Galileo D, Gray G, Owens G, Majors J and Sanes J (1990): Neurons and glia arise from a common progenitor in chicken optic tectum: demonstration with two retroviruses and cell type-specific antibodies. *Proc Natl Acad Sci (USA)* 87:458–462.

Gimlick R, Braun J (1985): Improved fluorescent compounds for tracing cell lineage. *Dev Biol* 109:509–514.

Gimlick R, Cook J (1983): Cell lineage and the induction of second nervous systems in amphibian development. *Nature* 306:471–473.

Golden JA, Cepko CL (1996): Clones in the chick diencephalon contain multiple cell types and siblings are widely dispersed. *Development* 122:65–78.

Golden J, Fields-Berry S, Cepko C (1995): Construction and characterization of a highly complex retroviral library for lineage analysis. *Proc Natl Acad Sci USA* 92:5704–5708.

Hamburger V, Hamilton H (1951): A series of normal stages in the development of the chick embryo. *J Morphol* 88:49–91.

Hughes S and Blau H (1990): Migration of myoblasts across lamina during skeletal muscle development. *Nature* 345:350–352.

Kasahara N, Dozy A, Kan Y (1994): Tissue-specific targeting of retroviral vectors through ligand–receptor interactions. *Science* 266:1373–1376.

Le Douarin N (1973): A biological cell labelling technique and its use in experimental embryology. *Dev Biol* 30:217–222.

Le Douarin N (1982): *The neural crest.* Cambridge: Cambridge University Press.

Levison S, Chuang C, Abramson B and Goldman J (1993): The migrational patterns and developmental fates of glial precursors in the rat subventricular zone are temporally regulated. *Development* 119:611–622.

Price J (1987): Retroviruses and the study of cell lineage. *Development* 101:409–419.

Price J, Turner D, Cepko C (1987): Lineage analysis in the vertebrate nervous system by retrovirus-mediated gene transfer. *Proc Natl Acad Sci USA* 84:156–160.

Roe T, Reynolds T, Yu G, Brown P (1993): Integration of murine leukemia virus DNA depends on mitosis. *EMBO J* 12:2099–2108.

Ryder EF, Cepko CL (1994): Migration patterns of clonally related granule cells and their progenitors in the developing chick cerebellum. *Neuron* 12:1011–1028.

Sambrook J, Fritsch E, Maniatis T (1989): *Molecular cloning: a laboratory manual*, 2nd ed, vol 1. Cold Springs Harbor, NY: Cold Springs Harbor Laboratory Press.

Sanes J (1989): Analysing cell lineage with a recombinant retrovirus. *Trends Neurosci* 12: 21–28.

Sanes J, Rubenstein J, Nicolas J-F (1986): Use of a recombinant retrovirus to study lineage in mouse embryos. *EMBO J* 5:3133–3142.

Sequence analysis software package, 7 ed. (1991): Madison, WI: Genetics Computer Group, Inc.

Stoker A, Bissell M (1987): Quantitative immunocytochemistry assay for infectious avian retroviruses. *J Gen Virol* 68:2481–2485.

Turner DL and Cepko CL (1987): A common progenitor for neurons and glia persists in rat retina late in development. *Nature* 328:131–6.

Turner DL, Snyder EY and Cepko CL (1990): Lineage-independent determination of cell type in the embryonic mouse retina. *Neuron* 4:833–45.

Walsh C, Cepko CL (1992): Widespread dispersion of neuronal clones across functional regions of the cerebral cortex. *Science* 255:434–440.

Walsh C, Cepko CL, Ryder EF, Church GM, Tabin C (1992): The dispersion of neuronal clones across the cerebral cortex [letter]. *Science* 258:317–320.

PART IV

EVOLUTION IN VARIABLE ENVIRONMENTS AND PHYSIOLOGICAL ADAPTATION

Osmoregulatory Gene Expression and Implications for Evolutionary Studies: Strategies in Identification of the Osmotic Response Element (ORE)

JOAN D. FERRARIS and ARLYN GARCÍA-PÉREZ

Laboratory of Kidney and Electrolyte Metabolism, National Heart, Lung and Blood Institute, National Institutes of Health, Bethesda, Maryland 20892-1598

CONTENTS

SYNOPSIS

The action of osmotic response elements (OREs) is proposed to form the molecular basis for the almost ubiquitous ability of organisms to adapt to a hyperosmotic environment (Ferraris et al., 1994). Diverse organisms, including bacteria, yeast, plants, and animals, accumulate organic osmolytes to adapt to an increase in osmolality. Organic osmolytes accumulate either metabolically or via transport through

Molecular Zoology: Advances, Strategies, and Protocols, Edited by Joan D. Ferraris and Stephen R. Palumbi.
ISBN 0-471-14461-4   © 1996 Wiley-Liss, Inc.

the action of osmoregulatory genes. In all osmoregulatory genes examined, hyperosmolality induces gene transcription but the molecular basis of gene osmoregulation remains obscure. To address this question, we cloned an osmotically regulated gene that codes for aldose reductase. Aldose reductase catalyzes the conversion of glucose to sorbitol, a nonperturbing organic osmolyte. To identify the transcription start site we used two independent methods: primer extension analysis by direct mRNA dideoxynucleotide sequencing and a PCR-enhanced method to determine 5'-end sequence. In each case we found the transcription start site to be 36 base pairs upstream of the initiator methionine codon. Reporter gene constructs of the 5' flanking region were used in transient transfection analyses to identify a functional promoter (base pair -208 to $+27$) and to provide the first evidence of OREs in a eukaryotic genome. Identification and characterization of OREs within the 5' flanking region (base pair ≈ -3429 to -208) and their associated *trans*-acting factors should reveal the molecular mechanisms of gene regulation in osmotic stress. Identification of the action of OREs as the molecular basis underlying adaptation to an anisosmotic environment has clear implications for speciation events where salinity presents a barrier to populations.

INTRODUCTION

Organisms occupy highly variable osmotic environments. *Dunaliella salina,* a green alga, is abundant in the Dead Sea but can live in pond water, marine invertebrates occupy tide pools, estuaries, and mangroves, whereas insects withstand remarkable desiccation. The mammalian extracellular fluid is regulated at about 300 mOsm but the osmolality of cells in the renal papilla varies over thousands of milliosmoles. As diverse as organisms are, virtually all (with the noted exception of the halophilic archaebacteria) share a common mechanism to adapt to a hyperosmotic environment. When confronted with an increase in osmolality, bacteria, yeast, plants, and animals have the same response; they accumulate intracellular organic osmolytes to adapt (Yancey et al., 1982). In the case of most osmolytes examined, an osmoregulatory gene has been identified that causes, in response to hyperosmotic stress, accumulation of the osmolyte. Some osmoregulatory genes code for enzymes in the metabolic pathway of an osmolyte. An example of this is aldose reductase (AR), which catalyzes the conversion of glucose to sorbitol. Other osmoregulatory genes code for the proteins that transport osmolytes such as betaine or inositol into the cell. A mechanism shared by the osmoregulatory genes examined to date is that hyperosmolality markedly induces transcription (Smardo et al., 1992; Uchida et al., 1993; Yamauchi et al., 1993). But what is it that confers osmotic sensitivity to these genes? Recently, we reported the discovery of sequences upstream of the rabbit AR promoter that confer hyperosmotic response capability (Ferraris et al., 1994). This finding represented the first evidence of an osmotic response element (ORE) within a specific DNA fragment in a eukaryotic genome.

The work leading up to and including cDNA cloning is briefly outlined below;

detailed strategies for genomic cloning and identification of the ORE(s) follow. Basic protocols are either referenced, for example, genomic DNA isolation (Maniatis et al., 1982), or the manufacturer of the appropriate kit (includes instructions) is cited, for example, radiolabeling DNA by random priming (Boehringer-Mannheim). Specialized protocols are referenced in the text (as Ferraris and García-Pérez.#) and provided in the Protocol section at the back of this volume.

Although the accumulation of osmolytes in response to osmotic stress was first discovered in a marine crab, *Eriochier sinensis* (Duchateau and Florkin, 1955; Florkin, 1956, 1962), osmotic regulation of genes is best understood for mammalian renal medullary cells. This was largely facilitated by the establishment of PAP-HT25 cells, a line of cells derived from the rabbit inner medulla (Uchida et al., 1987). One of the principal osmolytes accumulated by renal medullary cells and also by PAP-HT25 cells during adaptation to hyperosmotic stress is sorbitol (Bagnasco et al., 1987; Cowley et al., 1990). As aforementioned, sorbitol is synthesized from glucose in a reaction catalyzed by aldose reductase (AR) (EC 1.1.1.21). Using this cell line, we originally demonstrated that extracellular hyperosmolality induces transcription of the AR gene (Smardo et al., 1992), resulting in a rise in AR mRNA (García-Pérez et al., 1989), followed by increased AR protein synthesis rate (Moriyama et al., 1989) and, ultimately, increased sorbitol accumulation (Uchida et al., 1989). To understand the molecular mechanisms involved in the osmotic regulation of the AR gene, we cloned the rabbit AR gene and characterized its structure (Ferraris et al., 1994).

STRATEGIES

Overview

Since this is essentially a problem of gene regulation, the first step was to clone the gene for aldose reductase. This was done by screening a library of genomic DNA clones using the AR cDNA as a probe. The second step was to analyze and authenticate the genomic clone(s) obtained. To map the positions of introns and exons we used two independent methods, PCR mapping and restriction enzyme fragment analysis. Both the PCR and restriction fragments were sequenced. The next step, to determine the start site of transcription, also was performed using two independent methods. In the first, we sequenced the mRNA directly using dideoxynucleotide termination sequencing; for the second, we used a PCR-enhanced method to determine the 5' end sequence of the mRNA. Once we identified the transcription start site (by convention, this nucleotide is labeled +1), we could define the downstream limit (base pair −1) of the 5' flanking region of the gene. We could then test segments of the 5' flanking region of the gene for promoter activity and for osmotic sensitivity. This was done by inserting fragments of the 5' flanking region into plasmids containing the reporter gene luciferase, transfecting the luciferase reporter gene constructs into PAP-HT25 cells, and analyzing the cells for luciferase activity under different osmotic conditions.

Genomic Cloning

The first step is to obtain a clone of the gene of interest. Ideally, if the gene is small enough (20–25 kb), the entire gene—that is, 5' flanking region, all exons and introns, and 3' flanking region—will be contained in a single clone. The way that we attempted to achieve this was first to screen a λ genomic library with the most complete cDNA probe available, then to select among the positive clones with probes designed to contain just the 5' end or the 3' end of the cDNA. Rabbit spleen genomic DNA was isolated (Maniatis et al., 1982) and used to prepare a custom λFixII library (Stratagene, La Jolla, CA). Assuming you cannot borrow an appropriate library from a colleague, ready-made genomic libraries or kit components can be purchased from several companies including Stratagene and Clontech (Palo Alto, CA). Forty clones were identified by screening (Ferraris and García-Pérez.1) with a 1287-bp AR cDNA insert released from pAR10 (García-Pérez et al., 1989) by EcoRI digestion. The 1287-bp insert was subsequently radiolabeled with α-[^{32}P]dCTP (Random Primed DNA Labelling Kit, Boehringer-Mannheim, Indianapolis, IN). Since plaques are not pure when initially picked and since we wished to select for a complete gene, we performed the necessary plaque purifications (Ferraris and García-Pérez.1) with two probes derived from pAR10, a PstI restriction fragment containing 137 bp of 5' translated region and a 70-bp PvuII restriction fragment containing 3' untranslated region (UTR). A single clone, λgAR1 (Fig. 1), hybridized to both probes. After mapping, we discovered that λgAR1 contained 2000 bp of 3' untranslated region but only the last 26 bp of exon 1. Clearly, we now needed another clone. The new clone should not only contain 5' flanking sequence

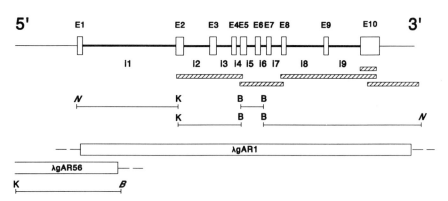

Figure 1. Schematic diagram of the rabbit AR gene (14.7 kb plus 3.5-kb 5' flanking region and 2-kb 3' flanking region). Exon (open boxes) lengths are E1, 102 bp; E2, 168 bp; E3, 117 bp; E4, 78 bp; E5, 123 bp; E6, 107 bp; E7, 82 bp; E8, 84 bp; E9, 83 bp; E10, 367. Intron (bold lines) lengths are i1, ~5383 bp; i2, ~1500 bp; i3, ~800 bp; i4, 213 bp; i5, 473 bp; i6, 310 bp; i7, 556 bp; i8, ~2335 bp; i9, ~1816 bp. Genomic clones λgAR1 and λgAR56 (open bars are AR gene regions, broken lines are λ arms) are shown in relation to the gene. Other representations are as follows: PCR fragments, hatched bars; restriction fragments, bracketed lines; B, BamHI; K, KpnI; N, NotI. Restriction sites in vector sequences are italicized. (From Ferraris et al., 1994, with permission.)

but should also substantially overlap λgAR1, so we designed two new probes. We also used a new library since we had performed multiple screens on the original library without finding a clone that had more upstream sequence than λgAR1.

To obtain the necessary upstream sequences, we screened a λDashII rabbit genomic library (Stratagene, La Jolla, CA) with two probes. They were a λgAR1, 375-bp intron 1-specific probe amplified by PCR (GeneAmp, Perkin Elmer, Foster City, CA) and a 465-bp BamHI cDNA restriction fragment, extending from the last 24 bp of exon 1 to exon 5, of the 1287-bp pAR10 cDNA insert. Plaque purifications were performed with both probes; a single clone, λgAR56, was characterized further.

Genomic Clone Mapping and Characterization

Intron positions in λgAR1 were determined using PCR with overlapping primer sets (Bruzdzinski and Gelehrter, 1989). With an unmapped gene, there is no way to tell where the introns will lie, so the strategy employed is to design sets of primers (sense and antisense) that overlap each other in the cDNA. All the sets taken together should span the total length of the cDNA. A PCR amplification is performed using a set of primers and no template (control), cDNA as template, and genomic clone as template. Ideally, the PCR will yield no product in the control, a single band of predicted X bp with the cDNA template, and with the genomic clone as template, either a band of X bp if there is no intron or a band of X + Y bp if there are one or more introns totalling Y bp between the sense and the antisense primers. We tried many different primer sets and most often did not get just a single band from the genomic clone (there is a lot more DNA available for mispriming). When we did get a predominant band from the genomic DNA, we verified the band by hybridization of a Southern transfer (Maniatis et al., 1982) to a nested end-labeled oligonucleotide probe. Since the cloned genomic DNA is bound on both sides by vector sequence, we also used vector primers on each end (paired with a cDNA primer) to determine the orientation of the gene. To confirm intron positions and because we were unable to amplify one of the regions of λgAR1 (between exons 1 and 2), we also characterized the clones by restriction analysis. We selected restriction enzymes that would release the clone from the lambda arms (NotI and BamHI) and empirically determined others that would cut the gene only once or twice. With λgAR1, the combination of BamHI, KpnI, and NotI yielded four fragments (Fig. 1), the relative positions of which were determined by hybridization to specific end-labeled oligonucleotide probes. In λgAR56 (insert size 13.1 kb) restriction digestion with BamHI, KpnI, and NotI yielded three fragments. To determine which of these fragments overlapped clone λgAR1 (Fig. 2), we determined which fragment hybridized to an upstream, exon 1-specific oligonucleotide. This identified the 5 kb KpnI-BamHI fragment (gAR56_5) (Fig. 2). After we sequenced gAR56_5, we found that it contained the first 1435 bp of intron 1, all of exon 1, and approximately 3500 bp of 5' flanking region. gAR56_5 was used as the template for all PCR amplifications for the principal reporter gene constructs (below). Restriction fragments (Figs. 1 and 2) were subcloned into pBluescript SK+II (Stratagene, La Jolla,

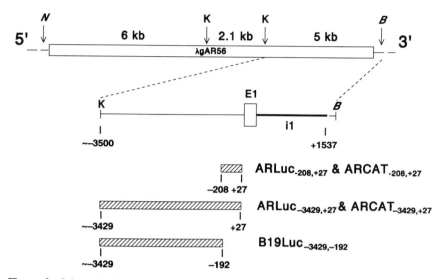

Figure 2. Schematic diagram of λgAR56 showing positions of KpnI restriction sites and an enlargement of the 5-kb restriction fragment (gAR56_5). Hatched bars represent PCR-amplified gene fragments used in the primary reporter gene constructs. ARLuc$_{-208,+27}$ and ARCAT$_{-208,+27}$ contain the promoter alone, as previously defined (Wang et al., 1993); ARLuc$_{-3429,+27}$ and ARCAT$_{-3429,+27}$ contain this promoter plus upstream sequence; and B19Luc$_{-3429,-192}$ contains upstream sequence alone. *B,* BamHI; K, KpnI; *N,* NotI. Restriction sites in vector sequences are italicized.

CA) and sequenced by primer-directed, double-stranded plasmid sequencing (Sequenase DNA Sequencing Kit, U.S. Biochemical Corp., Cleveland, OH). PCR fragments (Fig. 1) were sequenced directly (Bachmann et al., 1990).

The mapping and characterization of a gene should yield the length of the gene and confirmation that exon sequences are identical to the cDNA and to the results of the primer extension analysis (below). Typically, exon–intron boundaries should conform to the GT–AG rule; that is, introns begin with a GT and end with an AG.

Determination of the Transcription Start Site

Like most cloned cDNAs, pAR10 did not contain the transcription start site; it also did not contain the 5′-most 14 amino acids. We determined the transcription initiation start site using two independent methods. In both methods the template was poly(A)+ RNA isolated from PAP-HT25 cells (Uchida et al., 1987). To increase the abundance of AR mRNA, these cells were first exposed for 24 hours to medium made hyperosmotic (500 mOsm/kg H$_2$O) with NaCl (García-Pérez et al., 1989; Smardo et al., 1992). Primer extension analysis was performed by direct mRNA dideoxynucleotide sequencing (Geliebter, 1987; see Ferraris and García-Pérez.[2]) with three different antisense primers, each of which was expected to lie within 200 bp of the start site (Fig. 3A). All three reactions stopped at the same position but the precise identity of the transcription start site nucleotide ($+1$) was unclear. We then used a PCR-enhanced method to identify the nucleotide (Hofman and Brian, 1991; see Ferraris and García-Pérez.3) (Fig. 3A).

Reporter Gene Expression Analysis of Transient Transfectants

Reporter gene constructs and transient transfections (Ferraris and García-Pérez.4) were used for two purposes. One was to demonstrate that the promoter of the AR gene we had cloned was functional; this also defined basal promoter activity. The other was to examine segments of the 5' flanking region of the gene for evidence of an osmotic response element. We used several pUC8-derived expression vectors (Ferraris et al., 1994). 007Luc and 007CAT are promoterless, whereas B19Luc and B19CAT contain the B19 parvovirus-derived promoter in unique XhoI–HindIII sites immediately upstream of the luciferase and chloramphenicol acetyltransferase (CAT) genes, respectively. All AR fragments to be subcloned into expression vectors were PCR-amplified using primers synthesized with appropriate restriction enzyme sites toward their 5' end. We have found that using PCR-amplified fragments to which we have added appropriate restriction enzyme sites is more convenient than having to locate appropriately positioned restriction sites in the genomic sequence and then modify them to allow insertion into vector sequence. Transfections are performed like other experiments and require positive and negative controls; thus transfections are performed in groups, for example, transfection of AR-Luc promoter constructs were performed concomitantly with transfections of B19Luc (positive control for promoter activity), 007Luc (negative control for promoter activity), and transfection components only (no DNA; control for background). In transient transfections, the DNA is merely taken up by the cells. Unlike stable transfectants, the constructs are not incorporated into the genome of the cells. Thus it is important to normalize for transfection efficiency, that is, for the amount of DNA that enters the cells. For example, in transfections where the primary construct carries the luciferase gene, luciferase activity in relative light units (RLU) per microgram (μg) total cell protein is normalized by CAT protein in picograms (pg) per μg total cell protein (from the cotransfected B19CAT construct) (RLU luc/pg CAT). In transfections where the primary construct carries the CAT gene, CAT protein (pg per μg total cell protein) is normalized by luciferase activity (RLU per μg total cell protein) (from the cotransfected B19Luc) (pg CAT/RLU luc).

Since the promoter for the human AR gene had recently been characterized (Wang et al., 1993), we could determine whether the equivalent sequence in the rabbit gene would drive a downstream promoter. Figure 3B shows the alignment of the human sequence with that of the rabbit AR gene. *Cis*-elements previously shown to affect basal promoter activity were the TATA box, the CAAT box, and a GA element. The hAR sequence shown in Figure 3B (bp -192 to $+31$) demonstrated full promoter activity (Wang et al., 1993). So we subcloned the equivalent sequence in the rabbit gene, a fragment (rAR, bp -208 to $+27$) (Fig. 3B), directionally into

B

```
-208  CAGCGGTAGGAGAGACGGAATCCGGTGC        -180  rAR
      |||||| ||||| ||||| |||||| |||
-197  GAGCGTTGGGGCGGAAAGAATCCGCTGC        -169  hAR

-179  CGCTAGGACCTGGCGGAAGCTCCACCCGGGGCGGGGGTCGCCCGGGGCAAC   -120  rAR
      | |||||||| ||||||||||  | |||   || | ||| || || || |
-168  CACTAGGACCAGGCGGAAGAAGCATCCCGCGACCCTTGGGAAGGCCGCCGCGCACC   -109  hAR

-119  GCCCACCTCGGGAGTGCGGCCAATCAGATGGCGCTTTCGCACGCCAATCGC   -60  rAR
      | ||    |||   | ||||| |||   ||| |  |||||||  || || |
-108  CCC--------AGCGCAACCAATCAGAGAGCTCCTTCGCGCGCGCCAACCGC   -59  hAR

-59   GGGCGCCGATTCTGCAG--CCGCACGGGCTATTTAAAGGTGCGCGCCGAAGCT        -1  rAR
      ||||||| || ||||    |  | |||||||||||||||| | |||||   |
-58   AGGCGCCCTTTCTGCCGACCTCACGGGCTATTTAAAGGTACGCGCCGCGCGG---CCAAGGCC   -1  hAR

+1    ACAAACGGTTCTGGGCCTGGGTCTCTG-AGGG--GCGGCCATG   +39  rAR
      | | | ||| |||||||  |||||||| ||||    |||||||
+1    GCA--CCGTACTGGGCGGGGGTCTGGGAGGCGCAGCAGCCATG   +41  hAR
```

A

```
+40   GCGACCCACCTTGTGCTCTACACAACGGCGCCAAGATGCCGATCCTGGGGCTGGGCACCTGG   +99   ARmRNA
+100  AAGTCACCACCGGGCCAGGTGACCAGGCCGTGAAGACAGCCATCGACCTCGGGTACCGC     +159  ARmRNA
+160  CACATCGACTGCGCCCACGTGTACCAGAACGAGAACGAGAGGTCGGGTGGCCCTGCAGGAG   +219  ARmRNA
+220  AAGCTCAAGGAGCAGGTGGTGAAACGTGGTGAGGAGCTCTTCATCGTGCAGCAAGCTGTGGTGC  +279  ARmRNA
+280  ACGTCCCACGACAAGAGCCT                                            +299  ARmRNA
```

the XhoI–HindIII sites immediately upstream of the luciferase gene in 007Luc to produce $ARLuc_{-208, +27}$. Cells cotransfected with $ARLuc_{-208, +27}$ and B19CAT, the latter used for normalization of transfection efficiency, were maintained in isoosmotic medium (300 mOsm/kg H_2O) until harvested. $ARLuc_{-208, +27}$ showed an 11.3-fold increase in luciferase gene expression compared to 007Luc (mean expressed as RLU luc/pg CAT: $ARLuc_{-208, +27}$, 1455; 007Luc, 129.2). In contrast, another ARLuc construct containing a DNA fragment spanning from intron 3 to intron 7 was designed to control for artifactual promoter activity due to mere disruption of the 007Luc vector with a DNA fragment; this construct showed luciferase gene expression less than that of 007Luc. We were able to conclude that the upstream sequence cloned into $ARLuc_{-208, +27}$ can effectively drive the transcription of the downstream gene.

If the promoter of a given gene has not yet been characterized, then it is necessary to do so. The first step would be to identify putative promoter elements such as a CAAT box, TATA box, GC elements, Sp1 sites, and so on, using any of a variety of DNA analysis programs such as IBI Pustell or PCgene. These programs will only tell you that there are DNA sequences that are similar or identical to consensus sequences of promoter elements found in other genes. Computer identifications of cis-elements tend to be nonspecific; however, they provide an important starting point. Conservatively, it should only be necessary to analyze bases -600 to -1. The next step is to use transient transfections to determine the promoter activity generated by the largest DNA fragment, that is, one containing all possible computer-identified promoter elements. Once the largest fragment is shown to drive a downstream gene, this fragment becomes the positive control for narrowing down the promoter. Next, one systematically tests ever smaller DNA fragments, each fragment containing one less promoter element, for promoter activity.

Our next step was to test for the presence of a region that might respond to extracellular hyperosmolality. To do so, we prepared an additional construct, $ARLuc_{-3429, +27}$ (Figs. 2 and 4). $ARLuc_{-3429, +27}$ contains a fragment extending from bp ~ -3429 to $+27$ situated in the forward direction upstream of the luciferase gene in 007Luc. This construct is equivalent to $ARLuc_{-208, +27}$ plus ~ 3221 bp upstream. When cells cotransfected with $ARLuc_{-3429, +27}$/B19CAT were exposed to hyperosmotic (500 mOsm/kg H_2O) versus isoosmotic medium (300 mOsm/kg H_2O) for 24 hours, luciferase gene expression was more than 40-fold greater in the cells in

---

Figure 3. (A) Nucleotide sequence of rabbit renal AR mRNA from nt $+40$ to nt $+299$. Primers used in primer extension analysis and PCR-enhanced determination of 5'-end sequence are overlined. In the gene, introns 1 and 2 are positioned between nt 102 and 103 and nt 270 and 271, respectively. (B) Nucleotide sequence of the 5' upstream region of the rabbit AR gene (rAR). Numbering is positive downstream beginning with $+1$ as the transcription initiation start site. Sequence is aligned with hAR (human; Wang et al., 1993) through the first codon (ATG). Cis-elements previously shown to affect basal promoter activity (Wang et al., 1993) (TATA box, CAAT box, and GA element) and the first codon (ATG) are double-underlined. Primers used to PCR-amplify gene fragments for reporter gene expression analysis are overlined. (From Ferraris et al., 1994, with permission.)

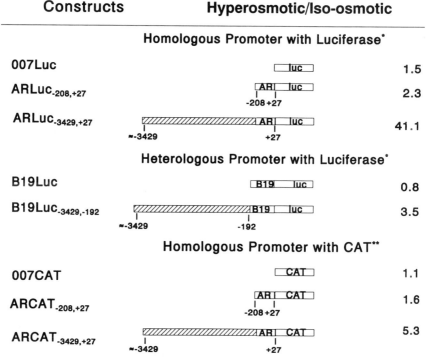

Figure 4. Effect of hyperosmolality on reporter gene expression in transient transfectants. Transfected PAP-HT25 cells were maintained in isoosmotic medium (300 mOsm/kg H$_2$O) or exposed to hyperosmotic medium (500 mOsm /kg H$_2$O) for 24 hours. Values are expressed as the ratio of the values in the two conditions. ARLuc$_{-208,+27}$ and ARCAT$_{-208,+27}$ contain the promoter alone (bp −208 to +27) (Figs. 2 and 3B), as previously defined (Wang et al., 1993); ARLuc$_{-3429,+27}$ and ARCAT$_{-3429,+27}$ contain this promoter plus upstream sequence (bp ~−3429 to +27); and B19Luc$_{-3429,-192}$ contains upstream sequence alone (bp ~−3429 to −192). All values represent the mean of ≥3 independent transfections. *Cells were cotransfected with a given luciferase construct and B19CAT. Luciferase activity (RLU) per μg cell protein was normalized by dividing by CAT protein (pg) per μg cell protein; calculation of the hyperosmotic/isoosmotic ratio cancels residual units. **Cells were cotransfected with a given CAT construct and B19Luc. CAT protein (pg) per μg cell protein was normalized by luciferase activity (RLU) per μg cell protein.

hyperosmotic medium (Fig. 4) (mean expressed as RLU luc/pg CAT: AR-Luc$_{-3429,+27}$ hyperosmotic, 70466; ARLuc$_{-3429,+27}$ isoosmotic, 1714). In contrast, in 007Luc and ARLuc$_{-208,+27}$ luciferase gene expression in hyperosmotically treated cells was 1.5-fold and 2.3-fold that of isoosmotically treated cells, respectively (mean expressed as RLU luc/pg CAT: 007Luc hyperosmotic, 188.1; 007Luc isoosmotic, 129.2; ARLuc$_{-208,+27}$ hyperosmotic, 3343.6; ARLuc$_{-208,+27}$ isoosmotic, 1455). We concluded that the region between bp ~−3429 and −208 contains one or more putative osmotic response elements (OREs) (Ferraris et al., 1994).

Typically, an enhancer should exert an effect (though not necessarily to the same degree) regardless of the promoter; for example, it should control the expression of a promoter belonging to a different gene. To test whether the region containing the putative OREs would have an effect on a heterologous promoter, $B19Luc_{-3429, -192}$ was examined. This construct contains a fragment extending from bp ~ -3429 to -192 placed in the forward direction upstream of the B19 promoter in B19Luc. When cells cotransfected with $B19Luc_{-3429, -192}/B19CAT$ were exposed to hyperosmotic medium, luciferase expression increased by 3.5-fold compared to cells maintained in isoosmotic medium (Fig. 4) (mean expressed as RLU luc/pg CAT: $B19Luc_{-3429, -192}$ hyperosmotic, 29486; $B19Luc_{-3429, -192}$ isoosmotic, 8458). In comparison, luciferase gene expression in cells transfected with B19Luc, containing the B19 promoter alone upstream of the luciferase gene, in hyperosmotic medium was 0.8-fold that in isoosmotic medium (mean expressed as RLU luc/pg CAT: B19Luc hyperosmotic, 20682; B19Luc isoosmotic, 26017). Thus the region in the rAR between bp ~ -3429 and -192 contains putative OREs capable of conferring hyperosmotic response to a heterologous promoter. Unfortunately, because that region appears to have a negative effect on basal B19 promoter activity (comparing the means of $B19Luc_{-3429, -192}$ and B19Luc in cells in isoosmotic medium), this may obscure some of the osmotic response expressed by $B19Luc_{-3429, -192}$. When we narrow down the ORE to the essential nucleotides, we will reexamine the effect on a heterologous promoter as well as examine the ORE for independence in orientation; that is, an enhancer should work not only in the forward orientation but also in the reverse orientation.

Another aspect to consider is the reporter gene to use. We selected luciferase for our primary reporter gene construct and CAT as our control for transfection efficiency. However, we also examined the reverse. To do this, we constructed AR-$CAT_{-3429, +27}$ (Figs. 2 and 4), containing the same AR gene fragment as AR-$Luc_{-3429, +27}$ but subcloned into 007CAT immediately upstream of the CAT gene. As a control, we constructed $ARCAT_{-208, +27}$; the same promoter fragment as described for $ARLuc_{-208, +27}$ was PCR-amplified and subcloned directionally into the XhoI–HindIII sites of 007CAT immediately upstream of the CAT gene. When cells cotransfected with $ARCAT_{-3429, +27}/B19Luc$ were exposed to hyperosmotic medium, CAT gene expression increased by 5.3-fold over that of cells in isoosmotic medium (Fig. 4) (mean expressed as pg CAT/RLU luc: $ARCAT_{-3429, +27}$ hyperosmotic, 6.93×10^{-5}; $ARCAT_{-3429, +27}$ isoosmotic, 1.32×10^{-5}). In contrast, CAT gene expression in cells transfected with 007CAT or $ARCAT_{-208, +27}$ (Figs. 2 and 4) and exposed to hyperosmotic medium was 1.1-fold and 1.6-fold that of cells exposed to isoosmotic medium, respectively (Fig. 4) (mean expressed as pg CAT/RLU luc: 007CAT hyperosmotic, 1.23×10^{-5}; 007CAT isoosmotic, 1.14×10^{-5}; $ARCAT_{-208, +27}$ hyperosmotic, 4.34×10^{-5}; $ARCAT_{-208, +27}$ isoosmotic, 2.71×10^{-5}). Thus the hyperosmotic response conferred by the ~ -3429 to -208 region is clearly demonstrable in transfected PAP-HT25 cells regardless of the reporter gene used; however, the degree of enhancement appears different. Results will depend on the strength of the enhancer, the sensitivity of the reporter gene assay, and the linearity of the reporter gene assay. The latter may be particularly important. The luciferase assay is linear for changes in luciferase activity that range over orders of magnitude; the CAT assay is not.

Previously, we had found that transcription of the aldose reductase gene was induced by extracellular hyperosmolality (Smardo et al., 1992). Now, using the techniques described above, we have cloned the gene and discovered a region of 5′ flanking sequence that confers osmotic response to the gene. Currently, we are characterizing this region to define the specific nucleotides of the OREs. This is done first by transient transfection of ever-smaller regions of DNA to identify the specific group of nucleotides required for osmotic response. The second step is to demonstrate evidence for the generation of *trans*-acting factors in hyperosmotically treated cells. *Trans*-acting factors are nuclear proteins whose binding may influence transcription of the gene. To demonstrate their existence we are using mobility shift assays. Mobility shift assays demonstrate a retardation of the mobility of a DNA fragment when it is bound by a protein. For both transient transfections and mobility shift assays we will also incorporate mutations into the DNA sequence to aid in analysis of essential nucleotides. The next step is to footprint the ORE to identify the nucleotides to which *trans*-acting factors bind. Each of these methods will allow us to continue to elucidate the molecular mechanisms underlying osmotic regulation of eukaryotic genes.

Inherent in the discovery of the existence of OREs and the recognition that the action of OREs forms the molecular basis for the ability of organisms to adapt to a hyperosmotic environment (Ferraris et al., 1994) are strong implications for the study of evolutionary biology. One of the most obvious instances is in processes of environmentally induced speciation (Ferraris, 1993). The presence of a consistent environmental challenge, such as variation or extreme in salinity, in combination with existing genetic variability leads to the selection of genetic variants. An example of salinity-induced isolation leading to speciation would include populations of organisms found at the extremes of a salinity cline. These populations could have altered protein structure or enzyme kinetics but more commonly mechanisms of adaptive evolution involve genetic change. Genetic change can be accomplished through mutation in the primary structure of proteins; however, the major mechanism is thought to be through change in regulatory genes (Wilson, 1976). Thus, by initiating the transcription of the gene into messenger RNA or by altering the rate at which it is transcribed, the amount of an enzyme can be increased; this will increase the activity of the enzyme without change in the primary structure or the properties of the enzyme. If a gene has a regulatory region that is responsive to the environmental stress, for example, an ORE, that gene will be adaptive. It should be noted that an ORE is not necessary for normal or constitutive expression; if it were absent the product of the gene could still be generated and life would go on as long as the salinity did not change. Various individuals in a species are likely to have the same genes, but genetic variants may occur at the level of an ORE. In an ORE, a single base change could make the difference between ability to respond by the required increase in expression and a lack of adaptation.

REFERENCES

Bachmann B, Luke W, Hunsmann G (1990): Improvement of PCR amplified DNA sequencing with the aid of detergents. *Nucleic Acids Res* 18:1309.

Bagnasco S, Uchida S, Balaban R, Kador P, Burg M (1987): Induction of aldose reductase and sorbitol in renal inner medullary cells by elevated extracellular NaCl. *Proc Natl Acad Sci USA* 84:1718–1720.

Britten R, Davidson E (1985): Hybridisation strategy. In: Hames B, Higgins S (eds). *Nucleic acid hybridisation: a practical approach.* Washington DC: IRL Press, pp 3–15.

Bruzdzinski C, Gelehrter T (1989): Determination of exon–intron structure: a novel application of the polymerase chain reaction technique. *DNA* 8:691–696.

Cowley BD Jr, Ferraris JD, Carper D, Burg MB (1990): *In vivo* osmoregulation of aldose reductase mRNA, protein and sorbitol in renal medulla. *Am J Physiol* 258:F154–F161.

Duchateau G, Florkin M (1955): Concentration du milieu exterieur et etat stationnaire du pool des acides amines non proteiques des muscles d'*Eriocheir sinensis*, Milne Edwards. *Arch Int Physiol Biochim* 63:249–251.

Ferraris JD (1993): Molecular approaches to the study of evolution and phylogeny of the Nemertina. *Hydrobiologia* 266:255–265.

Ferraris JD, Williams CK, Martin BM, Burg MB, García-Pérez A (1994): Cloning, genomic organization, and osmotic response of the aldose reductase gene. *Proc Natl Acad Sci USA* 91:10742–10746.

Florkin M (1956): Vergleichende Betrachtung des stationaren Zustandes der nichteiweissgebundenen Aminosauren der Tiere. *Colloq Ges Physiol Chem* 6:62–99.

Florkin M (1962): La regulation isosmotique intracellulaire chez les invertebres marins euryhalins. *Bull Acad Belg Cl Sci* 48:687–694.

García-Pérez A, Martin B, Murphy HR, Uchida S, Murer H, Cowley BD, Handler JS, Burg MB (1989): Molecular cloning of cDNA coding for kidney aldose reductase: regulation of specific mRNA accumulation by NaCl-mediated osmotic stress. *J Biol Chem* 264:16815–16821.

Geliebter J (1987): Dideoxynucleotide sequencing of RNA and uncloned cDNA. *Focus* 9:5–8.

Hofman M, Brian D (1991): A PCR-enhanced method for determining the 5' end sequence of mRNAs. *PCR Methods Appl* 1:44–45.

Maniatis T, Fritsch E, Sambrook J (1982): *Molecular cloning: a laboratory manual.* Cold Spring Harbor, NY: Cold Spring Harbor Laboratory Press.

Moriyama T, García-Pérez A, Burg MB (1989): Osmotic regulation of aldose reductase protein synthesis in renal medullary cells. *J Biol Chem* 264:16810–16814.

Smardo F, Burg M, García-Pérez A (1992): Kidney aldose reductase gene transcription is osmotically regulated. *Am J Physiol* 262:C776–C782.

Uchida S, Green N, Coon H, Triche T, Mims S, Burg M (1987): High NaCl induces stable changes in phenotype and karyotype of renal cells in culture. *Am J Physiol* 253:C230–C242.

Uchida S, García-Pérez A, Murphy H, Burg MB (1989): Signal for induction of aldose reductase in renal medullary cells by high external NaCl. *Am J Physiol* 256:C614–C620.

Uchida S, Yamauchi A, Preston A, Kwon H, Handler J (1993): Medium tonicity regulates expression of the Na^+- and Cl^--dependent betaine transporter in Madin-Darby canine kidney cells by increasing transcription of the transporter gene. *J Clin Invest* 91:1604–1607.

Wang K, Bohren K, Gabbay K (1993): Characterization of the human aldose reductase gene promoter. *J Biol Chem* 268:16052–16058.

Wilson, AC (1976). Gene regulation in evolution. In: Ayala, FJ (ed). *Molecular evolution.* Sunderland, MA: Sinauer, pp 225–234.

Yamauchi A, Uchida S, Preston AS, Kwon HM, Handler JS (1993): Hypertonicity stimulates transcription of the gene for the Na^+/myo-inositol cotransporter in MDCK cells. *Am J Physiol* 264:F20–F23.

Yancey P, Clark M, Hand S, Bowlus R, Somero G (1982): Living with water stress: evolution of osmolyte systems. *Science* 217:1214–1222.

CHAPTER 18

A Molecular Approach to the Selectionist/Neutralist Controversy

DENNIS A. POWERS and PATRICIA M. SCHULTE

Hopkins Marine Station, Stanford University, Pacific Grove, California 93950

CONTENTS

SYNOPSIS

Using the model teleost *Fundulus heteroclitus*, we illustrate the power of a multidisciplinary approach to address the classical "Selectionist/Neutralist" controversy. The first step was to establish the relationship between genetic variation, geography and environmental variables. Of over 50 protein coding loci studied, approximately half showed genetic variation. For many of these loci there were directional changes in gene frequency (*i.e.*, clines) between populations in relation to latitude. We used mitochondrial DNA analyses to delineate the phylogeography of this species. In

Molecular Zoology: Advances, Strategies, and Protocols, Edited by Joan D. Ferraris and Stephen R. Palumbi.

ISBN 0-471-14461-4   © 1996 Wiley-Liss, Inc.

addition, we conducted biochemical analyses on a variety of nuclear genes. As an example of our multidisciplinary approach, we have summarized some of our work with lactate dehydrogenase-B. We have reviewed the kinetic properties of the allelic isozymes, compared their cDNA sequences, and illustrated the use of site directed mutagenesis to determine the functional significance of specific amino acid differences. We have also shown that differences in enzyme concentration between populations are caused by variations in transcriptional regulation. The detailed molecular mechanisms underlying this phenomenon are currently under investigation.

INTRODUCTION

Few subjects in biology have been more strongly debated than the evolutionary significance, or lack thereof, of protein polymorphisms. Proponents of the selectionist school have asserted that natural selection maintains protein polymorphisms, whereas the neutralists have argued that the vast majority of such variation is selectively neutral; others have favored intermediate positions.

Classically, this conflict has been addressed by biogeographical analyses and/or the generation of mathematical models. After decades of heated debate, it became clear that the same biogeographic data could be used to support diametrically opposed theories, depending on the variables used in the model. Two decades ago, Lewontin (1974) summarized the failure of evolutionary biologists to resolve this important controversy by conventional approaches. As a consequence, some evolutionary biologists (e.g., Powers and Powers, 1975) began using an experimental approach with the assumption that the neutralist hypothesis could be rejected for a specific locus whenever functional nonequivalence could be established between allelic alternatives. Clarke (1975) formalized the general approach that these experimental biologists were using to address problems of genetic variation at enzyme synthesizing loci (reviewed later by Koehn, 1978). First, Clarke (1975) suggested that investigators make a detailed biochemical and physiological study of the allelic isozymes. Based on the nature of these observed biochemical and physiological differences, the function of the enzyme being studied, and the ecology of the organism, he suggested that investigators postulate one or more selective factors that might be operating on the allelic isozymes and then generate a hypothesis that establishes a potential mechanistic link between the selective factors and the gene product in question. The hypothesis could then be tested by experimental manipulation to produce a predictable response. Based on the experimental results, Clarke (1975) recommended that investigators reexamine the natural populations and develop a comprehensive theory for the observed gene frequency patterns.

During the past two decades, several research groups have used this general approach to study allelic isozymes of model organisms (reviewed by Powers et al., 1993, and references therein). While not all of the important studies discussed in the review by Powers et al. (1993) take advantage of the full power of the experimental approach, each has contributed important information about the adaptive significance of allelic isozymes and insight concerning the selectionist/neutralist controversy.

In 1972 we began using an experimental approach similar to that which was later formalized by Clarke (1975). Although we presented our initial results at scientific meetings as early as 1972, the first publication took an additional three years (Powers and Powers, 1975). Since then, the amount of information and level of sophistication have expanded exponentially (e.g., see review by Powers et al., 1993). We have reasoned that if biochemical analysis of allelic isozymes leads to predictable differences in cell physiology, then experiments designed to test those predictions should yield results that allow one to make other testable predictions at higher levels of biological organization. As each new cycle of predictions is followed by experimental validation, presumably, one would eventually be led to accepting either the selectionist or neutralist paradigm. If predictions can be followed by experimental validation, then the selectionist viewpoint would be supported; otherwise, the neutralist position would be favored. It is this cycle of *a priori* predictions, coupled with testing those predictions, that provides the real power of the experimental approach. In the course of following this approach, we have utilized a number of modern molecular techniques that have either not been used by others in the field or have not been incorporated into the interdisciplinary approach delineated above. These molecular approaches have allowed us to gain particular insight into the question of the role and mechanisms of natural selection in maintaining protein polymorphisms and have illuminated many aspects of the biology and physiology of our model teleost, *Fundulus heteroclitus*.

THE MODEL ORGANISM

Studies on the genus *Fundulus* have probably contributed more to the advancement of various branches of biological science in North America than any other group of fishes. A symposium on the biology of *Fundulus,* published in *American Zoologist* (1986), is a testament to the use of this model system by a variety of scientific disciplines.

   Fundulus heteroclitus range from the Matanzas River in Florida to Port au Port Bay in Newfoundland, Canada. Along this coast there is a 1°C change in annual mean water temperature per degree change in latitude (Fig. 1C), making it one of the steepest thermal gradients in the world. Fish in the southern marshes can experience summer temperatures in excess of 40°C, while northern populations are seldom exposed to such temperatures. Winter temperatures in northern regions result in extensive ice formation, while southern marshes don't even approach freezing temperatures (Fig. 1B).

   Although many environmental fluctuations exist in this fish's habitat, most are secondary effects brought about by changes in temperature, oxygen, and salinity. Since *F. heteroclitus* is a poikilotherm, temperature has a dramatic effect on its metabolism. The temperature difference of approximately 15°C in annual mean water temperature between the northern and southern extremes of the species' range (Fig. 1C) should result in a significant difference in the metabolic activity of these cold-blooded animals unless adequate thermal compensation can be achieved. There are two fundamental levels at which such thermal compensatory readjustments

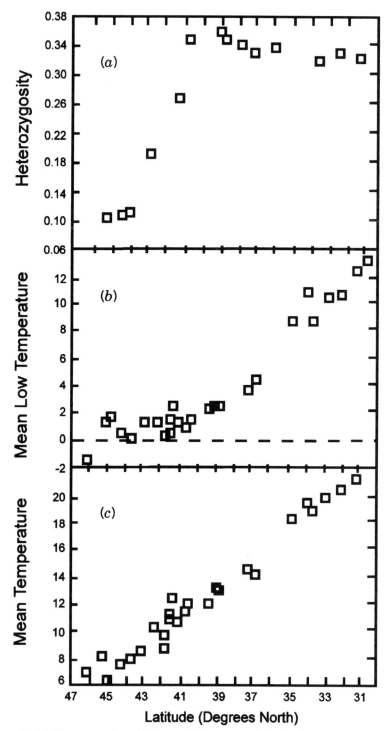

Figure 1. (a) Heterozygosity within populations versus latitude (°N). (b) Lowest average monthly surface water temperature (°C) versus latitude. (c) Mean surface water temperature versus latitude. (Adapted from Ropson et al., 1990, and reproduced with permission from the Annual Review of Genetics, Volume 25, © 1991 by Annual Reviews Inc.)

could take place: (1) at the individual level through either behavioral or physiological acclimation, and/or (2) at the population level by differential selection of allelic alternatives. In order for the second level of compensatory adjustments to take place, there has to be adequate genetic variation within the species.

BIOGEOGRAPHICAL VARIATION OF NUCLEAR
AND MITOCHODRIAL LOCI

Fundulus heteroclitus is highly polymorphic for a number of enzyme-encoding loci as determined by protein electrophoresis (e.g., Holmes and Whitt, 1970; Whitt, 1969, 1970a, 1970b; Massaro and Brooke, 1972; Avise and Kitto, 1973; Kempf and Underhill, 1974; Mitton and Koehn, 1975; Powers and Powers, 1975; Place and Powers, 1978; Van Beneden et al., 1981; Brown et al., 1988). Allele frequencies of the polymorphic loci have been determined for a large number of populations along the Atlantic Coast of North America (Powers and Powers, 1975; Powers and Place, 1978; Cashon et al., 1981; Ropson et al., 1990) and from the Chesapeake and Delaware Bays and their tributaries (Powers et al., 1986). Some of these loci have allelic isozymes that change in relation to latitude, while others do not (reviewed by Powers et al., 1993).

Examination of the polymorphic enzyme-encoding loci in *F. heteroclitus* along the Atlantic Coast has uncovered significant directional changes in gene frequency (i.e., clines) and in degree of genetic diversity (Fig. 1A) in relation to latitude (Powers and Place, 1978; Cashon et al., 1981; Ropson et al., 1990). The gene frequency patterns of coastal populations, in relation to latitude, can be grouped into four general types. The first type is represented by those loci that have two or three predominant alleles—one that dominates in northern populations, while an alternate allele (or alleles) dominates in southern populations (e.g., *Ldh-B, Mdh-A, Gpi-B, Pgm-B, Mpi-A,* and *H6pdh-A*). The second pattern involves loci that are multiallelic with one allele fixed at the northern end of the species range, but whose gene frequency drops to an intermediate value in southern populations with several other alleles being well represented (e.g., *Idh-A, Idh-B, Pgm-A, Fum-A, Est-S,* and *Est-B*). The third gene frequency pattern that has been observed involves the fixation of an allele with the same electrophoretic mobility at both the northern and southern extremes of the species range, but with genetic variability at middle latitudes (e.g., *6Pgdh-A* and *Aat-A*). The fourth gene frequency pattern does not show a directional change in gene frequency with latitude (e.g., *Aat-B, Ap-A, Adk-A,* and *Est-D*). It is difficult to rationalize this wide variety of patterns if all loci are behaving neutrally.

To begin to determine the effect of historical processes on the distribution of allelic isozymes, Gonzalez-Villasenor and Powers (1990) analyzed mtDNA restriction fragment length polymorphism (RFLP) patterns of four Atlantic Coast populations and identified two major mtDNA haplotype assemblages with a transition zone somewhere along the coast of New Jersey (Fig. 2). Previously, we suggested that the Wisconsin glaciation might have helped shape the present allelic isozyme clines because several showed sharp gene frequency changes near the Hudson River, which are associated with the edge of the last major glacial advance (Cashon et al.,

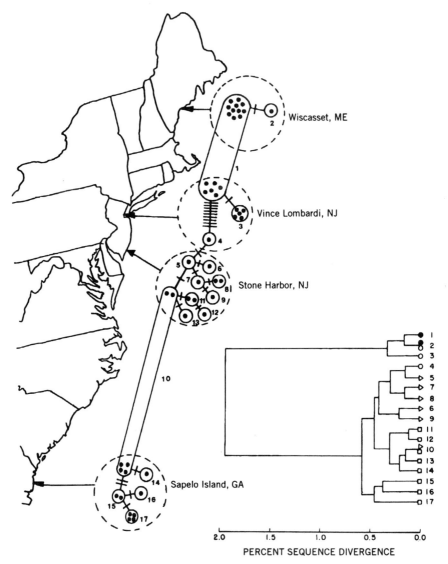

Figure 2. Phylogenetic network and phenogram of mtDNA haplotypes. The phylogenetic network is a composite of 17 mtDNA restriction phenotypes. Each circle indicates a different mtDNA clone. Dots inside circles identify the number of individual fish sharing the mtDNA phenotype. Clones are interconnected by branches with solid lines crossing the branches of the network indicating the minimum number of base substitutions required to account for the different mtDNA clones. Arrows indicate the collection areas. Clone 1 is shared between the populations of Maine and Vince Lombardi, New Jersey, while clone 10 is shared between Stone Harbor, New Jersey, and Georgia. The phenogram of mtDNA genotypes, illustrated as an insert in the figure, was generated by a UPGMA analysis of nucleotide sequence divergence (p) estimates. The numbers and symbols (circles, triangles, and squares) correspond to the mtDNA clones illustrated in the network. (From Gonzalez-Villasenor and Powers, 1990.)

1981). Although the mtDNA data showed a sharp disjunction consistent with that hypothesis, the number of nucleotide differences between the northern and southern mtDNA haplotypes suggested a divergence time significantly prior to the last glacial advance. Since the Chesapeake and Delaware Bays were only rivers during the last glacial advance, we examined the mtDNA haplotypes of 740 individual *Fundulus heteroclitus* from 29 populations along the Atlantic Coast and within the Chesapeake and four Delaware Bays to determine if remnant northern mtDNA haplotypes could be detected within the Bays (reviewed by Powers et al., 1993). Not only were northern mtDNA haplotypes detected, but they were distributed in a clinal fashion up the Bays and their tributaries (e.g., see Fig. 3). On the basis of these data, we have suggested that a single ancestral contact zone previously existed at least as far south as the mouth of the Chesapeake Bay prior to the last glaciation and that this zone has been unstable during the past several hundred thousand years.

Combining previous work on morphology (e.g., Morin and Able, 1983) and allelic isozymes with an analysis of the maternally inherited mitochondrial genome achieved a new synthesis and understanding of *F. heteroclitus* history and biogeography. Paleontological data for many flora and fauna indicate that local populations have frequently tracked environmental changes related to glaciations and sea level changes of these Milankovich cycles (Bennett, 1990). Morphological, physiologi-

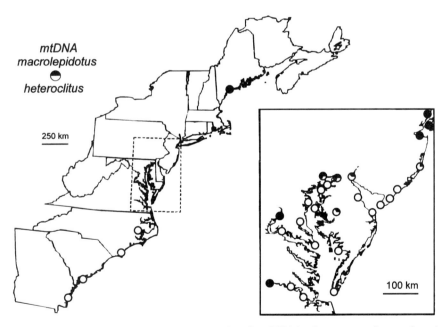

Figure 3. Frequencies of "northern" and "southern" mtDNA haplotypes superimposed as pie diagrams on a map of the Atlantic Coast and insert of the Mid-Atlantic region. Symbols: solid regions represent the fraction of a population represented by a "northern" mtDNA haplotype; open regions represent the fraction of a population represented by a "southern" mtDNA haplotype. (Adapted from Powers et al., 1993.)

cal, and genetic patterns characteristic of *F. heteroclitus* populations demonstrate the impact of frequent glaciations and sea level changes in shaping present day zoogeographic patterns.

While biogeographical studies, like those described above, are useful for placing genetic variations into a historical context, they do not provide the relative contributions of chance (e.g., genetic drift) and adaptive (e.g., natural selection) forces that shaped the genetic divergence between the allelic alternatives of specific loci prior to, during, and/or after a geographic isolation event, nor do they provide insights concerning the molecular mechanisms that drive these processes. Yet, it is the relative roles of chance and adaptive forces that are at the very heart of the neutralist/selectionist controversy. Thus additional approaches must be undertaken to determine the role of natural selection, if any, as a driving force for generating and/or maintaining gene diversity within and between populations of *F. heteroclitus*.

As pointed out earlier, *F. heteroclitus* is found in one of the steepest thermal gradients in the world (Fig. 1C) and, being a poikilotherm, must be profoundly influenced by this environmental parameter. The gene diversity of this species decreases at latitudes greater than 41°N but is unchanged at more southern latitudes (Fig. 1A). We have previously pointed out that this phenomenon is correlated with the amount of time that each population spends at or below freezing temperatures (Powers et al., 1983). Thus if temperature affects, or has affected, the differential survival of *F. heteroclitus* with specific allelic isozymes, then natural selection could be, or may have been, acting to change the gene frequency of populations that experience different thermal regimens along the East Coast. While these correlations are suggestive of a causal relationship, they are not definitive because they could be explained by stochastic events.

Since our work on the *Ldh-B* locus and its allelic alternatives has been most extensive to date, we shall use that locus as an example to illustrate our general strategy toward addressing the relative roles of selection and stochastic processes in maintaining protein polymorphisms. In the present chapter we emphasize our biochemical and molecular studies. The way that we have used this molecular information to make testable hypotheses at higher levels of biological organization has been reviewed elsewhere (Powers et al., 1991, 1993).

LACTATE DEHYDROGENASE: A GENERAL EXAMPLE

There are three vertebrate *Ldh* genes (*Ldh-A, Ldh-B,* and *Ldh-C*) that appear to be independently regulated and show tissue specificity. The products of the three *Ldh* genes are designated LDH-A_4, LDH-B_4, and LDH-C_4, respectively. While *Ldh-A* and *Ldh-B* are simultaneously expressed in the same tissues of many vertebrates, there is a remarkable tissue specificity and exclusivity of LDH expression in the tissues of many marine fishes. White skeletal muscle, whose metabolism is predominantly anaerobic, expresses the *Ldh-A* locus. Red muscle and liver that have significant aerobic metabolisms express almost exclusively *Ldh-B*. *Fundulus heteroclitus* erythrocytes, which have some aerobic capability (about 5–10%), also express *Ldh-*

B exclusively. The suggested functional significance of the difference in LDH isozymes is that LDH-A_4 is principally involved in the conversion of pyruvate to lactate (i.e., anaerobic glycolysis), while LDH-B_4 is principally involved in the conversion of lactate to pyruvate (i.e., gluconeogenesis and aerobic metabolism). The *Ldh-C* locus is expressed in the testes of mammals and in a variety of tissues in teleosts. While this locus is generally referred to as *Ldh-C* in mammals, birds, and fish, based on DNA sequence similarity, we have demonstrated that the *Ldh-C* of mammals and fish were derived independently (Quattro et al., 1993).

We have analyzed the isozyme patterns derived from the three *Ldh* loci of *F. heteroclitus* (Place and Powers, 1978). We have also cloned and sequenced the cDNAs of *Ldh-A* (Quattro et al., 1995), *Ldh-B* (Crawford et al., 1989), and *Ldh-C* (Quattro et al., 1993), sequenced *Ldh-B* cDNAs for a number of individuals from four different populations (Bernardi et al., 1993), and sequenced the *Ldh-B* promoter from several individual fish from a variety of different populations (Schulte et al., 1995b).

Although the *Ldh-A* and *Ldh-C* loci are relatively monomorphic between *Fundulus* populations, the *Ldh-B* locus has two codominant alleles: *Ldh-B^a* and *Ldh-B^b*. The relative proportions of these alleles vary with latitude. The *Ldh-B^b* allele predominates in the northern (i.e., colder) portions of the range (Fig. 4A). Since there are only two major alleles, *Ldh-B^a* is the predominant form in the southern (i.e., warmer) portion of the range (Powers and Place, 1978). We have used biochemical analyses to determine if these LDH-B_4 allelic isozymes differ in (1) catalytic efficiency, (2) substrate inhibition, and (3) enzyme stability (Place and Powers, 1979, 1984a, 1984b). We have also demonstrated enzyme concentration differences between northern (cold water) and southern (warm water) populations (Crawford and Powers, 1989, 1990; Crawford et al., 1990) and have begun to delineate the molecular mechanisms that regulate these differences in LDH-B_4 concentrations (Segal et al., 1995; Schulte et al., 1995b). We shall review some of those studies below.

Catalytic Differences Between LDH-B$_4$ Allelic Isozymes

At all temperatures and pH values, the k_{cat} values of the two LDH-B_4 allelic isozymes are identical, but the pseudo-first-order rate constant (V_{max}/K_m) and the related second-order rate constant (k_{cat}/K_m) are significantly different (Place and Powers, 1979, 1984b). At low temperatures, the LDH-B_4^b allelic isozyme, whose gene frequency is greatest in the northern colder waters (Fig. 4A), has a greater apparent catalytic efficiency at low substrate concentrations (V_{max}/K_m) than LDH-B_4^a (Place and Powers, 1979, 1984b). At higher temperatures the situation appears to be reversed, with LDH-B_4^a, whose gene frequency is greatest in the southern warm waters, having a greater catalytic efficiency than the LDH-B_4^b enzyme. The most dramatic differences occur at low temperatures and low pH values. For example, at 10°C and pH 6.5, the allelic isozyme that dominates in colder northern waters has a second-order rate constant that is twice that of the allelic isozyme that dominates at southern latitudes. These differences become less pronounced at higher pH values. At 25°C the two homozygous allelic isozymes are essentially equivalent, but at

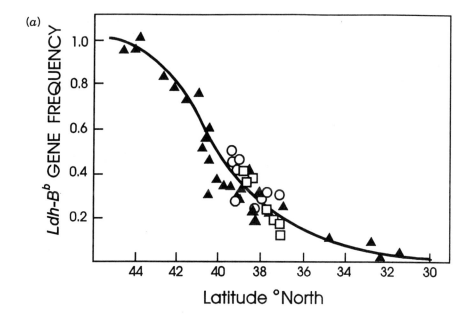

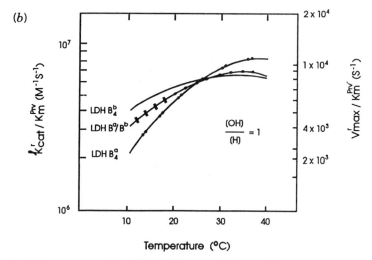

Figure 4. (A) Frequency of the *Ldh-B^b* allele versus latitude (°N). Solid triangles are from coastal populations, whereas the open symbols are from populations from the Chesapeake Bay and its tributaries (from Powers et al., 1991). (B) The effect of temperature on pyruvate reduction at a constant $[OH^-]/[H^+]$ ratio of unity. The values of k_{cat}/K_m and V_{max}/K_m are plotted against temperature at constant enzyme concentration, $[E_t]$. The Nernst equation was used to reduce the effect of temperature on pH in order to maintain a constant relative alkalinity. (Adapted from Place and Powers, 1979 and Powers et al., 1993.)

higher temperatures (e.g., 40°C), the allelic isozyme that predominates in populations from warmer southern latitudes has a greater catalytic efficiency than its northern cold water counterpart. A possible confounding factor is that the pH of cells decreases as their temperature increases. Since there is a physical relationship between temperature and pH of fluids, it is possible to transform the three-dimensional data into a two-dimensional representation that reflects the maintenance of relatively constant internal $\frac{[OH]}{[H]}$. After these pH changes are taken into account, the catalytic efficiency differences between LDH-B$_4$ allelic isozymes are illustrated in Figure 4B.

Very high concentrations of reaction products or substrates may inhibit an enzyme's function. For example, during the conversion of pyruvate to lactate, the LDH-B$_4^a$ allelic isozyme is much less susceptible to product (i.e., lactate) inhibition than the LDH-B$_4^b$ isozyme. For the conversion of lactate to pyruvate, the LDH-B$_4^b$ allelic isozyme is more susceptible to substrate (i.e., lactate) inhibition than the LDH-B$_4^a$ isozyme. Moreover, the magnitude of inhibition is greater at cool temperatures than at warm ones. The putative selective significance of this difference may relate to the accumulation of lactate during extreme swimming activity in fish, which can exceed 20 mM (Powers et al., 1979; DiMichele and Powers, 1982a; Place and Powers, 1984b), or during early embryonic development where lactate concentrations start somewhere between 40 and 50 mM (Paynter et al., 1991).

Structural Differences Between LDH-B$_4$ Allelic Isozymes

We determined the cDNA sequence of fish with each *Ldh-B* genotype. The amino acid sequences derived from the two cDNA clones indicated variation at amino acid residues 185 (Ser in *Ldh-B^b* and Ala in *Ldh-B^a*) and 311 (Ala in *Ldh-B^b* and Asp in *Ldh-B^a*). Since the lactate dehydrogenase structure has been highly conserved throughout evolution, structural comparisons can be made across taxa (Holbrook et al., 1975). Residue 311 is located at the exterior of the molecule pointing toward the solvent (Fig. 5). While substitution of Asp by Ala at residue 311 is responsible for the charge difference between the allelic isozymes, there is no obvious functional significance to this substitution. On the other hand, residue 185 is located at the interface between subunits (Fig. 5). Residues 182 to 185 form a hairpin loop that extends deep into the crevice of the other subunit. This structure may well be stabilized by hydrogen bonding between the hydroxy hydrogen of Ser 185 and the imidazolium nitrogen of His 182. The substitution at residue 185 of Ser in LDH-B$_4^b$ by Ala in LDH-B$_4^a$ should result in a loss of hydrogen bonding between subunits. We reasoned that this substitution might account for the increased susceptibility to heat denaturation of the latter. These predictions represent clearly testable hypotheses. Therefore site-directed mutagenesis studies were done (Lauerman and Powers, 1995) to ascertain the role, if any, that amino acid substitutions at positions 185 and/or 311 might play in the heat denaturation and/or other functional or structural differences. All possible combinations of amino acid substitutions were synthesized and their proteins produced in *Escherichia coli*. As predicted, residue 185 was found to affect the thermal stability of the LDH-B$_4$ isozymes. Although substituting Asp for Ala at residue 311 affected the charge of the LDH-B$_4$ electromorphs, there

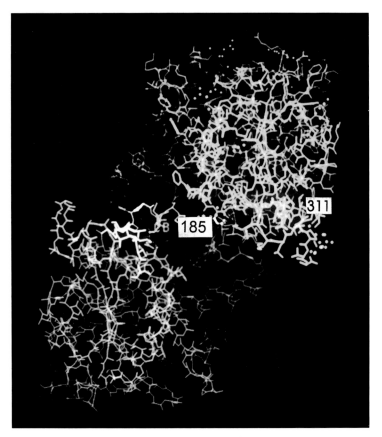

Figure 5. Computer simulated structure of LDH-B$_4$. Two subunits of the tetramer are shown with the subunit interaction involving residue 185 illustrated, and with residue 311 on the outside of the molecule. (Reproduced with permission from the Annual Review of Genetics, Volume 25, © 1991 by Annual Reviews Inc.)

was little, if any, effect on its thermal stability. Each of the four recombinant proteins (i.e., LDH-B$_4^{\text{Ala185/Asp311}}$; LDH-B$_4^{\text{Ser185/Asp311}}$; LDH-B$_4^{\text{Ala185/Ala311}}$; LDH-B$_4^{\text{Ser185/Ala311}}$) had K_m^{prv} and $K_l^{\text{-act}}$ values and rates of proteolysis (Lauerman and Powers, 1995) that were similar to those for the LDH-B$_4^b$ allelic isozyme isolated from a Maine population (Place and Powers, 1984a, 1984b). These results suggest several possibilities, including: (1) kinetically different but electrophoretically cryptic polymorphisms exist at the *Ldh-B* locus; (2) a high-affinity inhibitor was copurified during the purification of enzyme preparations from southern populations but not during purification of the enzyme from northern localities; or (3) there is differential post-translational modification (e.g., phosphorylation) of the LDH-B$_4$ allelic isozymes in the fish's cells that is not reproduced in *E. coli*. Recent evidence supports the latter.

Since only one cDNA from one individual was sequenced with an LDH-B$_4^a$

electrophoretic phenotype, it could not be said for certain that these two amino acids are the only differences between the LDH-B$_4$ allelic isozymes from Maine and Georgia populations. The fact that one cryptic amino acid substitution (position 185; Ser for Ala) was found suggested that there may have been others. There are a variety of data to support this hypothesis. For example, while substituting Ala for Ser at residue 185 affected the thermal stability of LDH-B$_4$, the thermal profiles obtained for the recombinant proteins did not completely match the profiles obtained from enzyme purified from several hundred individuals from Maine and Georgia, respectively. These data suggest that there may be differential post-translational modifications (e.g., phosphorylation) of the LDH-B$_4$ allelic isozymes in the fish, which are not reproduced in *E. coli,* or that there may be more than one allelic isozyme in the preparations purified from natural populations. Detailed zoogeographic studies of enzyme polymorphisms (Cashon et al., 1981; Powers and Place, 1978; Ropson et al., 1990), mtDNA restriction length polymorphisms (Gonzalez-Villasenor and Powers, 1990; Smith, 1989; Smith et al., 1992), and other data are concordant with southern populations having greater genetic diversity than northern populations, which is consistent with the latter hypothesis. On the other hand, we have sequenced six and seven *Ldh-B* cDNAs from northern and southern populations, respectively. Although we found a greater number of silent nucleotide substitutions in the southern population, other than those substitutions discussed above for amino acids 185 and 311, we have not yet found cryptic amino acid substitutions that affect either the kinetics or thermal stability of LDH-B$_4$ (Bernardi et al., 1993; G. Bernardi and D. A. Powers, *unpublished data*). The southern populations always had Ala at position 185 and Asp at position 311, while the *Ldh-B* cDNAs from the northern group always had Ser and Ala at positions 185 and 311, respectively. Six *Ldh-B* cDNA sequences obtained from a New Jersey population had slightly different amino acid sequences. For example, some had "northern" phenotypes at positions 185 and 311, others had "southern" phenotypes, and still others had a mixture at these amino acid positions (e.g., Ala at 185 and Ala at 311, or Ser at 185 and Asp at 311).

In order to ascertain the frequency of *Ldh-B* allelic variants at amino acids 185 and 311 in natural populations of *Fundulus,* we developed an allele-specific PCR method to rapidly determine the genotype at both positions (Powell et al., 1992). Even though the allelic alternatives varied by only a single nucleotide (Fig. 6A), we developed PCR primers (Fig. 6B) that would allow amplification of the DNA between sites 185 and 311 under defined experimental conditions. Since an amplification product only resulted when the correct primer was present, this method allows us to discriminate between the single nucleotide differences. Under those conditions, amplification between the two variable sites on the template DNA only took place when there was perfect complementarity between the primers and the target DNA.

The composition of the four reactions used to analyze each sample along with the template DNA that would give a product in each reaction are illustrated at the top of Figure 7 and a cartoon showing an example analysis of a template gene *Ldh-B^a* whose protein product would be LDH-B$_4$[Ala185/Asp311] is diagrammed in the bottom

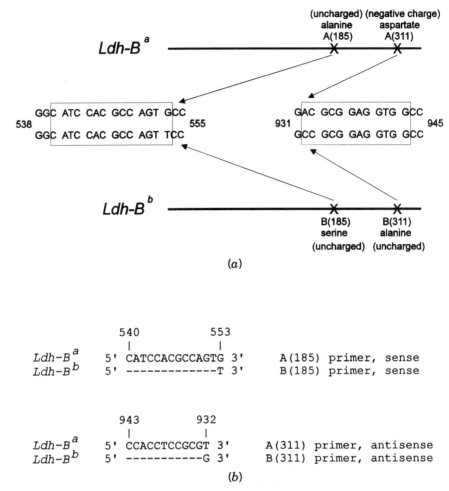

Figure 6. (a) The nucleotide and amino acid differences between *Ldh-B^a* and *Ldh-B^b* and their gene products, LDH-B$_4^a$ and LDH-B$_4^b$, respectively. The polymorphic amino acids that influence the electrophoretic mobility of LDH-B$_4$, and thus define the electrophoretic alleles *Ldh-B^a* and *Ldh-B^b*, are labeled as site 311, and the alternate amino acids are A(311) and B(311). The amino acids at the other polymorphic site, which do not affect electrophoretic mobility, are labeled sites A(185) and B(185). The nucleotide numbers refer to the first and last nucleotides shown for each region and are the numbers relative to the first nucleotide in the protein-coding portion of the gene. The boxes that outline sequences are used for PCR primers. (b) The PCR primers. The primers are specific for the regions shown in part (a). The numbers are relative to the first nucleotide in the protein-coding portion of the gene. (From Powell et al., 1992.)

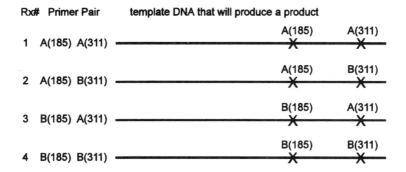

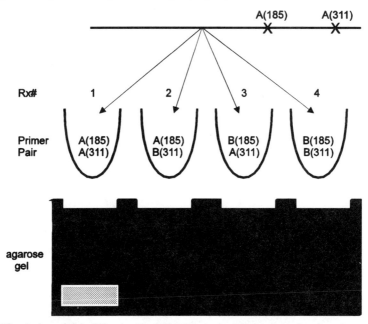

Figure 7. The design of the allele-specific PCR. The composition of the four reactions used to analyze each sample along with the template DNA that will give a product in each reaction are shown at the top. An example of an analysis of template gene A(185) A(311) is shown below. After temperature cycling, the products of the four reactions were loaded on an agarose gel. The presence or absence of a product was determined by the presence or absence of a band of the correct size. Each of the reactions (1–4) used one primer to each of the two polymorphic sites in the gene. A product was produced only when both primers exactly matched the template gene. For example, the A(185) and A(311) primer pair (reaction 1) produced a product only when the template DNA contained the A(185) A(311) sequences (see Fig. 8); reactions 2, 3, and 4 did not produce a product from the A(185) A(311) template DNA. (From Powell et al., 1992.)

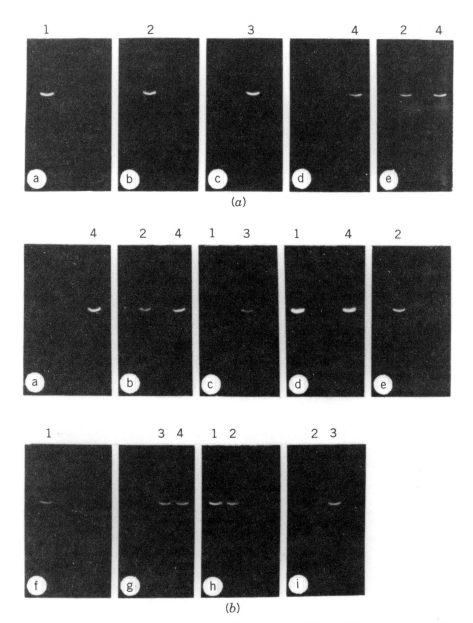

Figure 8. (a) Analysis by allele-specific PCR of template DNAs of known sequence using the primers and experimental approach illustrated in Figures 6 and 7. The numbers indicate which of the four lanes contain a PCR product for each template. Lanes 1 to 4 for each template are the results of reactions 1 to 4 for that template. Templates used were: a, A(185) A(311); b, A(185) B(311); c, B(185) A(311); d, B(185) B(311); e, two templates used— A(185) B(311) and B(185) B(311). Template DNAs were produced by *in vitro* mutagenesis of a cloned LDH-B cDNA. (b) Analysis of the LDH-B genotype of nine *Fundulus* by allele-specific PCR. Fish RNA was used to produce cDNA, which was then amplified by PCR, as

of Figure 7. Figure 8 illustrates the results of an actual analysis of DNA samples taken from an intermediate population and clearly illustrates how each of the homozygotes and heterozygotes could be identified by this single nucleotide PCR method. Analyses of the data shown in Figure 8 and other gels demonstrate that intermediate populations contained all possible combinations at amino acid positions 185 and 311 (Powell et al., 1992). Although starch gel electrophoresis had previously revealed only two alleles (i.e., *Ldh-B^a* and *Ldh-B^b*) and three phenotypes (i.e., two different homozygotes and one heterozygote), the allele-specific PCR assays identified four alleles and nine LDH-B$_4$ phenotypes. In addition to identifying base changes that cause amino acid substitutions, this technique can also be used to rapidly determine any single nucleotide variant (i.e., coding or noncoding) and can be used on DNA isolated from either homozygotes or heterozygotes. Our studies on genetic variation of *Ldh-B* cDNAs within and between populations in combination with site-directed mutagenesis and functional studies should help delineate the precise functional role of the amino acid variants in populations of *Fundulus heteroclitus*.

LDH-B$_4$ Enzyme Concentration Differences Between Populations

Near the extremes of its natural distribution (e.g., Maine and Georgia), we found that the specific activity of some liver enzymes in *F. heteroclitus* differed significantly. For example, the LDH-B$_4$ of fish liver collected from Maine had a specific activity approximately twice that of liver collected from Georgia (Crawford et al., 1985, 1990; Fig. 9A). These differences in specific activity remain even after extensive temperature acclimation, suggesting that this may represent an evolutionary adaptation (Crawford and Powers, 1989). Similar analyses of populations from intermediate latitudes indicate LDH-B$_4$-specific activities intermediate to Maine and southern fish (Crawford et al., 1990).

Maine fish are essentially only *Ldh-B^b* genotype, and fish from Georgia and Florida have the *Ldh-B^a* genotype. Since the LDH-B$_4$ allelic isozymes have the same rate constant (k_{cat}) at the same temperatures, the observed differences in enzyme-specific activity are most likely a function of different enzyme concentrations because $V_{max} = k_{cat}[E_t]$. Evidence to support this possibility comes from immunoprecipitation studies (Crawford and Powers, 1989, 1990) wherein enzyme activity differences were directly correlated with differences in LDH-B$_4$ protein concentrations (Fig. 9B). Dissimilar LDH-B$_4$ concentrations between populations

---

outlined. Lanes 1 to 4 represent the results of reactions 1 to 4 for each template. Fish cDNA that produced one product was scored as homozygous, and DNA that produced two products was scored as heterozygous. The genotypes were scored as follows: a, homozygous B(185) B(311); b, heterozygous A(185) B(311) and B(185) B(311); c, heterozygous A(185) A(311) (the band in lane 1 is faint) and B(185) A(311); d, heterozygous A(185) A(311) and B(185) B(311); e, homozygous A(185) B(311); f, homozygous A(185) A(311); g, heterozygous B(185) A(311) and A(185) B(311); h, heterozygous A(185) A(311) and A(185) B(311); i, heterozygous A(185) B(311) and B(185) A(311). (From Powell et al., 1992.)

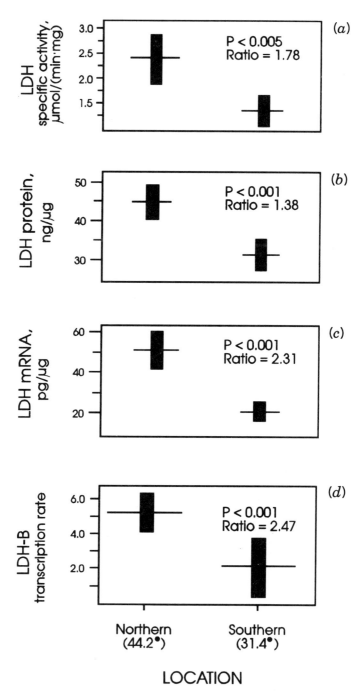

Figure 9. (a) LDH-specific activity of livers from Maine and Georgia populations collected from 44.2°N and 31.4°N, respectively. (b) LDH protein concentrations of the livers referred to previously. (c) LDH-specific mRNA concentrations of the liver samples. (d) LDH-B transcriptional rate for the same samples as the other measurements illustrated in this figure. (Reproduced with permission from the Annual Review of Genetics, Volume 25, © 1991 by Annual Reviews Inc.)

were apparently primarily genetic because the differences were present even after long-term acclimation (Crawford and Powers, 1990) and genetic transmission analyses are also consistent with that hypothesis (L. DiMichele and D. A. Powers, *unpublished data*).

The differences in LDH-B$_4$ concentrations could be the result of different *Ldh-B* gene copies, transcriptional factors, LDH-B mRNA concentrations, rates of mRNA degradation, ratios of active versus inactive enzyme, or rates of LDH-B$_4$ synthesis (or degradation). The cloned *Ldh-B^b* cDNA was used as a probe to determine if there were different numbers of *Ldh-B* gene copies and LDH-B mRNA concentrations between Maine and Georgia populations (Crawford and Powers, 1989). While the number of gene copies did not vary between populations, the LDH-B mRNA concentrations were found to be different (Fig. 9C). The LDH-B mRNA concentrations were approximately twofold higher in Maine fish than in their Georgia counterparts.

Because the differences in LDH-B mRNA could be due to transcriptional or post-transcriptional regulation, the *Ldh-B* transcriptional rates for Maine and Georgia fish were determined (Crawford and Powers, 1992) using nuclear run-on assays. Northern *F. heteroclitus*, acclimated to 20°C, had a significantly higher transcription rate from the *Ldh-B* locus than acclimated fish from the southern population (Fig. 9D). The greater LDH-B transcription rate was not due to a general increase in the rate of RNA synthesis *per se,* but was specific for *Ldh-B*. Thus populations of *F. heteroclitus* that live in different thermal regimes appear to compensate by the transcriptional regulation of the *Ldh-B* locus. Working out the detailed molecular mechanisms responsible for this difference in gene expression has begun to provide insight concerning the adaptive significance of regulatory events to natural populations in a changing environment and at the extremes of a species natural range (P. M. Schulte et al., 1995b).

Studies on enzymes from distantly related and congeneric species have shown that variations in kinetic rate constants and/or enzyme concentration are important for environmental adaptation (Siebenaller and Somero, 1978; Graves and Somero, 1982; Hochachka and Somero, 1984; Somero and Hand, 1990). In some cases, allelic variants within a species have evolved different kinetic constants that affect important physiological parameters and selection (Daly and Clarke, 1981; DiMichele and Powers, 1982a, 1982b, 1991; Vigue et al., 1982; Hilibish and Koehn, 1985; Zamer and Hoffman, 1989; Paynter et al., 1991). In other cases, enzyme concentrations have been altered by differential gene regulation (McDonald et al., 1977; McDonald and Ayala, 1978; Crawford and Powers, 1989). Since gene regulation allows enzyme activity to vary independent of amino acid changes in the gene product (McDonald and Ayala, 1978; Wilson, 1976), it is important to establish the relative role of differential gene expression as an evolutionary mechanism for population adaptation to changing environments, and the potential role, *if any,* of regulatory changes in the genesis of new species. This is one of the most exciting and important areas of research for population and evolutionary biologists for the coming decade and has become our major research focus. While our preliminary results were presented at the American Society of Zoology meeting, because those data are mostly unpublished they will not be detailed here. However, an overview of the results and how the general approach may be used to address similar problems in other systems will be summarized.

It is possible to use a combination of DNA sequence analysis, cell culture transfection assays, and *in vivo* DNA footprinting to investigate the molecular mechanisms underlying naturally occurring variation in transcription rate. We have used these approaches to study the molecular mechanisms regulating differential expression of the *Ldh-B* gene between northern and southern populations of *Fundulus heteroclitus*. The detailed results of those studies will be presented elsewhere (Schulte et al., 1995a, 1995b; Segal et al., 1995).

Since the transcription of genes is regulated by a complex interaction between regulatory proteins and specific DNA sequences in the regulatory regions of genes, it is possible to identify the specific regulatory proteins and their DNA-binding sites by a technique often referred to as "DNA footprinting." This technique utilizes the fact that specific DNA sequences that are bound to regulatory proteins are protected against digestion by DNase I. After digestion, the protected DNA will yield a different electrophoretic pattern than unprotected DNA. The resulting DNA "footprints" can be used to identify the specific DNA sequences to which regulatory proteins are bound or they can be used as a probe to isolate the regulatory proteins. Although the most common method employs an *in vitro* approach, *in vivo* techniques can provide information about accessibility *in vivo* that cannot always be assessed *in vitro*. We have developed an *in vivo* DNA "footprinting" technique that has been used to identify some of the sequences of the *Ldh-B* promoter that bind regulatory proteins (Schulte et al., 1995a; Segal et al., 1995). This technique can readily be adapted for similar problems in other marine organisms. Preliminary data from transfection studies utilizing the *Ldh-B* promoter from Maine and Georgia populations indicate that there are significant differences in their ability to drive expression of the reporter gene to which they are ligated (Schulte et al., 1995b). The molecular mechanisms underlying these regulatory differences are currently under investigation.

The Functional Role of Molecular Differences

Differences in enzyme kinetics, thermal stability, enzyme concentration, amino acid sequence, and other parameters that have been uncovered between the LDH-B$_4$ allelic isozymes and their genes indicate that they are structurally and functionally nonequivalent, both within and between populations. Moreover, the nonequivalence can be correlated with the temperatures at which each allelic isozyme is most common. The question arises, "Are these and other structural and functional differences reflected at higher levels of biological organization?" Ideally, one would like to predict physiological differences at the cellular and organismal level that should result from observations at the molecular level. However, such predictions assume that *in vitro* differences are large enough to produce significant variations *in vivo*. In other words, physiological differences predicted on the basis of *in vitro* molecular studies must be proved experimentally at the cellular and organismal level, which is where selection takes place. Since this chapter focused on molecular approaches, we refer the reader to other review papers that carefully document the linkage between our molecular results and higher levels of biological organization (i.e., cellular, physiological, organismic, and population; see Powers et al., 1991, 1993). In those reviews we not only summarize the structural, kinetic, thermodynamic, and enzyme concentration differences between alternative *Ldh-B* allelic isozymes, but we also demonstrate how these variations are correlated with differences in oxygen consumption, metabolic flux, developmental rate, hatching time, swimming performance, and survivorship in selection experiments (for primary studies see Place and Powers, 1979, 1984a, 1984b; DiMichele and Powers, 1982a, 1982b, 1984, 1991; DiMichele et al., 1986, 1991; Crawford and Powers, 1989, 1990, 1992.)

SUMMARY AND CURRENT RESEARCH DIRECTIONS

Using the model teleost, *Fundulus heteroclitus,* we have formulated a multi-disciplinary approach to help provide a better understanding of the evolutionary mechanisms that affect genetic variation at enzyme-encoding loci in natural populations. Starting with molecular methodologies, we have demonstrated structural, kinetic, thermodynamic, and enzyme concentration differences between alternative *Ldh-B* allelic isozymes and have begun to uncover some of the molecular mechanisms driving these differences. In other reviews (Powers et al., 1991, 1993) we have shown how these are correlated with differences in oxygen consumption, metabolic flux, developmental rate, hatching time, and swimming performance. Although we have provided considerable evidence consistent with the hypothesis that the *Ldh-B* locus directly affects these biological functions, the possibility remains that *Ldh-B* may only be a genetic marker associated with the genes actually causing the observed biological differences. The ideal way to determine whether genetic variation at a particular locus (e.g., *Ldh-B*) directly affects biochemical or metabolic functions, like those we have uncovered, would be to develop genetic strains with a common genetic background but with variation only at the locus in question. Although we have constructed *Fundulus* strains with restricted genetic backgrounds, we have not been able to develop strains that vary only at a single locus. Since it would take many years to achieve this goal, we have attempted a shorter term alternative approach by altering the organism's enzyme phenotype via exchanging the endogenous allozyme with that derived from a different genotype. Although these enzyme exchange studies (Paynter et al., 1991) were done on individuals from a single population, with a restricted genetic background (i.e., all individuals shared common alleles for most of the polymorphic enzymes and a single mtDNA haplotype), and the results were consistent with a direct effect of the *Ldh-B* locus on development rate and oxygen consumption, those results do not exclude the possibility that another locus or loci were responsible for the observed biological phenomenon.

Using transgenic technology, we are now in the process of determining if allelic alternatives of the *Ldh-B* gene are directly responsible for observed biological differences in oxygen consumption, metabolic flux, developmental rate, hatching time, and/or swimming performance that we have previously shown to be associated with, or marked by, allelic alternatives of *Ldh-B*. Using a similar approach, we intend to determine if LDH-B$_4$ enzyme concentrations have a direct biological effect by creating a series of transgenic *Fundulus* that express introduced constructs of the *Ldh-B* gene, including variants of both the promoter and coding regions. Biological performance of these transgenic animals can then be compared to their non-transgenic counterparts in order to determine the magnitude of the biological impact, if any, provided by the introduced *Ldh-B* genes. The null hypothesis is that transgenic fish that express the foreign *Ldh-B* will not differ significantly in biological function from their nontransgenic counterparts. These interesting experiments should provide a definitive answer as to whether differences in *Ldh-B* genotype alone can have a measurable impact on function at the whole-organism level.

REFERENCES

American Zoologist (1986): 26:109–288 Entire volume.

Ausubel FM, Brent R, Kingston RE, Moore DD, Seidman JG, Smith JA, Struhl K (1987): *Current protocols in molecular biology.* New York: Greene Publishing/Wiley.

Avise JC, Kitto GB (1973): Phosphoglucose isomerase gene duplication in the bony fishes: an evolutionary history. *Biochem Genet* 8:113–132.

Bennett KD (1990): Milankovich cycles and their effect on species in ecological and evolutionary time. *Paleobiology* 16:11–21.

Bernardi G, Sordino P, Powers DA (1993): Concordant mitochondrial and nuclear DNA phylogenies for populations of the teleost *Fundulus heteroclitus*. *Proc Natl Acad Sci USA* 99:9271–9274.

Brown D, Ropson I, Powers DA (1988): Biochemical genetics of *Fundulus heteroclitus* (L.). V. Inheritance of 10 polymorphic loci. *Heredity* 79(5):359–365.

Cashon RE, Van Beneden RJ, Powers DA (1981): Biochemical genetics of *Fundulus heteroclitus* (L). IV. Spatial variation in gene frequencies of *Idh-B*, *6-Pgdh-A*, and *Est-S*. *Biochem Genet* 19:715–718.

Clarke B (1975): The contribution of ecological genetics to evolutionary theory: detecting the direct effects of natural selection on particular *polymorphic loci*. *Genetics* 79:101–108.

Crawford DL, Powers DA (1989): Molecular basis of evolutionary adaptation in two latitudinally extreme populations of *Fundulus heteroclitus*. *Proc Natl Acad Sci USA* 86:9365–9369.

Crawford DL, Powers DA (1990): Molecular adaptation to different thermal environments: genetic and physiological mechanisms. In: Clegg, M.T., O'Brian, S.J. (eds). *Molecular evolution.* New York: Wiley-Liss, pp 213–222.

Crawford DL, Powers DA (1992): Evolutionary adaptation to different thermal environments via transcriptional regulation. *Mol Biol Evol* 9(5):806–813.

Crawford DL, Place A, Cashon R, Powers DA (1985): Thermal acclimation of differential regulation. *Am Zool* 25:122.

Crawford DL, Constantino HR, Powers DA (1989): Lactate dehydrogenase-B cDNA from the teleost *Fundulus heteroclitus:* evolutionary implications. *Mol Biol Evol* 6(4):369–383.

Crawford DL, Place AR, Powers DA (1990): Clinal variation in the specific activity of lactate dehydrogenase-B from the teleost *Fundulus heteroclitus. J Exp Zool* 255:110–113.

Daly K, Clarke B (1981): Selection associated with alcohol dehydrogenase locus in *Drosophila melanogaster:* differential survival of adults maintained on low concentration of ethanol. *Heredity* 46:219–226.

DiMichele L, Powers DA (1982a): Physiological basis for swimming endurance differences between *Ldh-B* genotypes of *Fundulus heteroclitus. Science* 216:1014–1016.

DiMichele L, Powers DA (1982b): *Ldh-B* genotype specific hatching times of *Fundulus heteroclitus* embryos. *Nature* 296:560–563.

DiMichele L, Powers DA (1984): Developmental and oxygen consumption differences between *Ldh-B* genotypes of *Fundulus heteroclitus* and their effect on hatching times. *Physiol Zool* 57:52–56.

DiMichele L, Powers DA (1991): Developmental heterochrony and differential mortality in the model teleost, *Fundulus heteroclitus. Physiol Zool* 64(6):1426–1443.

DiMichele L, Paynter K, Powers DA (1991): Lactate dehydrogenase-B allozymes directly affect development of *Fundulus heteroclitus. Science* 253:898–900.

Gonzalez-Villasenor LI, Powers DA (1990): Mitochondrial-DNA restriction-site polymorphisms in the teleost *Fundulus heteroclitus* support secondary intergradation. *Evolution* 44(1):27–37.

Graves JE, Somero GN (1982): Electrophoretic and functional enzyme evolution in four species of eastern Pacific barracudas from different thermal environments. *Evolution* 36:97–106.

Hilibish TJ, Koehn RK (1985): Dominance in physiological phenotypes and fitness at an enzyme locus. *Science* 229:52–54.

Hochachka PW, Somero GN (1984): *Biochemical adaptation.* Princeton, NJ: Princeton University Press.

Holbrook JJ, Liljas A, Steindel SJ, Rossman MG (1975): Lactate dehydrogenase. In: Boyer PE (ed). *The enzymes,* vol II, part A. New York: Academic Press, pp 191–292.

Holmes RS, Whitt GS (1970): Developmental genetics of the esterase isozymes of *Fundulus heteroclitus. Biochem Genet* 4:471–480.

Kempf CT, Underhill DK (1974): A serum esterase population in *Fundulus heteroclitus. Copeia* 1974:792–794.

Koehn RK (1978): Physiology and biochemistry of enzyme variation: the interface of ecology and population genetics. In: Brussard P.F. (ed). *Ecological genetics:* the *interface.* New York: Springer, pp 511–572.

Kunkel TA, Roberts JD, Zakour RA (1987): Rapid and efficient site specific mutagenesis without phenotypic selection. *Methods Enzymol* 154:367–382.

Lauerman T, Powers DA (1995): Stability and kinetic analysis of four recombinant LDH-B$_4$ allelic isozymes cloned from the teleost fish, *Fundulus heteroclitus* (in preparation).

Lewontin RC (1974): *The genetic basis of evolutionary change.* New York: Columbia University Press.

Massaro EJ, Brooke HE (1972): A mutant A-type lactate dehydrogenase subunit in *Fundulus heteroclitus* (Pisces Cyprinodontidae). *Copeia* 1972:298–302.

McDonald JF, Ayala FJ (1978): Genetic and biochemical basis of enzyme activity variation in natural populations. I. Alcohol dehydrogenase in *Drosophila melanogaster. Genetics* 89:371–388.

McDonald JF, Chambers GK, David J, Ayala FJ (1977): Adaptive response due to changes in gene regulation: a study with *Drosophila. Proc Natl Acad Sci USA* 74:4562–4566.

Mitton JB, Koehn RK (1975): Genetic organization and adaptive response of allozymes to ecological variables in *Fundulus heteroclitus. Genetics* 79:97–111.

Morin RP, Able KW (1983): Patterns of geographic variation in the egg morphology of the fundulid fish, *Fundulus heteroclitus. Copeia* xx:726–740.

Mueller PR, Wold B (1989): In vivo footprinting of a muscle specific enhancer by ligation mediated PCR. *Science* 246:780–786.

Mueller PR, Wold B (1991): Ligation mediated PCR: applications to genomic footprinting. *Methods (companion to Methods Enzymol)* 2:20–31.

Paynter KT, DiMichele L, Hand SC, Powers DA (1991): Metabolic implications of *Ldh-B* genotype during early development in *Fundulus heteroclitus. J Exp Zool* 257:24–33.

Place AR, Powers DA (1978): Genetic bases for protein polymorphism in *Fundulus heteroclitus.* I. Lactate dehydrogenase (*Ldh-B*), malate dehydrogenase (*Mdh-A*), phosphoglucoisomerase (*Pgi-B*), and phosphoglucomutase (Pgm-A). *Biochem Genet* 16:577–591.

Place AR, Powers DA (1979): Genetic variation and relative catalytic efficiencies: LDH-B allozymes of *Fundulus heteroclitus. Proc Natl Acad Sci USA* 76:2354–2358.

Place AR, Powers DA (1984a): The lactate dehydrogenase (LDH-B) allozymes of *Fundulus heteroclitus* (Lin.): I. Purification and characterization. *J Biol Chem* 259:1299–1308.

Place AR, Powers DA (1984b): The LDH-B allozymes of *Fundulus heteroclitus:* II. Kinetic analyses. *J Biol Chem* 259:1309–1318.

Powell M, Crawford DL, Lauerman T, Powers DA (1992): Analysis of cryptic alleles of *Fundulus* lactate dehydrogenase by a novel allele-specific polymerase chain reaction. *Mol Mar Biol Biotechnol* 1(6):391–396.

Powers DA, Place AR (1978): Biochemical genetics of *Fundulus heteroclitus.* I. Temporal and spatial variation in gene frequencies of *Ldh-B, Mdh-A, Gpi-B* and *Pgm-A. Biochem Genet* 16:593–607.

Powers DA, Powers DW (1975): Predicting gene frequencies in natural populations: a testable hypothesis. In: Markert C (ed). *Isozymes IV: genetics and evolution.* New York, Academic Press, pp 63–84.

Powers DA, Greaney GS, Place AR (1979): Physiological correlation between lactate dehydrogenase genotype and hemoglobin function in killifish. *Nature* 277:240–241.

Powers DA, DiMichele L, Place AR (1983): The use of enzyme kinetics to predict differences in cellular metabolism, developmental rate, and swimming performance between *Ldh-B* genotypes of the fish, *Fundulus heteroclitus.* In: Whitt G, Markert GC (eds). *Isozymes: current topics in biological and medical research,* Vol. 10. New York: Academic Press, p 171.

Powers DA, Ropson I, Brown D, Van Beneden R, Cashon R, Gonzalez-Villasenor I, DiMichele L (1986): Genetic variation in *Fundulus heteroclitus:* geographic distribution. *Am Zool* 26:131–144.

Powers DA, Lauerman T, Crawford D, DiMichele L (1991): Genetic mechanisms for adapting to a changing environment. *Annu Rev Genet* 25:629–659.

Powers DA, Smith M, Gonzalez-Villasenor I, DiMichele L, Crawford DL, Bernardi G, Lauerman TA (1993): Multidisciplinary approach to the selectionist/neutralist controversy using the model teleost *Fundulus heteroclitus.* In: Futuyma D, Antonovics J (eds). *Oxford surveys in evolutionary biology,* vol 9. Oxford: Oxford University Press, chap 2, pp 43–107.

Quattro JM, Woods HA, Powers DA (1993): Sequence analysis of teleost retina-specific lactate dehydrogenase-C: implications for the evolution of the vertebrate lactate dehydrogenase gene family. *Proc Natl Acad Sci USA* 90:242–246.

Quattro JM, Pollock D, Powell MA, Woods HA, Powers DA (1995): Evolutionary relationships among vertebrate muscle-type lactate dehydrogenases. *Mol Mar Biol Biotechnol* 4(3):224–231.

Ropson IJ, Brown DC, Powers DA (1990): Biochemical genetics of *Fundulus heteroclitus* (L.) VI. Geographical variation in the gene frequencies of 15 loci. *Evolution* 44(1): 16–26.

Schulte PM, Segal JA, Crawford DL, Powers DA (1995a): A rapid *in vivo* footprinting method for the detection of DNA–protein interactions in isolated nuclei. *Mol Mar Biol Biotechnol* 4(3):200–205.

Schulte PM, Gomez-Chiarri M, Powers DA (1995b): Geographical variation in the function and sequence of the regulatory region of the *Ldh-B* locus in *Fundulus heteroclitus* (Submitted).

Segal JA, Schulte PM, Powers DA, Crawford DL (1995): Variation in promoter sequence and transcription factor binding at the *Ldh-B* locus of *Fundulus heteroclitus. J. Exp. Zool.* (Submitted).

Siebenaller GN, Somero GN (1978): Pressure-adaptive differences in lactate dehydrogenases of congeneric fishes living at different depths. *Science* 201:225–257.

Seibenaller GN and Somero GN (1982): The maintenance of different enzyme activity levels in congeneric fishes living at different depths. *Physiol Zool* 55:171–179.

Smith MW (1989): *Mitochondrial and nuclear gene analysis of the killifish, Fundulus heteroclitus; a reconstruction of phylogenetic history.* PhD thesis. Johns Hopkins University, Baltimore, MD, p 183.

Smith MW, Glimcher MC, Powers DA (1992): Differential introgression of nuclear alleles between subspecies of the teleost *Fundulus heteroclitus. Mol Mar Biol Biotechnol* 1(3):226–238.

Somero GN, Hand SC (1990): Protein assembly and metabolic regulation: physiological and evolutionary perspectives. *Physiol Zool* 63:443–471.

Van Beneden RJ, Cashon RE, Powers DA (1981): Biochemical genetics of *Fundulus heteroclitus* III, inheritance of isocitrate dehydrogenase (*Idh-A* and *Idh-B*), 6-phosphogluconate dehydrogenase (*6-Pgdh-A*), and serum esterase (*Est-S*) polymorphisms. *Biochem Genet* 19:701–714.

Vigue CL, Weisgram PA, Rosenthal E (1982): Selection at the alcohol dehydrogenase locus of *Drosophila melanogaster:* effects of ethanol and temperature. *Biochem Genet* 20:681–688.

Whitt GS (1969): Homology of lactate dehydrogenase genes: E gene function in the teleost nervous system. *Science* 166:1156.

Whitt GS (1970a): Developmental genetics of the lactate dehydrogenase isozymes of fish. *J Exp Zool* 175:1–35.

Whitt GS (1970b): Genetic variation of supernatant and mitochondrial malate dehydrogenase isozymes in the teleost *Fundulus heteroclitus. Experientia* 26:734.

Wilson AC (1976): Gene regulation in evolution. In: Ayala FJ (ed). *Molecular evolution.* Sunderland, MA: Sinauer, pp 225–234.

Zamer WE, Hoffman RJ (1989): Allozymes of glucose-6-phosphate isomerase differentially modulate pentose-shunt metabolism in the sea anemone *Metritium senile. Proc Natl Acad Sci USA* 86:2737–2741.

CHAPTER 19

Molecular Studies of the Sequential Expression of a Respiratory Protein During Crustacean Development

NORA BARCLAY TERWILLIGER and GREGOR DURSTEWITZ

Oregon Institute of Marine Biology, Charleston, Oregon 97420; and Department of Biology, University of Oregon, Eugene, Oregon 97403

CONTENTS

Molecular Zoology: Advances, Strategies, and Protocols, Edited by Joan D. Ferraris and Stephen R. Palumbi.
ISBN 0-471-14461-4   © 1996 Wiley-Liss, Inc.

SYNOPSIS

Oxygenation properties of hemocyanin, the copper-containing oxygen transport protein found in arthropod hemolymph, are responsive to both the external milieu and the internal metabolism of the organism. During development from a swimming megalopa to a crawling crab, hemocyanin subunit composition and oxygen affinity change in *Cancer magister,* the Dungeness crab. To understand the molecular mechanisms responsible for these ontogenetic changes, we are investigating patterns of tissue-specific and developmental stage-specific hemocyanin transcription. To identify site of biosynthesis and onset of adult hemocyanin biosynthesis, we used PCR and a combination of both specific and universal degenerate primers to develop hemocyanin-specific cDNA probes. These probes were then used for both Northern blot analysis of mRNA transcripts and cDNA library screening. A developmentally regulated hemocyanin cDNA has been cloned and is being sequenced.

INTRODUCTION

Hemocyanin, like the other oxygen-transporting molecules, hemoglobin and hemerythrin, combines reversibly with oxygen at the respiratory surface of the organism and carries oxygen via the circulatory system to cells and tissues far from the animal's surface. Thus the hemocyanin molecule is poised between the external environment and the internal milieu. Its functional properties, including oxygen affinity and cooperativity, respond to changes in both external and internal parameters such as temperature, salinity, pH, and metabolic compounds (Van Holde and Miller, 1982, for review).

Hemocyanins occur in only two phyla, the Mollusca and the Arthropoda. We now know that these hemocyanins are two very different proteins, even though they share the functional property of combining reversibly with oxygen, and they each contain two copper atoms at their active sites. Indeed, one of the copper-binding sites of molluscan hemocyanin, CuB, shows clear sequence homology to the CuB-binding region of arthropodan hemocyanins (Drexel et al., 1987). There is no significant homology between the molluscan and arthropodan CuA sites, however, nor between the rest of the amino acid sequences as far as is known (Volbeda and Hol, 1989b; Lang and Van Holde, 1991). Not surprisingly, molluscan and arthropodan hemocyanin molecules show marked differences in quaternary structure and subunit size (for reviews see Markl and Decker, 1992; Van Holde et al., 1992).

Arthropodan hemocyanins, found in chelicerates, crustaceans, and one myriapod, are made up of individual polypeptide chains or subunits of about 75 kDa. Based on the two arthropodan hemocyanins whose crystal structures are known, *Panulirus interruptus* a and *Limulus polyphemus* II (Gaykema et al., 1984; Volbeda and Hol, 1989a; Hazes et al., 1993), each polypeptide chain is bean shaped and composed of three structural regions or domains. Domain 1 is made up of seven α-helices while domain 2, containing the two copper-binding sites, CuA and CuB, is composed of two pairs of antiparallel α-helices. The third domain is a seven-stranded β-barrel with two long loops that wrap around domain 2 to interact with domain 1. Domain 2, containing the functional copper sites where oxygen is bound, is the most conserved region of the subunit, a feature that has been helpful in the present study. The subunits self-assemble to form extracellular hexameric and multi-hexameric oligomers that circulate in the hemolymph. Subunit heterogeneity in the oligomers varies among the arthropodan hemocyanins, ranging from 2 to as many as 12 different polypeptide chains, depending on the species (Van Holde and Miller, 1982, for review). Functional studies on hemocyanins from a number of different arthropods indicate that subunit composition affects the oxygen-binding properties of the oligomer (Sullivan et al., 1974; Truchot, 1992, for review).

The hemocyanin of the Dungeness crab, *Cancer magister,* is an especially intriguing hemocyanin to study because it changes in both structure and function during development of the crab from megalopa through the early juvenile instars to the adult (Terwilliger and Terwilliger, 1982). Megalopa and early juvenile crab hemocyanins are composed of five different subunits, numbered in order of increasing mobility in sodium dodecyl sulfate–polyacrylamide gel electrophoresis (SDS-PAGE) as shown in Figure 1. Adult hemocyanin contains the same five subunits plus another, subunit 6, that is not present in the megalopa and young juvenile

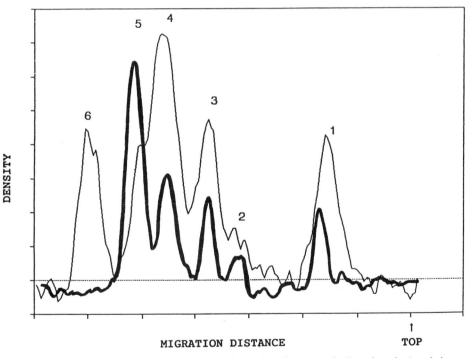

Figure 1. Gel scan comparing five subunits of megalopa hemocyanin (broad tracing) and six subunits of adult hemocyanin (thin tracing) from *Cancer magister* separated by electrophoresis on 7.5% SDS-PAGE.

stages. Furthermore, the relative amounts of two other subunits, 4 and 5, switch during the change from juvenile to adult. These latter three subunits–4, 5, and 6–appear to be developmentally regulated, in contrast to subunits 1, 2, and 3 whose stoichiometries are constant during development.

As the hemocyanin subunit composition changes from juvenile to adult pattern, so too does the oxygen affinity of the hemocyanin (Terwilliger et al., 1985; Terwilliger and Brown, 1993). Megalopa and juvenile hemocyanins have an intrinsically lower oxygen affinity than does adult hemocyanin when measured at the same pH and ionic composition. As subunit 6 appears and subunits 4 and 5 reverse their relative concentrations, the oxygen affinity of the hemocyanin increases to adult levels. In addition, the changes in hemocyanin structure and functional properties during development are integrated with the ontogeny of hemolymph ion regulation (Brown and Terwilliger, 1992; Terwilliger and Brown, 1993).

There are many examples of ontogenetic changes in protein expression from a variety of phyla. The fetal–maternal shift in mammalian hemoglobin is one of the classic examples (Bunn et al., 1977). This ontogenetic shift in Dungeness crab hemocyanin is the first documented change in a copper protein whose biochemical and physiological roles in both the adult and the juvenile organism have been well studied (McMahon et al., 1979; Graham et al., 1983; Morris and McMahon, 1989; Brown, 1991).

Cancer magister has a number of attributes that make it a particularly suitable organism for these biochemical, physiological, and molecular studies. First, wild megalopas can be collected easily in sufficient numbers when they return to Oregon coastal waters and estuaries in the spring after several months of oceanic larval life. Second, the megalopa stage of this species is markedly larger than any of the other Pacific Coast megalopas and thus can be identified at a glance rather than having to be laboriously sorted out from a swirl of similar beasts varying only in length of rostral spine or patterns of hairs on hairy little legs. Third, the megalopas can be raised through the different instars in a running seawater system at ambient temperature and salinity. In addition, hemocyanin protein is easily obtained by bleeding a crab, and changes in an individual crab's hemolymph proteins can be analyzed over time by repeated sampling (Otoshi, 1994; Terwilliger and Otoshi, 1994). Thus the crab fits the August Krogh principle: "For a large number of problems there will be some animal of choice on which it can be most conveniently studied" (Krogh, 1929).

As more information has been obtained on developmental changes in hemocyanin structure and function, more questions have arisen, ones that could best be approached using molecular techniques. This chapter asks the following questions and describes the molecular strategies we are using to answer them.

1. Where is *C. magister* hemocyanin synthesized? Hemocyanin circulates as an extracellular protein in the hemolymph, but where are the cells located that synthesize the hemocyanin? The hepatopancreas has been implicated as the site of synthesis in several studies of crustacean hemocyanin biosynthesis (Senkbeil and Wriston, 1981; Préaux et al., 1986; Hennecke et al., 1990). Hemocyanin synthesis has also been described in horseshoe crab blood cells within sinuses around the eye (Fahren-

bach, 1970; Wood and Bonaventura, 1981), scorpion endocuticle (Alliel et al., 1983), tarantula heart (Kempter, 1986; Markl et al., 1990), and crab reticular connective tissue around several organs (Ghiretti-Magaldi et al., 1973, 1977). Is the site of synthesis species specific, or are there multiple sites within an animal? Our approach to determine where *C. magister* hemocyanin is synthesized was to look for hemocyanin mRNA in various tissues.

2. When does synthesis of hemocyanin subunit 6 begin? We knew when hemocyanin containing subunit 6, "adult hemocyanin," was first detectable in the hemolymph, but we wished to know when synthesis of subunit 6 first occurred. Such information might provide clues about the mode of assembly of the multihexameric molecules as well as insights into the regulation of biosynthesis. Ultimately, this approach may reveal whether the onset of adult hemocyanin biosynthesis is solely regulated by an internal developmental program or is correlated to extrinsic environmental cues as well. To begin to answer this question, we need to investigate hemocyanin mRNA expression in specific developmental stages of *C. magister.*

3. What is the evolutionary relationship of crustacean hemocyanin to other arthropod hemolymph proteins, including chelicerate hemocyanin, arthropod cryptocyanin (Terwilliger and Otoshi, 1994), and insect hemolymph proteins? To answer this question, we decided to construct a *C. magister* cDNA library and determine the hemocyanin cDNA sequence.

WHERE IS *CANCER MAGISTER* HEMOCYANIN SYNTHESIZED?

We chose a combination of approaches to address this question. They included designing a hemocyanin-specific degenerate primer (a short DNA sequence complementary to hemocyanin mRNA) and also preparing *C. magister* cDNA from crab hepatopancreas. With the hemocyanin-specific primer and a commercially available oligo-dT primer plus the crab cDNA, we would try to amplify by the polymerase chain reaction (PCR) any crab cDNA complementary to hemocyanin mRNA. The amplified hemocyanin cDNA fragment could then be used as a probe to assay hemocyanin mRNA expression in different tissues of the crab using Northern blots.

Design of Hemocyanin-Specific Degenerate Primers

Since the CuA-binding site in domain 2 is highly conserved in all crustacean and chelicerate hemocyanin subunits thus far sequenced (Beintema et al., 1994), we expected it to be a conserved feature in *C. magister* hemocyanin subunits as well. Therefore a 32-bp oligonucleotide primer (CuA primer I) was designed based on a conserved sequence of 10 amino acids within the CuA site of another crustacean hemocyanin polypeptide, subunit a of the spiny lobster, *Panulirus interruptus* (Bak and Beintema, 1987). Due to the degeneracy of the genetic code, the "primer" actually consisted of a family of oligonucleotides that represented all possible ways of coding for the short sequence of amino acids. The primer (degeneracy = 16384) was synthesized in a Model 380B automated DNA synthesizer (Applied Biosystems, Inc.) at the University of Oregon Biotechnology Laboratory. Its sequence was:

```
    E   L   F   F   W   V   H   H   Q   L   T            AA sequence of
                                                         subunit a from
                                                         P. interruptus
5'  GAA-TTT-TTT-TTT-TGG-GTT-CAT-CAT-CAA-TTT-AC   3'      CuA primer I
    G C C   C   C           C   C   C   G C C
    A                               A       A
    G                               G       G
```

Another primer was designed to specifically amplify mRNA coding for hemocyanin subunit 6, the subunit present only in adult *C. magister*. We needed a short amino acid sequence from subunit 6 hemocyanin to develop this primer. All six subunits of adult *C. magister* hemocyanin were purified and the N-terminal amino acid sequence of each was determined (*unpublished data*). This 5' subunit 6 primer, a degenerate 26-bp oligonucleotide (degeneracy = 65536, see below), was based on the unique N-terminal amino acid sequence of subunit 6:

```
    S   A   G   G   A   F   D   A   Q            N-terminal AA
                                                 sequence of sub-
                                                 unit 6 from C.
                                                 magister
5'  TCT-GCT-GGT-GGT-GCT-TTT-GAT-GCT-CA   3'      5' subunit 6
    AGC   C   C   C   C   C   C   C              primer
    A   A   A   A   A           A
    G   G   G   G   G           G
```

Isolation of Total RNA

Adult crab hepatopancreas was chosen for the initial source of RNA for PCR, based on crab mRNA literature and the abundance of hepatopancreas tissue. Adult male Dungeness crabs were collected from Coos Bay and quickly killed. Tissue samples (1 g) were immediately dissected, thoroughly rinsed with *C. magister* saline buffer (Brown and Terwilliger, 1992), placed in liquid nitrogen, and ground to a fine powder with mortar and pestle. Total RNA was isolated with the guanidinium isothiocyanate method using the *RAPID* Total RNA Isolation Kit (5 Prime → 3 Prime, Inc.). This standard procedure quickly inactivates cellular RNases that were likely to be present in high concentrations in actively metabolizing hepatopancreas tissue.

Development of Probes: Reverse Transcription of mRNA, PCR Amplification, Cloning, and Sequencing of Hemocyanin cDNA

First strand cDNA was prepared from hepatopancreas total RNA using AMV reverse transcriptase to synthesize complementary DNA from the RNA template

(Terwilliger and Durstewitz.1). Using the CuA primer I plus an oligo-dT primer directed against the poly(A)+ tail of mRNA, the PCR reaction was carried out with hepatopancreas cDNA as template (Fig. 2 and Terwilliger and Durstewitz.1). PCR products were size analyzed on 1.2% agarose Tris-acetic acid–EDTA (TAE) minigels; several major fragments between 1200 and 2000 bp had been generated, as seen in Figure 3. These fall within the expected size range for a hemocyanin cDNA amplified from the CuA site to the 3' end, based on known hemocyanin sequences (Linzen et al., 1985). The PCR products of interest were cloned (Terwilliger and Durstewitz.1) into a Bluescript II SK+ phagemid vector (Stratagene Inc., La Jolla, CA). A positive clone containing a 1600-bp insert was selected, and sequencing grade DNA was prepared from the clone with a QIAGEN maxiprep kit according to manufacturer's instructions. The 1600-bp insert was then partially sequenced with a Sequenase kit (Version 2.0, U.S. Biochemical, Cleveland, OH) using T3 and T7 sequencing primers and the dideoxy sequencing method. The best-fit alignment, (FASTA algorithm, Devereux et al., 1984) of the derived amino acid sequence of the 5' end of the 1600-bp PCR fragment and the corresponding CuA-coding portion of *Panulirus interruptus* subunit a is shown in Figure 4. With 81% amino acid similarity (chemical similarity) and 76% amino acid identity between the two se-

Adult *Cancer magister* hepatopancreas cDNA

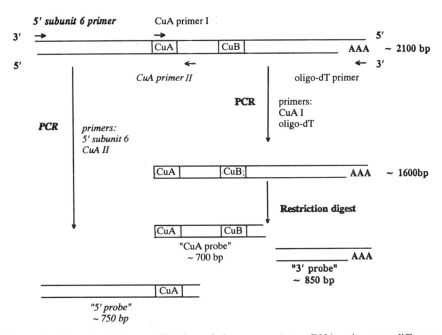

Figure 2. Schematic PCR amplification of *Cancer magister* cDNA using two different combinations of primers (CuA I and oligo-dT, *5' subunit 6* and *CuA II*) and an EcoRV restriction digest to obtain three hemocyanin-specific cDNA probes (CuA probe, 3' probe, and *5' probe*).

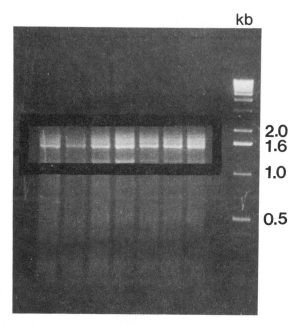

Figure 3. PCR products on 1.2% agarose TAE minigel. Target sequence: adult *Cancer magister* hepatopancreas cDNA; primers: CuA I and oligo-dT. From left to right, products of PCR reactions run at 1 mM, 2 mM, 3 mM, 4 mM, 5 mM, 6 mM, and 8 mM Mg^{2+}; right lane, 1-kb ladder (Gibco BRL). Box outlines 1200–2000 bp PCR fragments excised from gel and cloned.

quences, the PCR fragment is clearly a hemocyanin cDNA. Choosing the conserved portion, the CuA site, of the crustacean hemocyanin subunit as the basis for our primer was a successful strategy for obtaining hemocyanin cDNA. Which of the six possible hemocyanin subunits of *C. magister* this cDNA represented could not be identified at this point. The cDNA insert was mapped by restriction analysis and the 5' and 3' ends were sequenced with the dideoxy technique. Digestion of the 1600-

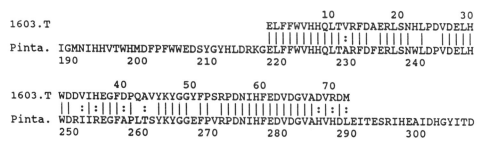

Figure 4. Amino acid sequence alignment of 5' end of 1600-bp PCR fragment from *Cancer magister* (1603.T) with *Panulirus interruptus* hemocyanin subunit a (Pinta). CuA portion of Pinta sequence shown is numbered after Linzen et al. (1985).

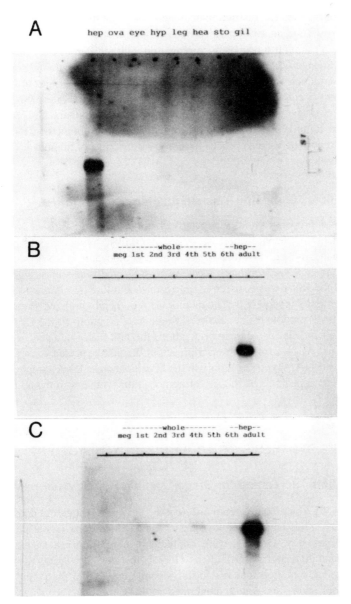

Figure 5. Northern blots of *Cancer magister* RNA hybridized with ^{32}P random prime labeled probes. (A) Total RNA from different tissues of adult *C. magister*. Probe, 750-bp 5' probe. Hep, hepatopancreas; ova, ovary; eye, eyestalk; hyp, hypodermis; leg, leg muscle; hea, heart; sto, stomach; gil, gill. (B) mRNA from different developmental stages of *C. magister*. Probe, 750-bp 5' probe. Meg, megalopa; 1st, 1st instar; 2nd, 2nd instar; 3rd, 3rd instar; 4th, 4th instar; 5th, 5th instar; 6th hep, 6th instar hepatopancreas; adult hep, adult hepatopancreas. (C) Same as (6B), but hybridized with 700-bp CuA probe.

bp cDNA fragment by the restriction enzyme EcoRV yielded two fragments, a 700-bp "CuA probe" and an 850-bp "3′ probe" (Fig. 2).

Using the 5′ subunit 6 primer in combination with a different CuA primer, CuA II, whose sequence was derived from the 1600-bp clone, we were able to amplify, clone, and sequence a 750-bp fragment corresponding to the 5′ end of the hemocyanin subunit 6 cDNA. Its 3′ end overlapped with the CuA site of the 1600-bp PCR fragment (Fig. 2). Thus with PCR and different combinations of primers, we were able to amplify different regions of hemocyanin subunit 6 cDNA. These regions could now be used as probes to identify transcripts of hemocyanin subunits in various tissues.

Northern Blots: Screening Tissues for Hemocyanin mRNA

Total RNA was prepared as described above from eight different tissues of adult *C. magister,* including hepatopancreas, ovary, eyestalks, hypodermis, leg muscle, heart, stomach, and gill in order to see where hemocyanin mRNA was located. Equal amounts of RNA from each tissue were subjected to denaturing gel electrophoresis and the RNA was transferred onto a nitrocellulose membrane as described in Terwilliger and Durstewitz.[2]. The membrane was then hybridized to the 5′ probe that had been ^{32}P-random prime labeled ("Probe-Eze" Random-Prime-Labeling Kit, 5 Prime→3 Prime, Inc.). After hybridization, the blot was evaluated by autoradiography (Fig. 5A). A 2100-bp transcript, about the size expected for a full-length hemocyanin mRNA, was present only in the hepatopancreas RNA sample. Thus the hepatopancreas appears to be the site of hemocyanin synthesis in the adult Dungeness crab.

WHEN DOES SYNTHESIS OF ADULT HEMOCYANIN BEGIN?

Northern Blots: Screening for Stage-Specific Hemocyanin mRNA

Messenger RNA was prepared from different developmental stages to determine the stage at which adult hemocyanin synthesis begins. Megalopa larvae collected in May from Coos Bay, Oregon, were maintained in running seawater aquaria at ambient temperature and salinity at the Oregon Institute of Marine Biology. The megalopas quickly molted into first instar juvenile crabs, and the juveniles, fed a diet of mussel, fish, and squid, continued molting through the instar stages in synchrony with field-caught juveniles. As each developmental stage reached intermolt, aliquots of megalopa through 5th instar juvenile were harvested by quick-freezing batches of whole animals in liquid nitrogen. The aliquots were stored at −80°C until the older instars had developed, at which point 1-g samples of frozen whole animals were ground to a fine powder with mortar and pestle in liquid nitrogen, and RNA was isolated. The larger stages, 6th instar and adult, were dissected as described above; 1-g samples of hepatopancreas were quickly rinsed and frozen. Total RNA was isolated with the guanidinium isothiocyanate method using a *RAPID* Total RNA Isolation Kit, and poly (A)+ mRNA was prepared with oligo-dT spin columns (5 Prime $\rightarrow$ 3 Prime, Inc.). The mRNA yield was quantified by measuring absorbance at 260 nm.

Eight developmental stages were assayed by Northern blots for the presence of hemocyanin subunit 6 mRNA. Equal amounts (0.3 μg) of mRNA from each stage were separated according to size by electrophoresis and blotted as described in Terwilliger and Durstewitz.2. Results of hybridizing the blot to the 750-bp 5' probe are shown in Figure 5B. Hemocyanin mRNA was detected only in the adult stage, where a 2.1-kb transcript hybridized strongly. When a duplicate blot was incubated with the 700-bp CuA probe (Fig. 5C), the adult sample again gave a strong signal. Some degree of hybridization occurred in the earlier stages as well. These results suggest that synthesis of hemocyanin subunit 6 occurs after the 6th instar stage. The low levels of hybridization of the earlier stages with the CuA probe probably indicate cross-reactions with mRNA transcripts coding for one or more of the other hemocyanin subunits, 1–5, that are present in megalopa and early instars or low levels of subunit 6 mRNA. Experiments in progress to further pinpoint the onset of adult hemocyanin synthesis will include Northern blots of mRNA from instars 7, 8, and 9 as the juveniles develop.

WHAT IS THE EVOLUTIONARY RELATIONSHIP OF HEMOCYANIN TO OTHER PROTEINS?

Construction of a *Cancer magister* cDNA Library

The third question we wished to investigate, the cDNA sequence of the developmentally regulated hemocyanin subunit 6, led us to construct a *C. magister* cDNA library. The cDNA library would eventually give us the complete sequences of all hemocyanin subunits including the 5' noncoding regions. Adult crab hepatopancreas mRNA was used to create the library in a lambda phage vector (Lambda ZAP cDNA Synthesis Kit, Stratagene). In this procedure, double-strand cDNA was synthesized from mRNA; each cDNA was then inserted into the multiple cloning site of a Bluescript SK− vector within a Lambda ZAP II phage. The library was amplified once on *Escherichia coli* XL-1 blue MRF' cells to a final titer of 1.3 $\times$ 10^7 plaque forming units/μl. An aliquot was plated out and screened with filter lifts using the random prime labeled 5' probe described above, according to standard procedure (Sambrook et al., 1989). Positive clones were excision rescued according to manufacturer's instructions. Plasmid DNA was isolated by alkaline lysis miniprep (Sambrook et al., 1989) and subjected to restriction enzyme analysis. The longest clones that had hybridized to the 5' probe, including an 1800-bp cDNA, were selected for sequencing.

Creating Nested Deletions for DNA Sequencing

Initial sequencing of the 5' and 3' ends of the 1800-bp cDNA showed that it was a hemocyanin cDNA. To sequence it entirely, overlapping nested deletions were created by cleavage with DNase I in the presence of Mn^{2+} (Lin et al., 1985; B.

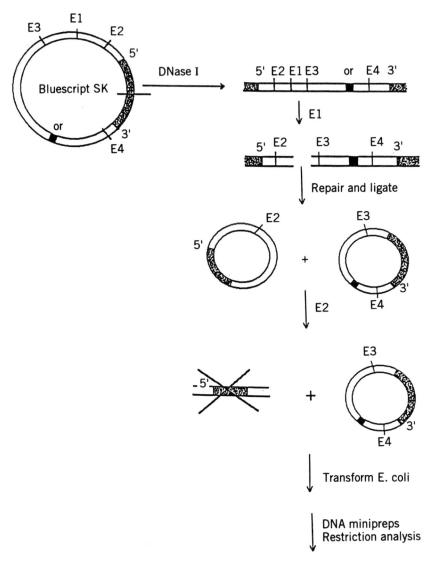

Figure 6. Formation of DNase I nested deletions for DNA sequencing. Bluescript SK−
carrying the 1800-bp hemocyanin cDNA insert (stippled) is randomly linearized by digestion
with DNase I in the presence of Mn^{2+}. The linearized fraction is cut in two by digestion with
SmaI (E1), polyethylene glycol precipitated, and repaired with Klenow fragment. Both
segments are recircularized with T4 DNA ligase. The construct carrying the 5′ end is recut
with EcoRI (E2) and then the mixture is used to transform competent *E. coli* XL-1 Blue cells.
Clones where DNase I had cut outside the insert and clones containing 5′ portions of the
hemocyanin cDNA insert are rendered incapable of transformation because they are linearized by E2. The remaining "useful" clones (those carrying cDNA inserts with 5′ end
deletions of various lengths) have no E2 site, remain circularized, and can transform. Insert
size is determined by restriction analysis with XbaI and XhoI (E3 and E4). Clones of
appropriate length are then selected for dideoxy sequencing.

kb

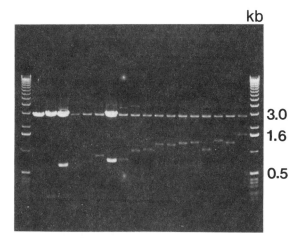

3.0

1.6

0.5

Figure 7. Restriction analysis of DNase I nested deletions of 1800-bp hemocyanin cDNA. Miniprep DNA (0.1 μg) from each clone was digested with 10 units restriction enzymes (XbaI/XhoI) in 10 μl total volume for 2 hours at 37°C and analyzed on 1.2% TAE minigel. Clones shown in lanes 3 to 19 were selected for dideoxy sequencing. Insert size increases from ~150 bp (left) to ~1700 bp (right). Lanes 1 and 20, 1-kb ladder (Gibco BRL); lane 2, Bluescript vector (2.9 kb) linearized with XbaI/XhoI.

Thisse and C. Thisse, *personal communication*) as described in Figure 6 and Terwilliger and Durstewitz.3. The resulting nested deletions derived from the 1800-bp cDNA clone provided a family of clones with overlapping staggered deletions extending from various sites in the sequence to the 5' end. Fifteen clones, ranging in size from 150 to 1800 bp (Fig. 7), were selected for sequencing by the dideoxy method, using a Sequenase kit as described above. To date, 1200 bp at the 5' end of the cDNA have been sequenced, along with 200 bp at the 3' end, and the full sequence is expected shortly.

CONCLUSIONS

We view hemocyanin as a model system for a developmentally regulated multi-subunit protein. Several molecular approaches to the study of hemocyanin ontogeny have been presented here. Most have led to more questions or opened up other avenues of research. How is hemocyanin gene expression controlled? How do individual subunits contribute to the functional properties of the whole protein? Does the cDNA library we constructed contain blueprints of other exciting crab hemolymph proteins? What insights into the evolution of arthropod hemolymph proteins and other oxygen-binding proteins can be derived from the cDNA sequence of this arthropodan hemocyanin? A better understanding of this hemolymph protein will lead to further understanding of how animals cope with the challenge of oxygen transport during development.

ACKNOWLEDGMENTS

We thank Dr. Yi Lin Yan for her expert technical advice and countless helpful suggestions and Kristin O'Brien for protein purification and enthusiastic arthropod husbandry. We also acknowledge the ideas offered by Drs. Ry Meeks-Wagner and John Postlethwaite. This work was supported by NSF Grant IBN-9217530 to N.B.T.

REFERENCES

Alliel PM, Dautigny A, Lamy J, Lamy J-N, Jolles P (1983): Cell free synthesis of hemocyanin from the scorpion *Androctonus australis*. Characterization of the translation products by monospecific antisera. *Eur J Biochem* 134:407–414.

Bak HJ, Beintema JJ (1987): *Panulirus interruptus* hemocyanin. The elucidation of the complete amino acid sequence of subunit a. *Eur J Biochem* 169:333–348.

Beintema JJ, Stam WT, Hazes B, Smidt MP (1994): Evolution of arthropod hemocyanins and insect storage proteins (hexamerins). *Mol Biol Evol* 11:493–503.

Brown AC (1991): *Effects of salinity and temperature on the respiratory physiology of the Dungeness crab, Cancer magister, during development.* PhD Thesis, University of Oregon.

Brown AC, Terwilliger NB (1992): Developmental changes in ionic and osmotic regulation in the Dungeness crab, *Cancer magister. Biol Bull* 182:270–277.

Bunn HF, Forget BG, Ranney HM (1977): *Human hemoglobins.* Philadelphia: Saunders.

Devereux J, Haeberli P, Smithies O (1984): A comprehensive set of sequence-analysis programs for the VAX. *Nucleic Acids Res* 12(1):387–395.

Drexel R, Siegmund S, Schneider H-J, Linzen B, Gielens C, Preaux G, Lontie R, Kellermann J, Lottspeich F (1987): Complete amino acid sequence of a functional unit from a molluscan hemocyanin (*Helix pomatia*). *Biol Chem Hoppe-Seyler* 368:617–635.

Fahrenbach WH (1970): The cyanoblast: hemocyanin formation in *Limulus polyphemus*. *J Cell Biol* 44:445–453.

Gaykema WPJ, Hol WGJ, Vereijken JM, Soeter NM, Bak HJ, Beintema JJ (1984): 3.2 Å structure of the copper-containing, oxygen carrying protein *Panulirus interruptus* hemocyanin. *Nature* 309:23–29.

Ghiretti-Magaldi A, Milanesi C, Salvato B (1973): Identification of hemocyanin in the cyanocytes of *Carcinus maenas. Experientia* 29:1265–1267.

Ghiretti-Magaldi A, Milanesi C, Tognon G (1977): Hemopoiesis in Crustacea Decapoda: origin and evolution of hemocytes and cyanocytes of *Carcinus maenas. Cell Differ* 6:167–186.

Graham RA, Mangum CP, Terwilliger RC, Terwilliger NB (1983): The effect of organic acids on oxygen binding of hemocyanin from the crab *Cancer magister. Comp Biochem Physiol* 74A:45–50.

Hazes B, Magnus KA, Bonaventura C, Bonaventura J, Dauter Z, Kalk K, Hol WGJ (1993): Crystal structure of deoxygenated *Limulus polyphemus* subunit II hemocyanin at 2.18 Å resolution: clues for a mechanism for allosteric regulation. *Protein Sci* 2:597–619.

Hennecke R, Gellissen G, Spindler-Barth M, Spindler K-D (1990): Hemocyanin synthesis in the crayfish, *Astacus leptodactylus*. In: Preaux G, Lontie R (eds). *Invertebrate dioxygen carriers*. Leuven: Leuven University Press, pp 503–506.

Kempter B (1986): Intracellular hemocyanin and the site of biosynthesis in the spider *Eurypelma californicum*. In: Linzen B (ed). *Invertebrate oxygen carriers*. Berlin: Springer, pp 489–494.

Krogh A (1929): Process of physiology. *Am J Physiol* 90:243–251.

Lang W, van Holde K (1991): Cloning and sequencing of *Octopus dofleini* hemocyanin cDNA: derived sequences of functional units Ode and Odf. *Proc Natl Acad Sci USA* 88:244–248.

Lin HC, Lei S, Wilcox G (1985): An improved DNA sequencing strategy. *Anal Biochem* 147:114–119.

Linzen B, Soeter NM, Riggs AF, Schneider H-J, Schartau W, Moore MD, Yokota E, Behrens PQ, Nakashima H, Takagi T, Nemoto T, Vereijken JM, Bak HJ, Beintema JJ, Volbeda A, Gaykema WPJ, Hol WGJ (1985): The structure of arthropod hemocyanins. *Science* 229:519–524.

Markl J, Decker H (1992): Molecular structure of the arthropod hemocyanins. In: Mangum CP (ed): *Advances in comparative and environmental physiology,* vol 13. Berlin: Springer-Verlag, pp 325–376.

Markl J, Stumpp S, Bosch FX, Voit R (1990): Hemocyanin biosynthesis in the tarantula *Eurypelma californicum*, studied by *in situ* hybridization and immuno-electron microscopy. In: Preaux G, Lontie R (eds). *Invertebrate dioxygen carriers*. Leuven: Leuven University Press, pp 497–502.

McMahon BR, McDonald DG, Wood CM (1979): Ventilation, oxygen uptake and haemolymph oxygen transport following enforced exhausting activity in the Dungeness crab *Cancer magister*. *J Exp Biol* 80:271–285.

Morris S, McMahon BR (1989): Potentiation of hemocyanin oxygen affinity by catecholamines in the crab *Cancer magister:* a special effect of dopamine. *Physiol Zool* 62:654–667.

Otoshi C (1994): *Distribution and function of the hemolymph proteins, hemoecdysin and hemocyanin, in relation to the molt cycle of the juvenile Dungeness crab, Cancer magister, and size-specific molting and reproductive capability of the adult female Cancer magister.* Masters thesis, University of Oregon.

Préaux G, Vandamme A, De Bethune B, Jacobs M-P, Lontie R (1986): Hemocyanin-mRNA-rich fractions of cephalopodan Decabrachia and of Crustacea, their *in vivo* and *in vitro* translation. In: Linzen B (ed). *Invertebrate oxygen carriers*. Berlin: Springer-Verlag, pp 485–488.

Sambrook J, Maniatis T, Fritsch EF (1989): *Molecular cloning,* 2nd ed. Cold Spring Harbor, NY: Cold Spring Harbor Laboratory Press.

Senkbeil EG, Wriston JC (1981): Hemocyanin synthesis in the American lobster, *Homarus americanus*. *Comp Biochem Physiol* 68B:163–171.

Sullivan B, Bonaventura J, Bonaventura C (1974): Functional differences in the multiple hemocyanins in the horseshoe crab, *Limulus polyphemus* L. *Proc Natl Acad Sci USA* 71:2558–2562.

Terwilliger NB, Brown AC (1993): Ontogeny of hemocyanin function in the Dungeness crab *Cancer magister:* the interactive effects of developmental stage and divalent cations on hemocyanin oxygenation properties. *J Exp Biol* 183:1–13.

Terwilliger NB, Otoshi C (1994): Cryptocyanin and hemocyanin: fluctuation and functions of crab hemolymph proteins during molting. *The Physiol* 37:A-67.

Terwilliger NB, Terwilliger RC (1982): Changes in the subunit structure of *Cancer magister* hemocyanin during larval development. *J Exp Zool* 221:181–191.

Terwilliger NB, Terwilliger RC, Graham R (1985): Crab hemocyanin function changes during development. In: Linzen B (ed). *Invertebrate oxygen carriers.* Berlin: Springer-Verlag, pp 333–335.

Truchot JP (1992): Respiratory function of arthropod hemocyanins. In: Magnum CP (ed). *Advances in comparative and environmental physiology,* vol 13. Berlin: Springer-Verlag, pp 377–410.

Van Holde KE, Miller KJ (1982): Hemocyanins. *Q Rev Biophys* 15:1–129.

Van Holde KE, Miller KI, Lang WH (1992): Molluscan hemocyanins: structure and function. In: Mangum CP (ed). *Advances in comparative and environmental physiology,* vol 13. Berlin: Springer-Verlag, pp 258–300.

Volbeda A, Hol WGJ (1989a): Crystal structure of hexameric hemocyanin from *Panulirus interruptus* refined at 3.2 Å resolution. *J Mol Biol* 209:249–279.

Volbeda A, Hol WGJ (1989b): Pseudo 2-fold symmetry in the copper-binding domain of arthropodan hemocyanins. Possible implications for the evolution of oxygen transport proteins. *J Mol Biol* 206:531–546.

Wood EJ, Bonaventura J (1981): Identification of *Limulus polyphemus* hemocyanin messenger RNA. *Biochem J* 196:653–656.

CHAPTER 20

The Role of P450$_{arom}$ in Sex Determination of Prototheria and Nonmammalian Vertebrates

PANCHARATNAM JEYASURIA and ROSEMARY JAGUS

University of Maryland Biotechnology Institute, Center of Marine Biotechnology, Baltimore, Maryland 21202

VALENTINE LANCE

Center of Reproduction of Endangered Species, Zoological Society of San Diego, San Diego, California 92112-0551

ALLEN R. PLACE

University of Maryland Biotechnology Institute, Center of Marine Biotechnology, Baltimore, Maryland 21202

CONTENTS

Molecular Zoology: Advances, Strategies, and Protocols, Edited by Joan D. Ferraris and Stephen R. Palumbi.
ISBN 0-471-14461-4 © 1996 Wiley-Liss, Inc.

SYNOPSIS

The mechanism of sex determination in egg-laying amniotes may be fundamentally different from that of the placental mammals. Differentiation of the mammalian ovary proceeds normally in the absence of estrogen, whereas in birds and reptiles estrogen is essential for ovarian development. Recent data have shown that it is possible to sex reverse female embryos of birds and reptiles simply by blocking estrogen synthesis in the undifferentiated gonad. Conversely, application of estrogen to male reptile embryos results in development of an ovary. These data suggest that the enzyme necessary for estrogen synthesis (CYP19, aromatase) in the developing gonad plays a critical role in sex determination in these vertebrates. We have begun an examination of the role and regulation of the aromatase gene in sex determination in two species of reptiles with TSD (temperature-dependent sex determination): the diamond-back terrapin, *Malaclemys terrapin,* and the American alligator, *Alligator mississippiensis.* These species were selected because (1) the sex of the embryos can be manipulated at will simply by setting incubation conditions, (2) extensive experience with successful handling of the eggs and embryos of these reptiles is at hand, and (3) the temperature regimes that give rise to males or females differ markedly between the two species. Three full-length cDNAs for the terrapin aromatase and a partial cDNA for the alligator aromatase have been obtained. The shortest of the cDNA constructs from the terrapin is capable of producing *in vitro* (coupled transcription/translation) as well as *in vivo* (transfected COS cells) a functional aromatase enzyme. *In situ* hybridization studies, as well as a competitive reverse transcription–polymerase chain reaction (RT-PCR) procedure, are being employed to discern the ontogeny of aromatase expression in these two reptiles. Sex determination in birds and reptiles may depend on the initiation of estrogen synthesis in the indifferent gonad, which inhibits male differentiation and stimulates ovarian development. In the absence of this estrogenic signal, a testis develops. Future studies will address whether it is simply the activation of the aromatase gene or a gene or genes acting upstream from aromatase that is the initial trigger of the sex-determining cascade in reptiles.

INTRODUCTION

A Comparison of Sex Determination in Mammals and Nonmammals

Eukaryotic animals have evolved a bewildering variety of chromosomal, genetic, and environmental mechanisms of sex determination (Bull, 1983). In placental mammals with heteromorphic chromosomes, sex is clearly dependent on the presence or absence of a Y chromosome. The picture is less clear in prototheria and nonmammalian vertebrates. How the mammalian system evolved from egg-laying amniotes has been the subject of much speculation (Jablonka and Lamb, 1990; Charlesworth, 1991; Kraak and De Looze, 1993; Graves and Foster, 1994; Hurst, 1994a, 1994b; Lucchesi, 1994), but little data exist to support any of the proposed theories. *The mechanism of sex determination in egg-laying amniotes may be fundamentally different from that of the placental mammals.* It is important, therefore, that we understand how sex is determined in reptiles and birds if we are to understand how sex-determining genes evolved in the Mammalia.

In mammals there is convincing evidence that sex determination is controlled by a gene on the Y chromosome that initiates testis differentiation (Gubbay et al., 1990; Sinclair et al., 1990). This gene, SRY in humans, Sry in mice, is a member of a large family of HMG box, or Sox genes (SRY-related HMG box containing genes) that code for protein with a DNA-binding motif (Laudet et al., 1993). The gene has been shown to be present on the short arm of the Y chromosome in all placental mammals examined, including marsupials (Foster et al., 1992), but, despite considerable effort, has not been found in the egg-laying mammals, echidna and platypus (J. M. Graves, *personal communication*). In mice Sry is expressed in the gonadal ridge at days 10.5–11.5 postcoitum (pc), coincident with induction of testis differentiation (Koopman et al., 1990). A number of XX human males were shown to have Y to X translocation of SRY, and XY female humans have been found either lacking SRY, or with mutations in the HMG box region of SRY (Hawkins et al., 1992; Wachtel, 1994; Wachtel and Simpson, 1994). Transgenic XX mice expressing a 14-kb region of Sry developed testes and male phenotype (Koopman et al., 1991). However, despite the overwhelming evidence implicating SRY as sex determining in mammals, there are a number of intriguing exceptions in which XX human males have no detectable SRY sequences in their genome (Wachtel, 1994) and the majority of human XY females have an apparently normal SRY sequence (Hawkins et al., 1992; Wachtel and Simpson, 1994). Wachtel and Simpson (1994) suggest that at least three genes are involved in testicular differentiation. Wagner et al. (1994) and Foster et al. (1994) provide evidence that SOX9 is one such gene in cases of campomelic displasia with XY sex reversal. The DAX-1 gene on the mammalian X chromosome may be yet another (Muscatelli et al., 1994; Zanaria et al., 1994).

It has been suggested that the mammalian SRY evolved from a SOX3 gene present on the marsupial X chromosome, and presumably on the primitive Y chromosome (Foster and Graves, 1994). This may explain why SRY is sex specific only in mammals, but it does not explain how SRY acquired a testis-determining role, as there is no evidence that SOX3 is involved in testis differentiation. Testicular differentiation in nonmammalian amniotes is morphologically similar to that of mammals and it is likely that the same genes are involved in testis formation, but there are no sex-specific SRY-like genes in birds or reptiles. Using a probe from the conserved HMG box region of SRY, Tiersch et al. (1991) found an equal signal in male and female DNA from birds, reptiles, amphibians, and fish. Similarly, attempts to clone an SRY-like gene from chicken, gecko, alligator, and turtle have revealed a whole family of Sox genes: 10 in the turtle (Spotila et al., 1994), 17 in the alligator, and 18 in the chicken (Coriat et al., 1994). None of these SRY-like genes appears to be sex specific. However, stage- and tissue-specific expression of these genes has yet to be studied. Graves and Foster (1994) suggest that sex determination in nonmammalian vertebrates may be controlled by quite different genes from those that determine sex in mammals. The model of Jost (1953) in which maleness is imposed on the neutral, or default, female phenotype may not apply. There is evidence in birds, for example, that femaleness is imposed on the default, or neutral, male phenotype (Graves and Foster, 1994).

Although Mullerian inhibiting substance (MIS) is a good marker for Sertoli cell differentiation, it is not necessary for male sex differentiation. Mice that were MIS-deficient developed testes and produced functional spermatozoa, but the Mullerian duct derivatives were also retained and interfered with insemination. The testes of these MIS-deficient mice exhibited Leydig cell hyperplasia (Behringer et al., 1994).

The extant crocodilians and the birds are believed to have evolved from a common archosauran ancestor (Gautier et al., 1988). Protein sequence data (deJong et al., 1985) and peptide hormone sequence data (Lance et al., 1984; Tohohiko et al., 1992; Rodriguez-Bello et al., 1993; Wang and Conlon, 1993) lend strong support to this relationship. There are no sex chromosomes in crocodilians (Cohen and Gans, 1970) and all crocodilians exhibit temperature-dependent sex determination, TSD (Lang and Andrews, 1994). All birds show genetic sex determination, GSD, but not all birds have heteromorphic sex chromosomes. The more primitive ratites have morphologically identical sex chromosomes (de Boer, 1980). Birds have a ZZ, ZW sex chromosome complement in which the female has the heteromorphic pair, but these chromosomes bear no genetic relationship to the sex chromosomes of mammals (Mizuno et al., 1993; Graves and Foster, 1994). Although the W chromosome of domestic fowl and most passerines is far smaller than the Z chromosome, there is no evidence of a dosage compensation mechanism (Baverstock et al., 1982; Solari, 1994), and no evidence that a gene similar to the SRY is present on these chromosomes, or that an SRY-like gene is involved in sex determination. A number of genes have been mapped to the Z chromosome of chicken, none of which are found on the mammalian X or Y (Levin et al., 1993), but only two genes have actually been sequenced and their position located on the Z chromosome (Mizuno et al., 1993). To date only a single gene of unknown function has been found on the W chromosome, but its precise location is unclear and it may be situated in the pseudoautosomal region (Dvorak and Smith, 1992). It is possible that sex determination in birds is a simple dosage phenomenon in which the presence of two Z chromosomes is sufficient to either suppress female differentiation or induce male differentiation. Triploid chickens with a ZZW complement develop a left ovotestis and a right testis, triploid ZZZ chickens develop normal looking testes but are infertile, ZWW triploids die early in embryogenesis (Thorne and Sheldon, 1993). How this system evolved from a crocodilian system in which the chromosomes are identical in males and females is unknown.

The Role of Estrogens

In both birds and reptiles, however, the synthesis of estrogen in the developing embryo plays an important role in sex differentiation, whereas in mammals it is clearly unimportant. Simply blocking estrogen synthesis in female chick embryos by injecting an aromatase inhibitor resulted in phenotypically male hatchlings that were capable of spermatogenesis when sexually mature (Elbrecht and Smith, 1992;

Wartenburg et al., 1992). Similar results have been found in turtles with temperature-dependent sex determination (TSD). Eggs incubated at a temperature known to produce female hatchlings produced male hatchlings when treated with aromatase inhibitors. Conversely, eggs incubated at a temperature known to produce male hatchlings produced female hatchlings when treated with estrogen (Crews and Bergeron, 1994; Lance and Bogart, 1994; Jeyasuria et al., 1994; Pieau et al., 1994; Wibbels et al., 1994).

It has not been possible to fully sex reverse genetically male chick embryos using hormonal or drug treatments (Samsel et al., 1982). Male chick embryos that are treated with estrogens exhibit feminized gonads, but these feminized gonads revert to normal male structures some weeks after hatching (Wolff and Ginglinger, 1935; Burns, 1961). Although some mammalian fetal ovaries are capable of synthesizing estrogen, female sex differentiation occurs normally in the absence of estrogen. A human female with a mutation in the aromatase gene and thus unable to synthesize estrogen had normal ovarian development at 17 months but exhibited infantile genitalia and polycystic ovaries at age 18 (Ito et al., 1993). A gene knockout experiment in the mouse in which a nonfunctioning estrogen receptor was introduced into the germline resulted in female mice that were totally unresponsive to estrogen; the gonad apparently differentiated normally as an ovary but, as in the human aromatase deficiency case, exhibited cystic follicles (Lubahn et al., 1993). Further evidence from a recent study on the ontogeny of steroidogenic enzyme gene expression showed that the fetal mouse ovary did not express three key enzymes, including aromatase, indicating again that estrogen is not necessary for ovarian differentiation or development (Greco and Payne, 1994).

A Role for Aromatase (CYP19)

Estrogens are produced normally in a steroid biosynthetic pathway that involves androgens as a substrate. The aromatization of androgens to estrogens takes place in the endoplasmic reticulum and is classified as a mixed-function oxidase reaction. The aromatase enzyme complex (CYP19) is the sole mediator for the conversion of androgens to estrogens in vertebrates. The "aromatase" complex consists of two membrane-bound P450 enzymes, which are $P450_{arom}$ (which binds the androgen substrate and inserts oxygen into the molecule) and an ubiquitous flavoprotein, NADPH-cytochrome P450 reductase. Reducing equivalents from NADPH are transferred via the flavoprotein to $P450_{arom}$. The overall reaction involves a three-step hydroxylation, release of formic acid, and a resulting spontaneous aromatization of the A ring (see Fig. 1).

The gene (CYP19) for aromatase has been cloned from several mammalian species (see Simpson et al., 1994), chicken (McPhaul et al., 1988), zebra finch (Shen et al., 1994), alligator and turtle (Jeyasuria et al., 1995), and two species of teleost fish (Tanaka et al., 1992; Trant, 1994), and the regulation of its tissue-specific expression in mammals has been the subject of considerable research (Simpson et al., 1994). In mammals the full sequence of the aromatase gene (CYP19) is not known, but it spans a distance of at least 75 kb and may be larger (Simpson et al.,

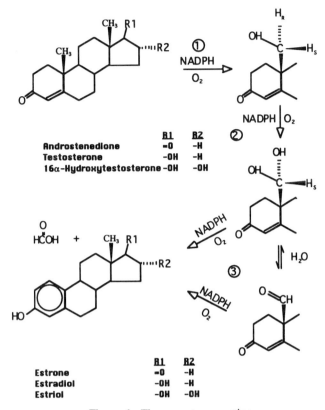

Figure 1. The aromatase reaction.

1994). In the medaka, the aromatase gene only spans 2.6 kb (Tanaka et al., 1995). The mammalian, avian, and fish genes contain 9 exons (McPhaul et al., 1993; Tanaka et al., 1995). The promoter region of human aromatase has been studied in great detail (Simpson et al., 1994), but only fragmentary information is available on the chicken promoter region (McPhaul et al., 1993) with slightly better information being available on the fish promoter (Tanaka et al., 1995).

Work from the laboratories of Keith Parker and Ken-ichiri Morohashi has shown that steroidogenic factor 1 (SF-1), also known as Ad4BP, an orphan nuclear receptor with a zinc finger DNA-binding domain, the mammalian homologue of *fushi tarazu* factor 1 (FTZ-F1) from the fruitfly (Lavorgna et al., 1991), is a key regulator of steroidogenic enzyme gene expression (Rice et al., 1991; Lynch et al., 1993; Morohashi et al., 1993). This nuclear protein of 52,000 daltons binds to the promoter region of all steroidogenic P450 genes including aromatase and has been shown to act as a transcriptional activator. SF-1 has also been shown to be important in MIS expression (Shen et al., 1994), and pituitary expression of LHβ, FSHβ, and the receptor for gonadotropin releasing hormone (Ingraham et al., 1994).

In rats, SF-1 was first detected in the primordial adrenal glands and testes of the

13.5-day pc fetus, but only trace amounts were detected in the fetal ovaries (Hatano et al., 1994). When the gene for this factor was disrupted in mice, the embryos developed without gonads or adrenal glands and died shortly after birth (Luo et al., 1994). Clearly this factor is a key determinant of gonadal development. However, given the widespread tissue distribution of SF-1 and its multiple roles in steroidogenesis and pituitary function, it is obvious that other factors must also be involved. In rat granulosa cells, for example, aromatase activity increases in response to FSH via an increase in cAMP. The promoter region of rat aromatase gene has a cAMP response element (CRE) that binds the transcription factor, CRE-binding protein (CREB), and an SF-1 binding site (Richards, 1994). However, as the authors point out, these two regulatory elements alone cannot account for the tissue-specific expression of aromatase in granulosa cells as both these elements are present in adrenal cells that do not express aromatase. At least four promoters have been identified in the human aromatase gene, including the SF-1, some of which are responsible for tissue-specific expression of the gene (Simpson et al., 1994).

The recent identification of yet another factor, DAX-1, a member of the nuclear hormone receptor superfamily, also involved in gonadal development adds a further complication (Muscatelli et al., 1994; Zanaria et al., 1994). DAX-1 shows a similar distribution to SF-1 and is transcribed in fetal gonads and in adult testis, ovary, and adrenal (Zanaria et al., 1994; A. Swain, *personal communication*). It is very likely that these two transcription factor genes are also important in bird and reptile gonad differentiation.

Mizuno et al. (1993) have shown that transcription of the aromatase gene in female chick embryos starts as early as day 5 of incubation. Aromatase activity in turtle gonads peaks in female embryos at the time of sex differentiation (Desvages and Pieau, 1992), whereas in alligator and crocodile embryonic gonadal aromatase activity is first detected after gonadal differentiation (Smith et al., 1994).

OUR STRATEGY

As virtually nothing is known on the mechanism of sex determination in birds or reptiles, we started with the premise that regulation of the gene for aromatase is of central importance. Accordingly, we set out to clone both the terrapin as well as the alligator aromatase cDNAs. Currently, we have succeeded in obtaining full-length cDNAs for the terrapin and only a partial cDNA for the alligator aromatase.

cDNA Cloning—A Partial Clone or "So, What's New?"

The first cDNA isolated from the turtle ovary cDNA library (Jeyasuria et al., 1994) was roughly 500 base pairs short of the start site of translation in the 5' region in other known P450 aromatases (chicken, rat, mouse, human, trout, and catfish). After rescreening the library twice using both an unamplified library and the original amplified library, it was not possible to find a clone larger than the original clone (pTA 1837) isolated.

Northern Analysis

To determine the mRNA transcript complexity for the terrapin aromatase and obtain a better estimate of the expected size for the cDNA, we performed a Northern analysis. A 7-year-old female, a 4-year-old female, and a 7-year-old male were euthanised by decapitation and tissues were removed by dissection and snap frozen in liquid nitrogen. The tissues were stored in liquid nitrogen prior to RNA extraction. RNA was extracted using a guanidinium isothiocyanate method with an acidic phenol extraction and a propanol precipitation (Chomczynski and Sacchi, 1987). Ten micrograms of total RNA were run on a 0.8% agarose/2.2 M formaldehyde gel and transferred to nylon (MSI Magnagraph) in 10× SSC. The RNA was fixed onto the nylon by UV crosslinking. The blot was prehybridized in hybridization solution [6× SSC, 0.5% SDS, 5× Denhardt's, 0.1% bovine serum albumin (Fraction V), and 5 µg/ml yeast tRNA] at 65°C for 2 hours. The 1.8-kb insert from PTA 1837 was labeled by Klenow extension of random primers (hexanucleotides) with 5′α-^{32}P dATP (Feinberg and Vogelstein, 1983). The membranes were then hybridized with 1×10^7 cpm/ml of turtle aromatase probe in fresh hybridization buffer at 65°C overnight. After washing and exposure to x-ray film, the image shown in Figure 2

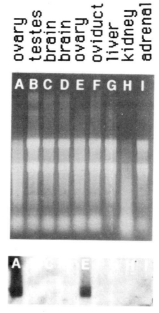

Figure 2. Northern blot analysis using the terrapin aromatase partial cDNA as a probe. Lanes A and E are RNA from ovaries of 7- and 4-year-old females, respectively. Lanes B and D are RNA from the testes and brain of a 7-year-old male. The rest of the lanes are tissue RNAs from the same individual as in lane A. Lane C/brain, F/oviduct, G/liver, H/kidney, and I/adrenal. Ten micrograms of total RNA were loaded per lane. Samples were run on a 0.8% agarose/2.2 M formaldehyde in MOPS buffer. *Top:* Ethidium bromide staining of the RNA gel. *Bottom:* Autoradiographic image. Only ovarian RNA provided a positive signal to the terrapin aromatase at this sensitivity. The terrapin ovarian transcript is calculated to be ∼ 3.2 kb.

was obtained. From the Northern analysis, we therefore expect to find aromatase cDNAs of approximately 3.5 kb for terrapin ovary.

Primer Extension Library

To acquire the remaining sequence at the 5′ end of the mRNA we decided to make a primer extension library using the known sequence from clone PTA 1837 rather than attempting to remake the cDNA library. The primer (5′GAGAGAGAAC-TAGT*CTCGAG*ATGGCATTTTCATCCAAAG3′) consists of a 5′ nonspecific tail (note that this tail ensures that the XhoI restriction enzyme has a sufficient number of bases on either side of its restriction site), a XhoI site (italicized), and a sequence-specific (bold lettering) reverse complement to a site 215 bp from a BamHI site of the original truncated cDNA (PTA 1837). First strand synthesis was performed using this primer and poly(A)+ selected mRNA from the ovary of an adult female terrapin using reverse transcriptase followed by second strand synthesis using DNA polymerase I and RNase H. The cDNA was then ligated to EcoRI linkers and then cut with XhoI and ligated to λ Zap Express XR DNA (λ Zap Express restricted with XhoI and EcoRI) and packaged into the bacteriophage. This strategy involves using Stratagene's cDNA Zap Express kit (Stratagene Cloning Systems, La Jolla, CA) with the exception of using the above 40 base pair oligonucleotide as the primer. The unamplified library was screened using the 215-bp EcoRI-BamHI fragment from the 5′ end of the turtle aromatase coding sequence (PTA 1837). Three unique primer extended clones were obtained (Fig. 3), each identical in coding sequence but different in the 5′ untranslated region. We are investigating the abundance of these transcripts in our poly(A)+ selected mRNA. Our initial interpretation is that the adult ovary may possess three aromatase transcripts (each around 3.2 kb) resulting from alternate 5′ untranslated splicing. Our Northern analysis (Fig. 2) is consistent with a population of aromatase transcripts in the 3.2-kb region. We are developing 5′ RACE procedures for examining whether these transcripts are found in the developing gonad.

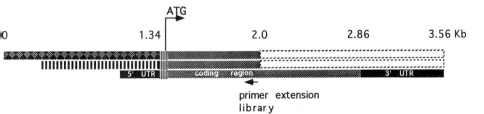

Figure 3. Schematic of cDNAs obtained from a primer extended terrapin ovary library. The largest insert is 2.0 kb in length. The smallest insert is shown contiguous with the original truncated clone. The stippled lines indicate an extrapolation of sequence based on the original aromatase cDNA. Similar patterns indicate homologous sequence and different patterns indicate nonhomologous sequence.

Gene Expression Analysis of Transient Transfectants

The shortest of the primer extended clones was ligated to our original partial clone and then inserted into the eukaryotic expression vector, pBK-CMV, for expression in COS-1 cells (see Jagus and Place.3). This construction proved to be extremely difficult as all the enzymes in the multiple cloning sites of the expression vector were also found internal to the aromatase sequence, thus requiring a three-fragment ligation to obtain the correct orientation. The three-way ligation consisted of cutting the vector with XhoI and EcoRI, the primer extended clone with EcoRI and BamHI, and the original truncated clone with XhoI and BamHI (see Fig. 4). These fragments were eluted from an agarose gel and ligated together.

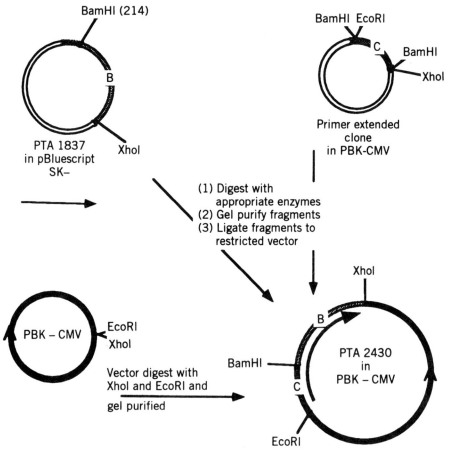

Figure 4. Schematic of the two plasmids rescued from (1) the original ovarian cDNA (PTA 1837) library and (2) the ovarian primer extended cDNA library and the final plasmid originating from the ligation of the two inserts to make a contiguous aromatase clone (PTA 2430) in an expression vector. (pBK-CMV) Stratagene

After ligation, the construct proved to be able to drive terrapin aromatase expression in COS-1 cells. Using Lipofectamine transfection, the COS-1 cells were able to aromatize androstenedione at a rate of approximately 3 pmol/hr · plate, (10 pmoles/hr · mg protein) which was completely inhibited by 4-hydroxyandrostenedione. When microsomes were prepared from these transfected cells, the activity at 27°C was greater than that measured with the same microsomes at 37°C. In our hands, the COS cell expressed terrapin aromatase is less stable at 37°C than at 27°C. We can find little information on the Q_{10} for the aromatase enzyme to interpret these findings.

In Vitro Aromatase Expression Using a Coupled Transcription–Translation Expression System

An increasingly common practice in molecular biology is the cloning of cDNAs in vectors carrying bacteriophage promoters, to allow *in vitro* transcripts to be prepared, which are then subsequently translated *in vitro*. This two-step procedure was made unnecessary when Craig et al. (1992) noted that with sufficient optimization for translation, the transcription and translation could be coupled to give a "single tube assay." Prior to transfection of COS cells with our terrapin pBK-CMV construct, we performed a coupled transcription–translation (Jagus and Place.1) in a reticulocyte message-dependent lysate (MDL) and obtained the results shown in Figure 5. When we used T3 polymerase (correct reading frame) in the presence of ^{35}S-methionine we obtained a 57-kDa protein, while T7 polymerase coupled transcription–translation (opposite reading frame) gave no incorporation above background (data not shown). This construct was capable of incorporating 3.8% of the methionine into aromatase protein. We calculate that a 10-μl transcription–translation reaction would provide approximately 1 ng of aromatase protein. When the reactions were performed in the presence of canine pancreatic microsomes (Jagus and Place.2), we observed an increase in molecular weight of the translated product, which we interpret as resulting from glycosylation after the translated product has been inserted into the membrane. Greater than 95% of the incorporated ^{35}S-methionine was found in microsomes recovered by ultracentrifugation. Only when we added rat recombinant P450 reductase (35 μmol/min · mg, a gift from Dr. Estabrook's laboratory) to the reaction mixture, could we recover aromatase activity (tritiated water assay, Lephart and Simpson, 1991) with an apparent turnover number of 128 nmol/hr · mg aromatase protein at 27°C ($\sim$35 min^{-1}). The activity was fully inhibited by 100 μM 4-hydroxyandrostenedione, a nonsteroidal inhibitor of aromatase. This coupled transcription–translation assay is a general procedure for examining structure–function relationships among cytochrome P450s. We are currently optimizing the assay in regard to reductase addition, pH, temperature, and lipid addition. We only examined recovered activity in the presence of canine pancreatic microsomes; however, we propose to perform these studies also with simple phospholipid emulsions (e.g., dilaurylphosphatidyl choline) as well as microsomes prepared from oviducts (terrapin oviducts do not express aromatase mRNA, see Fig. 2) of alligators and terrapins.

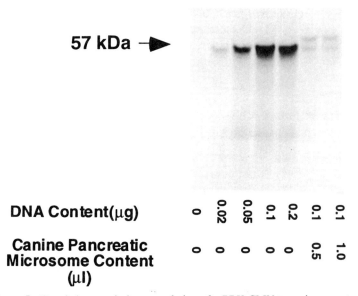

DNA Content(μg)	0	0.02	0.05	0.1	0.2	0.1	0.1
Canine Pancreatic Microsome Content (μl)	0	0	0	0	0	0.5	1.0

Figure 5. Coupled transcription–translation of a PBK-CMV terrapin aromatase construct in the presence of canine pancreatic microsomes. We suspect that glycosylation has resulted in the change of molecular weight observed when the coupled transcription–translation reactions are performed in the presence of the pancreatic microsomes.

We have begun a series of *in situ* hybridization studies with embryos to establish when aromatase is expressed in development, what tissues express aromatase, and if temperature shifts or steroid administrations affect this expression. Once these data are in hand we will begin examination of the promoter region(s) for the aromatase gene in order to determine potential regulatory elements in the reptilian sex cascade.

We believe that sex determination in birds and reptiles may depend on the initiation of estrogen synthesis in the indifferent gonad, which inhibits male differentiation and stimulates ovarian development. In the absence of this estrogenic signal a testis develops. Whether it is simply the activation of the aromatase gene or a gene or genes acting upstream from aromatase that is the initial trigger of the sex-determining cascade remains to be determined.

ACKNOWLEDGMENTS

This study was supported by the Maryland Sea Grant and NSF MCB93-17264 (R.J.). We would like to thank Mira Kautsky and Pam Terry for assistance in rearing diamond-back terrapin hatchlings and laboratory work. We would also like to thank Dr. Angela Brodie for supplying the 4-hydroxyandrostenedione; Dr. Charles W. Fisher of Dr. R. W. Estabrook's laboratory for supplying the recombinant rat P450 reductase; and Promega Corp. for providing samples of canine pancreatic microsomes. This is contribution No. 258 from the Center of Marine Biotechnology, University of Maryland Biotechnology Institute.

REFERENCES

Baverstock PR, Adams M, Polkinghorne RW, Gelder M (1982): A sex-linked enzyme in birds—Z-chromosome conservation but no dosage compensation. *Nature* 296:763–766.

Behringer RR, Finegold MJ, Cate RL (1994): Mullerian-inhibiting substance function during mammalian sexual development. *Cell* 79:415–425.

Bogan JS, Page DC (1994): Ovary? Testis?—a mammalian dilemma. *Cell* 76:603–607.

Bull JJ (1983): *The evolution of sex determining mechanisms.* Menlo Park, CA: Benjamin-Cummings.

Burns RK (1961): Role of hormones in the differentiation of sex. In: Young WC (ed). *Sex and internal secretions,* 3rd ed. Baltimore: Williams & Wilkins, pp 76–158.

Charlesworth B (1991): The evolution of sex chromosomes. *Science* 251:1030–1033.

Chomczynski P, Sacchi N (1987): Single step method of RNA isolation by acid guanidinium thiocyanate–phenol-chloroform extraction. *Anal Biochem* 162:156–159.

Ciccarone V, Hawley-Nelson P, Jessee J (1993): Lipofectamine reagent: a new higher efficiency polycationic liposome transfection reagent. *Focus (Life-Technol)* 15:73–79.

Clark BJ, Waterman MR (1991): Heterologous expression of mammalian P450 in COS cells. *Methods Enzymol* 206:100–108.

Cohen MM, Gans, C (1970): The chromosomes of the order Crocodilia. *Cytogenetics* 9: 81–105.

Coriat A-M, Valleley E, Ferguson MWJ, Sharpe PT (1994): Chromosomal and temperature-dependent sex determination: The search for a conserved mechanism. *J Exper Zool* 270:112–116.

Craig D, Howell MT, Gibbs CL, Hunt T, Jackson RJ (1992): Plasmid cDNA-directed protein synthesis in a coupled eukaryotic in vitro transcription–translation system. *Nucleic Acids Res* 20:4987–4995.

Crews D (1994): Temperature, steroids and sex determination. *J Endocrinol* 142:1–8.

Crews D, Bergeron JM (1994): Role of reductase and aromatase in sex determination in the red-eared slider (*Trachemys scripta*), a turtle with temperature-dependent sex determination. *J Endocrinol* 143:279–289.

Desvages G, Pieau C (1992): Aromatase activity in gonads of turtle embryos as a function of the incubation temperature of eggs. *J Steroid Biochem Mol Biol* 40:4783–4806.

Detera-Wadleigh SD, Fanning TG (1994): Phylogeny of the steroid receptor superfamily. *Mol Phylogenet Evol* 3:192–205.

de Boer LEM (1980): Do the chromosomes of the kiwi provide evidence for a monophyletic origin of the ratites. *Nature* 287:84–85.

de Jong WW, Zweers A, Versteeg M, Dessauer HC, Goodman M (1985): alpha-crystallin A sequences of Alligator mississippiensis and the lizard Tupinambis teguixin: Molecular evolution and reptilian phylogeny. *Molec Biol Evol* 2:484–493.

Dubin RA, Ostrer H (1994): SRY is a transcriptional activator. *Mol Endocrinol* 8:1184–1192.

Duzgures N, Felgner PL (1993): Intracellular delivery of nucleic acids and transcription factors by cationic liposomes. *Methods Enzymol* 221:303–306.

Dvorak Elbrecht AE, Smith RG (1992): Aromatase enzyme activity and sex determination in chickens. *Science* 255:467–470.

Elbrecht AE, Smith RG (1992): Aromatase enzyme activity and sex determination in chickens. *Science* 255:467–470.

Fitzpatrick SL, Richards JS (1994): Identification of a cyclic adenosine 3′,5′-monophosphate-response element in the rat aromatase promoter that is required for transcriptional activation in rat granulosa cells and R$_2$C Leydig cells. *Mol Endocrinol* 8:1309–1319.

Feinberg AP, Vogelstein B (1983): A technique for radiolabelling DNA restriction endonuclease fragments to high specific activity. *Anal Biochem* 132:6–13.

Felgner PL, Ringold GM (1989): Cationic liposome-mediated transfection. *Nature* 337:387–388.

Foster JW, Graves JAM (1994): An SRY-related sequence on the marsupial X chromosome: implications for the evolution of the mammalian testis-determining gene. *Proc Natl Acad Sci USA* 91:1927–1931.

Foster JW, Brennan FE, Hampikian GK, Goodfellow PN, Sinclair AH, Lovell-Badge R, Selwood L, Renfree MB, Cooper DW, Graves JAM (1992): Evolution of sex determination and the Y chromosome: SRY-related sequences in marsupials. *Nature* 359:531–533.

Foster JW, Dominguez-Steglich MA, Guiloi S, Kwok J, Weller PA, Stevanovic, Weissenbach J, Mansour S, Young ID, Goodfellow PN, Brook JD, Schafer AJ (1994): Campomelic displasia and autosomal sex reversal caused by mutations in an SRY-related gene. *Nature* 372:525–530.

Gauthier J, Kluge AG, Rowe T (1988): Amniote phylogeny and the importance of fossils. *Cladistics* 4:105–205.

Graves JAM, Foster JW (1994): Evolution of mammalian sex chromosomes and sex-determining genes. *Int Rev Cytol* 154:191–259.

Greco TL, Payne AH (1994): Ontogeny of expression of the genes for steroidogenic enzymes P450 side-chain cleavage, 3 β-hydroxysteroid dehydrogenase, p450 17-hydroxylase/C17 2o lyase, and p450 aromatase in fetal mouse gonads. *Endocrinology* 135:262–268.

Gubbay J, Collignon J, Koopman P, Capel B, Economou A, Munsterberg A, Vivian N, Goodfellow P, Lovell-Badge R (1990): A gene mapping to the sex-determining region of the mouse Y chromosome is a member of a novel family of embryonically expressed genes. *Nature* 346:245–250.

Haqq CM, King C-Y, Ukiyama E, Falsafi S, Haqq TN, Donahoe PK, Weiss MA (1994): Molecular basis of mammalian sexual determination: activation of mullerian inhibiting substance gene expression by SRY. *Science* 266:1494–1500.

Hatano O, Takayama K, Imai T, Waterman MR, Takakusa T, Morohashi K-I (1994): Sex-dependent expression of a transcription factor, Ad4BP, regulating steroidogenic P-450 genes in the gonads during prenatal and postnatal rat development. *Development* 120:2787–2797.

Hatano O, Takayama K, Imai T, Waterman MR, Takakusa T, Morohashi K-I (1994): Sex-dependent expression of a transcription factor, Ad4BP, regulating steroidogenic P-450 genes in the gonads during prenatal and postnatal rat development. *Development* 120:2787–2797.

Hawkins JR, Taylor A, Berta P, Levilliers J, Van der Auwera B, Goodfellow PN (1992): Mutational analysis of SRY: nonsense and missense mutations in XY sex reversal. *Hum Genet* 88:471–474.

Hinshelwood MM, Corbin CJ, Tsang PCW, Simpson ER (1993): Isolation and characterization of a complementary deoxyribonucleic acid insert encoding bovine aromatase cytochrome P450. *Endocrinology* 133:1971–1977.

Hurst LD (1994a): Embryonic growth and the evolution of the mammalian Y chromosome. I. The Y as an attractor for selfish growth factors. *Heredity* 73:223–232.

Hurst LD (1994b): Embryonic growth and the evolution of the mammalian Y chromosome. II. Suppression of selfish Y-linked growth factors may explain escape from X-inactivation and rapid evolution of Sry. *Heredity* 73:233–243.

Ingraham HA, Lala DS, Ikeda Y, Luo X, Shen W-H, Nachtigal MW, Abbud R, Nilson JH, Parker KL (1994): The nuclear receptor steroidogenic factor 1 acts at multiple levels of the reproductive axis. *Genes Dev* 8:2302–2312.

Ito Y, Fisher CR, Conte FA, Grumbach MM, Simpson ER (1993): Molecular basis of aromatase deficiency in an adult female with sexual infantilism and polycystic ovaries. *Proc Natl Acad Sci USA* 90:11673–11677.

Jablonka E, Lamb MJ (1990): The evolution of heteromorphic sex chromosomes. *Biol Rev* 65:249–276.

Jagus R (1987): Translation in cell-free translation systems. *Methods Enzymol* 152:296–306.

Jeyasuria P, Roosenburg WM, Place AR (1994): The Role of P-450 Aromatase In Sex Determination Of The Diamondback Terrapin, *Malaclemys terrapin*. *J Exper Zool* 270:95–111.

Jeyasuria P, Place AR, Lance VA, Blumberg B (1995): The role of p450 aromatase in temperature-dependent sex determination in reptiles. *J Cell Biochem Suppl* 19B:51.

Jost A (1953): Studies on sex differentiation in mammals. *Rec Prog Horm Res* 8:379–418.

Kain KC, Orlandi PA, Lanar DE (1991): Universal promoter for gene expression without cloning: expression PCR. *Biotechniques* 10:366–370.

Koopman P, Munsterberg A, Capel B, Vivian N, Lovell-Badge R (1990): Expression of a candidate sex-determining gene during mouse testis differentiation. *Nature* 348:450–452.

Koopman P, Gubbay J, Vivian N, Goodfellow PN, Lovell-Badge R (1991): Male development of chromosomally female mice transgenic for Sry. *Nature* 351:117–121.

Kraak SBM, De Looze EMA (1993): A new hypothesis on the evolution of sex determination in vertebrates: big females ZW, big males XY. *Netherlands J Zool* 43:260–273.

Krust A, Green S, Argos P, Kumar V, Walter P, Bornert JM, Chambon P (1986): The chicken oestrogen receptor sequence: homology with v-erbA and the human oestrogen and glucocorticoid receptors. *EMBO J* 5:891–897.

Lala DS, Rice DA, Parker KL (1992): Steroidogenic factor 1, a key regulator of steroidogenic enzyme expression, is the mouse homolog of fushi tarazu-factor 1. *Mol Endocrinol* 6:1249–1258.

Lance VA, Bogart MH (1991): Tamoxifen "sex reverses" alligators at male producing, but is an antiestrogen in female hatchlings. *Experientia* 47:263–266.

Lance VA, Bogart MH (1992): Disruption of ovarian development in alligator embryos treated with an aromatase inhibitor. *Gen Comp Endocrinol* 86:59–71.

Lance VA, Bogart MH (1994): Studies on sex determination in the American alligator *Alligator mississippiensis*. *J Exp Zool* 270:79–85.

Lance V, Hamilton JW, Rouse JB, Kimmel JR, Pollock HG (1984): Isolation and characterization of reptilian insulin, glucagon, and pancreatic polypeptide: Complete amino acid sequence of alligator (*Alligator mississippiensis*) insulin and pancreatic polypeptide. *Gen Comp Endocrinol* 55:112–124.

Lang JW, Andrews HV (1994): Temperature-dependent sex determination in crocodilians. *J Exp Zool* 270:28–44.

Laudet V, Stehelin D, Clevers H (1993): Ancestry and diversity of the HMG box superfamily. *Nucleic Acids Res* 21:2493–2501.

Lephart ED, Simpson ER (1991): Assay of aromatase activity. *Methods Enzymol* 206:477–483.

Levin I, Crittenden LB, Dodgson JB (1993): Genetic map of the chicken Z chromosome using random amplified polymorphic DNA (RAPD) markers. *Genomics* 16:224–230.

Lavorgna G, Ueda H, Clos J, Wu C (1991): FTZ-FI, a steroid hormone receptor-like protein implicated in the activation of fushi tarazu. *Science* 252:848–851.

Longmire JL, Maltbie M, Pavelka RW, Smith LM, Witte SM, Ryder OA, Ellsworth DL, Baker RJ (1993): Gender identification in birds using microsatellite DNA fingerprint analysis. *The Auk* 110:378–381.

Lubahn DB, Moyer JS, Golding TS, Couse JF, Korach KS, Smithies O (1993): Alteration of reproductive function but not prenatal sexual development after insertional disruption of the mouse estrogen receptor gene. *Proc Natl Acad Sci USA* 90:11162–11166.

Lucchesi JC (1994): The evolution of hetermorphic sex chromosomes. *Bioessays* 16:81–83.

Luo X, Ikeda Y, Parker K (1994): A cell-specific nuclear receptor is essential for adrenal and gonadal development and sexual differentiation. *Cell* 77:481–490.

Lynch JP, Lala DS, Peluso JJ, Luo W, Parker KL, White B (1993): Steroidogenic factor 1, an orphan nuclear receptor, regulates the expression of rat aromatase gene in gonadal tissues. *Mol Endocrinol* 7:776–786.

Matsumine H, Herbst MA, Ignatius Ou S-H, Wilson JD, McPhaul MJ (1991): Aromatase mRNA in the extragonadal tissues of chickens with the Henny-feathering trait is derived from a distinct promoter structure that contains a segment of a retroviral long terminal repeat. *J Biol Chem* 266:19900–19907.

McPhaul MJ, Noble JF, Simpson ER, Mendelson CR, Wilson JD (1988): The expression of a functional cDNA encoding the chicken cytochrome P4540arom (aromatase) that catalyzes the formation of estrogen from androgen. *J Biol Chem* 263:16358–16363.

McPhaul MJ, Herbst MA, Matsumine H, Young M, Lephart ED (1993): Diverse mechanisms of control of aromatase gene expression. *J Steroid Biochem Mol Biol* 44:341–346.

Mizuno S, Saitoh Y, Nomura O, Kunita R, Ohtomo K, Nishimori K, Ono H, Saitoh H (1993): Sex-specific DNA sequence in galliformes and their application to the study of sex differentiation. In: Etches RJ, Verrinder AM (eds). *Manipulation of the avian genome*. Boca Raton, FL: CRC Press, pp 257–274.

Morohashi K-I, Zanger UM, Honda S-I, Hara M, Waterman MR, Omura T (1993): Activation of CYP 11A and CYP 11B gene promoters by the steroidogenic cell-specific transcription factor, Ad4BP. *Mol Endocrinol* 7:1196–1204.

Morohashi K-I, Iida H, Nomura M, Hatano O, Honda S, Tsukiyama T, Niwa O, Hara T, Takakusu A, Shibata Y, Omura T (1993): Functional difference between Ad4BP and ELP, and their distributions in steroidogenic tissues. *Molec Endocrinol* 8:643–653.

Muscatelli F, Strom TM, Walker AP, Zanaria E, Recans D, Meindi A, Bardoni B, Guioli S,

Zehetner G, Rabi W, Schwartz HP, Kaplan J-C, Camerino G, Meitlinger T, Monaco AP (1994): Mutations in the DAX-1 gene give rise to both X-linked adrenal hypoplasia congenita and hypogonadotropic hypogonadism. *Nature* 372:672–676.

Pelham HRB, Jackson RJ (1976): An efficient mRNA-dependent translation system from reticulocyte lysate. *Eur J Biochem* 67:247–256.

Pieau C, Girondet M, Richard-Mercier N, Desvages G, Dorizzi M, Zborski P (1994): Temperature sensitivity of sexual differentiation of gonads in the European pond turtle: hormonal involvement. *J Exp Zool* 270:86–94.

Rice DA, Mouw AR, Bogerd AM, Parker KL (1991): A shared promoter element regulates the expression of three steroidgenic enzymes. *Molec Endocrinol* 5:1552–1561.

Richards JS (1994): Hormonal control of gene expression in the ovary. *Endocr Rev* 15:725–751.

Rodrigues-Bello A, Kah O, Tramu G, Conlon JM (1993): Purification and primary structure of alligator neurotensin. *Peptides* 14:1055–1508.

Safer R, Jagus R, Kemper WM (1978): Analysis of initiation factor function in highly fractionated and unfractionated reticulocyte lysate. *Methods Enzymol* 60:61–87.

Saitoh Y, Ogawa A, Hori T, Kunita R, Mizuno S (1993): Identification and localization of two genes on the chicken Z chromosome: expression of rat aromatase gene in gonadal tissues. *Mol Endocrinol* 7:776–786.

Samsel J, Zeis A, Weninger J-P (1982): Feminisation du testicule embryonnaire de poulet par le diethylstilboestrol et action antagoniste du tamoxifene. *Biochimie* 64:369–376.

Shen P, Campagnoni CW, Kampf K, Schlinger BA, Arnold AP, Campagnoni AT (1994): Isolation and characterization of a zebra finch aromatase cDNA: in situ hybridization reveals high aromatase expression in brain. *Mol Brain Res* 24:227–237.

Simpson ER, Mahendroo MS, Means G, Kilgore MW, Hinshelwood MM, Graham-Lorence S, Amarneh B, Ito Y, Fisher CR, Michael MD, Mendelson CR, Bulun S (1994): Aromatase cytochrome P450, the enzyme responsible for estrogen biosynthesis. *Endocr Rev* 15:342–355.

Sinclair AH, Berta P, Palmer MS, Hawkins JR, Griffiths BL, Smith MJ, Foster JW, Frischauf AM, Lovell-Badge R, Goodfellow PN (1990): A gene from the human sex determining region encodes a protein with homology to a conserved DNA-binding motif. *Nature* 346:240–244.

Smith CA, Elf PK, Joss JMP (1994): Aromatase enzyme activity during gondal sex differentiation in alligator embryos. *Differentiation* 58(4):281–290.

Solari AJ (1994): *Sex chromosomes and sex determination in vertebrates.* Boca Raton, FL: CRC Press.

Spotila LD, Kaufer NF, Theriot E, Ryan KM, Penick D, Spotila JR (1994): Sequence analysis of the ZXY and Sox genes in the turtle, *Chelydra serpentina*. *Mol Phylogen Evol* 3:1–9.

Spotila JR, Spotila LD, Kaufer NF (1994): Molecular mechanisms of TSD in reptiles: a search for the magic bullet. *J Exp Zool* 270:117–127.

Tanaka M, Telecky TM, Fukada S, Adachi S, Chen S, Nagahama Y (1992): Cloning and sequence analysis of the cDNA encoding P450 aromatase (P450arom) from a rainbow trout (*Oncorhynchus mykiss*) ovary; relationship between the amount of P450arom mRNA and the production of oestradiol-1713 in the ovary. *J Mol Endocrinol* 8:53–61.

Tanaka M, Fukada S, Matsuyama M, Nagahama Y (1995): Structure and promoter analysis of the cytochrome P-450 aromatase gene of the teleost fish, medaka (*Oryzias latipes*). *J Biochem* 117:719–725.

Tereba A, McPhaul MJ, Wilson JD (1991): The gene for aromatase (P450arom) in the chicken is located on the long arm of chromosome 1. *J Hered* 82:80–81.

Thorne MH, Sheldon BL (1993): Triploid intersex and chimeric chickens: useful models for studies of avian sex determination. In: Reed KC, Graves JAM (eds). *Sex chromosomes and sex Determining genes*. Chur, Switzerland: Harwood Academic Publishers, pp 199–205.

Tiersch TR, Mitchell MJ, Wachtel SS (1991): Studies on the phylogenetic conservation of the SRY gene. *Hum Genet* 87:571–573.

Tohohiko N, Swanson P, Lance VA, Kawauchi H (1992): Isolation and characterization of glycosyated and non-glycosylated prolactins from two reptiles, alligator and crocodile. *Int J Peptide and Protein Res* 39:250–257.

Trant JM (1994): Isolation and characterization of the cDNA encoding the channel catfish (*Ictalurus punctatus*) form of cytochrome P450$_{arom}$. *Gen Comp Endocrinol* 95:155–168.

Wachtel SS (1994): XX sex reversal in the human. In: Wachtel SS (ed). *Molecular genetics of sex determination*. San Diego: Academic Press, pp 267–285.

Wachtel SS, Simpson JL (1994): XY sex reversal in the human. In: Wachtel SS (ed). *Molecular genetics of sex determination*. San Diego: Academic Press, pp 287–309.

Wagner T, Wirth J, Meyer J, Zabel B, Held M, Zimmer J, Pasantes J, Bricarelli FD, Keutel J, Hustert E, Wolf U, Tommerup N, Schempp W, Scherer G (1994): Autosomal sex reversal and campomelic dysplasia are caused by mutations in and around the SRY-related gene SOX9. *Cell* 79:1111–1120.

Walter P, Blobel G (1983): Signal recognition particle: a ribonucleoprotein particle required for co-translation of proteins, isolation and properties. *Methods Enzymol* 96:84–93.

Wang Y, Conlon JM (1993): Neuroendocrine peptides (NPY, GRP, VIP, somatostatin) from the brain and stomach of the alligator. *Peptides* 14:573–579.

Wartenberg H, Lenz E, Schweikert H-U (1992): Sexual differentiation and the germ cell in sex reversed gonads after aromatase inhibition in the chicken embryo. *Andrologia* 24:1–6.

Wibbels T, Bull JJ, Crews D (1994): Temperature dependent sex determination: a mechanistic approach. *J Exp Zool* 270:71–78.

Willier BH, Gallagher TF, Koch FC (1935): Sex modifications in the chick embryo resulting from injections of male and female hormones. *Proc Natl Acad Sci USA* 21:625–631.

Wolff E, Ginglinger A (1935): Sur la transformation des poulets males en intersexues par injection d'hormone femelle (folliculine) aux embryos. *Arch Anat Histol Embryol* 20:219–278.

Zanaria E, Muscatelli F, Bardoni B, Strom TM, Guioli S, Guo W, Lalli E, Moser C, Walker AP, McCabe ERB, Meitlinger T, Monaco AP, Sassone-Corsi P, Camerino G (1994): An unusual member of the nuclear hormone receptor superfamily responsible for X-linked adrenal hypoplasia congenita. *Nature* 372:635–641.

CHAPTER 21

Molecular Approaches to Control of Reproductive Behavior and Sexual Differentiation of Brain in Rodents

MARGARET M. MCCARTHY

Department of Physiology, University of Maryland School of Medicine,
Baltimore, Maryland 21201

CONTENTS

SYNOPSIS
INTRODUCTION
MOLECULAR APPROACHES TO CONTROL OF REPRODUCTIVE BEHAVIOR
    Quantitative *in Situ* Hybridization Histochemistry
    Lysate Version of RNase Protection Assay
    Gene Knockouts with Antisense Oligonucleotides
MOLECULAR APPROACHES TO SEXUAL DIFFERENTIATION OF THE BRAIN
ACKNOWLEDGMENTS
REFERENCES

SYNOPSIS

Steroid hormones are characterized by their binding to an intracellular receptor, which then acts as a transcription factor capable of enhancing or inhibiting gene expression in a highly specific manner. The fact that steroid hormones act in the brain is well known but many of the mechanistic aspects of these effects are not. During a temporally limited critical neonatal period, gonadal steroids induce permanent differentiation of the brain that subsequently influences steroid actions in the adult brain and the regulation of sexually dimorphic behaviors. Using the laboratory

Molecular Zoology: Advances, Strategies, and Protocols, Edited by Joan D. Ferraris and Stephen R. Palumbi.
ISBN 0-471-14461-4   © 1996 Wiley-Liss, Inc.

rodent as a model allows for examination of the molecular mechanisms regulating steroid-induced behaviors in a controlled and manipulatable setting. Quantification of gene expression at the cellular level can be accomplished with *in situ* hybridization histochemistry for individual mRNAs. Use of the lysate version of the RNase protection assay allows for simultaneous quantification of multiple mRNAs from exceedingly small tissue samples. Lastly, cause and effect relationships between steroid-induced gene expression and behavioral endpoints can be investigated with the use of synthetic antisense oligonucleotides designed to block the translation of a targeted mRNA into protein.

INTRODUCTION

The concept that steroid hormones act during two different phases in the life of an organism, and that the characteristics of the second phase are dependent on those of the first, originated with the work of Phoenix, Goy, Gerall, and Young (1959). Now known as the classic "organizational versus activational" action of steroid hormones, this tenet has provided a framework on which to base all investigations into the sexual differentiation of the brain and control of sexually dimorphic behaviors in adults. The basic principle, that the female brain is the intrinsic phenotype that is then differentiated or masculinized by neonatal exposure to steroid hormones of testicular origin during a restricted critical period, is generalizable across the mammalian order. In general, a masculinized brain results in an animal that does not exhibit female sexual behavior or an ovulation-inducing gonadotropin surge in response to estrogen. This is not to suggest that the mechanisms of differentiation are so conserved as to be redundant across species. On the contrary, there are substantial interspecies and intraspecies differences in the timing and hormonal agents of organizational events and questions remain regarding whether or not the female brain is truly undifferentiated (see McCarthy, 1994a) or what degree of plasticity is maintained in the adult CNS (Arnold and Breedlove, 1985; Tobet and Fox, 1992). Nonetheless, we can now exploit the organizational/activational concept of steroid hormone action to ask about the molecular bases of these events, both during development and in adulthood.

In many ways the sexual behavior of the female laboratory rat is an ideal system for investigating molecular mechanisms of steroid hormone action in the brain. A number of features make this behavioral reflex a useful tool. First, the lordosis response, a characteristic posture adopted by the female rat involving dorsiflexion of the spine, elevation of the rump, and lateral deflection of the tail, is both steroid and stimulus dependent. In the absence of at least 24 hr of prior exposure to circulating estrogen, the female will not assume this posture no matter what the stimulus. Alternatively, with even the most effective of steroid hormone regimes, the lordosis response will not be evoked without physical stimulation of the flanks and perineal region by either the male rat or investigator. These constraints on the behavioral response confer a tremendous degree of control to the investigator such that a behavioral "assay" can be conducted at an opportune time. Furthermore, the lordosis response is highly quantifiable. Commonly, the behavior is reported as an index or lordosis quotient (LQ), in which the male is allowed to mount the female a predetermined number of times and the number of lordoses recorded. The LQ = (# of lordosis/10) × 100, so that a fully receptive female has an LQ of 80–100% and a weakly receptive female exhibits an LQ of 10–40%. The intensity of the reflex is also quantified such that increasing scores are attributed to increasing degrees of dorsiflexion of the spine. Thus both quantitative and qualitative data can be garnered and used as an indicator of steroid action in the brain, although the contribution of steroid modulation of peripheral sensory nerve endings cannot be excluded.

A second desirable feature of this behavioral response is that the neural circuitry has been extremely well mapped. Identified populations of interconnected neurons in the preoptic area, dorsomedial hypothalamus, midbrain, hindbrain, and spinal cord have been characterized in regard to their neurochemical identity, afferent and efferent projections, and steroid hormone sensitivity (see Pfaff et al., 1994). This allows for manipulation of the pathway at various points and determination of what components are critically required for the response, versus those playing a modulatory role. For example, by intracerebral implantation it has been determined that estrogen acting exclusively in the ventromedial nucleus of the hypothalamus is both necessary and sufficient for a full lordosis response (Rubin and Barfield, 1980). This treatment will be ineffective, however, if protein synthesis is prevented or if the efferent projections of the hypothalamus are blocked. Given that activated steroid receptors act as transcription factors, this raises the question of how gene expression is regulated by estrogen and how this in turn alters neuronal activity such that a behavior is now expressed that previously was not.

MOLECULAR APPROACHES TO CONTROL OF REPRODUCTIVE BEHAVIOR

Steroid hormone receptors are characterized by possessing a DNA-binding domain that recognizes and interacts with sequence-specific hormone response elements (HREs) incorporated into a particular gene (for review see Beato, 1989). Most commonly, a steroid receptor is "activated" by binding of its ligand (although there are exceptions, see below), which then transduces the receptor to act as a positive or negative allosteric modulator of gene expression. In the case of estrogen receptor-induced lordosis, the temporal characteristics have clearly been delineated and have led to the Cascade Hypothesis of estrogen action (Pfaff, 1989) in which early genomic events that are mediated by estrogen then activate subsequent gene expression, which is essential for the appropriate induction of behavior.

Several points along this "cascade" have been investigated in attempts to characterize the molecular underpinnings of steroid-induced behavior. Beginning with the steroid receptor itself, estrogen was found to down-regulate the expression of its own receptor in female (Lauber et al., 1990; Simerly and Young, 1991) and male rat forebrain (Lisciotto and Morrell, 1993). While this apparent feedback mechanism of steroids on the expression of their own receptors is of interest, it does not address the question of what gene products are in fact responsible for the neuronal events regulating behavior. One of the most obvious gene products induced by activated estrogen receptor is the progesterone receptor. As little as 24 hr of estrogen exposure greatly increases progesterone receptor mRNA (Romano et al., 1989). This is presumably a direct effect of estrogen receptor on the progesterone receptor gene since in the brain progesterone receptors are virtually always expressed in the same neurons as estrogen receptor (Blaustein and Turcotte, 1993). Of particular interest is the observation that estrogen treatment in males is ineffective at inducing progesterone receptor expression (Lauber et al., 1991), suggesting the regulation of gene expression itself has been sexually differentiated. But again, the predominant function of an activated progesterone receptor is to positively or negatively regulate the expression of other genes, and so what are these?

Numerous "downstream" gene products induced by estrogen have been identified. For example, when quantified by slot blots or *in situ* hybridization histochemistry, estrogen increased mRNA for preproenkephalin in as little as 1 hr in brain areas related to control of lordosis (Romano et al., 1988) and may be related to the ability of specific opioid receptor subtype agonists to facilitate lordosis (Pfaus and Pfaff, 1992). Other genes in the brain regulated by estrogen and related to the control of behavior include vasoactive intestinal peptide (Gozes et al., 1989), substance P (Brown et al., 1990), and preprocholecystokinin (Micevych et al., 1994) to name but a few.

A number of techniques can be used to quantify gene expression in the brain, including slot blot hybridization, Northern blots, *in situ* hybridization histochemistry, and RNase protection assay, each having its own advantages and disadvantages. In my experience, when addressing questions of steroid induction of gene expression in the brain, the approaches of quantitative *in situ* hybridization and the lysate version of RNase protection offer the considerable advantage of high sensitivity while retaining tissue resolution. However, neither of these techniques is trivial to establish in the laboratory. Both require substantial amounts of specific reagents, nuclease-free working conditions, and constant vigilance to quality control. Therefore an investigator should only consider using these approaches if it is expected that they will become a routine part of the laboratory repertoire.

Quantitative *in Situ* Hybridization Histochemistry

Probably the single most important advantage of *in situ* hybridization histochemistry over other techniques is the ability to quantify signal at the individual cell level. This is a particularly important feature when dealing with a rare message expressed in a highly heterogeneous tissue such as the brain. Frequently, mRNAs expressed in the brain are restricted to a small region and interspersed among other cells that do not express the message. Substantial numbers of studies have successfully quantified steroid-induced changes in gene expression in discrete populations of neurons in the brain.

A common criticism leveled at *in situ* hybridization studies is that mechanisms of functional significance are bound to involve changes in protein levels and verification of changes in mRNA may or may not relate to changes in the gene product. However, attempts at cellular-level quantification of protein by immunocytochemical methods have been problematic, most likely due to inherently high variance and lack of linearity of the signal. Receptor proteins can be localized by *in vitro* receptor autoradiography and satisfactorily quantified, but cellular resolution is lost. Therefore the quantification of an autoradiographic signal generated by hybridization of a radiolabeled DNA or RNA probe to a specific mRNA in tissue can be a satisfactory compromise.

Important methodological issues to consider when using this approach include demonstrating that the probe is present in excess compared to the target and that comparisons between groups are made in the linear response range of the photographic emulsion. Both these parameters can be optimized by varying the amount of probe and duration of exposure of the emulsion prior to developing (for a more detailed discussion of these issues see McCabe and Pfaff, 1989; McCabe et al., 1993). In general, it is not advisable to attempt to make comparisons between assays and so individual runs should be designed to include samples from all animals and all groups. A manageable assay involves 70–120 slides.

In some instances there is a further advantage to quantification at the mRNA level as opposed to the protein. For example, the enzyme that synthesizes the inhibitory neurotransmitter GABA actually comes in two different forms coded by two separate genes. The two forms of glutamic acid decarboxylase are now referred to as GAD_{65} and GAD_{67} because of a 2-kDa difference in their size (Erlander et al., 1991). The functional significance of these two forms is largely unknown. By exploiting the specificity of the genetic code, we used PCR-generated single-strand DNA probes (Fig. 1) specific to each form of GAD and performed *in situ* hybridization on adjacent sections of rat brain in estrogen- and non-estrogen-treated females (McCarthy et al., 1995a). In this way we were able to establish a complex pattern of estrogen regulation of GAD expression that was variable in different brain regions and select populations of cells.

It has been established that both forms of GAD are expressed in the same cells (Kaufman et al., 1991) and one limitation of this technique is that we could not simultaneously measure mRNA of two different types in the same cell. Fluorescent or nonisotopic labeling procedures would allow for distinction of the two forms of message in one cell but would lose the criteria for quantification (i.e., linearity and saturability of signals). Therefore questions of whether estrogen is modulating the two GADs in opposite directions in the same cell versus separate populations of cells that exhibit differential sensitivity will have to await further technical advances. However, we have found it possible to quantify mRNA for the two forms of GAD in the same *sample* by using a lysate version of RNase protection assay.

Lysate Version of RNase Protection Assay

The basic principle of the RNase protection assay is a simple one. A radiolabeled riboprobe is incubated with tissue RNA and allowed to hybridize. The riboprobe is a sequence-specific strand of ribonucleotides designed to hybridize to a tar-

1) Reverse transcribe total RNA using Random Hexamers

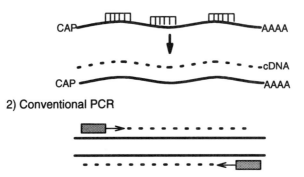

2) Conventional PCR

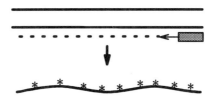

3) Purify DNA template by spin column chromatography
4) Repeat PCR with only the antisense primer for 50-75 cycles
5) Include 33P-dCTP, limit or exclude cold dCTP

6) Purify radiolabeled ssDNA probe by column chromatography

Figure 1. Diagrammatic representation of the use of asymmetric PCR to generate a single-strand DNA radiolabeled probe for *in situ* hybridization as based on the protocol of Brooks et al. (1993).

geted mRNA. RNase that selectively degrades all single-strand RNA (probe and endogenous) is added, and only the specifically formed RNA:RNA hybrids (riboprobe:mRNA) are protected from degradation. Separation by gel electrophoresis and visualization by autoradiography complete the process. This technique is extremely sensitive, detecting rare mRNAs easily, and is reliably quantitative. However, the standard RNase protection assay requires the prior extraction of total or polyA-RNA and therefore requires fairly substantially sized tissue samples. To obtain accurate regionally specific information in a tissue such as the rodent brain would require the pooling of samples from several different animals. Recent advances in this technique allow for direct hybridization of riboprobe to exceedingly small tissue samples without the prior extraction of RNA (Strauss and Jakobowitz, 1993).

We have been using this technique with good effect in tissue samples microdissected from adult and neonatal brain. Across the estrus cycle, message for both forms of GAD was observed to significantly decline in the nucleus of the diagonal band of Broca in the medial preoptic area selectively on the afternoon of proestrus

(Grattan et al., 1995). This is the time at which a gonadotropin surge (LH) is elicited from the pituitary to induce ovulation and it has been speculated that steroids act to reduce GABAergic inhibition on the release of LH. These data suggest one mechanism by which this is accomplished is via a down-regulation of mRNA for GAD. A significant sex difference in both forms of GAD has also been observed on postnatal day 1 using this technique (McCarthy et al., 1995), suggesting GABA may be involved in the mechanism of sexual differentiation of the brain.

While characterizing changing patterns of gene expression in response to hormonal exposure based on behavioral responses is an informative and useful approach, it often leaves unanswered the question of functional significance of observed changes. Teasing apart the various contributing regulatory components has traditionally been accomplished in behavioral pharmacology with the use of selective receptor agonists, antagonists, enzyme inhibitors, or other neurochemical agents. Recent advances in the use of synthetic antisense oligodeoxynucleotides has placed a new tool in the repertoire of behavioral neuroscientists and has been used with considerable success to temporarily and selectively knock out gene expression.

Gene Knockouts with Antisense Oligonucleotides

The basic principle of gene knockouts by synthetic antisense oligonucleotides is appealing in its simplicity and easy availability to the general researcher. An oligonucleotide is a short strand of synthetic DNA. All that is required to begin an investigation involving the use of antisense oligonucleotides is knowledge of the sequence you wish to target, some general aspects of the protein product in question, and access to a computer that can run sequence comparison against the GenBank or EMBL databases. This is necessary in order to establish that your antisense oligonucleotide does not exhibit significant homology with other known mRNAs. Once a suitable sequence has been identified, synthesized, and purified, the researcher need only administer it and hope for the desired effect. Of course, this is where it gets complicated.

There have been several mechanisms proposed for the action of antisense oligonucleotides in biological systems (Fig. 2), the most obvious being that the synthetic oligonucleotide hybridizes with the target mRNA and the resulting duplex either prevents the ribosomal complex from forming or blocks its movement down the mRNA strand, an action known as "hybridization arrest" (Toulme, 1992). An additional but not exclusive mechanism involves the action of the nuclease RNase-H, which selectively degrades the RNA component of an RNA:DNA duplex, leaving the DNA (the antisense oligonucleotide) intact and capable of forming additional hybrids (Walder and Walder, 1988). When active, this nuclease can greatly increase the efficacy of antisense oligonucleotide action, but it can also increase the probability of nonspecific effects, particularly in poikilotherms (Woolf et al., 1992). Additional potential mechanisms of action of antisense oligonucleotides include the blocking of splicing sites, disruption of mRNA stability, and prevention of transport of heteronuclear RNA out of the nucleus. None of these mechanisms has been definitively proved to occur in an *in vivo* situation and so to-date there remains only circumstantial

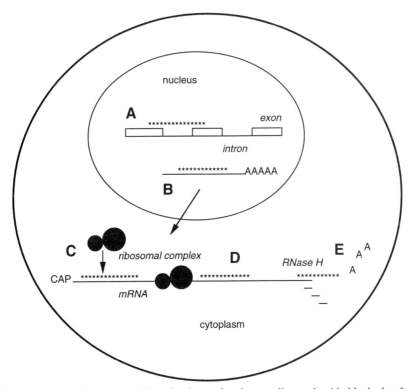

Figure 2. Some of the potential mechanisms of antisense oligonucleotide blockade of gene expression in cells: (A) hybridization of splice sites to prevent formation of mature mRNA, (B) prevention of transport of RNA out of the nucleus, (C) blocking of the formation of the ribosomal complex on the mRNA, (D) prevention of the movement of the ribosomal complex down the mRNA by "hybridization arrest," and (E) activation of the nuclease RNase-H, which degrades the RNA portion of an RNA:DNA duplex.

evidence that antisense oligonucleotides are indeed acting in the predicted fashion (see Stein and Krieg, 1994). Furthermore, reports of unexpected and nonspecific effects of antisense oligonucleotides continue to increase with increasing use of the technique (see McCarthy, 1994b).

All of these caveats aside, this approach has been used quite effectively in the study of hormonal control of female reproductive behavior. As mentioned previously, one of the first and most prominent gene products induced by estrogen is the progesterone receptor. Progesterone has long been known to synergize with estrogen to facilitate lordosis and proceptive behaviors (Pfaff et al., 1994). However, there has been a consistently unresolved issue as to whether progesterone is acting in a classic receptor-mediated genomic fashion, or whether there are membrane receptor non-genomic actions of progesterone that contribute to the regulation of behavior as well (see Delville, 1991). Blocking the synthesis of estrogen-induced progesterone receptor by intracerebral infusions of antisense oligonucleotides designed to be complementary to, and thereby hybridize with, progesterone receptor mRNA and block its translation into protein has helped to establish that most of progesterone's effects on lordosis are in fact mediated by the progesterone receptor (Pollio et al., 1993; Mani et al., 1994; Ogawa et al., 1994). A further insite into the role of this receptor in lordosis behavior was also provided by use of antisense oligonucleotides. It previously had been observed that agonists of the dopamine receptor activate the progesterone receptor independently of progesterone, including binding to DNA and induction of gene expression (Power et al., 1991). A physiological significance of this observation was established by blocking dopaminergic facilitation of lordosis with antisense oligonucleotides against the progesterone receptor (Mani et al., 1994).

A likely component of success in the case of antisense oligonucleotides targeting the progesterone receptor was that the induction of gene expression by estrogen treatment could be temporally correlated with the administration of synthetic oligonucleotides. This approach has also been exploited in the use of antisense oligonucleotides against oxytocin receptor mRNA, which is strongly induced by estrogen (McCarthy et al., 1994a). In general it has proved difficult, but not impossible, to significantly block with antisense oligonucleotides any mRNAs that are either highly abundant, constitutively expressed, or extremely stable (see McCarthy, 1994b). These and other criteria that must be considered when beginning a study involving the use of antisense oligonucleotides can be found in protocol McCarthy.3. As with any new methodology, the potential advantages must be weighed against the uncertainties in interpretation. A commonly asked question in neuroendocrine and behavioral neuroscience studies that involve the use of antisense is: "Is this really better than the already existing antagonists, blockers, and so on?" In some cases the answer is definitively "yes." For example, after determining by *in situ* hybridization that estrogen induced mRNA for both forms of GAD in a particular brain area, and knowing that GABA in this area facilitated lordosis, we used antisense oligonucleotides specific to one or the other form of GAD to determine which form of GAD was synthesizing behaviorally relevant GABA (McCarthy et al., 1994b). In this way, the use of antisense oligonucleotides to temporarily knock out or reduce gene expression can be a complementary tool to those techniques designed to quantify changes in gene expression and hopefully aid in the determination of cause and effect relationships.

The issues of appropriate controls and demonstration of the specificity of antisense oligonucleotide gene knockout are paramount when using this approach. Controls can be divided into two categories: (1) oligonucleotide controls and (2) response controls. The purpose of the oligonucleotide controls is to demonstrate that observed results are due to sequence specificity and not the result of exposure of the cells to large amounts of synthetic DNA. Appropriate controls include sense strand oligonucleotides, scrambled oligonucleotides, and missense oligonucleotides. The details of each are discussed in protocol McCarthy.3. It is optimal to use more than one oligonucleotide control; that is, one group receives scrambled oligonucleotide and another receives sense oligonucleotide. The prediction is that only the group receiving antisense oligonucleotide will exhibit a specific response. The response controls must include demonstration that the targeted protein has been decreased only by the antisense oligonucleotide and not the controls. In addition, it is desirable to demonstrate that a related protein has not been altered by the antisense oligonucleotide administration. The demonstration that the neuropeptide Y-Y1 receptor was specifically reduced by antisense oligonucleotide, but the related neuropeptide Y-Y2 receptor was not, is an excellent example of this type of control (Wahlstedt et al., 1993). Additional response controls are dependent on the specific system being investigated. In the case of behavioral studies it is always important to demonstrate that the animal has not been generally impaired by the treatment and that observations of behavioral changes are specific. For example, antisense oligonucleotide against tryptophan hydroxylase, the rate-limiting enzyme in serotonin synthesis, was found to increase anxiety levels in mice but had no effect on their locomotor ability (McCarthy et al., 1995b).

In addition to being useful approaches in understanding the molecular basis of steroid-induced behavior in the adult animal, all these techniques can also be applied to the question of sexual differentiation of the brain.

MOLECULAR APPROACHES TO SEXUAL DIFFERENTIATION OF THE BRAIN

An understanding of the mechanisms of sexual differentiation of the rodent brain can fairly be characterized as rudimentary at best. The majority of effort to-date has focused on the role of the estrogen receptor, but what lies beyond that is largely a black box. When a neonatal female is administered exogenous testosterone, it reaches the brain where it is aromatized into estrogen, presumably binds to the estrogen receptor, regulates transcription of some unknown genes, and thereby permanently differentiates the brain. The approaches that work well in the adult, that is, administering hormone and monitoring gene expression based on preconceived concepts of likely candidates, has not been generally applied to the process of sexual differentiation. This is in part due to the fact that there are few obvious "likely candidates" that would be specific to this developmental process. One notable exception is the growth associated protein, GAP-43, which is a membrane-bound protein found in axonal growth cones. *In situ* hybridization revealed a regionally specific sex difference in mRNA for GAP-43 in neonatal rat brain (Shughrue and Dorsa, 1994a) and further demonstrated that message levels were regulated by both estrogen- and androgen-receptor activation in a region-specific manner (Shughrue and Dorsa, 1994b).

In regard to the estrogen receptor, both the mRNA for the receptor itself and the mRNA for the enzyme involved in ligand synthesis, aromatase, have been mapped and quantified in the neonatal brain. Levels of both these proteins appear to be regulated transcriptionally and have a regulatory influence on each other. Specifically, estrogen receptor mRNA is significantly higher in neonatal female brain and changes developmentally, peaking at postnatal day 2 and declining to male levels by day 10. Levels of estrogen receptor mRNA in males remain constant throughout this period (DonCarlos and Handa, 1994). The mRNA for aromatase, on the other hand, is high in the neonatal male brain during this critical period of sexual differentiation of the brain (Lauber and Lichtensteiger, 1994). As mentioned previously, in the adult, estrogen down-regulates the mRNA for its own receptor; therefore in the neonatal brain the synthesis of masculinizing estrogen by local aromatase may regulate the level of gene expression for the estrogen receptor.

Evidence for a causal relationship between estrogen receptor mRNA and differentiation was provided by the use of antisense oligonucleotides to the estrogen receptor mRNA administered directly into the brain during the critical period (McCarthy et al., 1993b). Using the model of the androgenized female (genetic female neonatally administered testosterone and thereby "masculinized"), female rats were infused directly into the hypothalamus with antisense oligonucleotides to estrogen receptor mRNA, scrambled control oligonucleotides, or vehicle. All animals were subsequently raised to adulthood and various behavioral and physiological measures of a masculinized versus feminized brain were assessed. When females had been treated with testosterone neonatally, and received antisense to the estrogen receptor (remember that testosterone is converted to estrogen), they were protected from the masculinizing effects of the neonatal testosterone. There was no protective effect of the scrambled control oligonucleotide or the vehicle infusions. In contrast, normal females that did not receive testosterone exhibited only a few effects in response to the estrogen receptor antisense oligonucleotide infusions.

Based on these findings, it was concluded that the presence of estrogen receptor is in fact required for masculinization of the brain in response to testosterone. While comforting, this finding is hardly surprising. However, the more important implication of the study is the potential this technique offers for establishing the functional significance of new gene products if and when they are identified. Because of the tight temporal and stimulus constraints on estrogen-induced gene expression to differentiate the brain, this should be an ideal system for methods to identify rare and novel genes. These would include subtractive hybridization (Rubenstein et al., 1990) and differential display PCR (Liang and Pardee, 1992). In the near future studies should begin to identify the gene products regulating this important developmental event.

ACKNOWLEDGMENTS

This work was supported by a Special Research Initiative Support Grant from the University of Maryland School of Medicine. I would like to thank Dr. David Grattan and Dr. Gregory Ball for helpful comments on the manuscript.

REFERENCES

Arnold AP, Breedlove SM (1985): Organizational and activational effects of sex steroids on brain and behavior: a reanalysis. *Horm Behav* 19:469–498.

Beato M (1989): Gene regulation by steroid hormones. *Cell* 56:335–344.

Blaustein JD, Turcotte JC (1993): Immunocytochemical localization of midbrain estrogen receptor- and progestin receptor-containing cells in female guinea pigs. *J Comp Neurol* 328:76–87.

Brooks PJ, Kaplitt MG, Kleopoulos SP, Funabashi T, Mobbs CV, Pfaff DW (1993): Cell-detection of messenger RNA and low abundance of heteronuclear RNA with single-stranded DNA probes produced by amplified primer extension labeling. *J Histochem Cytochem* 41:1761–1766.

Brown ER, Harlan RE, Krause JE (1990): Gonadal steroid regulation of substance P (SP) and SP-encoding messenger ribonucleic acids in the rat anterior pituitary and hypothalamus. *Endocrinology* 126:330–340.

Delville Y (1991): Progesterone-facilitated sexual receptivity: a review of arguments supporting a nongenomic mechanism. *Neurosci Biobehav Rev* 15:407–414.

DonCarlos L, Handa RJ (1994): Developmental profile of estrogen receptor mRNA in the preoptic area of male and female neonatal rats. *Dev Brain Res* 79:283–289.

Erlander MG, Tillakaratne NJK, Feldblum S, Patel N, Tobin AJ (1991): Two genes encode distinct glutamate decarboxylases. *Neuron* 7:91–100.

Gozes I, Werner H, Fawzi M, Abdelatty A, Shani Y, Fridkin M, Koch Y (1989): Estrogen regulation of vasoactive intestinal peptide mRNA in rat hypothalamus. *J Mol Neurosci* 1:55–61.

Grattan DR, Selmanoff M, Rocca MS, Strauss KI, McCarthy MM (1995): GABAergic neuronal activity and mRNA levels for both forms of glutamic acid decarboxylase are reduced in the diagonal band of Broca during the afternoon of proestrus. Brain Res. Submitted.

Kaufman DL, Houser CR, Tobin AJ (1991): Two forms of the gamma-aminobutyric acid synthetic enzyme glutamate decarboxylase have distinct intraneuronal distributions and co-factor interactions. *J Neurochem* 56:720–723.

Lauber ME, Lichtensteiger W (1994): Pre- and postnatal ontogeny of aromatase cytochrome P450 messenger ribonucleic acid expression in the male rat brain studied by *in situ* hybridization. *Endocrinology* 135:1661–1668.

Lauber AH, Romano GJ, Mobbs CV, Pfaff DW (1990): Estradiol regulation of receptor messenger ribonucleic acid in rat mediobasal hypothalamus: an *in situ* hybridization study. *J Neuroendocrinol* 2:605–611.

Lauber AH, Romano GJ, Pfaff DW (1991): Sex difference in estradiol receptor mRNA in rat mediobasal hypothalamus as demonstrated by *in situ* hybridization. *Neuroendocrinology* 53:608–613.

Liang P, Pardee AB (1992): Differential display of eukaryotic messenger RNA by means of the polymerase chain reaction. *Science* 257:967–971.

Lisciotto CA, Morrell JI (1993): Circulating gonadal steroid hormones regulate estrogen receptor mRNA in the male rat forebrain. *Mol Brain Res* 20:79–90.

Mani SK, Allen JMC, Clark JH, Blaustein JD, O'Malley BW (1994): Convergent pathways for steroid hormone- and neurotransmitter-induced rat sexual behavior. *Science* 265:1246–1249.

McCabe JT, Pfaff DW (1989): *In situ* hybridization: a methodological guide. In: Conn PM (ed). *Methods in neuroscience,* vol 1. Orlando, FL: Academic Press, pp 98–126.

McCabe JT, Kao T-C, Volkov ML (1993): An assessment of the efficacy of *in situ* hybridization as a quantitative method by variance components estimation. *Microsc Res Tech* 25:61–67.

McCarthy MM (1994a): Molecular aspects of sexual differentiation of the rodent brain. *Psychoneuroendocrinology* 19:415–427.

McCarthy MM (1994b): Use of antisense oligonucleotides to block gene expression in the central nervous system. In: de Kloet ER, Sutano W (eds). *Methods in neurosciences: neurobiology of steroids,* vol 22. Orlando, FL: Academic Press, pp 342–356.

McCarthy MM, Brooks PJ, Pfaus J, Brown HE, Flanagan LM, Schwartz-Giblin S, Pfaff DW (1993a): Antisense technology in behavioral neuroscience. *Neuroprotocols* 2:67–74.

McCarthy MM, Schlenker EH, Pfaff DW (1993b): Enduring consequences of neonatal treatment with antisense oligodeoxynucleotides to estrogen receptor mRNA on sexual differentiation of rat brain. *Endocrinology* 133:433–439.

McCarthy MM, Kleopolous SP, Mobbs CV, Pfaff DW (1994a): Infusion of antisense oligodeoxynucleotides to the oxytocin receptor in the ventromedial hypothalamus reduces estrogen-induced sexual receptivity and oxytocin receptor binding in the female rat. *Neuroendocrinology* 59:432–440.

McCarthy MM, Masters DB, Rimvall K, Schwartz-Giblin S, Pfaff DW (1994b): Intracerebral administration of antisense oligodeoxynucleotides to GAD_{65} and GAD_{67} mRNAs modulates reproductive behavior in the female rat. *Brain Res* 636:209–220.

McCarthy MM, Grattan DR, Davis AM, Selmanoff M (1995): Sex differences in glutamic acid decarboxylase mRNA content in the neonatal brain. *Soc Neurosci Abstract,* Vol 21, #791.2. San Diego, CA.

McCarthy MM, Kaufman LC, Brooks PJ, Pfaff DW, Schwartz-Giblin S (1995a): Estrogen modulation of mRNA for the two forms of glutamic acid decarboxylase (GAD) in female rat brain. *J Comp Neurol* 360:685–697.

McCarthy MM, Nielsen DA, Goldman D (1995b): Antisense oligonucleotide inhibition of tryptophan hydroxylase activity in mouse brain. *Regul Pept* 59:163–170.

Micevych PE, Abelson L, Fok H, Ulibarri C, Priest CA (1994): Gonadal steroid control of preprocholecystokinin mRNA expression in the limbic–hypothalamic circuit: comparison of adult with neonatal steroid treatments. *J Neurosci Res* 38:386–398.

Ogawa S, Olazabal UE, Parhar IS, Pfaff DW (1994): Effects of intrahypothalamic administration of antisense DNA for progesterone receptor mRNA on reproductive behavior and progesterone receptor immunoreactivity in female rat. *J Neurosci* 14:1766–1774.

Pfaff DW (1989): Patterns of steroid hormone effects on electrical and molecular events in hypothalamic neurons. *Mol Neurobiol* 3:135–154.

Pfaff DW, Schwartz-Giblin S, McCarthy MM, Kow L-M (1994): Cellular mechanisms of female reproductive behaviors. In: Knobil E, Neil E, et al (eds). *The physiology of reproduction,* 2nd ed, vol 2. New York: Raven Press, pp 107–220.

Pfaus JG, Pfaff DW (1992): μ-, δ-, and κ-Opioid receptor agonists selectively modulate sexual behaviors in the female rat: differential dependence on progesterone. *Horm Behav* 26:457–473.

Phoenix CH, Goy RW, Gerall AA, Young WC (1959): Organizing action of prenatally administered testosterone proprionate on the tissues mediating mating behavior in the female guinea pig. *Endocrinology* 65:369–382.

Pollio G, Zue P, Zanisi M, Nicolin A, Maggi A (1993): Antisense oligonucleotide blocks progesterone-induced lordosis behavior in ovariectomized rats. *Mol Brain Res* 19:135–139.

Power RF, Mani SK, Codina J, Conneely OM, O'Malley BW (1991): Dopaminergic and ligand-independent activation of steroid hormone receptors. *Science* 254:1636–1639.

Romano GJ, Harlan RE, Shivers BD, Howells RD, Pfaff DW (1988): Estrogen increases proenkephalin messenger ribonucleic acid levels in the ventromedial hypothalamus of the rat. *Mol Endocrinol* 2:1320–1328.

Romano GJ, Krust A, Pfaff DW (1989): Expression and estrogen regulation of progesterone receptor messenger RNA in neurons of rat hypothalamus. *Mol Endocrinol* 3:1295–1300.

Rubenstein JLR, Brice AEJ, Ciaranello RD, Denny D, Porteus MH, Usdin TB (1990): Subtractive hybridization system using single-stranded phagemids with directional inserts. *Nucleic Acids Res* 18:4833–4842.

Rubin BS, Barfield RJ (1980): Priming of estrous responsiveness by implants of 17β-estradiol in the ventromedial hypothalamic nucleus of female rats. *Endocrinology* 106:504–509.

Shughrue PJ, Dorsa DM (1994a): The ontogeny of GAP-43 (neuromodulin) mRNA in postnatal rat brain: evidence for a sex dimorphism. *J Comp Neurol* 340:174–184.

Shughrue PJ, Dorsa DM (1994b): Estrogen and androgen differentially modulate the growth-associated protein GAP-43 (neuromodulin) messenger ribonucleic acid in postnatal rat brain. *Endocrinology* 134:1321–1328.

Simerly RB, Young BJ (1991): Regulation of estrogen receptor messenger ribonucleic acid in rat hypothalamus by sex steroid hormones. *Mol Endocrinol* 53:424–432.

Stein CA, Krieg AM (1994): Problems in interpretation of data derived from *in vitro* and *in vivo* use of antisense oligodeoxynucleotides. *Antisense Res Dev* 4:67–70.

Strauss KI, Jakobowitz DM (1993): Quantitative measurement of calretinin and β-actin mRNA in rat brain micropunches without prior isolation of RNA. *Mol Brain Res* 20:229–239.

Tobet SA, Fox TO (1992): Sex difference in neuronal morphology influenced hormonally throughout life. In: Gerall AA, Moltz H, Ward IL (eds). *Handbook of behavioral neurobiology.* New York: Plenum Press, pp 41–83.

Toulme JJ (1992): Artificial regulation of gene expression by complementary oligonucleotides—An Overview. In: Antisense RNA and DNA, Wiley-Liss Inc., pp 175–194.

Wahlstedt C, Pich EM, Koob GF, Yee F, Heilig M (1993): Modulation of anxiety and neuropeptide Y-Y1 receptors by antisense oligonucleotides. *Science* 259:528–531.

Walder RY, Walder JA (1988): Role of RNase H in hybrid-arrested translation by antisense oligonucleotides. *Proc Natl Acad Sci* 85:5011–5015.

Whitesell L, Rosolen A, Neckers LM (1991): *In vivo* modulation of N-myc expression by continuous perfusion with an antisense oligonucleotide. *Antisense Res Dev* 1:343–350.

Woolf TM, Melton DA, Jennings CGB (1992): Specificity of antisense oligonucleotides *in vivo. Proc Natl Acad Sci USA* 89:7305–7309.

CHAPTER 22

Transgenic Fish: Ideal Models for Basic Research and Biotechnological Applications

THOMAS T. CHEN, JENN-KAN LU, MICHAEL J. SHAMBLOTT,
CLARA M. CHENG, and CHUN-MEAN LIN

Center of Marine Biotechnology, University of Maryland Biotechnology Institute,
Baltimore, Maryland 21202; and Department of Biological Sciences, University of Maryland
Baltimore County, Baltimore, Maryland 21202

JANE C. BURNS

Department of Pediatrics, School of Medicine, University of California–San Diego,
San Diego, California 92093-0609

RENATE REIMSCHUESSEL

Aquatic Pathobiology Group, Department of Pathology, University of Maryland at Baltimore,
Baltimore, Maryland 21201

NAGARAJ CHATAKONDI and REX A. DUNHAM

Department of Fisheries Science and Allied Aquaculture, Auburn University,
Auburn, Alabama 36849

CONTENTS

Molecular Zoology: Advances, Strategies, and Protocols, Edited by Joan D. Ferraris and Stephen R.
Palumbi.
ISBN 0-471-14461-4 © 1996 Wiley-Liss, Inc.

SYNOPSIS

Organisms into which heterologous DNA has been artificially introduced and stably integrated in their genomes are termed *transgenic*. Since 1985, a wide variety of transgenic fish species have been produced. They are produced by microinjecting, electroporation, or infection with pantropic defective retroviral vectors. These transgenic fish can serve as excellent experimental models for basic scientific investigations as well as biotechnological applications. In this chapter, using research results generated in our laboratories and those of others as examples, we review the current status of the transgenic fish research and illustrate the potential application of this technology in both basic research and biotechnological applications.

INTRODUCTION

Animals or plants into which heterologous DNA has been artificially introduced and integrated in their genomes are called *transgenic*. Since the early 1980s, transgenic plants (Gasser and Fraley, 1989), nematodes (Stinchcomb et al., 1985), fruitflies (Rubin and Spradling, 1982), sea urchins (McMahan et al., 1984, 1986), frogs (Etkin and Pearman, 1987), laboratory mice (Gordon, 1989; Jaenisch, 1990), and farm mammals such as cows, pigs, and sheep (Pursel et al., 1989) have been successfully produced. In plant systems, the DNA is introduced into cells by infection with *Agrobacterium tumefaciens* or by physical means. In animal systems, the DNA is injected into the pronuclei of fertilized eggs and the injected embryos are incubated *in vitro* or implanted into the uterus of a pseudopregnant female for subsequent development. In these studies, multiple copies of transgenes are integrated at random locations in the genome of the transgenic individuals. If the transgenes are linked with functional promoters, expression of transgenes as well as display of change in phenotype is expected in some of the transgenic individuals. Furthermore, the transgenes in many transgenic individuals are also transmitted through the germline to subsequent generations. These transgenic animals play important roles in basic research as well as applied biotechnology. In basic research, transgenic animals provide excellent models for studying molecular genetics of early vertebrate development, actions of oncogenes, and the biological functions of hormones at different stages of development. In applied biotechnology, transgenic animals offer unique opportunities for producing animal models for biomedical research, improving the genetic background of broodstock for animal husbandry or aquaculture, and designing bioreactors for producing valuable proteins for pharmaceutical or industrial purposes.

Since 1985, a wide range of transgenic fish species have been produced (Chen and Powers, 1990; Fletcher and Davis, 1991; Hackett, 1993) by microinjecting or electroporating homologous or heterologous transgenes into newly fertilized or unfertilized eggs. Several important steps are routinely taken to produce a desired transgenic fish. First, an appropriate fish species must be chosen, depending on the nature of the studies and the availability of the fish holding facility. Second, a specific gene construct must be prepared. The gene construct contains the structural gene encoding a gene product of interest and the regulatory elements that regulate the expression of the gene in a temporal, spatial, and developmental manner. Third, the gene construct has to be introduced into the developing embryos in order for the transgene to be integrated stably into the genome of every cell. Fourth, since not all instances of gene transfer are efficient, a screening method must be adopted for identifying transgenic individuals.

Although remarkable progress has been made in producing transgenic fish by gene transfer technology, a critical review of the published results has shown that a majority of the research effort has been devoted to confirming the phenomenon of foreign gene transfer into various fish species. Very few attempts have been made to explore the application of transgenic fish technology in basic as well as applied research. Recently, we have devoted a substantial amount of our research effort to this problem, with promising results. In this chapter, we will discuss the potential application of this technology using results generated in our laboratories as examples.

PRODUCTION OF TRANSGENIC FISH

Selection of Fish Species

Gene transfer studies have been conducted in several different fish species including channel catfish, common carp, goldfish, Japanese medaka, loach, northern pike, rainbow trout, salmon, tilapia, walleye, and zebrafish (for review see Chen and Powers, 1990; Hackett, 1993). Depending on the purpose of the transgenic fish studies, the embryos of some fish species are more suited for gene transfer studies than the others. For example, Japanese medaka (*Oryzias latipes*) and zebrafish (*Brachydanio rerio*) have short life cycles (3 months from hatching to mature adults), produce hundreds of eggs on a regular basis without exhibiting a seasonal breeding cycle, and can be maintained easily in the laboratory for 2–3 years. Eggs from these two fish species are relatively large (diameter: 0.7–1.5 mm) and possess very thin, semitransparent chorions, features that permit easy microinjection of DNA into the eggs if appropriate glass needles are used. Furthermore, inbred lines and various morphological mutants of both fish species are available. These fish species are thus suitable candidates for conducting gene transfer experiments for (1) studying developmental regulation of gene expression and gene action, (2) identifying regulatory elements that regulate the expression of a gene, (3) measuring the activities of promoters, and (4) producing transgenic models for environmental toxicology. However, a major drawback of these two fish species is their small body size that makes them unsuitable for some endocrinological or biochemical analyses.

Channel catfish, common carp, rainbow trout, and salmon are commonly used, large-body-size model fish species in transgenic fish studies. Since the endocrinology, reproductive biology, and physiology of these fish species have been well worked out, they are well suited for conducting studies on comparative endocrinology as well as in aquaculture applications. However, the long maturation time of these fish species and a single spawning cycle per year will hamper rapid research progress in this field.

Loach, killifish, goldfish, and tilapia are the third group of model fish species suitable for conducting gene transfer studies since their body sizes are large enough for most biochemical and endocrinological studies. Furthermore, shorter maturation times as compared to catfish, rainbow trout, or salmon allow easier manipulation of transgenic progeny. Unfortunately, the lack of a well-defined genetic background and asynchronous reproductive behavior of these fish species render them less amenable to gene transfer studies.

Transgene Constructs

A transgene used in producing transgenic fish for basic research or biotechnological applications is a recombinant gene construct that will produce a gene product at appropriate levels in the desired tissue(s) at the desired time(s). The prototype of a transgene is usually constructed in a plasmid to contain an appropriate promoter/enhancer element and the structural gene.

Depending on the purpose of the gene transfer studies, transgenes can be grouped into three main types: (1) *gain-of-function*, (2) *reporter function*, and (3) *loss-of-function*. The *gain-of-function* transgenes are designed to add new functions to the transgenic individuals or to facilitate the identification of the transgenic individuals if the genes are expressed properly in the transgenic individuals. Transgenes containing the structural genes of mammalian and fish growth hormones (GH, or their cDNAs) fused to functional promoters such as chicken and fish β-actin gene promoters are examples of the gain-of-function transgene constructs. Expression of the GH transgenes in transgenic individuals will result in growth enhancement (Zhang et al., 1990; Du et al., 1992; Lu et al., 1992; Chen et al., 1993). Bacterial chloramphenicol acetyl transferase (CAT), β-galactosidase, or luciferase genes fused to functional promoters are one type of the *reporter function* transgenes. These reporter genes are commonly used to identify the success of gene transfer effort. A more important function of a reporter gene is used to identify and measure the strength of a promoter/enhancer element. In this case, the structural gene of the CAT, β-galactosidase, or luciferase gene is fused to a promoter/enhancer element in question. Following gene transfer, the expression of the reporter gene activity is used to determine the transcriptional regulatory sequence of a gene or the strength of a promoter (Moav et al., 1992).

The *loss-of-function* transgenes are constructed for interfering with the expression of host genes. These genes might encode an antisense RNA to interfere with the post-transcriptional process or translation of endogenous mRNAs. Alternatively, these genes might encode a catalytic RNA (a ribozyme) that can cleave specific mRNAs and thereby cancel the production of the normal gene product (Cotten and Jennings, 1989). Although these genes have not yet been introduced into a fish model, they could potentially be employed to produce disease-resistant transgenic broodstocks for aquaculture or transgenic model fish defective in a particular gene product for basic research.

Methods of Gene Transfer

Techniques such as calcium phosphate precipitation, direct microinjection, lipofection, retrovirus infection, electroporation, and particle gun bombardment have widely been used to introduce foreign DNA into animal cells, plant cells, and germlines of mammals and other vertebrates. Among these methods, direct microinjection and electroporation of DNA into newly fertilized eggs have proved to be the most reliable methods of gene transfer in fish systems.

Microinjection of Eggs or Embryos. Microinjection of foreign DNA into newly fertilized eggs was first developed for the production of transgenic mice in the early 1980s. Since 1985, the technique of microinjection has also been adopted for introducing transgenes into Atlantic salmon, common carp, catfish, goldfish, loach, medaka, rainbow trout, tilapia, and zebrafish (Chen and Powers, 1990; Fletcher and Davis, 1991). The gene constructs that were used in these studies include human or rat growth hormone (GH) gene, rainbow trout or salmon GH cDNA, chicken δ-crystalline protein gene, winter flounder antifreeze protein gene, *Escherichia coli* β-galactosidase gene, and *E. coli* hygromycine resistance gene (Chen and Powers, 1990; Fletcher and Davis, 1991). In general, gene transfer in fish by direct microinjection is conducted as follows. The parameters for microinjection are summarized in Table 1. Eggs and sperm are collected in separate, dry containers. Fertilization is initiated by adding water and sperm to the egg, with gentle stirring to enhance fertilization. Eggs are microinjected within the first few hours after fertilization. The injection apparatus consists of a dissecting stereo microscope and two micromanipulators, one with a micro-glass-needle for injection and the other with a

**TABLE 1. Parameters of Gene Transfer in Fish
by Microinjection and Electroporation**

Parameters	Gene Transfer Method	
	Microinjection	Electroporation
Developmental stage	1 to 2 cells	1 to 2 cells
DNA size	<10 kb	<10 kb
DNA concentration	$10^6–10^7$ molecules/embryo	100 μg/ml
DNA topology	Linear	Linear
Chorion barrier	Dechorionated/micropyle	Intact chorion
Electrical field strength	N/A	500–3000 V
Pulse shape	N/A	Exponential/square
Pulse duration	N/A	ms to seconds
Temperature	RT	RT
Medium	PBS/saline	PBS/saline

Abbreviations: N/A, not applicable; RT, room temperature (25°C); ms, millisecond; PBS, phosphate-buffered saline (10 mM, pH 7.5, 0.15 M NaCl).

micropipette for holding fish embryos in place. Routinely, about $10^6–10^8$ molecules of a linearized transgene in about 20 nl are injected into the egg cytoplasm. Following injection, the embryos are incubated in water until hatching. Since natural spawning in zebrafish or medaka can be induced by adjusting photoperiod and water temperature, precisely staged newly fertilized eggs can be collected from the aquaria for gene transfer. If the medaka eggs are maintained at 4°C immediately after fertilization, the micropyle on the fertilized eggs will remain visible for 2 hours. The DNA solution can easily be delivered into the embryos by injecting through this opening.

Depending on the species, the survival rate of injected fish embryos ranges from 35% to 80% while the rate of DNA integration ranges from 10% to 70% in the survivors (Table 2; Chen and Powers, 1990; Fletcher and Davis, 1991). The tough chorions of the fertilized eggs in some fish species (e.g., rainbow trout and Atlantic salmon) can make insertion of glass needles difficult. This difficulty has been overcome by any one of the following methods: (1) inserting the injection needles through the micropyle, (2) making an opening on the egg chorions by microsurgery, (3) removing the chorion by mechanical or enzymatic means, (4) preventing chorion hardening by initiating fertilization in a solution containing 1 mM glutathione, or (5) injecting the unfertilized eggs directly.

Electroporation. Electroporation is a successful method for transferring foreign DNA into bacteria, yeast, and plant and animal cells in culture. This method has become popular for transferring foreign genes into fish embryos in the past 3 years (Lu et al., 1992; Powers et al., 1992). Electroporation utilizes a series of short electrical pulses to permeate cell membranes, thereby permitting the entry of DNA molecules into embryos. The patterns of electrical pulses can be emitted in a single pulse of exponential decay form (i.e., exponential decay generator) or high-frequen-

TABLE 2. Transfer of Foreign DNA into Medaka Embryos by Different Gene Transfer Methods

	Microinjection[a]	Electroporation		Pantropic Retroviral Vector	
		I[b]	II[c]	Electroporation[d]	Incubation[e]
Viability (at hatching)	50%	70%	90%	50%	70%
Integration[f] rate	20%	15%	25%	50%	70%
Transgene expression	Yes	Yes	Yes	Yes	Yes
Efficiency (eggs/min)	1–2	200	200	200	200

[a]Injecting is carried out via micropyle prior to blastodisc formation.
[b]Exponential-decay impulse mode.
[c]Square-wave impulse mode.
[d]Electroporation with square-wave mode at 3.5 kV.
[e]Fertilized eggs are exposed to a mixture of medaka hatching enzyme and pancreatin for 2 hr. The dechorinated embryos are incubated with the pantropic pseudotyped retrovirus overnight at room temperature.
[f]Integration rate is calculated from the surviving embryos after gene transfer.

cy multiple peaks of square waves (i.e., square wave generator). The basic parameters are summarized in Table 1. Studies conducted in our laboratory (Lu et al., 1992; Powers et al., 1992) and those of others (Buono and Linser, 1992) have shown that the rate of DNA integration in electroporated embryos is on the order of 20% or higher in the survivors (Table 2). Although the overall rate of DNA integration in transgenic fish produced by electroporation was equal to or slightly higher than that of microinjection, the actual amount of time required for handling a large number of embryos by electroporation is orders of magnitude less than the time required for microinjection. Recently, several reports have also appeared in the literature describing successful transfer of transgenes into fish by electroporating sperm instead of embryos (Symonds et al., 1994; Tseng et al., 1994). Electroporation is therefore considered as an efficient and versatile massive gene transfer technology.

Transfer of Foreign Genes by Pantropic Retroviral Infection. Although transgenes can readily be introduced into the fish species studied to date by microinjection or electroporation at high efficiency, the resulting P_1 transgenic individuals are mosaics as a result of delayed transgene integration. Furthermore, these methods are not effective or not successful in transferring foreign DNA into embryos of some marine fish and invertebrates. Recently, a new gene transfer vector, a highly defective pantropic retroviral vector, has been developed (Burns et al., 1993). This vector contains the long terminal repeat (LTR) sequences of Moloney murine leukemia virus (MoMLV) and transgenes in a viral envelop with the G-protein of vesicular stomatitis virus (VSV). Since entry of VSV into cells is mediated by interaction of the VSV G-protein with a phospholipid component of the cell, this pseudotyped retroviral vector has a very broad host range and is able to transfer transgene into many different cell types. Using the pantropic pseudotyped defective retrovirus as a gene transfer vector, transgenes containing neo^R or β-galactosidase have been introduced into zebrafish (Lin et al., 1994) and medaka (Table 2; Lu et al., 1994). Recently, the feasibility of using a pantropic pseudotyped retroviral vector for introducing genes into marine invertebrates has been tested in dwarf surf clams and the results showed that transgenes can readily be transferred into clams at high efficiency (J. K. Lu, J. C. Burns, S. Allen, and T. T. Chen, *unpublished results*).

CHARACTERIZATION OF TRANSGENIC FISH

Identification of Transgenic Fish

The most time-consuming step in producing transgenic fish is the identification of transgenic individuals. Traditionally, dot blot and Southern blot hybridization of genomic DNA were common methods used to determine the presence of transgenes in the presumptive transgenic individuals. These methods involve isolation of genomic DNA from tissues of presumptive transgenic individuals, digestion of DNA samples with restriction enzymes, and Southern blot hybridization of the digested DNA products. Although these methods are expensive, laborious, and insensitive, they offer a definitive answer whether a transgene has been integrated into the host genome. Furthermore, they also reveal the pattern of transgene integration if appropriate restriction enzymes are employed in the Southern blot hybridization analysis. In order to handle a large number of animals efficiently and economically, a polymerase chain reaction (PCR) based assay has been adopted (Lu et al., 1992; Chen et al., 1993). The strategy of the assay is outlined in Figure 1. It involves isolation of genomic DNA from a very small piece of fin tissue, PCR amplification of the transgene sequence, and Southern blot analysis of the amplified products. Although this method does not differentiate whether the transgene is integrated in the host genome or remains as an extrachromosomal unit, it serves as a rapid and sensitive screening method for identifying individuals that contain the transgene at the time of analysis. In our laboratory, we use this method as a preliminary screen for transgenic individuals when screening thousands of the presumptive transgenic fish.

Expression of Transgenes

An important aspect of gene transfer studies is the detection of transgene expression. Depending on the levels of transgene products in the transgenic individuals, the following listed methods are commonly used to detect transgene expression: (1) RNA Northern or dot blot hybridization, (2) RNase protection assay, (3) reverse transcription/polymerase chain reaction (RT/PCR), (4) immunoblotting assay, and (5) other biochemical assays for determining the presence of the transgene protein

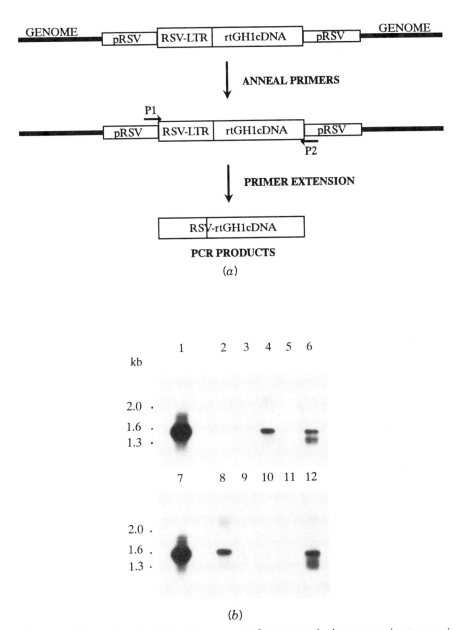

Figure 1. Strategy for identifying the presence of transgenes in the presumptive transgenic fish by PCR and Southern blot hybridization. DNA samples were isolated from pectoral fin tissues of presumptive transgenic fish and subjected to PCR amplification. The amplified products were analyzed by electrophoresis on agarose gels and Southern blot hybridization. (A) Strategy of PCR amplification, (B) Southern blot analysis of PCR-amplified products. Lanes 2–6 and 8–12, DNA samples from presumptive transgenic fish; lanes 1 and 7, transgene construct (RSVLTR-rtGH1 cDNA. (From Chen et al., 1993, with permission.)

products. Among these assays, RT/PCR is the most sensitive method and only requires a small amount of sample. The strategy of this assay is summarized in Figure 2. Briefly, it involves the isolation of total RNA from a small piece of tissue, synthesis of single-strand cDNA by reverse transcription, and PCR amplification of the transgene cDNA by employing a pair of oligonucleotide primers specific to the transgene product. The resulting products are resolved on agarose gels and analyzed by Southern blot hybridization using a radiolabeled transgene as a hybridization probe. Transgene expression can also be quantified by a quantitative RT/PCR method (Ballagi-Pordany et al., 1991).

Pattern of Transgene Integration

Studies conducted in many fish species have shown that following injection of linear or circular transgene constructs into fish embryos, the transgene is maintained as an extrachromosomal unit through many rounds of DNA replication in the early phase of the embryonic development. At later stages of embryonic development, some of the transgenes are randomly integrated into the host genome while others are degraded, resulting in the production of mosaic transgenic fish (for review see Hackett, 1993). In many fish species studied to date, multiple copies of transgenes were found to integrate in a head-to-head, head-to-tail, or tail-to-tail form, except in transgenic common carp and channel catfish, where single copies of transgenes were integrated at multiple sites on the host chromosomes (Zhang et al., 1990).

Inheritance of Transgenes

Stable integration of the transgenes is an absolute requirement for continuous vertical transmission to subsequent generations and establishment of a transgenic fish line. To determine whether the transgene is transmitted to the subsequent generation, P_1 transgenic individuals are mated to nontransgenic individuals and the progeny are assayed for the presence of transgenes by the PCR assay method described earlier (Lu et al., 1992; Chen et al., 1993). Although it has been shown that the transgene may persist into the F_1 generation of transgenic zebrafish as extrachromosomal DNA (Stuart et al., 1988), detailed analysis of the rate of transmission of the transgenes to the F_1 and F_2 generations in many transgenic fish species indicates true and stable incorporation of the constructs into the host genome (for review see Chen and Powers, 1990; Hackett, 1993). If the entire germline of the P_1 transgenic fish is transformed with at least one copy of the transgene per haploid genome, at least 50% of the F_1 transgenic progeny will be expected in a backcross involving a P_1 transgenic with a nontransgenic control. In many such crosses, only about 20% of the progeny are transgenic (Stuart et al., 1988, 1990; Zhang et al., 1990; Shears et al., 1991; Lu et al., 1992; Chen et al., 1993). When the F_1 transgenic is backcrossed with a nontransgenic control, however, at least 50% of the F_2 progeny are transgenics. These results clearly suggest that the germlines of the P_1 transgenic fish are mosaic as a result of delayed transgene integration during embryonic development.

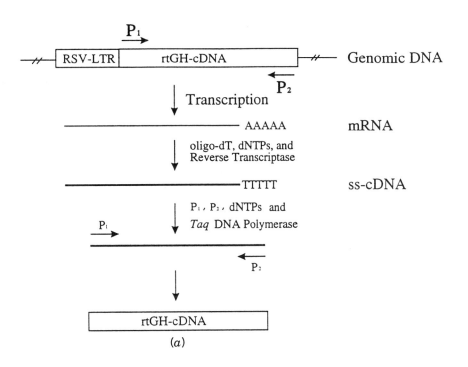

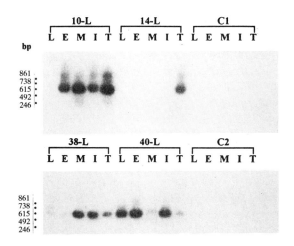

Figure 2. Strategy of detecting rtGH transgene expression by reverse transcription (RT)/PCR assay. (A) Strategy of RT/PCR. (B) Detection of rtGH transgene expression in transgenic carp by RT/PCR. Total RNA was isolated from liver, muscle, eye, gut, and testis of F_1 transgenic carp and controls following the acid guanidinium thiocyanate–phenol–chloroform method. Single-strand cDNA was prepared by reverse transcription from each total RNA and used as a template for PCR amplification of rtGH using synthetic oligonucleotides as amplification primers. the resulting products were analyzed by Southern blot analysis using radiolabeled rtGH cDNA as a hybridization probe. (From Chen et al., 1993, with permission.)

411

APPLICATION OF TRANSGENIC FISH IN BASIC RESEARCH

Transgenic fish, like transgenic mice, can serve as excellent experimental models for a wide variety of basic scientific investigations. These studies include (1) identifying the regulatory elements of a gene, (2) examining the molecular genetics of early vertebrate development, (3) studying the functions of a gene product, (4) identifying the biological actions of hormones, (5) developing models for biomedical research, and (6) establishing models for environmental toxicant analysis.

In higher vertebrates, growth is primarily modulated by the availability of growth hormone (GH) and insulin-like growth factors (IGFs) to their respective receptors. The secretion of GH from the pituitary gland and the binding of GH to its receptor signal the production of IGF I mRNA and the corresponding polypeptide by the liver (endocrine production) and other tissues (autocrine/paracrine function). Although the influence of GH on IGF induction and the molecular mechanism that underlies the GH-controlled IGF I gene expression have been under intensive investigation in higher vertebrates for many years, very little is known in lower vertebrates such as fish. Using rainbow trout as experimental animals, we are interested in studying the mechanism by which GH and IGF control growth in lower vertebrates.

Age-Dependent, Tissue-Specific, and Growth Hormone-Dependent Expression of Fish IGF Genes

As a step toward understanding the regulation of growth in fish by GH, we initiated work to identify the presence of IGF I and IGF II in rainbow trout by PCR and screening of a liver cDNA library. Two unique cDNA sequences have been identified. On the basis of a 98.7% nucleotide sequence homology to coho salmon IGF I, one cDNA sequence was identified as rainbow trout IGF I. The second cDNA sequence shared 43.3% identity with trout IGF I at the predicted amino acid level and 53.6% identity with human IGF II and was identified as trout IGF II (Shamblott and Chen, 1992). This was the first time that an IGF II was identified in a fish species.

As a result of differential splicing in the 5' untranslated region, signal peptide, E-domain, and 3' untranslated region, as well as transcription initiated from more than one promoter, multiple size forms of IGF I and II mRNA have been detected in mammals (Bell et al., 1985; Rotwein et al., 1988). To detect the presence of multiple size forms of IGF I and II mRNA in rainbow trout, an RT/PCR method was adopted (Shamblott and Chen, 1993). This assay employed two sets of primers each for IGF I and II so that small size differences of PCR products could be resolved on high concentration (e.g., 3%) agarose gels and the identity of each product could be confirmed by nucleotide sequence determination. The primer sets were designed to separately amplify the 5' region (predicted start codon to C-domain) or 3' region (C-domain to approximately 100 bp beyond the predicted stop codon) of both IGF I and II. While only one size form of IGF I and II mRNA resulted from RT/PCR with the 5' IGF I and both 5' and 3' IGF II primer sets, four size forms of IGF I mRNA resulted from the 3' IGF I primer set. Results of nucleotide sequence determination of the four size forms of IGF I mRNA showed that the size differences were due to insertions or deletions in the E-domain. These four forms of IGF I mRNA (Fig. 3), in increasing nucleotide length, are designated as IGF IEa-1, -2, -3, -4 in accordance with suggested revisions of IGF I nomenclature (Holthuizen et al., 1991; Duguay et al., 1992). The predicted amino acid residues of the E-domain are 35, 47, 62, and 74, respectively (Shamblott and Chen, 1993). The entire nucleotide sequences for IGF IEa-2 and Ea-3 mRNA have been determined from their respective intact cDNA clones. Duguay et al. (1992) recently detected three forms of IGF I mRNA for coho salmon by using an RT/PCR assay and these three mRNA forms are equivalent to rainbow trout IGF IEa-1, Ea-3, and Ea-4. By using the same approach, Wallis and Devlin (1993) also detected three size forms of IGF I mRNA for chinook salmon. These three size forms correspond to rainbow trout IGF IEa-1, Ea-2, and Ea-4. The reasons for the absence of rainbow trout IGF IEa-2 and IEa-3 in the livers of coho salmon and chinook salmon, respectively, are unknown. It is conceivable that the missing forms were not resolved and therefore not recognized after agarose gel electrophoresis. Alternatively, the IGF I mRNA form absent in these two reports may not have been present or detectable in these fish; in which case it is surprising that the two salmonid species lack different analogues of rainbow trout IGF I.

An RNase protection assay (RPA) was established to determine the mRNA levels of each of the four IGF I forms and IGF II in the liver, skeletal muscle, spleen, pyloric caecum (pancreatic tissue), heart, brain, and gill tissues of rapidly growing juvenile (7–8 months old) rainbow trout and sexually mature adults as well as in testes and ovaries from sexually mature adults (Shamblott and Chen, 1993). Probe templates for the RPA were constructed by cloning the 3' region (from the C-domain to approximately 100 bp into the 3' untranslated region) of each IGF I or

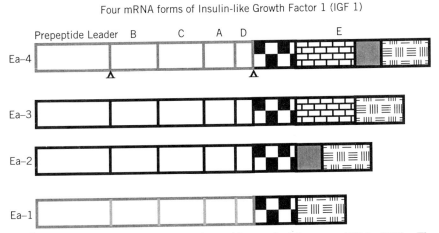

Figure 3. Schematic representation of the four subforms of rainbow trout IGF I mRNAs. The shaded line open box in Ea-2 and Ea-4 indicates that the nucleotide sequence of the molecule has not been confirmed from the cDNA clones yet. B, C, A, D, and E indicate different domains of the IGF prepropeptide. (From Shamblott and Chen, 1993, with permission.)

IGF II into the Bluescript plasmid vector in order to generate a radiolabeled antisense cRNA probe and unlabeled sense cRNA concentration standards, respectively, by *in vitro* transcription with T7 or T3 RNA polymerase. The level of 18S rRNA in each tissue was also determined in order to serve as an internal standard for normalization. Each of the four IGF I mRNA forms and the IGF II mRNA were readily distinguishable and determined by the RPA. The protected fragments for the four IGF I mRNA forms are 354 bp, 390 bp, 438 bp and for IGF II mRNA 496 bp. The results of the IGF mRNA RPA are summarized in Figure 4. At least one form of IGF I and one of IGF II mRNA are expressed in all the tissues examined in both developmental stages. Liver is the site of the greatest IGF mRNA abundance ($P <$ 0.01), and the levels of total IGF I and II mRNA are one to two orders of magnitude higher than in other tissues examined. Furthermore, it is interesting to note that the levels of total IGF I and II mRNA are twofold higher in the adult liver than the juvenile liver ($P < 0.01$). In mammals, IGF I mRNA has been detected primarily in the postnatal liver, kidney, spleen, pancreas, lung, and testis of the mouse (Mathews et al., 1986), the brain and several other regions of the central nervous system of rat (Rotwein et al., 1988), and the placenta and whole premenopausal ovary of human (Hernandez et al., 1992). In chicken, IGF I mRNA has been detected in the eye, skeletal muscle, and brain prior to hatching and the liver only after hatching (Kikuchi et al., 1991). IGF II mRNA has been detected in muscle, skin, lung, intestine, thymus, heart, kidney, brain, and spinal cord of fetal/neonatal rats and in the brain and spinal cord of adult rats. However, it is interesting to note that, except in the liver, levels of rtIGF II mRNA are much higher than those of the total rtIGF I mRNA in gill, kidney, heart, spleen, brain, muscle, pyloris, testis, and ovary. These results suggest that, in addition to IGF I, IGF II may play an important role in the fish growth as well as in maintenance of osmotic balance.

An *in vivo* study was conducted to determine the dependency of IGF mRNA accumulation upon GH treatment. In this study, yearling rainbow trout of about 150 g each maintained at 15°C were fasted for 5 days and each fish was injected with 10 μg/g fish body weight of bovine GH or carrier solution as control. Levels of IGF I and IGF II mRNA in different tissues were determined by an RNase protection assay at different periods post hormone treatment. Levels of liver IGF I mRNA significantly increased 6 hr post bovine GH treatment and remained significantly elevated at 12 hr, while liver IGF II mRNA levels were significantly elevated at 3 and 6 hr post hormone treatment (Shamblott et al., 1995). Both IGF I and IGF II mRNA levels responded with a three- to fourfold increase over mock injected controls. Although the levels of IGF I mRNA did not increase significantly in the pyloric caecum in response to bovine GH treatment, the levels of IGF II mRNA elevated at 12, 24, and 48 hr by about four-, two-, and fourfold, respectively. To determine whether the response of IGF mRNA induction by GH is dose-dependent, Shamblott et al. (1995) conducted further *in vitro* studies in a rainbow trout primary hepatocyte culture maintained in a serum-free medium supplemented with bovine GH. The results showed that both IGF I and IGF II mRNA levels responded to bovine GH

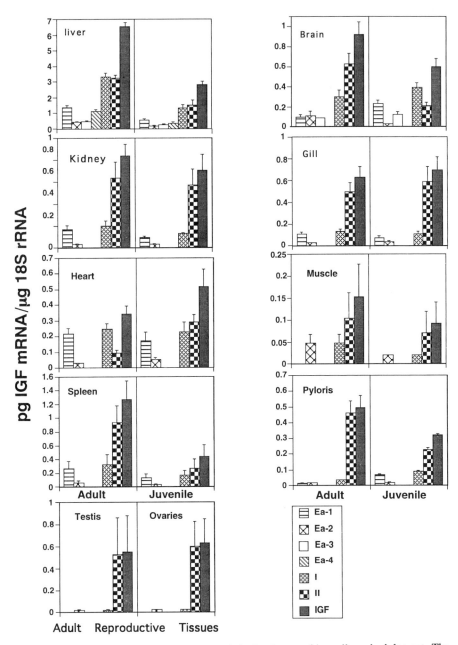

Figure 4. Levels of five forms of IGF mRNA in the tissues of juvenile and adult trout. The forms of IGF I mRNA are abbreviated as Ea-1, Ea-2, Ea-3, and Ea-4. Total IGF I, IGF II, and total IGF levels are abbreviated as I, II, and IGF, respectively. (From Chen et al., 1994, with permission.)

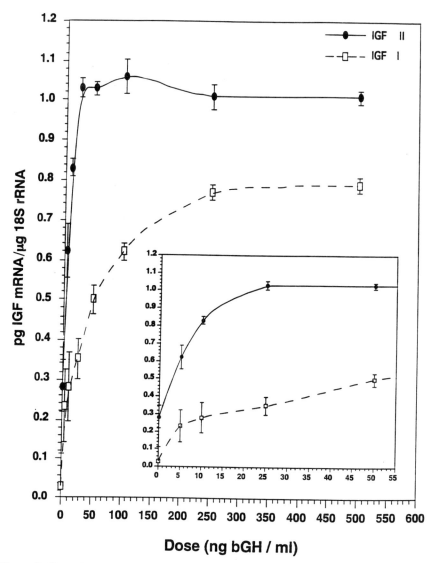

Figure 5. Dose-dependent accumulation of IGF I and II mRNA by *in vitro* cultured trout primary hepatocytes in response to bGH treatment. The levels of IGF mRNA were determined simultaneously in an RNase protection assay and normalized to levels of 18S rRNA. Mean levels (n + 4) and standard error of IGF II and I mRNA are indicated by a solid line and a dashed line, respectively. Graph inset uses identical symbols and axes to the larger figure. The 50% effective dosages for IGF I and II are approximately 45 ng/ml and 6 ng/ml, respectively. (From Shamblott et al., 1995, with permission.)

treatment in a dose-dependent fashion with ED_{50} values of about 45 and 6 ng/ml, respectively (Fig. 5). These results clearly showed that the synthesis of IGF I and IGF II mRNA in the liver of rainbow trout is under the modulation of GH.

IGF I Transgenic Fish

Although continuing investigation with the use of molecular biological approaches will shed light on the biological actions of IGFs, construction of transgenic fish producing elevated levels of these polypeptides by gene transfer technologies may generate alternative models for determining the effect of IGFs on fish growth and development. Toward this end, we have recently produced transgenic medaka by electroporating rtIGF I cDNA fused to a functional carp β-actin gene promoter (kindly donated by P. Hackett). Preliminary results have shown that more than 20% of the surviving embryos integrated the rtIGF I transgene in their respective genomes. Although the number of the P_1 transgenic individuals is still small, they are significantly larger than their nontransgenic controls ($P < 0.01$). Furthermore, it is interesting to note that both P_1 and F_1 IGF I transgenic fish hatched, on the average, 2 days earlier than their nontransgenic controls (Fig. 6). Unlike those studies conducted in mouse and rat, our results in medaka suggest that, in addition to regulating postembryonic somatic growth, IGF I may play an important role in embryonic development.

As shown in Figure 7A, a good number of the P_1 and F_1 transgenic female

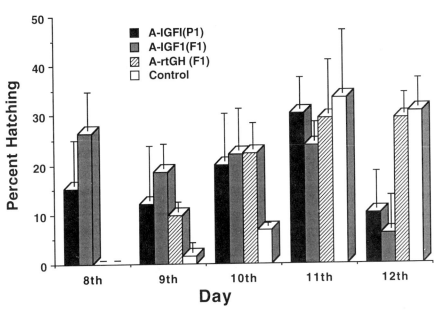

Figure 6. Effect of IGF I transgene on medaka embryonic development. Embryos were incubated at 25°C and the hatched fry were collected each day for observation.

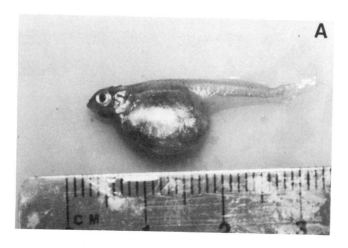

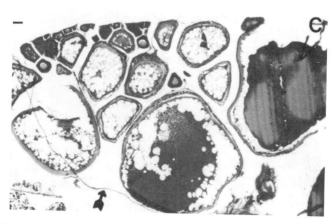

Figure 7. Histological examination of ovary from IGF I transgenic medaka. (A) Female IGF I transgenic medaka. (B) Female IGF I transgenic medaka showing an enlarged ovary. (C) Histological section of a nontransgenic ovary. (D–F) Histological sections of ovaries from IGF I transgenic medaka. (G) Histological section of ovary from GH transgenic medaka. Arrow indicates tunica albuginea. D, degenerating follicles; N, necrotic follicles; C, space in the ovary filled with a large amount of fluid. Solid bar = 100 μm.

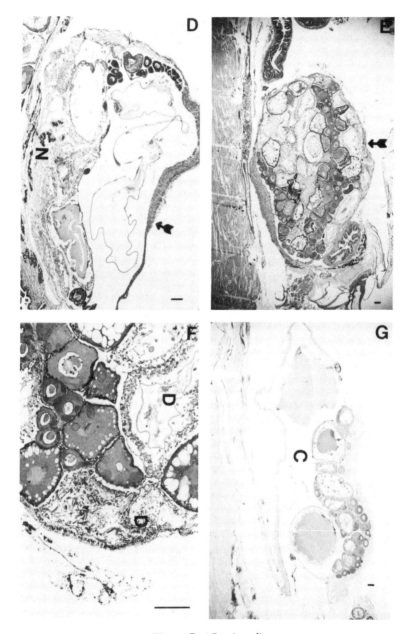

Figure 7. (*Continued*)

medaka exhibit unusual enlarged abdomens and these animals failed to spawn even after 6 months. Necropsy of these animals and histological examination have shown that the ovaries are filled with a gelatinous fluid and no mature eggs were observed (Fig. 7B).

A normal medaka ovary is surrounded by a thin tunica albuginea that is covered by mesothelium (Fig. 7C). Primary oocytes appear small and intensely basophilic and contain a prominent nucleus. As oocytes mature, they become larger and more eosinophilic and contain yolk granules. In the IGF I transgenic fish, several lesions in the ovary are noted (Fig. 7D–F). The tunica albuginea is thickened and numerous degenerating follicles are present in these ovaries. In some specimens, an inflammatory infiltrate of primarily macrophages surrounds necrotic follicles and extends into the tunica albuginea. It is interesting to note that similar ovary abnormality has also been observed in some of the GH transgenic female medaka (Fig. 7G). These histological findings are compatible with disruptions in normal ovarian function that may have been caused by overexpression of GH or IGF genes in the transgenic individuals.

One IGF I P_1 transgenic female further developed an adenocarcinoma on the maxilla. This female was initially tumorless and spawned several batches of eggs. When the tumor appeared, it ceased spawning. Histological studies have shown that the lesion was comprised of epithelial cells forming small cysts (Fig. 8). These cysts contained a pale eosinophilic material and some necrotic cells. The lesion was quite aggressive and replaced facial cartilage and bone, although it had not invaded the eye.

Although the above described observations are still preliminary and more detailed studies are required, it is clear that IGF transgenic fish can serve as models for studying the involvement of IGFs on (1) normal growth and development, (2) reproduction, and (3) tumor development.

APPLICATION OF TRANSGENIC FISH IN BIOTECHNOLOGY

The initial drive for transgenic fish research came from attempts to increase production of economically important fish for human consumption. The worldwide harvest of fishery products traditionally depends on natural populations of finfish, shellfish, and crustaceans in fresh and marine water. In recent years, however, the level of total worldwide annual harvest of fish products has approached or even surpassed the maximal potential level of about 150 million metric tons (as calculated by the U.S. Department of Commerce and the U.S. National Oceanic and Atmospheric Administration). In order to cope with the worldwide demand for fish products and the escalating increase in price, many countries have turned to aquaculture for increasing production of fish products. In 1985, the world production of finfish, shellfish, and macroalgae by aquaculture/mariculture reached 10.6 million metric tens, or approximately 12.3% of the worldwide catch generated by the international fishery efforts. Although aquaculture clearly has the potential for increasing world-

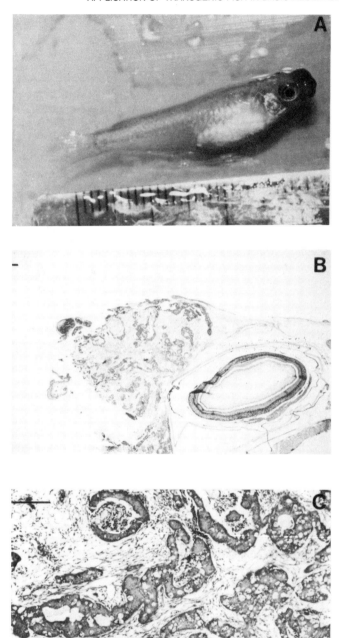

Figure 8. Histological examination of adenocarcinoma in IGF I transgenic medaka. (A) Gross morphology medaka bearing an adenocarcinoma. (B, C) Histological sections of adenocarnoma. Solid bar = 100 μm.

wide fish production, innovative strategies are needed to improve efficiency. What can transgenic technology offer?

Success in aquaculture depends on six factors: (1) complete control of the reproductive cycle of the fish species in culture; (2) excellent genetic background of the broodstock; (3) efficient prevention and detection of disease infection; (4) thorough understanding of the optimal physiological, environmental, and nutritional conditions for growth and development; (5) sufficient supply of excellent quality water; and (6) application of innovative management techniques. By improving these factors, the aquaculture industry has developed to a remarkable extent during the last decade. To sustain this growth, however, newly developed technologies in molecular biology and transgenesis will have to be increasingly applied by the aquaculture industry. These technologies can be employed to enhance growth rates, control reproductive cycles, improve feed compositions, produce new vaccines, and develop disease-resistant and hardier genetic stocks. In the last several years we have been searching for strategies to increase fish production by manipulating fish growth hormone and growth factor genes. The feasibility of this approach is evaluated below.

Biosynthetic Growth Hormone and Growth Enhancement

In recent years, cDNAs and their genomic DNA of growth hormone (GH) have been isolated and characterized for several fish species (for review see Chen et al., 1994). Expression of rainbow trout or striped bass GH cDNA in *E. coli* cells results in production of a large quantity of recombinant GH polypeptide (Agellon et al., 1988; Cheng et al., 1995). Since the GH polypeptide is highly hydrophobic and contains four cysteine residues, the newly synthesized recombinant GH polypeptide forms insoluble inclusion bodies in *E. coli* cells, rendering the hormone inactive. In an attempt to regain the biological activity of the recombinant hormone, Cheng et al. (1995) developed a procedure for renaturing the protein. It involves dissolving the insoluble recombinant hormone in a buffer containing 8 M urea and renaturing the polypeptide by slowly removing the urea from the protein solution. The biological activity of the renatured protein was then assessed by an *in vitro* sulfation assay (Fig. 9; Cheng and Chen, 1995).

In a series of *in vivo* studies, Agellon et al. (1988) showed that application of the recombinant trout GH to yearling rainbow trout resulted in a significant growth enhancement. After treatment of yearling rainbow trout with the recombinant GH for 4 weeks at a dose of 1 µg/g body weight/week, the weight gain among the individuals of the hormone-treated group was two times greater than that of the controls (Fig. 10). Significant length gain was also evident in hormone-treated animals. When the same recombinant hormone was administered to rainbow trout fry (Table 3) or small juveniles by immersing the fish in a GH-containing solution, the same growth-promoting effect was also observed (Agellon et al., 1988; J. Leong and T. T. Chen, *unpublished results*). These results are in agreement with those reported by Sekine et al. (1985), Gill et al. (1985), and many others (Sato et al.,

TABLE 3. Effect of GH Treatment on the Growth of Rainbow Trout Fry

Treatment	Weight[a] (g)		
	Initial	Final	% Gain
Saline control	1.33 ± 0.6[b]	3.94 ± 1.8[c]	196
GH (50 μ/liter)	1.29 ± 0.7[b]	5.51 ± 1.6[d]	327
GH (500 μ/liter)	1.35 ± 0.7[b]	5.30 ± 1.3[d]	293

[a]Values presented as mean ± SD. Groups of rainbow trout fry ($n = 15$) were subjected to osmotic shock in the presence or absence of GH. Weight was measured prior to and 5 weeks post-treatment. Differences between mean weights of GH-treated and control groups were evaluated using Student's t-test ($\alpha = 0.01$).

[b]No significant difference between these groups.

[c]Significantly different from the GH-treated groups ($P < 0.01$).

[d]No significant difference between these two treatments.

Source: Agellon et al. (1988), with permission.

1988; Moriyama et al., 1990). However, it is important to mention that the growth enhancement effect of the biosynthetic hormone was markedly reduced when a dose of more than 2 μg/g body weight of the hormone was applied to the test animals (Agellon et al., 1988). Recently, Paynter and Chen (1991) have observed that administration of recombinant trout GH polypeptide to spats of juvenile oysters

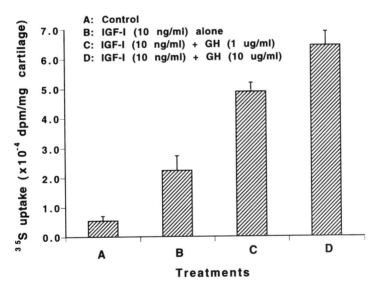

Figure 9. Effect of recombinant striped bass GH on uptake of [35S]sulfate into common carp gill cartilage maintained in culture.

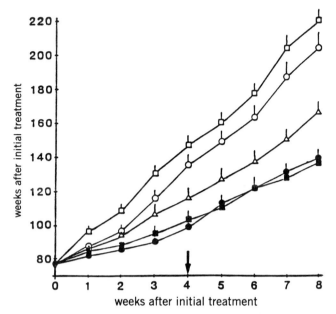

Figure 10. Effect of recombinant trout GH on growth of yearling rainbow trout. Groups of yearling rainbow trout received intraperitoneal injection of recombinant GH or control extract for 5 weeks. Wet weights of GH-treated and control fish are shown (mean ± SE). Open symbols, GH-treated fish: -○-, 0.2 μg/g body weight; -□-, 1.0 μg/g body weight; -△-, 2 μg/g body weight. Closed symbols, control fish: -●-, mock-treated fish; -■-, untreated fish. The arrow indicates the time of the last hormone treatment. (From Agellon et al., 1988, with permission.)

(*Crassostrea virginica*) by the "dipping method" referred to above also resulted in significant increases in shell height, shell weight, wet weight, and dry weight (Table 4). Furthermore, they also showed that oysters treated with recombinant trout GH, native bovine GH, or bovine insulin consumed more oxygen per unit time than controls. The results summarized above clearly suggest that exogenous application of recombinant fish growth hormone can enhance the somatic growth of finfish and shellfish.

GH Transgenic Fish

Although exogenous application of biosynthetic GH results in a significant growth enhancement in fish, it may not be cost effective because of the following reasons: (1) producing purified biosynthetic GH is costly; (2) treating individual fish with the hormone is labor intensive; (3) the optimal hormone dosage for each fish species is difficult to identify; and (4) GH uptake into fish from an exogenous source is inefficient. If new strains of fish producing elevated but optimal levels of GH can be produced, it would bypass all the problems associated with exogenous GH treat-

TABLE 4. Effect of Exogenously Applied Recombinant Rainbow Trout Growth Hormone on Oyster Growth

Treatment	Initial Height	Final Height	Total Weight	Shell Weight	Dry Weight
Control	8.14 (0.25)	11.68 (0.27)	206 (11)	136 (8)	6.10 (0.66)
10^{-9} M	8.04 (0.27)	11.74 (0.23)	199 (9)	131 (6)	6.87 (0.66)
10^{-8} M	8.72 (0.18)	12.79 (0.27)[b,c]	244 (20)	171 (11)[c]	9.42 (0.41)[b,c]
10^{-7} M	8.65 (0.32)	13.00 (0.36)[b,c]	252 (13)[c]	189 (13)[b,c]	9.41 (0.74)[b,c]

[a]Initial height represents mean size at the beginning of the experiment and final height, total weight, shell weight, and dry weight are mean values determined after the 5-week treatment cycle was concluded. Height was measured in millimeters (mm) from the umbo to the ventral shell margin; weight was measured in milligrams (mg). Standard errors of the mean (SEM) are in parentheses.
[b]Significantly larger than the control group (t-test; $P < 0.05$).
[c]Significantly larger than 10^{-9} M treatment group (t-test; $P < 0.05$).

ment. Moreover, once these fish strains have been generated, they would be far more cost effective than their ordinary counterparts because these fish would have their own means of producing and delivering the hormone and they could transmit their enhanced growth characteristics to their offspring.

Three aspects of fish growth characteristics that could be improved for aquaculture are (1) initial growth rate so that they reach maturation earlier, (2) enhanced somatic growth rate as adults to provide larger body size for market, and (3) fish with improved feed conversion efficiencies. Among these three, enhanced somatic growth rates via manipulation of the GH gene show considerable promise. Zhu et al. (1985) reported the first successful transfer of a human GH gene fused to a mouse metallothionein (MT) gene promoter into goldfish and loach. According to Zhu (*personal communication*), the F_1 offspring of these transgenic fish grew twice as large as their nontransgenic siblings. Unfortunately, Zhu and his colleagues failed to present compelling evidence for integration and expression of the foreign genes in their transgenic fish studies. Recently, many laboratories throughout the world have successfully confirmed Zhu's work by demonstrating that human or fish GH and many other genes can readily be transferred into the embryos of a number of fish species and integrated into the genome of the host fish. While a few groups have demonstrated expression of foreign genes in transgenic fish, only Zhang et al. (1990), Du et al. (1992), and Lu et al. (1992) have documented that a foreign GH gene could be (1) transferred to the target fish species, (2) integrated into the fish genome, and (3) genetically transmitted to the subsequent generations. Furthermore, the expression of the foreign GH gene may result in enhancement of growth rates of both P_1 and F_1 generations of transgenic fish (Zhang et al., 1990; Du et al., 1992; Lu et al., 1992).

In gene transfer studies conducted in common carp and channel catfish (Chen et al., 1990; Zhang et al., 1990; Dunham et al., 1992; Chen et al., 1993), about 10^6 molecules of a linearized recombinant plasmid containing the long terminal repeat (LTR) sequence of avian Rous sarcoma virus (RSV) and the rainbow trout GH cDNA were injected into the cytoplasm of one-cell, two-cell, and four-cell embryos. Genomic DNA samples extracted from the pectoral fins of presumptive transgenic fish were analyzed for the presence of RSVLTR-rtGH1-cDNA by PCR amplification and Southern blot hybridization of the amplified DNA products using radiolabeled LTR of RSV and/or trout GH1-cDNA as hybridization probes. In the case of transgenic carp studies (Zhang et al., 1990; Chen et al., 1993), about 35% of the injected embryos survived at hatching, of which about 10% of the survivors had stably integrated the RSVLTR-rtGH1-cDNA sequence. A similar percentage of transgenic fish was also obtained when RSVLTR-csGH-cDNA construct was injected into catfish embryos (Powers et al., 1991; Dunham et al., 1992). Southern blot analysis of genomic DNA samples of several transgenic carp revealed that a single copy of the RSVLTR-rtGH1-cDNA sequence was integrated at multiple chromosomal sites (Zhang et al., 1990).

The patterns of inheritance of RSVLTR-rtGH1 cDNA in the transgenic common carp were studied by fertilizing eggs collected from nontransgenic females or P_1 transgenic females with sperm samples collected from several sexually mature P_1 male transgenic fish. DNA samples extracted from the resulting F_1 progeny were assayed for the presence of RSVLTR-rtGH1-cDNA sequence by PCR amplification and dot blot hybridization (Chen et al., 1993). The percentage of the transgenic progeny resulting from nine matings were 0%, 32%, 26%, 100% (four progeny only), 25%, 17%, 31%, 30%, and 23%, respectively. If each of the transgenic parents in these nine matings carries at least one copy of the transgene in the gonad cell, about 50–75% transgenic progeny would have been expected in each pairing. Out of these nine matings, two siblots, both control × P_1, gave transgenic progeny numbers as large as or larger than expected ($P < 0.05$) and the remaining had lower than expected numbers of transgenic progeny. These results indicate that though most of these P_1 transgenic fish had RSVLTR-rtGH1 cDNA in their germline, they might be mosaics. Similar patterns of mosaicism in the germline of P_1 transgenic fish have been observed in many fish species studied to date (Ozato et al., 1986; Dunham et al., 1987, 1992; Stuart et al., 1988; Zhang et al., 1990; Lu et al., 1992).

If the transgene carries a functional promoter, some of the transgenic individuals are expected to express the transgene activity. According to Zhang et al. (1990) and Chen et al. (1993), many of the P_1 and F_1 transgenic common carp produced rtGH and the levels of rtGH produced by the transgenic individuals varied about tenfold. Chen et al. (1993) recently confirmed these results by detecting the presence of rtGH mRNA in the F_1 transgenic carp using an assay involving reverse transcription (RT)/PCR amplification. They have found that different levels of rtGH mRNA were detected in liver, eye, gonad, intestine, and muscle of the F_1 transgenic individuals.

Since the site of transgene integration differs among the individuals in any population of P_1 transgenic fish, they should be considered as totally different transgenic individuals and thus inappropriate for direct comparison of the growth performance among these animals. Instead, the growth performance studies should be conducted in F_1 transgenic and nontransgenic siblings derived from the same family. Recently, Chen et al. (1993) conducted studies to evaluate the growth performance of F_1 transgenic carp in seven families. In these experiments, transgenic and nontransgenic full-siblings were spawned, hatched, and reared communally under the same environment. Results of these studies showed that growth response by families of F_1 transgenic individuals in response to the presence of rtGH1 cDNA varied widely. When compared to the nontransgenic full-siblings, the results of four out of seven growth trials showed 20%, 40%, 59%, and 22% increase in growth, respectively (Table 5). In three of the four families where F_1 transgenics grew faster than their nontransgenic full-siblings, the maximum and minimum body weights of the transgenics were larger than those of the nontransgenics. In the fourth family, the minimum, but not the maximum, body weight of the transgenics was larger than that of the nontransgenics. In two of those three transgenic families in which transgenics did not grow faster than their nontransgenic full-siblings, the maximum and minimum body weights of the transgenics were smaller than those of the nontransgenics. In the third family, however, one of the F_1 transgenics was the largest fish in the family. The same extent of growth enhancement was also observed in F_2 offspring derived from crossing F_1 transgenics with nontransgenic controls (Fig. 11). Since the response of the transgenic fish to the insertion of the RSVLTR-rtGH1 cDNA appears to be variable, as a result of random integration of the transgene, the fastest growing genotype will likely be developed by utilizing a combination of family selection and mass selection of transgenic individuals following the insertion of the foreign gene. More dramatic growth enhancement in transgenic fish was obtained by introducing chinook salmon GH cDNA driven by the promoter of ocean pout antifreeze protein gene into Atlantic salmon embryos (Du et al., 1992). Some of these transgenic animals grew several times faster than their controls. In the studies of transgenic medaka carrying chicken β-actin gene promoter/human GH gene construct, the F_1 transgenic individuals also grew significantly faster than the nontransgenic siblings (Lu et al., 1992).

Manipulation of GH gene is just one of many examples of improving the genetic traits of fish for aquaculture. Other important traits such as increased tolerance to lower oxygen concentration, increased resistance to bacterial, fungal, viral, or parasitic infection, improved food conversion efficiency, and increased tolerance to low or high temperature may also be altered by transgenic fish technology provided that the genes responsible for each of these traits are determined.

Other Biotechnological Applications

Another important application of the transgenic fish technology will be the generation of novel animals for producing high economical value pharmaceuticals. Although no real example is available now, like transgenic cow or sheep, transgenic fish may be used as bioreactors for large-scale production of proteins such as human hemoglobin (Swanson et al., 1992), human tissue plasminogen (Pittius et al., 1988), human antihemophilic Factor IX (Clark et al., 1989), and human α-1-antitrypsin (Wright et al., 1991).

TABLE 5. Mean, Standard Deviation, Coefficient of Variation, and Percent Difference in Body Weight of Transgenic Common Carp, *Cyprinus carpio*, and Their Nontransgenic Full-Siblings

Family	Mating	Genotype	N	Mean Body Weight (SD)	Coefficient of Variation	% Difference	Range in Body Weight (g)
1	P_1 × control	T	31	120.6 (17.4)	14.4	20.8	95–173
		NT	65	99.3 (14.7)	14.8		65–129
2	P_1 × control	T	11	206.0 (45.2)	21.9	40.1	115–283
		NT	15	147.0 (48)	32.6		67–228
3	P_1 × P_1	T	7	5.8 (3.4)	58.6	−26.6	1.8–11.3
		NT	21	7.9 (3.1)	39.2		3.3–17.9
4	P_1 × P_1	T	28	66.1 (36.9)	55.8	58.5	18.5–338
		NT	65	41.7 (27.8)	66.6		8.3–141
5	P_1 × P_1	T	17	14.7 (6.8)	46.3	21.5	6.5–30.4
		NT	82	12.1 (8.4)	69.4		3.9–56.1
6	P_1 × P_1	T	97	114.2 (81.6)	71.5	−14.5	18.3–565.1
		NT	215	133.6 (83.6)	62.5		20.9–416.2
7	P_1 × P_1	T	15	72.2 (58.0)	80.3	−1.5	7.1–214.4
		NT	48	73.3 (47.6)	64.5		8.7–203.3

T, transgenic; NT, nontransgenic; *N*, number of fish; SD, standard deviation.

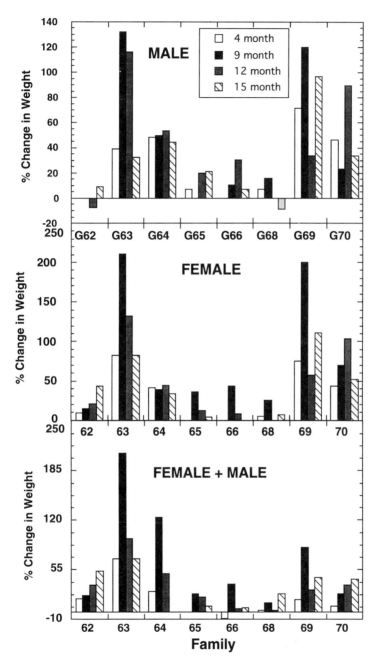

Figure 11. Body weight changes in F_2 transgenic and nontransgenic common carp carrying RSVLTR-rtGH transgene. Each family of fish is derived from a cross of F_1 transgenic and nontransgenic fish. All eight F_1 transgenics are derived from the same P_1 transgenic individual. Percent weight change in each F_2 individual is determined by comparing to the nontransgenic siblings.

ACKNOWLEDGMENT

The authors would like to thank Dr. Nick Vrolijk for critically reading the manuscript. This work was supported by grants from NSF (DCB-91-05719, IBN-93-17132) to T.T.C. and USDA (93-37205-9073) and BARD (US-2305-93RC) to T.T.C. and R.A.D.

REFERENCES

Agellon LB, Emery CJ, Jones JM, Davies SL, Dingle AD, Chen TT (1988): Growth hormone enhancement by genetically engineered rainbow trout growth hormone. *Can J Fish Aqua Sci* 45:146–151.

Ballagi-Pordany A, Ballagi-Pordany A, Funa K (1991): Quantitative determination of mRNA phenotypes by polymerase chain reaction. *Anal Biochem* 196:88–94.

Bell GI, Gerhardt DS, Fong NM, Sanchez-Pescador R, Rall LB (1985): Isolation of human insulin-like growth factor genes: insulin-like growth factor II and insulin genes are contiguous. *Proc Natl Acad Sci USA* 82:6450–6455.

Buono RJ, Linser PJ (1992): Transient expression of RSVCAT in transgenic zebrafish made by electroporation. *Mol Mar Biol Biotechnol* 1:271–275.

Burns JC, Friedmann T, Driever W, Burrascano M, Yee JK (1993): VSV-G pseudotyped retroviral vector: concentration to very high titer and efficient gene transfer into mammalian and nonmammalian cells. *Proc Natl Acad Sci USA* 90:8033–8037.

Chen TT, Powers DA (1990): Transgenic fish. *Trends Biotechnol* 8:209–215.

Chen TT, Lin C-M, Zhu Z, Gonzalez-Villasenor LI, Dunham RA, Powers DA (1990): Gene transfer, expression and inheritance of rainbow trout growth hormone genes in carp and loach. In: Church R (ed). *Transgenic models in medicine and agriculture.* New York: Wiley-Liss, pp 127–139.

Chen TT, Powers DA, Lin CM, Kight K, Hayat M, Chatakondi N, Ramboux AC, Duncan PL, Dunham RA (1993): Expression and inheritance of RSVLTR-rtGH1 cDNA in common carp, *Cyprinus carpio. Mol Mar Biol Biotechnol* 2:88–95.

Chen TT, Marsh A, Shamblott MJ, Chan KM, Tang YL, Cheng CM, Yang BY (1994): Structure and evolution of fish growth hormones and insulin-like growth factor genes. In: Hew CL, Sherwood N (eds). *Fish physiology.* Orlando, FL: Academic Press, pp 179–209.

Cheng CM, Chen TT (1995): Synergism of growth hormone (GH) and insulin-like growth factor-I (IGF-I) in stimulation of sulphate uptake by teleostean branchial cartilage *in vitro. J Endocrinol* 147:67–73.

Cheng CM, Lin CM, Shamblott M, Gonzalez-Villasenor LI, Powers DA, Woods C, Chen TT (1995): Production of a biologically active recombinant teleostean growth hormone in *E. coli* cells. *Mol Cell Endocrinol* 108:75–85.

Clark AJ, Bessos H, Bidhop JO, Harris S, Lathe R, McClenaghan M, Prowse C, Simons JP, Whitelaw CBA, Wilmut X (1989): Expression of human antihemophilic factor IX in the milk of transgenic sheep. *Biotechnology* 7:487–492.

Cotten M, Jennings P (1989): Ribozyme mediated destruction of RNA in vivo. *EMBO J* 8:3861–3866.

Du SJ, Gong GL, Fletcher GL, Shears MA, King MJ, Idler DR, Hew CL (1992): Growth enhancement in transgenic Atlantic salmon by the use of an "all fish" chimeric growth hormone gene construct. *Biotechnology* 10:176–181.

Duguay SJ, Park LK, Samadpour M, Dickhoff WW (1992): Nucleotide sequence and tissue distribution of three insulin-like growth factor I prohormones in salmon. *Mol Endocrinol* 6:1202–1210.

Dunham RA, Eash J, Askins J, Townes TM (1987): Transfer of the metallothionein-human growth hormone fusion gene into channel catfish. *Trans Am Fish Soc* 116:87–91.

Dunham RA, Ramboux AC, Duncan PL, Hayat M, Chen TT, Lin C-M, Kight K, Gonzalez-Villasenor LI, Powers DA (1992): Transfer, expression and inheritance of salmonid growth hormone genes in channel catfish, *Ictalarus punctatus,* and effects on performance traits. *Mol Mar Biol Biotechnol* 1:380–389.

Etkin LD, Pearman B (1987): Distribution, expression and germ line transmission of exogenous DNA sequences following microinjection into *Xenopus laevis* eggs. *Development* 99:15–23.

Fletcher GL, Davis PL (1991): Transgenic fish for aquaculture. In: Setlow JK (ed). *Genetic engineering,* vol 13. New York: Plenum Press, pp 331–370.

Gasser CS, Fraley RT (1989): Genetically engineering plants for crop improvement. *Science* 244:1293–1299.

Gill JA, Stumper JP, Donaldson EM, Dye HM (1985): Recombinant chicken and bovine growth hormone in cultured juvenile Pacific salmon. *Biotechnology* 3:4306–4310.

Gordon JW (1989): Transgenic animals. *Int Rev Cytol* 155:171–229.

Hackett PB (1993): The molecular biology of transgenic fish. In: Hochachka P, Mommsen T (eds). *Biochemistry and molecular biology of fish,* Vol 2. Amsterdam: Elsevier Science Publishers, pp 207–240.

Hernandez ER, Hurwitz A, Vera A, Pellicer A, Adashi EY, Leroith D, Roberts CT Jr (1992): Expression of the genes encoding the insulin-like growth factors and their receptors in the human ovary. *J Clin Endocrinol Metab* 74:419–425.

Holthuizen E, LeRoith D, Lund PK, Roberts C, Rotwein P, Spencer EM, Sussenbach JS (1991): Revised nomenclature for the insulin-like growth factor genes and transcripts. In: Spencer EM (ed). *Modern concepts of insulin-like growth factors.* New York: Elsevier, pp 733–736.

Jaenisch R (1990): Transgenic animals. *Science* 240:1468–1477.

Kikuchi K, Buonomo FC, Kajimoto Y, Rotwein P (1991): Expression of insulin-like growth factor I during chicken development. *Endocrinology* 128:1323–1328.

Lin S, Gaiano N, Culp P, Burns JC, Friedmann T, Yee J-K, Hopkins N (1994): Integration and germ-line transmission of a pseudotyped retroviral vector in zebrafish. *Science* 265:666–668.

Lu JK, Chrisman CL, Andrisani OM, Dixon JE, Chen TT (1992): Integration expression and germ-line transmission of foreign growth hormone genes in medaka, *Oryzias latipes. Mol Mar Biol Biotechnol* 1:366–375.

Lu JK, Burns JC, Chen TT (1994): Retrovirus-mediated transfer and expression of transgenes in medaka. In: Proceedings of the Third International Marine Biotechnology Conference at Tromoso, Norway, p 72.

McMahan AP, Novak JJ, Britten RJ, Davidson EH (1984): Inducible expression of a cloned heat shock fusion gene in sea urchin embryos. *Proc Natl Acad Sci USA* 81:7490–7494.

McMahan AP, Flytzanis CN, Hough-Evans BR, Wakamatsu Y, Okasa TS (1986): Production of cloned DNA in sea urchin egg cytoplasm: replication and persistence during embryogenesis. *Dev Biol* 108:420–430.

Mathews LS, Norsted G, Palmiter RD (1986): Regulation of insulin-like growth factor I gene expression by growth hormone. *Proc Natl Acad Sci USA* 83:9343–9347.

Moav B, Liu Z, Groll Y, Hackett PR (1992): Selection of promoters for gene transfer into fish. *Mol Mar Biol Biotechnol* 1:338–345.

Moriyama S, Takahashi A, Hirano T, Kawauchi H (1990): Salmon growth hormone is transported into the circulation of rainbow trout (*Oncorhynchus mykiss*) after intestinal administration. *J Comp Physiol B* 160:251–260.

Ozato K, Kondoh H, Inohara H, Iwamatsu T, Wakamatsu Y, Okada TS (1986): Production of transgenic fish: Introduction and expression of chicken δ-crystallin gene in medaka embryo. *Cell Differ* 19:237–244.

Paynter K, Chen TT (1991): Biological activity of biosynthetic rainbow trout growth hormone in the eastern oyster (*Crassostrea virginica*). *Biol Bull* 181:459–462.

Pittius CW, Hennighausen L, Lee E, Westphal H, Nicols E, Vitale J, Gordon K (1988): A milk protein gene promoter directs the expression of human tissue plasminogen activator cDNA to the mammary gland in transgenic mice. *Proc Natl Acad Sci USA* 85:5874–5878.

Powers DA, Gonzalez-Villasenor LI, Zhang P, Chen TT, Dunham RA (1991): Studies on transgenic fish gene transfer, expression and inheritance. In: First ML, Haseltine FP (eds). *Transgenic animals* (Biotechnology Series No 16). Boston: Butterworth-Heinmann, pp 307–324.

Powers DA, Hereford L, Cole T, Creech K, Chen TT, Lin CM, Kight K, Dunham RA (1992): Electroporation: a method for transferring genes into gametes of zebrafish (*Brachydanio rerio*), channel catfish (*Ictalurus punctatus*), and common carp (*Cyrinus carpio*). *Mol Mar Biol Biotechnol* 1:301–308.

Pursel VG, Pinkert CA, Miller KA, Bolt DA, Campbell RG, Palmiter RD, Brinster RL, Hammer RE (1989): Genetic engineering of livestock. *Science* 244:1281–1288.

Rotwein P, Burgess SK, Milbrandt JD, Krause JE (1988): Differential expression of insulin-like growth factor genes in rat central nervous system. *Proc Natl Acad Sci USA* 85: 265–269.

Rubin GM, Spradling AC (1982): Genetic transformation of *Drosophila* with transposable element vectors. *Science* 1218:348–353.

Sato N, Murata K, Watanabe K, Hayami T, Kuriya Y, Sakaguchi M, Kimura S, Nonak M, Kimura A (1988): Growth-promoting activity of tuna growth hormone and expression of tuna growth hormone cDNA in *Escherichia coli*. *Biotechnol Appl Biochem* 10:385–392.

Schulte PM, Down NE, Donaldson EM, Souza LM (1989): Experimental administration of recombinant bovine growth hormone to juvenile rainbow trout (*Salmo gairdneri*) by injection or immersion. *Aquaculture* 76:145–152.

Sekine S, Miizukzmi T, Nishi T, Kuwana Y, Saito A, Sato M, Itoh H, Kawauchi H (1985): Cloning and expression of cDNA for salmon growth hormone in *E. coli*. *Proc Natl Acad Sci USA* 82:4306–4310.

Shamblott M, Chen TT (1992): Identification of a second insulin-like growth factor in a fish species. *Proc Natl Acad Sci USA* 89:8913–8917.

Shamblott MJ, Chen TT (1993): Age-related and tissue-specific levels of five forms of insulin-like growth factor mRNA in a teleost. *Mol Mar Biol Biotechnol* 2:351–361.

Shamblott MJ, Cheng C, Bolt D, Chen TT (1995): Insulin-like growth factor (IGF) mRNA is accumulated in the liver and pyloric caeca of a teleost in a growth hormone-dependent manner. *Proc Natl Acad Sci USA* (*in press*).

Shears MA, Fletcher GL, Hew CL, Gauthier S, Davies PL (1991): Transfer, expression, and stable inheritance of antifreeze protein genes in Atlantic salmon (*Salmo salar*). *Mol Mar Biol Biotechnol* 1:58–63.

Stinchcomb DT, Shaw JE, Carr SH, Hirsh D (1985): Extrachromosal DNA transformation of *Caenorhabditis elegans*. *Mol Cell Biol* 5:3484–3496.

Stuart GW, McMurry JV, Westerfield M (1988): Replication, integration, and stable germ-line transmission of foreign sequence injected into early zebrafish embryos. *Development* 109:403–412.

Stuart GW, Vielkind JV, McMurray JV, Westerfield M (1990): Stable lines of transgenic zebrafish exhibit reproduction patterns of transgene expression. *Development* 109: 293–296.

Swanson ME, Martin MJ, O'Donnell JK, Hoover K, Lago W, Huntress V, Parsons CT, Pinkert CA, Pilder S, Logan JS (1992): Production of functional human hemoglobin in transgenic swine. *Biotechnology* 10:557–559.

Symonds JE, Walker SP, Sin FYT (1994): Development of mass gene transfer method in chinook salmon: optimization of gene transfer by electroporated sperm. *Mol Mar Biol Biotechnol* 3:104–111.

Tseng FS, Lio IC, Tsai HJ (1994): Introducing the exogenous growth hormone cDNA into lach (*Misgurnus anguillicaudatus*) eggs via electroporated sperms as carrier. In: Proceedings of the Third International Marine Biotechnology Conference, Tromso, Norway, abstract p 71.

Wallis AE, Devlin RH (1993): Duplicate insulin-like growth factor I gene in salmon display alternative splicing pathways. *Mol Endocrinol* 7:409–422.

Wright G, Garver A, Cottom D, Reeves D, Scott A, Simons P, Wilmut I, Garner I, Colman A (1991): High level expression of active human a-antitrypsin in the milk of transgenic sheep. *Biotechnology* 9:830–834.

Zhang P, Hayat M, Joyce C, Gonzales-Villasenor LI, Lin C-M, Dunham R, Chen TT, Powers DA (1990): Gene transfer, expression and inheritance of pRSV-Rainbow Trout-GH-cDNA in the carp, *Cyprinus carpio* (Linnaeus). *Mol Reprod Dev* 25:3–13.

Zhu Z, Li G, He L, Chen SZ (1985): Novel gene transfer into the goldfish (*Carassius auratus* L 1758). *Angew Ichthyol* 1:31–34.

.

Protocols

BURKE.1

WHOLE-MOUNT *IN SITU* HYBRIDIZATIONS
Modified after Wilkinson, 1992

Solutions
(If indicated, filter through 0.45 μm.)

DePC H₂O: 0.1% diethyl pyrocarbonate (DePC, Sigma) in H_2O, shake for approximately 30 sec and autoclave.

PBT: Filter before use.

PBS: 8 g NaCl; 0.2 g KCl; 1.44 g Na_2PO_4; 0.24 g KH_2PO_4 in 1 liter DePC H_2O. Make to 0.1% or 1.0% Tween-20.

RNA Transcription buffer, 10X: 400 mM Tris HCl, pH 8.25; 60 mM $MgCl_2$; 20 mM spermidine (Boehringer-Mannheim).

Ribonucleotide Mix: 10 mM GTP; 10 mM ATP; 10 mM CTP; 6.5 mM UTP; 3.5 mM digoxigenin-UTP (Boehringer-Mannheim).

RIPA: Make with DePC H_2O. Filter before use.

150 mM NaCl; 1% Nonidet-P-40; 0.5% sodium deoxycholate; 0.1% SDS; 1 mM EDTA; 50 mM Tris, pH 8.0.

Prehybridization and Hybridization Solutions: Make with DePC H_2O. Filter before use.

50% Formamide; 5X SSC (Sambrook et al., 1989), pH 4.5 (use citric acid to pH); 50 μg/ml yeast RNA; 1% SDS.

Molecular Zoology: Advances, Strategies, and Protocols, Edited by Joan D. Ferraris and Stephen R. Palumbi.
ISBN 0-471-14461-4 © 1996 Wiley-Liss, Inc.

50 μg/ml Heparin.

For hybridization, add approximately 1 μg probe/ml, usually about 10 μl/ml. Prehyb. and Hyb. can be stored in orange-topped tubes at −20°C.

Solution 1: Make up fresh and filter.

50% Formamide; 5X SSC, pH 4.5; 1% SDS.

Solution 2: Make up fresh and filter.

0.5 M NaCl; 10 mM Tris HCl, pH 7.5; 0.1% Tween-20.

Solution 3: Make up fresh and filter.

50% Formamide; 2X SSC, pH 4.5.

RNase A Stock: Prepare away from *in situ* area!

Dilute powder to a concentration of 10 mg/ml in 10 mM Tris HCl, pH 7.5; 15 mM NaCl.

Heat to 100°C, 15 min, cool slowly, aliquot, and store at −20°C.

Sheep Serum: Inactivate by heating to 65°C for 30 min. Can then be stored in aliquots at −20°C.

TBST: Make up fresh or from 10X stock of TBS then add Tween and levamisole, and filter.

0.14 M NaCl; 2.7 mM KCl; 25 mM Tris HCl, pH 7.5; 0.1% Tween-20; 2 mM levamisole.

NBT Stock:

0.075g NBT / 1 ml 70% dimethyl formamide.

BCIP Stock:

0.050 g/1 ml H_2O.

Reaction Mix: Make at last minute.

6.75 μl NBT/DMF and 5.25 μl BCIP/H_2O in 2 ml NTMT.

NTMT: Make up fresh and filter.

100 mM NaCl; 100 mM Tris HCl, pH, 9.5; 50 mM $MgCl_2$; 0.1% Tween-20; 2 mM levamisole.

Embryo Powder:

Sacrifice stage 32 ($\pm$) chick embryos, trim membranes, and remove heart at room temperature (RT) in PBS.

Homogenize in a minimum volume of ice-cold PBS. Add 4 × volume of ice-cold acetone, vortex, and put on ice for 30 min. Spin at 10,000g, 10 min at 4°C. Discard floating stuff; rinse pellet in ice-cold acetone. Allow to air dry completely on filter paper; grind to powder in mortar and pestle. Store at 4°C.

Protocol

Note: All washes are in volumes of 5–10 ml, in scintillation vials, at RT, rocking for 5 min unless otherwise noted.

Embryo Preparation:

1. Dissect embryos into PBS, remove membranes, and dissect as necessary. Using forceps or a needle, make a hole in the roof of the hindbrain and/or open the forebrain.
2. Fix in 15–30 ml fresh 4% paraformaldehyde, 4°C, overnight with gentle rocking. Wash in PBT, 2× for 5 min at 4°C, gently rocking, in volume of 5–10 ml.
   Wash in 25%, 50%, 75% PBT:MethOH, 5 min at RT, rocking.
   Wash 2× in 100% MethOH, 5 min at RT, rocking.
3. Store in 100% methanol at −20°C for up to 1 month.

Making Probe:

1. Combine the following reagents:
   dH$_2$O, 13 μl; 10X transcription buffer, 2 μl; ribonuclease inhibitor, 0.5 μl; nucleotide mix, 2 μl; linearized template DNA, 1 μl (1μg); RNA polymerase (commonly Sp6, T3, or T7), 1 μl (10 U).
2. Incubate for 2 hr at 37°C.
3. Electrophorese 1 μl of the reaction on a 1% TBE agarose gel with 0.5 μg/ml ethidium bromide to visualize the transcription product. There should be two visible bands, the larger is the remaining DNA template, and the smaller is the RNA transcription product. If RNA band is ~10× more intense than the template band, approximately 10 μg of probe have been synthesized. Occasionally, the RNA band is multiple, or appears as a smear, but can still give positive results.
4. Add 2 μl of DNase 1 (Boehringer: RNase-free) to transcription reaction, incubate for 15 min at 37°C.
5. Add 100 μl of TE (pH 8.0), 10 μl of 4 M LiCl, 300 μl ethanol and incubate for 30 min at −20°C.

6. Spin in a microcentrifuge at 4°C for 10 min, wash pellet with 80% ethanol, and air dry.

7. Resuspend in TE (pH 8.0) at an estimated concentration of 1 μg/μl. Store at −20°C.

Day 1: Pretreatment and hybridization.

1. Transfer embryos to clean scintillation vials (optional, may be taken to Prehyb. in 50 ml tubes).

2. Rehydrate samples in 75%, 50%, 25% MethOH:PBT, 1 wash each for 5 min. Wash 2× in PBT for 5 min at RT.

3. Bleach with 6% hydrogen peroxide in PBT for 1 hr at RT, rocking. Wash 3× in PBT for 5 min at RT.

Alternative treatment for embryos less than stage 15:

3a. Wash embryos with three changes of RIPA, 3× 30 min at RT Wash 3× in PBT, 5 min at RT; proceed to step 6.

4. Treat with 10 μg/ml proteinase-K in PBT for 15 min at RT, rocking.

5. Wash with 2 mg/ml glycine in PBT for 10 min at RT, make fresh and filter. Wash 2× in PBT for 5 min at RT.

6. Postfix with 4% paraformaldehyde and 0.2% gluteraldehyde in PBT for 20 min at RT. Wash 2× with PBT for 5 min at RT.

7. Add 1 ml Prehyb. to 1.5 ml vials; mix gently by inversion. Remove and add another 1 ml (optional, may add 2 ml directly). Prehybridize at 70°C for 1 hr. (Can store in Prehyb. solution at −20°C, before or after heating.)

8. Make up Hyb. solution (Prehyb. with approximately 1 μg/ml digoxigenin probe); keep liquid at 37°C. Remove Prehyb. Add 0.5 ml Hyb. Mix gently. Remove (also optional), add another 0.5 ml Hyb. Adjust volumes for scintillation vials. Hybridize overnight at 70°C.

Day 2: Posthybridization washes and antibody incubation.

Omitting the RNase treatment:

1. Wash with Solution 1: 3× for 30 min at 70°C, rocking.

2. Wash with Solution 3: 3× for 30 min at 65°C, rocking.

3. Wash with fresh TBST (1% Tween-20 and 2 mM levamisole) 3× for 5 min at RT, rocking.

With the RNase treatment:

1. Wash with Solution 1: 2× for 30 min at 70°C, rocking.
2. Wash with prewarmed Solution 1: Solution 2, 1:1 for 10 min at 70°C, rocking.
3. Wash with Solution 2: 3× for 5 min at RT, rocking.
4. Wash with 100 μg/ml RNase A in Solution 2: 2× for 30 min at 37°C, rocking. Prewarm Solution 2 to 37°C before adding RNase. g10.
   Mix separate, fresh batch of RNase/sol.2 for second wash.
5. Wash with Solution 2, then Solution 3, 5 min at RT, rocking.
6. Wash with Solution 3, 2× for 30 min at 65°C, rocking.
7. Wash with fresh TBST (1% Tween-20 and 2 mM levamisole) 3× for 5 min at RT, rocking.

Blocking and Antibody Preparation:

1. Wash with fresh TBST (1% Tween-20 and 2 mM levamisole) 3× for 5 min at RT, rocking.
2. Preblock embryos with 10% sheep serum in TBST, minimum volume for 2.5 hr, at RT, rocking.
3. Absorb antibody: In an appropriate tube, add approximately 3 mg of embryo powder and 0.5 ml TBST per embryo vial. Rock at 70°C for 30 min. Vortex for 10 min. Cool on ice.
4. Add 5 μl sheep serum and 1 μl anti-dig. antibody (Boehringer-Mannheim) per embryo vial. Shake gently at 4°C for 1 hr.
5. Spin for 10 min at 4°C. Collect supernatant and dilute with 1% sheep serum/TBST to a volume of 2 ml per embryo vial. Spin mixture again, at 4000 rpm at 4°C for 10 min.
6. Remove blocking serum from embryos. Add 1 ml Ab mix (or approximately 5× volume of embryos), mix gently, and remove (optional). Add another 1 ml Ab mix. Rock gently overnight at 4°C.

Day 3: Post-Ab hybridization washes.

1. Wash 3× for 5 min in TBST (1% Tween-20 and 2 mM levamisole), rocking at RT.
2. Wash 5× for 1–1.5 hr in TBST, rocking at RT.
3. Wash overnight in TBST, rocking at 4°C.

Day 4: Detection.

1. Wash 3× in NTMT (1% Tween-20 and 2 mM levamisole), rocking 10 min at RT.
2. Remove NTMT, add reaction mix (NBT and BCIP in NTMT), at least 2 ml per embryo vial. Cover tubes with foil, place on rocker for 20 min. Monitor about once an hour to check on progress.

Stopping the reaction

1. Wash 2× in NTMT for 10 min at RT rocking in dark (may be unnecessary).
2. Wash in PBT, pH 5.5, for 10 min at RT.
3. Postfix with 4% paraformaldehyde, 0.1% gluteraldehyde in PBS for 1 hr, RT.
4. Wash 2× in PBT; can store at 4°C in PBT.

To clear:

Wash in increasing concentrations of glycerol, 30–80%. Add azide 0.03% for long-term storage.

CAMERON.1

WHOLE-MOUNT *IN SITU* HYBRIDIZATION
References: Harkey et al., 1992; Ransick et al., 1993

1. Fix for 1 hr at room temperature in Streck Tissue Fixative (Streck Laboratories Inc., Omaha, NE). Handle embryos in flexible microtiter plates or microcentrifuge tubes. Embryos may be stored from 12 hr to 7 days in fixative.
2. Use 200–500 embryos per sample in microtiter well. Remove as much fixative as possible. Rinse with millipore-filtered (0.45 μm) seawater.
3. Remove seawater, rinse twice with phosphate-buffered saline (100 mM, pH 7.4) plus 0.1% Tween-20 (PBST).
4. Prepare hybridization buffer (HB): 50% formamide, 10% PEG 8000, 0.6 M NaCl, 5 mM EDTA, 20 mM Tris pH 7.4, 500 μg/ml yeast tRNA, 2X Denhardt's solution, 0.1% Tween 20.
5. Transfer embryos to hybridization buffer in three steps: 25% HB, 50% HB, 75% HB diluted in distilled water.
6. Prehybridize in 100% HB without probe at 45°C for 1 to several hours.
7. Prepare hybridization mix with 100% HB and RNA probe synthesized with digoxigenin-labeled uridine. Denature at 95°C for 10 min, then equilibrate for 2 min at 45°C.
8. Suspend embryos in hybridization mix, incubate at 45°C overnight.
9. Add one drop of PBST to hybridization mix containing embryos and mix well. Repeat twice.
10. Remove solution and wash twice with PBST at 45°C for 10 min.
11. Wash 3 times with 0.15 M sodium chloride, 0.17 M sodium citrate, pH 7.0 (1X SSC) at 65°C for 30 min each.
12. Rinse twice with PBST at room temperature.
13. Incubate for 30 min in PBST plus 5% goat serum.

14. Incubate 1 hr in PBST plus 5% goat serum and anti-digoxigenin antibody (Boehringer-Mannheim Biochemicals, Indianapolis, IN) at 1:500 dilution.

15. Rinse 3 times with PBST. Rinse twice with alkaline phosphate buffer pH 8.0 (100 mM Tris pH 8.0, 100 mM NaCl, 50 mM $MgCl_2$, 0.1% Tween-20).

16. Rinse once in alkaline phosphate buffer pH 9.5 (100 mM Tris pH 9.5, 100 mM NaCl, 50 mM $MgCl_2$, 0.1% Tween-20).

17. Prepare staining solution fresh daily: to 992 µl of alkaline phosphate buffer, pH 9.5, add 4.5 µl of nitro blue tetrazolium (75 mg/ml in 70% dimethylformamide) and 3.5 µl of 5-bromo-4-chloro-3-indoylphosphate (50 mg/ml in 100% dimethylformamide).

18. Add staining solution and check at 30-min intervals for sufficient staining. Stop staining reaction in PBST plus 1 mM EDTA.

19. Rinse once in PBST and dehydrate through graded ethyl alcohols beginning with 35%. Remove last 100% alcohol and add terpineol, mix well and mount.

CAMERON.2

MICROINJECTION OF SEA URCHIN EGGS
References: McMahon et al., 1985; Flytzanis et al., 1985

1. Protamine plate preparation: 60 mm, "Falcon" plastic petri-dish lids are filled with a 1% protamine sulfate solution for 1 min. The timing is critical: too short a period and the eggs stick poorly; too long and the eggs are unable to undergo normal elevation of the vitelline membrane on fertilization. Following protamine coating, dishes are rinsed with distilled water to remove unadhered protamine sulfate solution, and air dried.

2. Sea urchins are injected with 0.5 M KCl and shed eggs are collected in a beaker filled with millipore filtered seawater (MPFSW); then washed twice. Sperm are collected dry on ice.

3. The eggs are now handled in small quantities in small bore pipettes controlled by mouth pressure. The egg jelly is removed by treatment with pH 5 seawater for about 30 sec, after which the eggs are transferred in neat rows to a protamine sulfate-coated dish in 10 ml of MPFSW containing penicillin (20 units/ml) and streptomycin (50 µg/ml). The eggs are fertilized just before injection by adding small amounts of a 1–5% sperm suspension immediately over the eggs.

4. Microinjection needle preparation: Needles for microinjection are prepared from 1.0-mm outside diameter, 0.75-mm inside diameter, borosilicate, omega dot glass supplied by Frederick Haer and Co., Brunswick, Maine (No. 30-30-0). The glass is thoroughly cleaned by placing in a hot, 35% nitric acid solution for 1 hr, washing through several times with distilled water, and finally distilled water filtered through a 0.2-µm filter. Cleaned glass is placed in a screw-cap test

tube and dried in an oven. Fine-tipped microinjection needles are pulled on a Brown-Flaming micropipette puller (Sutter Instrument Co., Novato, CA) and, just prior to injection, are broken against the ridge formed at a surface scratch to open the tip to a diameter of ~0.4–0.9 μm.

5. The microinjection needle is filled from the rear, as near as possible to the tip with a DNA solution in 0.2 M KCl. Injection pressure can be supplied by a Picospritzer (General Valve Co., Fairfield, NJ). With the timed flow of the Picospritzer, the interval is adjusted to one that produces a bolus in the egg with a diameter about ¼ of the egg diameter.

6. Following injection, the seawater level is adjusted in the petri dish and the embryos are incubated at the appropriate temperature until the desired stage.

Note: The above procedure is presented in greater detail in McMahon et al., 1985; Flytzanis et al., 1985; Colin, 1986; and Summers et al., 1993.

CAMERON.3

PREPARATION OF NUCLEAR EXTRACT
Reference: Calzone et al., 1988, 1991

Note: This method can be used for amounts of embryos ranging from 1 ml up.

1. Pellet embryos by centrifugation at 1000g for 5 min at 4°C. Wash pellet once with 10–20 volumes of ice-cold 1 M glucose, then repellet by centrifugation at 2000g for 5 min at 4°C.

2. Resuspend embryo pellet in 10–20 volumes of Buffer A (10 mM Tris-HCl, pH 7.4, 1 mM EDTA, 1 mM EGTA, 1 mM spermidine trihydrochloride, 1 mM DTT, 0.36 M sucrose). Freeze in liquid nitrogen and store at −70°C until ready to process for nuclear extract.

3. Thaw frozen embryos in a beaker or bucket set in a luke-warm water bath, with vigorous stirring (we begin with a Teflon or hard rubber mallet, followed by a stiff PVC or glass stir rod until the embryos become slushy, then use a stainless steel stirring propeller attached to a variable speed motor) to lyse the embryos.

4. Pellet the nuclei by centrifugation at 2500g for 10 min at 4°C. Resuspend pellet (nuclei) in 20–30 volumes of ice-cold Buffer A (shake vigorously to resuspend), and repeat centrifugation. Repeat Buffer A washes in this manner 2–3×.

5. Resuspend nuclear pellet from final Buffer A wash in 20–30 volumes of ice-cold Buffer B (= Buffer A + 0.1% NP-40), and repellet by centrifugation (as above). Repeat this wash step 2–3×, until the nuclear pellet is colorless, thus removing the yolk. We generally find it helpful to enlist the aid of a blender to resuspend the nuclei in Buffer B (slow speed for about one-half minute).

6. Resuspend nuclear pellet from final Buffer B wash in an accurately measured volume of Buffer D (10 mM Hepes-KOH, pH 7.9, 1 mM EDTA, 1 mM EGTA, 1 mM spermidine trihydrochloride, 1 mM DTT, 10% glycerol). The volume of Buffer D should be about 2–3× the nuclear pellet volume (see step 9 below for determination of optimal volume of Buffer D); for example, if the nuclear pellet is about 50 ml, use 100–150 ml Buffer D to resuspend the nuclei. We use a blender (as described above) to resuspend the nuclei. The volume of the nuclear suspension is then accurately determined in a graduated cylinder, if needed, more Buffer D is added to ensure that the volume is 2–3× the original nuclear pellet volume. For example, if the measured volume of the nuclear suspension is 170 ml after adding 100 ml of Buffer D, add another 40 ml of Buffer D (total Buffer D added is then 2× the nuclear pellet volume).

7. Aliquot the nuclear suspension to Oak Ridge centrifuge tubes. The tubes should not be filled more than ¹/₃ to ¹/₂ full; for example, when using 30-ml Oak Ridge tubes, we generally do not add more than 10–12 ml nuclear suspension to each tube. The nuclei are now lysed by the addition of 0.1 volume of 4 M ammonium sulfate, pH 7.9, while vortexing, followed by vigorous shaking. The nuclear suspension should become highly viscous upon lysis of the nuclei and release of the chromatin.

8. Incubate the lysed nuclei on ice for 1 hr with occasional vigorous shaking.

9. Spin the lysed nuclei in a fixed angle rotor (Beckman Ti60, Ti70, or equivalent) at 35,000 rpm for 2–8 hr to remove the chromatin. The length of centrifugation time needs to be determined empirically; ideally, you should recover a supernatant volume that is at least ²/₃ the original volume of the nuclear suspension. Centrifugation times of over 8 hr are probably not advisable, as nuclear proteins of interest may begin to be lost to the chromatin pellet. If the volume of supernatant is not large enough after the centrifugation, try adding more Buffer D to the nuclear pellet next time.

10. Recover and combine the supernatants from the ultracentrifuge spin, and accurately determine the volume. Add 0.3 g ammonium sulfate per ml of supernatant, invert several times until the ammonium sulfate is completely dissolved, and incubate at 4°C overnight in order to precipitate proteins.

11. Recover the precipitated nuclear proteins by centrifugation at 16,000g for 30 min at 4°C. Resuspend the pellet in ¹/₂ the original nuclear pellet volume of ice-cold Buffer C (20 mM Hepes-KOH, pH 7.9, 40 mM KCl, 0.1 mM EDTA, 1 mM DTT, 20% glycerol, 0.1 mM PMSF, 0.1% NP-40) and let dissolve.

12. Dialyze the resuspended nuclear proteins against 50–100 volumes Buffer C overnight. Following dialysis, there is generally some precipitate, which should be removed by centrifugation at 16,000g for 30 min at 4°C. The supernatant is now ready for use in gel mobility shift assays or affinity chromatography, and should be stored in aliquots at −70°C until ready for use.

CAMERON.4

GEL MOBILITY-SHIFT ASSAYS
Reference: Calzone et al., 1988

1. Probes are double-strand DNA (either oligonucleotides or restriction fragments) with 5' overhangs that can be labeled with ^{32}P either by end-fill using Klenow or by kinasing. Typical specific activities of probes are on the order of 10^5 cpm/ng DNA.

2. Set up protein–DNA binding reactions in 10-μl final volume by the addition of the following components: 1 μl 10X binding buffer (0.1 M Hepes-KOH, pH 7.9, 5 mM DTT), 1 μl 50 mM $MgCl_2$, 1 μl 5 mg/ml poly dIdC/dIdC (or dAdT/dAdT—the quantity and type of this nonspecific competitor, which reduces the nonspecific binding of proteins to the probe, need to be optimized for the particular extract and probe being tested), 0.3 μl 2.5 M KCl (final concentration should be about 0.1 M), 1 μl probe (about 5×10^4 cpm), 4.7 μl Buffer C (see Cameron.3), and 1 μl nuclear extract. The amount of nuclear extract needed to produce a good signal will vary with different probes and different extracts and should be optimized by doing a dilution series in Buffer C. Similarly, the KCl and $MgCl_2$ concentrations may need optimization for different DNA-binding proteins.

3. Incubate the binding reactions on ice for 10 min, then load on a nondenaturing polyacrylamide gel (5–10% polyacrylamide in 0.5–1X TBE—the percentage of polyacrylamide and buffer concentration need to be optimized for each factor). Load 1–2 μl of DNA gel loading buffer (0.25% bromophenol blue, 0.25% xylene cyanol, 30% glycerol in H_2O) in one or more unused lanes as a way of tracking the progress of the gel. Run the gel at 200–300 V for 1.5–2 hr, dry gel, and expose autoradiographic film to visualize the gel mobility shift.

CAMERON.5

DNA BINDING SITE-SPECIFIC AFFINITY CHROMATOGRAPHY
References: Calzone et al., 1991; Coffman et al., 1992

1. Affinity columns are composed of double-strand concatemers of oligonucleotide binding sites (typically 25–30 nt long, in concatemers of >5) covalently linked to Sepharose CL4B (Sigma) via standard cyanogen bromide activation of the resin (for construction of such resins see protocols in Kadonaga and Tjian, 1986). Affinity resins can be stored for several years without loss of activity in storage buffer (10 mM Tris-HCl, pH 7.6, 0.3 M NaCl, 1 mM EDTA, 0.02% NaN_3) at 4°C.

2. Pack 0.5–1 ml affinity resin in a column (for automated affinity chromatography we typically use 1.5-ml disposable syringe type columns; for manual chroma-

tography, standard 0.5×10 cm columns such as those available from Biorad are suitable). The bed volume of the column should be determined by estimating the number of binding sites (assuming that $1/3$ of the sites linked to the resin are available for binding) and the number of specific protein molecules in the volume of nuclear extract to be processed, such that enough extract is passed over the column to saturate the binding sites. This will reduce the number of non-specific proteins that bind to the column.

3. Condition the column by passing 10 column volumes (cv) Buffer 0.1 (= Buffer C with 0.1 M KCl) through the resin, followed by 10 cv Buffer 1.0 (= Buffer C with 1.0 M KCl), followed again by 10 cv Buffer 0.1.

4. The following should be performed at 4°C. Bring an appropriate amount of nuclear extract (i.e., just enough to saturate the binding sites on the column) to 0.1 M with KCl. Gently load the extract on top of the column resin bed and allow to flow through by gravity. Collect the flowthrough for analysis and/or subsequent purifications. In practice, extract that has been passed through one affinity column is suitable for passing through another column containing a different site—in our lab, we use a robotic device to pass a volume of extract (typically 10–20 ml, containing extract from $1-2 \times 10^9$ embryos) over a tandem series of up to 12 columns, each bearing a different binding site.

5. Wash the column with 30–50 cv of Buffer 0.1.

6. Elute the column with a step gradient (3 cv each step) of Buffers 0.2, 0.3, 0.4, 0.5, 0.6, 0.7, 0.8, 0.9, and 1.0 (the buffer compositions are all the same as Buffer C except for the KCl concentration, which varies from 0.2 to 1.0 molar, respectively). Collect each elution step in a separate tube.

7. Regenerate the column by washing with 20–30 cv Buffer 1.0, followed by 5–10 cv storage buffer. Columns may be used repeatedly if regenerated and stored at 4°C in storage buffer.

8. Analyze the elution fractions from the column by gel mobility shift (Cameron.4) to locate the fractions in which the specific activity of interest elutes. The conditions for the binding reactions are essentially the same as for crude nuclear extract, except the concentration of nonspecific competitor DNA should be lower (200–500 ng/reaction) since the specific DNA-binding proteins are highly enriched in the elution fractions with respect to nonspecific DNA-binding proteins.

CAMERON.6

SOUTHWESTERN ASSAYS
Reference: Calzone et al., 1991

1. Southwestern blot assays are generally much less sensitive than gel mobility-shift assays and thus usually only work with purified or semipurified DNA-

binding proteins. Dilute affinity column elution fractions containing the peak activity (in our hands, typically 20 μl of the fraction) with an equal volume of 2X SDS sample buffer (0.2 M Tris-HCl, pH 6.8, 4% SDS, 10% β-mercaptoethanol, 20% glycerol, 0.1% bromophenol blue), heated to 95°C for 5 min, then run out on a standard Laemmli SDS gel (Laemmli, 1970). Since the molecular weight of the DNA-binding protein of interest is not known, ideally one should run a gradient gel; however, most DNA-binding proteins are between 30 and 80 kDa, and an 8% or 10% polyacrylamide gel will usually suffice.

2. Transfer the gel to nitrocellulose by your favorite electroblotting transfer method.

3. All of the following washes are performed at 4°C. Denature the blot by incubating in denaturation buffer (20 mM Hepes-KOH, pH 7.9, 0.1 mM EDTA, 1 mM DTT, 1 mM $MgCl_2$, 6M Guanidinium-HCl for 10 min. Renature by removing $1/2$ of the denaturation buffer and replacing with an equal volume of Buffer D (20 mM Hepes-KOH, pH 7.9, 0.1 mM EDTA, 1 mM DTT, 1 mM $MgCl_2$, 0.1% NP-40, 150 mM KCl, 20% glycerol); for example, if you initially added 100 ml of denaturation buffer, remove 50 ml and replace with 50 ml Buffer D. Incubate with agitation for 5 min, remove $1/2$ of the buffer, and replace with an equal volume of Buffer D. Repeat this step four more times.

4. Block the blot in Buffer D + 5% nonfat dry milk for at least 1 hr.

5. Wash the blot $3\times$ (5 min each) in Buffer D + 0.25% nonfat dry milk.

6. Prepare a ^{32}P-labeled binding site probe by nick translation using 400-ng double-strand concatenated oligonucleotide binding sites (e.g., the same concatemers that were used to prepare the affinity column in Cameron.5; generally these preparations are nicked enough already so that they may be directly labeled using *E coli* DNA polymerase I). Typical specific activities are on the order of 10^6 cpm/ng probe.

7. Incubate the blot with probe in Buffer D + 0.25% nonfat dry milk (10^6 cpm/ml) for 4 hr or overnight.

8. Wash the blot $3-5\times$ (2–3 min each) in Buffer D + 0.25% dry milk (use vigorous agitation and/or rubbing—don't forget to use gloves!). Wrap in Saran wrap (do not let dry) and expose autoradiographic film.

9. Wash the blot as above in Buffer D + 0.25% dry milk + 0.3 M KCl. Expose film as above.

10. Repeat step 9 using Buffer D + 0.25% dry milk + 0.6 M KCl. These washing steps generally remove nonspecific signals.

11. The blot may be regenerated by washing in Buffer D + 0.25% dry milk + 1.0 M KCl, and stored (wet) at −20°C for future use.

CAMERON.7

TWO-DIMENSIONAL GEL MOBILITY-SHIFT ASSAYS
Reference: Coffman et al., 1992

1. Two-dimensional gel mobility shifts consist of a preparative mobility shift run in a tube gel, the contents of which are then resolved by SDS-PAGE. Binding reactions for preparative gel mobility shifts are essentially the same as for analytical gel shifts, except the amounts of protein, probe, and nonspecific competitor are increased, which allows for visualization of the resolved complexes by silver staining of the second dimension SDS gel. Prepare the binding reaction by assembling the following ingredients (40 μl final volume): 4 μl 10X binding buffer (see Cameron.4), 4 μl 50 mM $MgCl_2$, 1 μl containing 1 μg poly dIdC/dIdC or other appropriate nonspecific competitor, 1 μl containing 0.2–1 μg probe (labeled or unlabeled; we generally use unlabeled probe, since the resolved protein complexes are visualized by silver staining); 10 μl Buffer E (= Buffer C without KCl; see Cameron.3), and 20 μl of affinity column fraction containing the activity of interest. Note that the final KCl concentration will be $^{1}/_{2}$ that of the concentration at which the DNA-binding protein of interest elutes from the column. In our experience this has not been a problem, but in some cases the volume of the binding reaction may need to be increased if the final KCl concentration is too high. Incubate the binding reactions on ice for 10 min.

2. Run the binding reactions out on 10% polyacrylamide tube gels (2–3 mm × 10 cm) in 1X TBE, 200–300 V, 1.5–2 hr; 2–3 μl of DNA loading buffer (see Cameron.4) may be added to the reaction just before loading so that the progress of the electrophoresis may be tracked.

3. Extrude the tube gel into 0.5 ml 2X SDS sample buffer (see Cameron.6) in a screw-cap test tube. Equilibrate for 30 min at room temperature with agitation, then boil for 5 min to denature the proteins and dissociate the protein–DNA complexes.

4. Layer the denatured tube gel across a standard Laemmli SDS gel (again, the optimal polyacrylamide percentage of the SDS gel must be guessed as for the Southwestern assay described above; most DNA-binding proteins will be resolved by a 10% polyacrylamide gel), and run at an appropriate voltage and current until the bromophenol blue is about 1 cm from the bottom of the gel.

5. Subject the gel to silver staining by your favorite method.

Note: It is advisable to run a control where no specific probe, or a heterologous probe, is substituted for the specific probe in the binding reaction. We have found that in most cases in such controls the DNA-binding proteins of interest do not enter the first dimension gel and show up as a vertical stack of bands on the side of the SDS gel corresponding to the top of the tube gel. However, some proteins do enter the first dimension even without probe DNA—comparison of the control gel with that containing specific probe allows assessment of which bands are resolved from the complex of interest. While we generally do not use labeled probe DNA, inclusion of label might help with location of the complex in the first dimension by autoradiographic exposure, especially when using oligonucleotide probes that comigrate with the dye front in the second dimension and are thus not visible by silver staining.

FERRARIS AND GARCÍA-PÉREZ.1

GENOMIC DNA LAMBDA LIBRARY SCREEN
Modified after Maniatis et al., 1982

1. This protocol was developed for use with a lambda library (e.g., Stratagene Lambda DASH II) and host bacteria that are antibiotic resistant (e.g., Stratagene SRB cells/kanamycin). Since contamination of host bacteria is one of the factors that can most severely interfere with effective library screening, we suggest the use of antibiotic-resistant host bacteria and invariably working from isolated single colonies. All solutions and materials should be sterile.

2. Preparation of single-colony plates of host bacteria.

   a. Prepare serial 10^{-2}, 10^{-4}, 10^{-5}, and 10^{-6} dilutions of host bacteria. Spread 50 μl of 10^{-4}, 10^{-5}, and 10^{-6} dilutions on NZY Kan plates.

   b. Grow overnight at 37°C; select plates containing well separated single colonies and store sealed with parafilm, inverted, at 4°C. Select single isolated colonies for all procedures.

3. Preparation of plating cells.

   a. Inoculate a single colony into a culture flask ($> 5\times$ volume of medium volume used) containing NZY medium (10 ml) plus 50 μg/ml kanamycin, plus 0.2% maltose and 10 mM $MgSO_4$. Grow at 37°C, 250 rpm.

   b. When OD_{600} of culture $= 1$, centrifuge at 4000g, 10 min.

   c. Resuspend in 10 mM $MgSO_4$ at $OD_{600} = 1$; store at 4°C.

   d. When ready to use, dilute an aliquot to $OD_{600} = 0.5$ (you will need 100 μl/plate).

4. Infection of host by lambda.

   a. Warm NZY Kan agar plates (10 cm) to 37°C; label, date, and number the plates.

   b. Boil NZY top agarose (0.7% Sea Plaque, FMC, Rockville, ME), mix thoroughly, cool to 47–52°C before use.

   c. You should screen the library at $\sim 2.5 \times 10^3$ pfu/plate (pfu's should be so dense as to fully cover the plate without spaces between units). If you do not know the titer of the library, it will have to be determined. Remove an aliquot of the library and prepare 10^{-5}, 10^{-6}, and 10^{-7} dilutions in SM (50 mM Tris, pH 7.5; 0.1 M NaCl; 8 mM $MgSO_4$; 0.01% gelatin); use 100 μl host cells at $OD_{600} = 0.5$: 100 μl λ/plate; plate in duplicate according to steps f–i; count plaque forming units (pfu)/plate; calculate titer of library in pfu/ml.

   d. Screen library at ~2.5×10^3 pfu/plate (pfu's should be so dense as to fully cover the plate without spaces between units); for example, for 20 plates of a library having a titer = 2.5×10^9 pfu/ml, dilute 10 µl with 990 µl SM (= 2.5×10^7 pfu/ml); dilute 10 µl of 2.5×10^7 pfu/ml with 990 µl SM (= 2.5×10^5 pfu/ml); dilute 21 µl of 2.5×10^5 pfu/ml with 2.08 ml SM (= 2.5×10^3 pfu/100 µl). Use a 50-ml tube for the last dilution.

   e. Add 2.1 ml host plating cells at OD_{600} = 0.5 to the library at 2.5×10^3 pfu/100 µl.

   f. Incubate at 37°C, 70 rpm for 20 min.

   g. Into each of 20, 5-ml polystyrene tubes pipette 200 µl of host/library mixture.

   h. Add 2.5 ml NZY top agarose (at 47–52°C) to a tube, vortex gently but quickly, pour without bubbles onto an NZY Kan agar plate; repeat for all tubes.

   i. Allow top agarose to gel completely at RT, invert, incubate at 37°C for 16–24 hr.

5. Prepare replicas, in duplicate (A and B) for each plate.

   a. Cool plates for 1–2 hr at 4°C; allow plates to come to room temperature.

   b. Number membranes (NEN, Boston, MA, Colony/Plaque Screen Hybridization Transfer Membrane) on the nonbinding side (check manufacturer's instructions as to sidedness of membrane).

   c. Place membrane on plate (binding side down onto agarose), make holes (1, 2, and 3 holes, asymmetrically arranged) near the edge with a 19-gauge needle. Remove the first membrane after 2 min (leave the second membrane, B, for 4 min). Lay the membrane DNA side up onto a piece of Whatman paper. For B membranes, use a lightbox to make holes in the same places as those for membrane A.

   d. Denature and neutralize as follows placing membranes on blotting paper between each transfer: 2 min denaturing solution (0.5 NaOH; 1.5 M NaCl) × 2 (in separate trays); 2 min or more neutralizing solution (0.5 M Tris pH 8.0; 1.5 M NaCl) × 2 (in separate trays). DNA side up throughout.

   e. Allow membranes to dry completely at room temperature; store in seal-a-meal bags, 4/bag, at 4°C until ready to hybridize.

   f. If you have a clone, such as a cDNA, to which your probe will bind, prepare it as a positive control by dotting 1 µl of a 1-ng/µl solution onto a membrane. Include the positive control in all steps to act as a control for random priming and hybridization.

6. Hybridization with random-primed DNA probe. Hybridization conditions should be selected based on the length of the probe, specificity of the probe, and the nature of the probe, for example, cDNA, RNA (not recommended), etc. (Britten and Davidson, 1985). The stringency (temperature and salt concentration) of hybridization solutions and washes should first be tested empirically for each probe using a Southern transfer (Maniatis et al., 1982) of restriction di-

gested genomic DNA. The higher the temperature or the lower the salt concentration, the more stringent the wash. The idea is to determine the stringency that will allow specific binding but wash off nonspecific binding. The following is the method we used for a 1287-bp cDNA.

a. Preheat all solutions to 42°C.

   Prehybridization solution: 5X SSC (20X stock = 3 M NaCl, 0.3 M Na citrate, pH 7.0); 5X Denhardt's [100X stock = 20% Ficoll 400, 20% polyvinylpyrrolidone, 20% bovine serum albumin (Pentax Fraction V)]; 10 mM Tris, pH 7.5; 250 µg/ml denatured salmon sperm DNA; 0.5% SDS (sodium dodecyl sulfate)

   Hybridization solution: 40% deionized formamide; 5X SSC; 5X Denhardt's; 10 mM Tris, pH 7.5; 100 µg/ml denatured salmon sperm DNA; 10% dextran sulfate; 0.5% SDS

b. Add prehybridization solution to each seal-a-meal bag (5 ml/membrane); do not seal. Arrange upright in temperature-regulated shaking H_2O bath with gentle circulation. The water level must be above the top of the membranes. Incubate 2 hr or more at 42°C.

c. Replace prehybridization solution with the same volume hybridization solution.

d. Denature the double-stranded probe at 95°C, 10 min, place on ice. Add 1–2 × 10^6 cpm random-primed probe/ml of hybridization solution. Seal bags and place horizontally; weigh down the corners of the bags. Incubate overnight at 42°C with gentle shaking.

e. Open bag, drain hybridization solution, gently remove membranes, place in 1-liter plastic box containing 500 ml of first wash solution (2X SSC, 0.5% SDS). Swirl gently, pour off liquid, and replace immediately. Cover and leave for 30 min at 42°C with gentle agitation. This is the first wash.

   (i) 2X SSC, 0.5% SDS, 42°C, 30 min (repeat of first wash)

   (ii) 0.1X SSC, 0.5% SDS, 42°C, 30 min (repeat)

f. Remove membranes from wash container and air dry. Mark the holes on the membranes with autoradiography ink or a very dilute ^{35}S solution.

g. Mount membranes on firm support (x-ray film from which the emulsion has been removed with Clorox works well) with DNA surface up. Cover with plastic wrap.

h. Expose to x-ray film for autoradiography.

i. Positives must appear on both A and B membranes. Because there is more DNA on the A membrane, a positive will be larger on the A membrane than on the B membrane. Negatives should appear as pale "doughnuts." Match positives on film with plaques on plate by aligning the holes. Pick a positive plaque by lifting a plug of top agarose with a P1000 tip (enlarge bore by cutting off end with sterile razor). Deposit into 1 ml SM; add 50 µl chloroform, vortex, let stand overnight at room temperature. Store at 4°C. For plaque purifications, assume one plaque in 1 ml SM contains 10^6–10^7 pfu;

prepare one plate at $0.5-1 \times 10^3$ pfu/plate. Proceed with plaque purifications (usually three) until no negative clones are present.

FERRARIS AND GARCÍA-PÉREZ.2

TRANSCRIPTION START SITE DETERMINATION BY DIDEOXY-TERMINATION SEQUENCING OF mRNA
Modified after Geliebter, 1987

1. All materials and solutions must be sterile and RNAse-free; H_2O must be DEPC-treated (Maniatis et al., 1982).
2. End-labeling the primer
   a. Select an antisense primer (~50% G-C content) that is within 200 nucleotides of the estimated transcription start site. If the primer turns out to be more than 200 nt away from the start site, sequential reactions can be used to "walk" through 5' end sequence.
   b. Add to a 1.5-ml tube: 1 μl primer (20 pmol/μl in H_2O); 1.5 μl H_2O; 6 μl 5X Forward Reaction Buffer (Gibco-BRL Life Technologies, Gaithersburg, MD; this comes with the kinase); 1.5 μl T4 polynucleotide kinase (Gibco-BRL 10 U/μl); 20 μl ^{32}P-γ-ATP (200 μCi, 40 pmol, >5000 Ci/mmol, Amersham, Arlington Heights, IL).
   c. Mix, centrifuge just to bring mixture to the bottom of the tube, incubate in a heat block at 37°C for 45 min. Inactivate enzyme at 65°C for 10 min.
3. Annealing primer to template
   a. Pipette 5 μg mRNA (as ethanol precipitate) into a 1.5-ml tube; centrifuge 15,000g, at 4°C for >30 min. Discard ethanol, dry pellet in a Speed-Vac for 5 min or to just evaporate remaining EtOH. Dissolve pellet in 10 μl annealing buffer (250 mM KCl; 10 mM Tris HCl, pH 8.3).
   b. Add 1 μl end-labeled primer, incubate at 80°C for 3 min then at 50–60°C for 45 min. Store at −80°C until ready for use. Determine annealing temperature empirically; estimate temperature at $T = [4(G+C)+2(A+T)] - 5$.
4. Sequencing reaction
   a. Prepare AMV Reverse Transcriptase Buffer [36 mM Tris HCl, pH 8.3; 24 mM $MgCl_2$; 12 mM DTT (dithiothreitol); 0.6 mM dATP; 0.6 mM dCTP; 0.6 mM dTTP; 1.2 mM dGTP; 150 μg/ml actinomycin-D].
   b. Label 1.5-ml reaction tubes G, A, T, C, or X. The X tube contains no ddNTP and is a good indicator of artificial stops. To each tube add: 2.2 μl of AMV RT Buffer; 1 μl of either 1 mM ddGTP, 1 mM ddATP, 2 mM ddTTP, 2 mM ddCTP, or H_2O; 2 μl primer:template; 1 μl AMV Reverse Transcriptase (20–25 U/μl Gibco-BRL). The higher the ratio of ddNTP:dNTP the more intense will be the bands close to the primer.

   c. Mix gently, centrifuge to bring contents to the bottom, incubate at 50°C for 45 min. Stop reactions with 2 μl stop solution (formamide, 0.37% EDTA, 0.3% bromophenol blue, 0.3% xylene cyanole).

   d. Store reactions at −80°C. When ready to load, boil 3 min, load 4 μl per lane in a sequencing gel.

FERRARIS AND GARCÍA-PÉREZ.3

TRANSCRIPTION START SITE DETERMINATION USING A PCR-ENHANCED METHOD FOR DETERMINING 5′ END SEQUENCE
Modified after Hofman and Brian, 1991

1. All materials and solutions must be sterile and RNAse-free; H_2O must be DEPC-treated (Maniatis et al., 1982).

2. End-labeling the primer

   a. Select an antisense primer (primer 1; ~50% G-C content) that is between 20 and 200 nucleotides downstream of the estimated transcription start site. If the primer turns out to be more than 200 nt away from the start site, sequential reactions can be used to "walk" through 5′ end sequence.

   b. Add to a 1.5-ml tube: 1 μl primer (20 pmol/μl in H_2O); 1.5 μl H_2O; 6 μl 5X Forward Reaction Buffer (Gibco-BRL Life Technologies, Gaithersburg, MD; this comes with the kinase); 1.5 μl T4 polynucleotide kinase (Gibco-BRL 10 U/μl); 20 μl ^{32}P-γ-ATP (200 μCi, 40 pmol, >5000 Ci/mmol, Amersham, Arlington Heights, IL).

   c. Mix, centrifuge just to bring mixture to the bottom of the tube, and incubate in a heat block at 37°C for 45 min. Inactivate enzyme at 65°C for 10 min.

3. Annealing primers to templates. Two reactions are run in parallel. In one, the primer is end-labeled; in the other, the same primer is not. The annealed end-labeled primer/template acts as a marker for the position of the annealed unlabeled primer/template.

   a. If possible, use mRNA in which the desired mRNA is induced. Pipette 5 μg mRNA (as ethanol precipitate) into two 1.5-ml tubes; centrifuge 15,000g, at 4°C for >30 min; discard ethanol; dry pellets in a Speed-Vac for 5 min or to just evaporate remaining EtOH. Dissolve pellets in 5 μl annealing buffer (250 mM KCl; 10 mM Tris HCl, pH 8.3).

   b. To one, add 1 μl end-labeled primer. To the other, add 1 μl unlabeled primer (50 pmol/μl), incubate at 80°C for 3 min, then at 50–60°C for 45 min. Determine annealing temperature empirically; estimate temperature at T = [4(G+C)+2(A+T)] − 5 (Hofman and Brian, 1991). Store at −80°C until ready for use.

4. First strand synthesis

   a. Prepare AMV Reverse Transcriptase Buffer [72 mM Tris HCl, pH 8.3; 48 mM MgCl$_2$; 24 mM DTT (dithiothreitol); 300 μg/ml actinomycin-D].

   b. To one 1.5-ml tube add: 1.1 μl RT Buffer; 2.1 μl 10 mM dNTP mix (each dNTP at 10 mM); 2 μl labeled primer/template; 1 μl AMV Reverse Transcriptase (20–25 U/μl).

   c. To a second 1.5-ml tube add: 3.3 μl RT Buffer; 6.3 μl 10 mM dNTP mix (each dNTP at 10 mM); 6 μl unlabeled primer/template; 3 μl AMV Reverse Transcriptase (20–25 U/μl).

   d. Mix gently, centrifuge briefly to bring contents to bottom of tube. Incubate at 42°C for 60 min.

   e. Stop both reactions with 0.1 vol 100 mM EDTA. Precipitate with 0.1 volume 3 M Na acetate plus 2 volumes ice-cold absolute EtOH; 30 min ice-water; 15,000g at 4°C for 1 hr. Resuspend labeled reaction in 2 μl loading buffer (formamide, 0.37% EDTA, 0.3% bromophenol blue, 0.3% xylene cyanol) and unlabeled rxn in 6 μl loading buffer. Store at −80°C.

5. Isolation of first strand cDNA from gel

   a. Boil reactions 3 min, load maximum volume per lane in a standard sequencing gel ~21 × 40 cm, 6% acrylamide/8.3 M urea, siliconized back plate, NON-wedge spacers, single load. Use a single lane for end-labeled reaction; three consecutive lanes for unlabeled reaction; leave several blank lanes between labeled and unlabeled reactions. Run bromophenol blue to bottom of gel.

   b. Do not fix or dry the gel. Place gel assembly horizontally on lab bench, front plate facing up. Remove comb. Pull front plate up by "working" on only one of the upper corners of the assembly only until vacuum is broken. Do not completely separate the plates.

   c. Flip over the entire assembly so that front plate is now directly on the bench. Gently and slowly pull plates apart (gel should stick to front plate since this is not siliconized.) Wrap the gel, together with the front plate, in plastic wrap. Avoid bubbles or folds in the plastic.

   d. Mark the gel (with an autoradiography pen on the plastic) for precise orientation of the film. In the dark, lay a piece of film (cut to the size of the glass) on top of the gel and tape it in place at the corners. Wrap the film, gel, and plate in aluminum foil.

   e. Expose at −80°C (exposure time will have to be determined empirically). After film is developed, realign film and gel. Tape film in place; flip gel so that glass faces up. Position of the full-length first strand cDNA should be the most slowly migrating, most intense band in the labeled primer reaction lane. Mark glass for position of band to be excised from the unlabeled primer reaction lanes.

   f. With a clean razor and forceps, excise strip of acrylamide containing cDNA and place in a 1.5-ml tube. Add 350 μl elution buffer (0.5 M NH$_4$OAc/1 mM EDTA/0.1% SDS); incubate at 37°C overnight.

   g. Remove gel slice; add 2 volumes ice-cold absolute EtOH; ice-water 1 hr; 15,000g at 4°C for 1 hr. Remove EtOH. Wash pellet, without resuspension, with 200 μl 70% EtOH and air dry. Resuspend in 20 μl ddH$_2$O.

6. First strand head-to-tail ligation

   a. To a 1.5-ml tube, add: 5 mg/ml salmon sperm DNA (first denature at 95°C for 5 min, ice until cool); 10 μg/ml bovine serum albumin; 50 mM Tris HCl, pH 8.0; 10 mM MgCl$_2$; 10 mM DTT (dithiothreitol); 1 mM ATP; 10 Units T4 RNA ligase; water to 20 μl.

   b. Incubate at 37°C for 4 hr; store at −20°C.

7. Second strand synthesis/PCR

   a. To a PCR tube, add: 1 μl ligation reaction, 5 μl 10X Perkin Elmer PCR Buffer; 0.1% Triton X-100; 200 μM dATP; 200 μM dCTP; 200 μM dTTP; 200 μM dGTP; 1.0 μM primer 2 (reverse complement of primer 1); 1.0 μM primer 3 (antisense just 3′ of primer 1); 1.25 units Taq polymerase; water to 50 μl. Mix, centrifuge just to bring contents to bottom of tube, overlay with 100 μl mineral oil. PCR cycle: 94°C for 1 min, 55°C for 1 min, 72°C for 10 sec for 40 cycles).

8. Visualize aliquot of PCR reaction in 4% Metaphor gel (FMC, Rockland, ME). Sequence the PCR product by direct sequencing (TAQuence Cycle Sequencing Kit, U.S. Biochemical, Cleveland, OH) or the PCR product can be subcloned (Prime PCR Cloner Kit, 5 Prime 3 Prime, Inc., Boulder, CO) and sequenced.

FERRARIS AND GARCÍA-PÉREZ.4

TRANSIENT TRANSFECTION OF CULTURED CELLS
(We use epithelial cells, other cells may require modifications.)

1. Preparation of reporter gene constructs. Two reporter gene constructs are required. The primary construct containing the DNA fragment to be tested for activity is the luciferase reporter gene construct. If the DNA fragment is to be tested for promoter activity, then the fragment is inserted immediately upstream of the luciferase gene. If some other activity is to be tested, such as osmotic response, then the DNA fragment and the homologous promoter are inserted upstream of the luciferase gene. The second reporter gene construct contains a strong promoter (usually viral) such as SV40 (for nonepithelial cells) or the parvovirus-derived B19 promoter upstream of the CAT (chloramphenicol acetyl transferase) gene.

2. All media must be prewarmed to 37°C before use; all solutions must be sterile. Both the chloroquine and DMSO (dimethylsulfoxide) treatments (below) are

optional. Enhancement of transfection by each of these treatments should be tested empirically.

3. Several days before transfection (depending on cell type), prepare actively growing cells: split one 150-mm 3–4 day postconfluent dish into six 150-mm dishes. Plan on one 150-mm dish for each DNA cotransfection and one for a negative control (treat identically but substitute water for DNA in $CaPO_4$/DNA precipitation step). Grow cells to 75% confluence. Feed 24 hr before transfection.

4. Prepare enough 35-mm dishes for *n* # of treatments plus negative control in triplicate. Aliquot 1 ml medium plus 2 μl of 100 mM chloroquine per 35-mm dish; incubate at 37°C.

5. $CaPO_4$/DNA coprecipitation

   a. For each cotransfection add to a 1.5-ml tube: 30 μl of luciferase construct (100 ng/μl); 120 μl CAT construct (100 ng/μl); 90 μl H_2O; 240 μl Buffer A (Cell-Phect, Pharmacia, Piscataway, NJ). Mix well; do not vortex; incubate at room temperature for 10 min.

   b. Add 480 μl Buffer B; mix by hand. Incubate at room temperature for 15 min.

6. Harvest cells (as you normally do since the method will be cell-type specific; below, we provide the method we use).

   a. Wash cells in 20 ml room temperature PBS (phosphate buffered saline without Ca^{2+} or Mg^{2+}) 2×, leave second wash on for 1 min.

   b. Add 3 ml trypsin/EGTA solution (0.1% trypsin/1 mM EGTA in PBS), incubate at 37°C until cells are lifted. Inactivate with 2 ml serum-containing growth medium.

   c. Using a 5-ml pipette, transfer cells from a 150-mm dish to a Falcon 2059 tube, centrifuge 2500 rpm for 1 min, aspirate medium.

7. Transfect

   a. Add 0.96 ml $CaPO_4$/DNA precipitate; resuspend 10× gently with a P1000 (Pipetman); incubate at room temperature for 15 min. Resuspend again; incubate at room temperature for 15 min.

   b. For each cotransfection, add 5.04 ml of prewarmed medium to a 50-ml tube. Using a P1000, transfer cell suspension to the medium. Resuspend well with 5-ml pipette; aliquot 1 ml per 35-mm dish (prewarmed and containing 1 ml medium plus chloroquine). Incubate for 2.5 hr at 37°C.

   c. Aspirate medium. Add 1.5 ml prewarmed medium containing 20% DMSO; incubate for 5 min at room temperature. Aspirate; add 1.5 ml medium; incubate for 5 min at room temperature. Aspirate; add 2 ml medium; incubate at 37°C.

   d. After 18–24 hr, change medium to control or experimental. This step can be eliminated if the activity examined is not one that is modified experimentally. Similarly, the time until harvest can be varied. We use 24 hr because expression of the AR gene is maximal 12–24 hr after hyperosmotic stimulus.

e. After 24 hr, harvest cells by adding 150 μl of Lysis buffer (Enhanced Luciferase Assay Kit, Analytical Luminescence Laboratory, San Diego, CA).

8. Analyze lysates for total protein (Bio-Rad Protein Assay Kit; Bio-Rad, Melville, NY), luciferase activity (Enhanced Luciferase Assay Kit; Monolight 2010 Luminometer; Analytical Luminescence Laboratory, San Diego, CA), and CAT protein (CAT Elisa Kit; Boehringer Mannheim, Indianapolis, IN, Titertek Multiskan plate reader at 405 nm). Express data as RLU (Relative Light Units) per μg total cell protein divided by CAT protein (pg per μg total cell protein).

LOEW AND FLEISCHER.1

MULTILOCUS DNA FINGERPRINTING

A more detailed protocol is available from the authors.

Purpose

Protocol for multilocus DNA fingerprinting from genomic DNA. The protocol is adaptable to virtually all species from which large amounts (2–10 μg) of clean, high-molecular-weight DNA can be extracted.

Comments on Isolation of Genomic DNA

DNA samples should be high-molecular-weight and very clean for multilocus fingerprinting to provide clear resolution and reliable results. In most cases DNA is extracted from soft tissues and blood; feather pulp, hair, and feces do not usually provide sufficient amounts of DNA for this method (for these DNA sources microsatellites are more suitable; see Fleischer and Loew.2). DNA extracted from small amounts of nonmammalian vertebrate blood (e.g., 25–100 μl) yields sufficient amounts of DNA for several fingerprints. For mammals, which have enucleate erythrocytes, a standard amount of whole blood needed for a single fingerprint is about 500 μl (but less may be required for specialized forensic methods). A number of protocols for extraction are available that involve organic or inorganic solutions; different protocols may be optimal for different tissues or taxa (e.g., Hoelzel 1992). Various buffers are available for storage of blood and soft tissue that do or do not require freezing or refrigeration (e.g., see Dessauer et al., 1990; Seutin et al., 1991); however, lysis buffers developed for storage of DNA at room temperatures do not always result in fingerprint-quality DNA. Lastly, we have found that dialysis of genomic DNA removes most contaminants that would otherwise interfere with restriction digestion. This protocol assumes that you have sufficient amounts of clean, high-molecular-weight genomic DNA that has been quantified via spectrophotometry or ethidium bromide (EtBr) staining.

1. *Restriction digestion of genomic DNA*

a. General comments: We usually load 4–5 μg of DNA per lane in a fingerprint gel but often digest 2–3 times that amount for reruns. The DNA must be digested to completion; thus we use a 5–10-fold excess of restriction enzyme. DNA fingerprinting almost always utilizes tetranucleotide recognizing restriction enzymes. *Hae*III usually provides good separation and resolution of fragments for a variety of vertebrate taxa, but other enzymes (e.g., *Hinf*I and *Alu*I) are often equally good. We recommend running a few test gels to determine optimal enzyme/probe combinations, running times, and hybridization temperature to produce fingerprints with well-resolved, variable fragments. Note that some enzymes do not digest methylated DNA and can result in incomplete digestion (Doefler, 1983). Place the following for each digest into a sterile eppendorf tube:

(i) 2–10 μg of genomic DNA sample.

(ii) $^1/_{10}$th of total volume of appropriate 10X buffer.

(iii) X units of restriction enzyme (at 10 units per μg of genomic DNA).

(iv) 1 μl of 5 mg/ml ribonuclease A (especially if DNA is isolated from tissue).

(v) Y μl of sterile water to bring up to no more than 70 μl total volume.

Spin for 5 sec in a microcentrifuge and then incubate 6 hr to overnight at manufacturer's recommended temperature.

2. *Electrophoresis of Digested Genomic DNA*

a. To determine degree of digestion and to quantify and adjust concentrations, we usually run a 1.0% agarose minigel in 1X TBE (0.05 M Tris, 0.05 M boric acid, 0.001 M EDTA, pH 8.0).

(i) Load 2–5 μl of sample per lane along with a known quantity of size marker (e.g., 1 μg of λ-*Hind*III) and run for 0.5–1.0 hr at 75 V.

(ii) Stain the gel with 1 or 2 drops of EtBr (10 mg/ml) for 10–20 min, destain in buffer for 5–10 min, and place on the transilluminator. Photograph.

(iii) Check to make sure all samples have digested to completion and that the amount of DNA per lane is of equal concentration. Complete digestion is generally indicated by an even smear of fragments and no high-molecular-weight DNA at the top of the lane.

b. Make a large 0.8–1.0% agarose gel. We use a 20 × 30 cm gel tray for an IBI® electrophoresis apparatus. The gel is made with 300–400 ml of fresh 1X TBE buffer (using double distilled water). A variety of high-quality agarose is available; we often use I.D.NA® agarose (FMC Bioproducts), especially for small DNA samples (e.g., from blood samples of mammalian infants).

(i) Boil the agarose and buffer and then cool to about 50°C. Add water to restore to original volume. Pour into the gel mold and place the comb (20–25 slots) into the gel.

(ii) Let the gel cool and harden. Pour a small amount of 1X TBE buffer onto gel and carefully remove the comb.

(iii) Put the gel into the gel rig and pour 1X TBE buffer (about 2.5 liters) over the gel and into the electrode tanks. Bring the buffer level to about 0.5–1 cm of buffer over the gel surface. The gel is now ready to load.

c. Loading the gel:
  (i) Mix 10X loading buffer/tracking dye (1% bromophenol blue, 1% xylene cyanol, 50% glycerol, 0.05 M EDTA) with size marker (e.g., λ-*Hind*III) to a concentration of 1 ng marker/μl dye (to later be probed as a within-lane size standard).
  (ii) Mix loading dye into digested DNA sample to 1X concentration and load carefully into the well. Load 3–4 μg of size marker into at least one lane.
  (iii) Putative parents should be run in lanes adjacent to offspring, preferably with parents flanking the offspring. Additional possible parents should be run in lanes next to those of the assumed parents.
  (iv) Run the gel for the optimal time and voltage as determined by the test runs. We typically run gels slowly, for example, 50 V for 40 hr, and change the 1X TBE buffer halfway through the run to obtain good resolution of fragments.

3. *Vacublotting the Gel*

a. To depurinate, denature, neutralize, and transfer the DNA from the gel onto a nylon membrane. Alternatively, DNA can be transferred through capillary action via Southern blotting (see Sambrook et al., 1990).
  (i) Remove the gel from the mold and stain with a few drops of EtBr (10 mg/ml) in about 500 ml of 1X TBE for 15–30 min. Destain for 10 min in buffer.
  (ii) Transfer the gel onto the transilluminator. Use a pipet tip to mark the size marker bands in the gel. Do not poke through the gel. Photograph the gel.
  (iii) Gently shake the gel on rotary shaker in 500 ml of depurination solution (0.25 N HCl) for 20 min.
  (iv) Remove depurinating solution and add 500 ml of the denaturing solution (1.5 M NaCl:0.5 M NaOH). Shake for 20 min. Repeat with fresh denaturing solution.
  (v) Remove denaturing solution and add 500 ml of the transfer solution (20X SSC; 3 M NaCl, 0.3 M Na_3-citrate, pH 7.0). Shake for 15 min.
  (vi) Assemble vacuum blotter apparatus and operate according to manufacturer's instructions. We use a Pharmacia Vacugene XL blotting system and transfer the DNA in 20X SSC for 2–3 hr. Transfer time depends on the thickness and concentration of the gel, and the size of the DNA fragments.
  (vii) After transfer is complete, mark the bands of the size standard by poking through the pipet tip indentations into the membrane with a pin. Do not shift gel.
  (viii) Remove the filter, gently rinse in 5X SSC for 5 min and UV crosslink and/or bake at 80°C for 1–2 hr. Label the edge of the filter with a ballpoint pen. Mark the size marker bands with soft pencil (probe binds to pencil).
  (ix) Re-stain the gel with EtBr to ascertain if all DNA has been transferred.

4. *Hybridization of Radioactively Labeled Probe to Blotted Filter.* This protocol can be used for a variety of minisatellite probes suitable for multilocus DNA fingerprinting, such as M13 bacteriophage DNA, and Jeffreys 33.15 or 33.6 probes. We generally perform hybridizations in hybridization tubes that are continually rotated in a hybridization oven (Robbins Scientific), but hybridization in seal-a-meal bags works equally well.

a. Prehybridization: To prevent nonspecific binding of the probe to the membrane, hence reduce "background" hybridization and increase the signal-to-noise ratio.

   (i) Place the filter into the hybridization tube with the DNA facing inward. Soak the filter in 5X SSC for 2 min. Avoid trapping large air bubbles between the filter and the glass walls or between layers of filter. Pour out the SSC.

   (ii) Add to the tube about 10 ml of prehybridization fluid (Westneat et al., 1988; 0.25 M Na_2HPO_4, pH 7.2; 7% sodium dodecyl sulfate [SDS]; 1 mM Na-EDTA, pH 8.0; 1% BSA, fraction 5). Prehybridize at optimal temperature as determined by test gels (e.g., 58–65°C) for 1–4 hr.

b. Labeling the probe via random priming: Several methods for labeling probes are available; we recommend the random priming method for minisatellite probes. For microsatellite probes we generally use nick translation (for long repeat arrays) or end-labeling methods. Here we provide our protocol for using the *Stratagene Prime-It Kit:*

   (i) Place the following in a sterile eppendorf tube:

      25–100 ng of probe DNA (purified insert or M13 whole)

      10 µl Primer mixture (random nonamers)

      Sterile water to total volume of 34 µl

   (ii) *Denature* the probe DNA in the above solution by heating to 95°C for 5–10 min, centrifuge for 5 sec, and then immediately chill on ice.

   (iii) *Add* the following to the eppendorf tube on ice:

      10 µl of 5X "dCTP" primer buffer (includes dATP, dGTP, and dTTP)

      5 µl of [^{32}P]dCTP, 3000 Ci/mmol

      1 µl of Klenow enzyme (5 units/µl)

   (iv) Incubate for 10 min at 37°C.

c. Checking the probe for incorporation: We check to see if the radionucleotides have been incorporated into the DNA by comparing a small aliquot of probe that has had the excess nucleotides washed from it with a control that has not been washed.

   (i) Combine 19 µl of water in a small eppendorf tube with 1 µl of hot probe solution.

   (ii) Place one DE81 (Whatman) filter in the filtration manifold on top of an erlenmeyer side arm flask.

   (iii) Put 1 µl of the diluted probe on the manifold DE81 filter. Place another 1 μl on a control filter and place it in a scintillation vial marked "C" for control.

(iv) Wash filter on manifold with about 10 ml of the 0.4 M Na_2HPO_4 solution, followed by a wash with 5 ml of 95% ethanol. Remove the filter and place it into a scintillation vial marked "W" for wash. Put 3 ml of aqueous scintillation cocktail into each vial and count with scintillation counter.

(v) Determine the percentage of radionucleotide incorporation; divide the number of counts of the washed sample by the number of counts of the control. Greater than 30–60% incorporation is expected, depending on the type of probe.

(vi) Determine the cpm/μg of probe as follows: [(cpm for the washed sample)(total volume of probe)(volume of the dilution)] / (amount of probe DNA in μg). For example: 246,567 cpm × 50 × 20 / 0.1 μg = 2.47×10^9 cpm/μg of probe DNA. Specific activities above 5×10^8 cpm/μg of DNA are considered satisfactory.

d. Cleaning the probe: Excess radionucleotides from the probe solution need to be removed before hybridization to reduce potential background on the autoradiogram. We use "Nick Columns" from Pharmacia, which contain Sephadex G-50. Follow the manufacturer's recommendations and the probe will be contained in about 900 μl.

e. Hybridization:

(i) Poke a pinhole in the cap of the eppendorf tube containing the cleaned probe so it will not pop open during heating (or use a screw-cap tube). Heat to 95°C for 10 min and immediately chill on ice.

(ii) Remove hybridization tubes from oven and reduce the prehybridization solution to 6 ml final volume. Add the probe directly to the bottom of the hybridization tube with a pipettor. Replace caps securely on hybridization tubes.

(iii) Return hybridization tubes to oven and hybridize overnight at the appropriate temperature. Hybridization stringency can vary and one should optimize conditions for each organism of study. Hybridization time can vary from about 15 to 24 hr, depending on the specific activity of the probe and amount of DNA on the filter.

f. Washing the filter:

(i) Remove the probe and hybridization fluid from the tube and save in a labeled, 50-ml centrifuge tube in the radiation freezer for up to 1 week. We have successfully reused probes up to two times.

(ii) Place 250 ml of wash solution (1X SSC: 0.1% SDS) into the tube and gently shake on rotary shaker for 15 min at room temperature. Pour off the wash solution into a liquid radioactive waste container.

(iii) Repeat the wash. Check radiation level of the filter with a geiger counter. If background appears high, repeat the wash up to 15 min at hybridization temperature or slightly lower.

(iv) Rinse the filter twice with distilled water and place onto a piece of 3MM paper. Do not let the filter dry completely or the probe will permanently bind to the filter.

(v) Wrap filter with a piece of Saran wrap and place it facing upward into cassette on top of an intensifying screen (e.g., DuPont Cronex Lightning Plus).

(vi) Place x-ray film on top of filter and additional intensifying screen on top of film. Expose at $-80°C$. Develop film within 24–72 hr (or longer if needed).

g. Stripping the filter: If the filter is to be rehybridized with another or the same probe, then it should be stripped of the current probe. Follow the instructions for stripping provided by the manufacturer of the nylon membrane (e.g., MSI, GeneScreen). Be very careful, as rough treatment of the filter can cause loss of digest DNA as well as probe.

FLEISCHER AND LOEW.2

CONSTRUCTION AND SCREENING OF MICROSATELLITE-ENRICHED GENOMIC LIBRARIES
Modified after Armour et al., 1994

A more detailed protocol is available from the authors.

Purpose

To increase the proportion of genomic library clones with inserts that contain microsatellite arrays. Thus a greater number of microsatellite flanking regions can be detected, sequenced, and subsequently used to design species-specific primers for microsatellite amplification.

1. *Processing Insert DNA*

a. Digest 5 μg of clean genomic DNA (from individual of heterogametic sex) to completion with restriction enzyme *Mbo*I, *Dpn*II, *Sau*3AI or other isoschizomer that leaves a 5'GATC overhang. Use the manufacturer's conditions and buffer and digest in a total volume of 30 μl.

b. Check completeness of digestion and requantify concentration by running 2 μl of digest in a 1.5% agarose gel in 1X TBE (0.05 M Tris, 0.05 M boric acid, 0.001 M EDTA, pH 8.0).

c. If digestion is complete, run the remainder of the digest on a 1.5% low-melting-point agarose gel in 1X TAE (0.04 M Tris, 0.02 M Na-acetate, 0.001 M EDTA, pH 7.2 with acetic acid). Excise the gel containing the 200–800 bp size fragments and proceed to isolating the DNA via Gene Clean (Bio101), GELase

(Epicentre Technologies) digestion, or electroelution. Resuspend DNA in 20 μl of 1X TE.

2. Processing of SAU Linkers

a. Synthesize complementary oligonucleotides so that when they are double stranded they have a 5′ overhang of CTAG and can serve as *SAU linkers:*

SAULA: 5′-GCG GTA CCC GGG AAG CTT GG-3′

SAULB: 5′-GAT CCC AAG CTT CCC GGG TAC CGC-3′

for example, 5′-GCG GTA CCC GGG AAG CTT GG-3′
            3′-CGC CAT GGG CCC TTC GAA CCC TAG-5′

b. Rehydrate each single-strand oligonucleotide to a molarity of 100 μM.

c. Phosphorylate the 5′ end of the SAULB oligonucleotide (*not* the SAULA):

(i)	10X kinase buffer (provided by manufacturer)	5 μl
(ii)	Unphosphorylated *SAULB*, 25 μg	36 μl
(iii)	10 mM ATP, pH 7.4 (*fresh*)	5 μl
(iv)	T4 Polynucleotide Kinase (12 U at 3 U/μl)	4 μl
	(dilute with PNK dilution buffer if included)	
	Total Volume:	50 μl

Incubate at 37°C for 1 hr; heat to 65°C to inactivate the PNK.

d. Allow equimolar amounts of SAULA and phosphorylated SAULB to anneal at room temperature for a minute or two and dilute to a final concentration of 0.5 μg/μl. Phosphorylated linkers are generally stable at −20°C for several months.

3. Linker-Ligation and Amplification of Digested Insert

a. Ligate SAU linkers to digested insert DNA at about 250:1 molar ratio (i.e., 2.0 μg:200 ng) in 20 μl total volume:

(i)	Insert DNA (200 ng)	x μl
(ii)	Linkers from above (2.0 μg)	4 μl
(iii)	sterile water(11 - x)	μl
(iv)	10X ligation buffer (supplied)	2 μl
(v)	10 mM ATP, pH 7.4 (*fresh*)	2 μl
(vi)	1.5 Weiss unit T4 DNA ligase	~1 μl

Incubate overnight at 16°C in water bath or thermocycler.

b. Run the 20-μl ligation mix in a 2.0% low-melting-point agarose gel in 1X TAE with a øX-*Hinc*II marker and visualize with EtBr. Linker dimers and tetramers (i.e., bands at about 44 and 88 bp) confirm successful ligation. In addition, linker-ligated insert DNA should be visible (250–850 bp).

c. Excise the linker-ligated inserts from the gel and recover the DNA by electroelution, Gene Clean, or GELase. Resuspend in 20 μl of sterile water.

d. Amplify the linker-ligated inserts in a 50-μl PCR reaction with the *SAULA* oligonucleotide as primer:

(i) Linker-ligated insert DNA	4 μl
(ii) SAULA primer (*10 μM stock*)	5 μl
(iii) dNTP mix (8 mM)	5 μl
(iv) PCR buffer (standard 10X, incl. MgCl$_2$)	5 μl
(v) *Taq* Polymerase (5 units)	0.5 μl
(vi) Sterile water	30.5 μl

Thermal profile is: 94°C for 1 min; 65°C for 1 min; and 70°C for 2 min; 30 cycles.

e. Run a 5-μl aliquot of the amplified insert on a 1.5% agarose gel in 1X TBE to check amplification. A bright smear between 250 and 850 bp indicates a successful amplification. If successful, remove the PCR components by centrifugal dialysis (e.g., Centricon 30) and dilute or concentrate to 25 μl total volume. Heat to 60°C for 10 min to inactivate possible nucleases. Run 2 μl on a 1.5% agarose gel in 1X TBE to quantify. Remove 500 ng for later comparison with enriched products (see 6.b below).

4. *Construction of Microsatellite Repeats by Ligation and Amplification*

a. Long sequences of microsatellite repeats are constructed and used to select large microsatellite repeat arrays from among the amplified inserts. Some long microsatellite oligomers are available commercially (e.g., CA/GT from Pharmacia). Others need to be assembled via oligonucleotide ligation followed by a self-priming PCR reaction. There is a large number of di-, tri-, and tetra-nucleotide repeats to choose from; it is a good idea to consult the literature to identify those that are generally found in genomes of the taxonomic group on which you work. For example, we have had good success with CAC$_n$/GTG$_n$ in mammals.

b. Oligonucleotides should be phosphorylated for ligation (see step 2.c above; alternatively phosphorylated oligonuclotides can be synthesized and obtained commercially).

c. To concatemerize and hence elongate each oligonucleotide, follow the ligation protocol (step 3.a above), but *substitute 5 μg of a particular phosphorylated oligonucleotide* (e.g., CAC$_n$/GTG$_n$) for the insert *and* linker DNA in 3.a.

d. Run half of the ligation in a 2% low-melting-point agarose gel in 1X TAE and excise the ligated concatemers from the gel (visible as discrete bands that are multiples of the original oligonucleotide). Isolate the DNA from the gel slice as in step 3.c above. Resuspend in about 20 μl of water.

e. To further increase the size of the oligonucleotide concatemer, set up a self-priming PCR reaction in 50 μl as follows:

(i) Concatemerized oligonucleotide from step 4.d	5 μl
(ii) dNTP mix (8 mM)	5 μl

    (iii) PCR buffer (standard 10X, incl. MgCl$_2$)   5 μl

    (iv) *Taq* Polymerase (5 units)   0.5 μl

    (v) Sterile water   34.5 μl

    Thermal profile is: 94°C for 1 min; 65°C for 1 min; and 70°C for 2 min; 30 cycles.

f. Run 5 μl of these cleaned, amplified repeats on a 1.5% agarose gel in 1X TBE to document the size and relative amount of the product. Remove PCR components by centrifugal dialysis as in step 3.e above. Requantify via spectrophotometry or gel assay.

5. Hybridization of Amplified Inserts to Amplified Repeats

a. Selection of inserts that contain microsatellites is accomplished by hybridization of the amplified inserts to amplified repeats bound to nylon filters. Different amplified repeats (e.g., CAC$_n$/GTG$_n$ and CATG$_n$/GTAC$_n$) can be pooled or used alone in hybridizations. If pooled, try to match repeats with similar annealing temperatures.

b. Denature 2 μg of amplified repeats (pooled or single) by addition of 1.0 M KOH to a final concentration of 150 mM for 5 min. Then neutralize with 0.25 volumes of 1 M Tris-HCl, pH 4.8. Alternatively can heat denature (95°C for 5 min).

c. Spot 1 μg of the denatured amplified repeats (pooled or single) onto a 3 mm × 3 mm piece of nylon filter. (Repeat each blot for optional insertless controls; see below.) Alternatively, use a dot blotter apparatus and cut later. Mark side of filter without DNA. UV crosslink DNA to the filters and allow to air dry and (optional) bake at 80°C for 1 hr.

d. For each filter, cut about 3 mm from the top of a yellow pipet tip (100 μl) and slit the tip. Put the filter in the slit to secure the filter in place within the tube and to allow easy transfer among tubes.

e. Prehybridize each filter in 1 ml of hybridization buffer (0.25 M Na$_2$HPO$_4$, pH 7.2; 7% SDS; 1 mM Na-EDTA, pH 8.0; 1% BSA, fraction 5) in a 1.5-ml microfuge tube at 65°C for 60 min and transfer the filter on the pipet tip to a new tube containing 100 μl of the same preheated buffer. (Do the same for the optional insertless control filters.)

f. Denature about 1 μg of amplified inserts by heating to 100°C for 5 min and placing in an ice/ETOH bath for 5–10 min.

g. Hybridize by adding the denatured amplified inserts to the tube containing the filter-bound amplified repeats in 100 μl buffer. Mix gently. (No insert DNA is added to the insertless control filters!)

h. Submerge each filter (*from here on* treat insertless control filters exactly as filters with amplified insert) completely in the buffer. Incubate at 65°C *overnight* in incubator or oven.

i. Wash filters twice for 5 min in 0.2X SSC, 0.01% SDS at 65°C to remove excess unbound insert DNA. Heat wash solutions in advance.

j. Remove and precipitate the bound microsatellite-enriched insert DNA from the filters as follows:

  (i) Carefully Pipet 100 μl of denaturing solution (50 mM KOH/0.01% SDS) over the filters 5–10 times at room temperature and let sit for 5 min. *Save each denaturing solution in a 1.5-ml eppendorf tube.*

  (ii) Pipet 100 μl of neutralizing solution (50 mM Tris-HCl, pH 7.5/0.01% SDS) over the filters 5–10 times. *Add* to the tubes from step a.

  (iii) Add 20 μl of 3.0 M Na-acetate (pH 7.0). Mix gently.

  (iv) Add 2 μl of *100 μM* SAULA primer as carrier DNA.

  (v) Add 2 volumes ice-cold 100% ethanol and place in ultracold freezer for 30 min.

  (vi) Spin at maximum velocity in refrigerated microfuge for 10 min to pellet DNA. Discard ethanol and rinse pellet in 400 μl of 70% ethanol.

  (vii) Discard 70% ethanol and speed-vac pellet until dry. Resuspend in 20 μl of sterile water. Place at 60°C for 10 min to inactivate DNAses.

6. *Amplification of Microsatellite-Enriched Inserts and Ligation Into Plasmid*

a. The microsatellite-enriched inserts (and the optional control, which may have oligo arrays that will self-prime) are reamplified in a *50-μl* PCR reaction with the SAULA primer as in 3.d *except* that 10 μl of insert from step 5.j.vii is used and 24 μl of sterile water. Quantify on 1.5% agarose gel in 1X TBE.

b. Dot blot on a nylon filter: 500 ng of unenriched PCR product from step 3.e and 500 ng of microsatellite-enriched PCR product (and the same volume of optional control amplification products as used for the enriched product).

  (i) Radioactively label (α-^{32}P) amplified repeats from step 4.f by nick translation, or end-label original oligonucleotides with terminal deoxynucleotidyl transferase (kits are available for both methods from U.S. Biochemical and other vendors).

  (ii) Hybridize probe to filter at 60–65°C in hybridization solution (see step 5.e) overnight. Wash twice for 10 min at room temperature in blotwash solution (1X SSC/0.1% SDS), and, if high background, repeat wash at 65°C.

  (iii) Expose overnight to x-ray film. Enrichment is indicated by a darker dot for the enriched component versus the unenriched. If the dot of the insertless control is as dark as that of the enriched component, a false positive is implied (i.e., repeat arrays washed off the filter during hybridization and amplified).

c. Clean the amplified inserts and remove the SAU linkers to form cohesive ends for cloning as follows:

  (i) Keep 25 μl as backup. Wash 25 μl of amplified inserts twice with 300 μl of sterile water in a Millipore ultra-free MC unit.

  (ii) Resuspend product in 20 μl of sterile water. Heat to 60°C for 10 min to inactivate nucleases.

(iii) Digest the 20 μl product with *Dpn*II or isoschizomer to remove linkers in 30 μl total volume as in 1.a above (incubate 1–2 hr at 37°C).

(iv) Add 70 μl of sterile water to reaction and 200 μl of phenol-chloroform (1:1). Mix and spin for 2 min.

(v) Remove supernatant to new, labeled tube. Add 10 μl of 3 M Na-acetate and 200 μl of ice-cold ethanol. Incubate at −70°C for 20 min and spin at full speed in refrigerated microfuge for 10 min.

(vi) Wash and dry pellet as in step 5.j.vi–vii, and resuspend in 10 μl of sterile water. Deactivate nucleases at 60°C. Requantify on a 1% agarose minigel in 1X TBE with EtBr staining.

d. Digestion and dephosphorylation of vector DNA:

(i) Digest 5 μg of pBluescript with 20 units of *Bam*HI in 20 μl total volume to completion according to manufacturer's conditions. Check for complete digestion by running 1 μl (250 ng) of digest on a 1% agarose minigel in 1X TBE adjacent to 250 ng of uncut pBluescript.

(ii) Mix the following:

(1) Digested vector DNA	19 μl
(2) 10X alkaline phosphatase buffer (supplied)	5 μl
(3) Sterile water	25 μl
(4) Alkaline phosphatase	1 μl

Incubate for 30 min at 37°C.

(iii) Add 50 μl of sterile water. Phenol-chloroform extract and ethanol precipitate the digested plasmid as in steps 6.c.iv–vi. Resuspend in 10 μl of sterile 1X TE.

(iv) Requantify amount of vector DNA.

e. Ligation of vector and insert:

(i) Do up to three ligations each in 20 μl: at a 1:3, 1:1, and 3:1 molar ratio of insert to vector DNA:

(1) pBluescript (~1 μg)	X ul
(2) Digested inserts (~50 ng)	Y ul
(3) Sterile water	16 − (X+Y) μl
(4) 10X ligation buffer (supplied)	1.5 μl
(5) 10 mM ATP, pH 7.4 (*if not in buffer*)	1.5 μl
(6) 1.5 Weiss unit T4 DNA ligase	≈1 μl

Incubate overnight at 16°C in water bath or thermocycler. A ligation control with digested øX-174 marker is suggested.

7. *Transformation of Competent E. coli Cells and Selection of Colonies Containing Microsatellite Inserts*

a. We have used a manufactured kit for transformation of plasmids into *E. coli* with great success (Stratagene Epicurian Coli® XL1-Blue supercompetent cells), but alternative kits and protocols are available. We exactly followed the manufacturer's instructions and obtained 30–60% recombinants from among hundreds of colonies per 100-mm plate. We typically inoculate two plates per library (i.e., 100 and 200 μl of transformation mixture).

b. Screening of library for positive clones containing microsatellites (see Bruford et al., 1992).

(i) Add 100 μl Luria-Bertani (LB) broth containing 50 μg/ml ampicillin and 15% glycerol into each well of a 96-well microtitre plate. Use highly sterile conditions and prevent cross-contamination between wells. With sterile pipette tip and a 200-μl micropipettor set at 20 μl transfer each white colony into a separate well. Incubate at 37°C to grow for several hours and then store at −20°C.

(ii) Cut a piece of nylon filter to the size of replica plater ("hedgehog," Sigma R-2508) or use circular nylon filters (100 mm size), and place on LB agar plates containing 50 μg/ml ampicillin. Rinse the prongs of the hedgehog in 70% ethanol and then flame off the ethanol to sterilize prongs. Allow to cool before use.

(iii) Dip hedgehog into plate wells to remove aliquots of cells, agitate gently and transfer cells to nylon filters on LB/amp plates. Place at 37°C and incubate overnight.

(iv) Once colonies are visible, pull filter off LB plate and allow to air dry. Lyse cells, and denature and neutralize DNA on filter by placing filter cell side up on Whatman paper soaked with:

(1) 10% SDS for 3 min.

(2) 1.5 M NaCl/0.5 N NaOH (fresh) for 5–10 min.

(3) 1.5 M NaCl/0.5 M Tris pH 7–8 for 5–10 min.

(4) 2X SSC for 1 min.

Crosslink DNA to filter and (optional) bake for 1 hr at 80°C.

(v) Hybridize ³²P-labeled repeat probe to filter, wash and autoradiograph (as in step 6.b).

(vi) Clones that contain microsatellites are usually much darker on the autoradiograph than those that do not. In our successful libraries 10–50% of the clones on a tray are darker. These positive clones are miniprepped as below. Alternatively, inserts can be PCR-amplified for sequencing directly from the colony with plasmid-specific primers (see Potts.1).

(vii) Isolation of plasmid/insert DNA. We use an alkaline-lysis/PEG precipitation procedure for this step, growing cells in 5 ml of LB (Applied Biosystems Bulletin 18, 1991). This procedure has yielded large amounts of plasmid/insert template of high quality ready for automated sequencing using pBluescript primer pairs (e.g., SK/KS). Alternative miniprep protocols are available from a number of sources (e.g., Bruford et al., 1992; Sambrook et al., 1989). Inserts can also be amplified using plasmid-specific primers, gel purified and then sequenced.

8. *Sequencing of Positive Clones and Primer Design*

a. We typically sequence plasmids or insert amplification products using a standard double-stranded protocol. We had good success with both manual (Sequenase 2.0, U.S. Biochemical) and automated (*Taq* Cycle Sequencing, Applied Biosystems Incorporated) methods. We refer the reader to these vendors or others for detailed protocols.

b. Primers are designed from sequences flanking the microsatellite arrays identified within the inserts. Occasionally a microsatellite array begins on the edge of the plasmid or is incomplete or too large for continued development. In addition, some clones contain the same insert. In general only a fraction (>30%?) will yield useful microsatellite primers.

c. Design rules are available from a number of sources (e.g., Hoelzel and Green, 1992).

d. Primers must be tested and optimized in order to guarantee faithful and consistent amplification. In addition, microsatellites should be assessed with pedigreed families to ensure mendelian inheritance and independent assortment.

GOLDEN.1

IN SITU HISTOCHEMICAL STAINING

1. Fix tissue in 4% paraformaldehyde [made in 1X PBS (phosphate buffered solution, 137 mM NaCl, 2.7 mM KCl, 8 mM Na_2HPO_4, 2.6 mM KH_2PO_4)].

2. Whole brains or cryostat sections were washed 30 min in PBS, changing the buffer every 10 min.

3. Heat tissue to 65°C for 30–40 min in PBS to inactivate endogenous alkaline phosphatase.

4. Place tissue in Buffer 3 (100 mM Tris pH 8.5, 100 mM NaCl, and 10 mM $MgCl_2$) with 1X Nitro Blue Tetrazolium (NBT) (Sigma) and 1X 5-bromo-4-chloro-3-indolyl-phosphate (X-phos) (Sigma). NBT is made as a 50X stock (50 mg/ml in 70% dimethylformamide and 30% H_2O), which can be stored for many months in glass covered with foil at −20°C. X-phos is made as a 100X stock (10 mg/ml in water) and also stored at −20°C. (*Note:* The concentrations of both NBT and X-phos have been varied to improve staining in some tissues.)

5. Allow staining for both whole brains and cut sections to proceed 30 min to overnight. Each reaction must be monitored to obtain the desired level of staining.

6. Stop reaction by washing in PBS several times followed by storage in 10 mM Tris pH 8.0 with 1 mM EDTA. Storage in the dark at 4°C will also help prevent a slow increase in background levels.

GOLDEN.2

PROTEINASE K DIGESTION OF CELLS PRIOR TO PCR

1. Remove coverslips from slides by immersion in H_2O.
2. Using a dissecting microscope, scrape single cells or small clusters of cells containing purple NBT precipitate with surrounding unlabeled tissue (approximately 0.5- to 2-mm tissue fragments) using a heat-pulled glass micropipette.
3. Transfer cells to a 96-well PCR (Hybaid) plate with 10 µl of a proteinase K solution (50 mM KCl, 10 mM Tris·Cl pH 7.5, 2.5 mM MgCl, 0.02% Tween-20, 200 µg/ml proteinase K).
4. Overlay each well with 1 drop of light mineral oil (Sigma) and heat to 60°C for 2 hr, 85°C for 20 min, and 95°C for 10 min in a thermocycler (Hybaid OmniGene).

GOLDEN.3

AMPLIFICATION OF DNA USING NESTED PCR

1. Add 0.15 µl Taq polymerase (Boehringer-Mannheim), 0.15 µl dNTP mix (Boehringer-Mannheim), 0.75 µl each of 10 µM oligonucleotide 0 (5'TGTG-GCTGCCTGCACCCCAGGAAAG3') and 10 µM oligonucleotide 5 (5'GTGT-GCTGTCGAGCCGCCTTCAATG3'), 2 µl PCR buffer with 15 mM $MgCl_2$ (Boehringer-Mannheim) and 16.2 µl of H_2O to each well of the 10 µl proteinase K solution (final volume 30 µl).
2. Thermocycle at 93°C for 2.5 min; [(94°C for 45 sec)(72°C for 2 min)] for 40 cycles; 72°C for 5 min.
3. Transfer 1 µl of reaction product from the first PCR to a new 96-well PCR (Hybaid) plate.
4. Add 0.15 µl Taq polymerase (Boehringer-Mannheim), 0.15 µl dNTP mix (Boehringer-Mannheim), 1 µl each 10 µM oligonucleotide 2 (5'GCCACCAC-CTACAGCCCAGTGG3') and 10 µM oligonucleotide 3 (5'GAGAGAGTGC-CGCGGTAATGGG3'), 2 µl PCR buffer with 15 mM $MgCl_2$ (Boehringer-Mannheim) and 14.5 µl of H_2O (final volume 30 µl).
5. Thermocycle at 93°C for 2.5 min; [(94°C for 45 sec)(70°C for 2 min)] for 30 cycles; 72°C for 5 min.
6. Load a 10-µl aliquot of the second PCR onto a 1.5% agarose gel (0.75% Seakem, 0.75% NuSeive) to ensure that the appropriate insert was amplified (Fig. 1).

GOLDEN.4

SEQUENCING OF THE PCR PRODUCT WITH CYCLE SEQUENCING

Sequencing was performed using the Cyclist™ Exo-*Pfu* DNA sequencing kit from Stratagene.

1. Add 5 μl of each d/ddNTP mix to each of four wells on a 96-well PCR (Hybaid) plate.

2. Add 5 μl from: 1 μl of the nested PCR product, 1 μl of 10 μM Oligo 3, 3 μl 10X sequencing buffer, 1 μl Exo-*Pfu*, 0.75 μl ^{35}S (10 μCi), 4 μl DMSO, and 11.25 μl H_2O to each well.

3. Thermocycle at 95°C for 5 min; [(95°C for 30 sec)(60°C for 30 sec)(72°C for 1 min)] for 30 cycles.

4. Add 5 μl of stop solution to the reaction and heat to 95°C for 2 min and then place on ice.

5. Load 2 μl onto a 6% acrylamide denaturing gel and run for 1.5 hr at 70 watts.

6. Develop with standard x-ray film.

GROSBERG.1

SIMPLE EXTRACTION AND RAPD-PCR PROTOCOLS

What follows are the basic protocols we use to prepare DNA and amplify RAPD markers from hydrozoans in the genus *Hydractinia,* and ascidians in the genera *Botryllus* and *Botrylloides*. These procedures seem to work well with all kinds of mucousy organisms, including anemones, large sea squirts, bryozoans, and even plants. Others have successfully used the same protocols to analyze zooplankton. Just play around with the amount of tissue you need, quantify the amount of DNA you extract from a given amount of tissue, and be sure to run the extracted DNA on an agarose minigel to see whether you get lots of undegraded (i.e., high molecular weight) DNA.

1. *Sample Preparation*

a. Large amounts of tissue

(i) If the zooids or polyps are small (<2–3 mm each), such as in many compound ascidians or hydroids, dissect about 4–8 polyps or zooids (or their equivalents) from a colony. If the zooids or polyps are large, one or two should do. If the sampled species is a brooder, be sure to remove all brooded embryos or larvae.

(ii) Blot excess water from the sample, and remove as much debris as possible.

(iii) Samples can either be ground into a fine powder after freezing in liquid nitrogen, or simply chopped up with a clean razor blade on an autoclaved glass slide. Try both ways and see what works best. Very mucousy organisms are virtually impossible to chop up while alive or defrosted. Once frozen in liquid nitrogen, the mucous loses much of its tenacity and you can usually grind the sample into a powder miscible in an aqueous extraction buffer.

b. Larvae and other small samples

(i) Mash a single frozen or fresh larva in 20 μl 2X CTAB and 1 μl proteinase-K on the side of a 0.5-μl eppendorf tube with a steel pin.

(ii) Incubate at 65°C for 60 min.

2. DNA Extraction

a. Large amounts of tissue

(i) Transfer tissue to 1.5-ml eppendorf tubes containing 500 μl 2X CTAB buffer (recipe below) and 5 μl proteinase-K (25 mg/ml stock solution).

2X CTAB Solution

- 2 g CTAB
- 35 ml 4 M NaCl
- 4 ml 0.5 M EDTA
- 10 ml 1.0 M Tris-HCl pH 8.0
- Glass-distilled H_2O to 100 ml
- 0.2 ml 2-mercaptoethanol
- Autoclave buffer first, then add mercaptoethanol

(ii) Mix with an autoclaved steel pin (or any small swizzle stick, previously sterilized in an appropriate medium) and incubate 2 hr at 65°C (shake tubes every 30 min or so).

(iii) Add 500 μl PCI (25 parts phenol: 24 parts chloroform: 1 part isoamyl alcohol), and hand mix by inverting tube a few times. Be careful to vent tubes once or twice to keep the sample from leaking.

(iv) Spin in an eppendorf centrifuge at 4°C at 8000 rpm for 18 min.

(v) Transfer supernate and repeat steps 3 and 4 with 500 μl PCI.

(vi) Transfer supernate to a tube filled with 500 μl CI, spin at 4°C at 8000 rpm for 18 min.

(vii) Transfer supernate and add 1 ml 95% EtOH (ice cold).

- Mix gently by hand.
- Store at −20°C for 20 min.
- Spin sample at 8000 rpm for 20 min at 4°C.

(viii) Decant EtOH and wash twice with 300–500 μl 70% EtOH.

- For each wash, spin 3–5 min at 3000 rpm at 4°C.

(ix) After second wash, pour off EtOH and air dry "pellet" overnight (don't worry if it's tinted).

(x) Resuspend in 50 μl 1X TE at 37°C for 2–3 hr. Flick the tube occasionally with your finger.

b. Larvae and other small samples

(i) Repeat all steps as for adults, but use only one PCI extraction, followed by one CI extraction.

(ii) After precipitating, washing, and drying sample, resuspend in ~15–20 μl 1X TE.

3. *Amplification Cocktails*

a. Large amounts of DNA

(i) Quantify DNA, preferably in a fluorometer. Follow the instructions; don't cut corners, be neat, and quit after a few hours (the fluorometer will tell you when it's ready for a nap). Do not try to do this after cocktail hour.

(ii) Adjust the DNA concentration in your sample to about 5 ng/μl.

(iii) Add sample DNA to first six ingredients in the PCR cocktail shown below, then overlay 1 drop of mineral oil on samples.

Amplification Cocktail for Large Amounts of Tissue
- 14.55 μl H_2O (glass distilled, autoclaved, and UV treated)
- 1.25 μl $MgCl_2$ (10 mM)
- 2.50 μl Reaction Buffer (from Parkin-Elmer)
- 2.50 μl of each dNTP (1 mM stock) → final concentration = 100 μm
- 0.20 μl Taq DNA polymerase (Perkin-Elmer, 5 units/μl)
- 3.00 μl Primer (10 μM) (from Operon Technologies)
- <u>1.00 μl</u> Sample DNA (adjusted to 5 ng/μl)
  25.00 μl final volume

b. Larvae and other small samples

(i) Don't bother to quantify the amount of DNA after you've determined it is somewhere between 1 and 5 ng/μl.

(ii) Add sample DNA to first six ingredients in the PCR cocktail shown below, then overlay 1 drop of mineral oil on samples.

Amplification Cocktail for Larvae and Other Small Samples
- 4.22 μl H_2O (glass distilled, autoclaved, and UV treated)
- 0.50 μl $MgCl_2$ (10 mM)
- 1.00 μl Reaction Buffer
- 1.00 μl dNTP's (1 mM stock)
- 0.08 μl Taq DNA polymerase (Perkin-Elmer, 5 units/μl)
- 1.20 μl Primer (10 μM)
- <u>2.00 μl</u> Sample DNA (we don't bother to measure concentration)
  10.00 μl final volume

4. *Thermocycler program (for all samples)*

- 94°C: 2 min, 30 sec ▶ 35°C: 1 min ▶ 72°C: 2 min   **1 Cycle**
- 94°C: 1 min   ▶ 35°C: 1 min ▶ 72°C: 2 min   **44 Cycles**
- 4°C soak

5. *Electrophoresis*

(i) Add 4 μl of Maniatis Type II loading buffer to each amplified sample.

(ii) Mix and load 10 μl into well of a 15 cm × 25 cm (for a double run) agarose gel made up of 0.6% agarose and 1.0% Synergel in 0.5X TBE (pH 8.0).
- Gel should be submerged in 0.5X TBE (pH 8.0)

(iii) Run gel at 100 V (4 V/cm) for 5 hr.

(iv) Stain for 8–15 min in 2.5 mg/500 ml ethidium bromide; destain for 30–60 min in distilled water.

(v) View (with proper face protection) and photograph (with appropriate filter) on UV lightbox.

HOLLAND.1

A SIMPLE METHOD FOR GENOMIC DNA PURIFICATION

1. *Materials*

Note: Autoclave all solutions unless otherwise noted.

a. GuSCN (5 M guanidinium isothiocyanate, 50 mM Tris, 25 mM EDTA pH to 7.4 with HCl). Make 100 ml, sterilize through a 0.45-μm filter. Just before use add 0.8 ml β-mercaptoethanol/100 ml.

b. 3 M NaAc (40 g Na acetate plus dH_2O to about 60 ml, add glacial acetic acid to pH 5.2, bring volume up to 100 ml).

2. *Method*

a. Homogenize 2 g tissue in 20 ml GuSCN *very gently* with a Dounce homogenizer with a loose-fitting pestle. Let the homogenate sit at room temperature for 1 hr to completely dissociate tissue. (*Note:* It is important not to shear the DNA.)

b. Centrifuge for 5 min at 2000 *g*.

c. Layer about 8 ml of supernatant onto 4 ml 5.7 M CsCl in an ultracentrifuge tube for the Beckman SW41 rotor. Repeat with the remaining supernatant. Fill tubes to within 2–3 mm of top. Top up with additional GuSCN if necessary. Mark interface with a permanent marker.

d. Centrifuge 11–12 hr or overnight at 20°C without braking.

e. Use a hypodermic syringe with an 18-gauge needle to *gently* pull out the DNA (above the white carbohydrate band and below the CsCl–GuSCN interface.

f. Transfer to a sterile 15-ml disposable tube and add 1 volume dH$_2$O, $^1/_{10}$ volume 3 M NaAc pH 5.2 and 2 volumes room temperature 100% ethanol. Rock gently until a precipitate appears.

g. Spool out DNA with a heat-sealed pasteur pipette. Transfer to a tube with 70% ethanol; let sit 5 min. Transfer to another tube with 70% ethanol; let sit another 5 min.

h. Stand pipette upright to evaporate ethanol. Dissolve pellet in 5–6 ml TE.

i. Extract twice with phenol equilibrated with TE/chloroform 1:1, once with chloroform alone.

j. Precipitate DNA with 1/10 volume 3 M NaAc and 2 volumes 100% ethanol.

k. Wash with 70% ethanol and dry in SpeedVac.

HOLLAND.2

PURIFYING DNA FROM RECOMBINANT BACTERIOPHAGE LAMBDA

1. *Materials*

Note: Autoclave all solutions unless otherwise noted.

a. TM buffer (50 mM Tris, 10 mM MgSO$_4$, pH 7.5).

b. NZY plates (per liter add 5 g NaCl, 2 g MgSO$_4$7H$_2$O, 5 g yeast extract, 10 g NZ amine (casein hydrolysate), 15 g agar, pH 7.5 with NaOH; autoclave; cool to about 55°C; pour into 9-cm petri dishes.

c. NZY top agarose: same as NZY plates except for 0.7 g agarose/100 ml instead of agar.

d. TB medium (5 g NaCl, 10 g bacto-tryptone/l).

e. LB medium (10 g bacto-tryptone, 5 g bacto-yeast extract, 10 g NaCl/liter; adjust to pH 7.5 with NaOH).

f. 5 M K acetate (11.5 ml glacial acetic acid, 28.5 ml dH$_2$O, 60 ml 5 M K acetate. The solution is 3 M with respect to K and 5 M with respect to acetate).

g. DNase I (1 mg/ml in sterile dH$_2$O; store frozen; do not autoclave).

h. RNase A (1 mg/ml in sterile dH$_2$O; store frozen; do not autoclave).

i. Tris-EDTA-SDS (0.5 M Tris pH 8, 0.25 M EDTA, 2.5% SDS).

j. 10 mg/ml proteinase K (in sterile dH$_2$O; do not autoclave; store frozen).

2. *Make a high-titer phage stock*

Note: To obtain enough lambda phage particles for DNA purification, it is necessary to infect a vigorous strain of *Escherichia coli,* such as LE392, at a ratio such that the infected bacteria replicate for several hours, reach stationary phase, and lyse completely. The initial ratio of infecting phage to bacteria is critical.

a. Day 1

    (i) Select a colony of a suitable bacterium such as *E. coli* strain LE392 from a fresh plate, add to 5 ml of TB medium supplemented with 10 mM $MgSO_4$ and 0.2% maltose; incubate overnight at 37°C with shaking.

b. Day 2

    (i) Centrifuge bacteria for 10 min at 4000g and resuspend pellet in 2.5 ml sterile 10 mM $MgSO_4$.

    (ii) Plate out phage stock at an approximate density of 50–1000 plaques on 9-cm NZY plates, using NZY top agarose. See instructions for plating out library supplied with the vector. Refrigerate plates when plaques are about 2 mM in diameter.

    (iii) Add single colony of *E. coli* strain LE392 to 5 ml sterile TB medium supplemented with 10 mM $MgSO_4$ (no maltose). Incubate overnight with shaking at 37°C.

c. Day 3

    (i) Centrifuge bacteria for 10 min at 4000g and resuspend pellet in 2.5 ml sterile 10 mM $MgSO_4$. To five tubes add 50 μl cells and 0, 1, 2, 3, 4 plaques from the plate from Day 2. (*Note:* Core with a 1-ml pipetter tip with the end cut off; save remaining bacteria at 4°C for use on Day 4).

    (ii) Mix. Let sit at room temperature for 5 min.

    (iii) Add 2 ml LB medium and 20 μl 1 M $MgSO_4$; incubate with shaking at 37°C for 4–6 hr until bacteria are lysed (solution becomes clear). As lysis occurs add 100 μl chloroform; vortex and store on ice.

    (iv) Centrifuge supernatants at 4500g for 10 min. Store at 4°C.

d. Day 4

    (i) Titer phage stocks from Day 3 by plating out bacteria saved from Day 3 infected with serial dilutions of the phage stocks to titer. (See Day 2.) The spot-titer method works well. Mix 0.1 ml plating bacteria with 3.0 ml NZY top agar at 42–45°C and spread on an NZY plate. Spot with 7.5 μl of 1 × 10^{-6} and 1 × 10^{-7} dilutions of the phage stock. Incubate overnight at 37°C. Select a lysate with $\gg 1 \times 10^{10}$ phage/ml for subsequent use.

    (ii) Add 1 colony of LE392 to 10 ml TB medium supplemented with 10 mM $MgSO_4$ and incubate overnight at 37°C with shaking.

e. Day 5

    (i) Centrifuge overnight culture of LE392, 4000g for 10 min, and resuspend in 5 ml of 10 mM $MgSO_4$.

    (ii) Add 1 μl high-titer phage stock to 625 μl cells in 10 mM $MgSO_4$.

    (iii) Incubate at room temperature for 5 min and add to a sterile 50-ml flask containing 25 ml LB medium, 10 mM $MgSO_4$. Shake 4–6 hr until lysis occurs.

Note: If lysis does not occur within 6 hr or if yields of DNA are low, it may be necessary to vary the ratio of phage to bacteria. For example, a ratio of 5 μl phage stock to 300 μl bacteria has given good results.

(iv) Add 1 ml chloroform and centrifuge at 4500*g* for 10 min.

(v) Decant supernatant and centrifuge in a glycerol step gradient in SW41 ultracentrifuge rotor as follows: To centrifuge tube add 3 ml 40% glycerol in TM buffer. Add 3 ml 5% glycerol in TM buffer. Layer on about 6 ml phage supernatant (fill tube to within 1–2 mm of top). Centrifuge for 1 hr at 35,000 rpm. Remove the top 6 ml from gradient and add a second 6 ml. Repeat centrifugation. Remove the top 6 ml from the gradient and add a third 6 ml. Repeat centrifugation. Resuspend pellet in 1 ml TM buffer.

(vi) Add 10 μl RNase A and 1 μl DNase I. Incubate for 30 min at 37°C.

(vii) Add 200 μl Tris-EDTA-SDS and 3 μl 10 mg/ml proteinase K. Incubate at 56°C for 30 min.

(viii) Add 400 μl 5 M K acetate/1.2 ml solution. Let sit on ice for 10 min. Centrifuge at 4000 rpm for 30 min at 4°C. Save supernatant and split into two tubes. Add 0.6 volume isopropanol.

(ix) Spool out DNA quickly with a heat-sealed Pasteur pipette. (*Note:* There may not be enough DNA to spool. If not, microfuge at top speed for 5 min at room temperature). Rinse pellet in 70% ethanol. Dry in SpeedVac, and resuspend in 50 μl TE. Incubate for 10 min at 65°C to disperse DNA. (*Note:* DNA is suitable for sequencing and restriction digests.)

HOLLAND.3

WHOLE-MOUNT *IN SITU* HYBRIDIZATION APPLICABLE TO *AMPHIOXUS* AND OTHER SMALL LARVAE

1. *General materials*

a. Lots of deionized or distilled water (dH$_2$O) treated with (diethylpyrocarbonate). (*Note:* For DEPC H$_2$O add 0.5% DEPC to H$_2$O in a bottle, shake, let sit a minimum of 2 hr and autoclave.)

Note: All solutions should be treated with DEPC and autoclaved except for solutions containing Tris, which should be made up in DEPC H$_2$O and then autoclaved.

b. Very good dissecting microscope with transmitted and incident light. (*Note:* Fiber optics are especially useful.)

c. 5X TBE (54 g Tris base, 27.5 g boric acid, 3.7 g Na$_2$EDTA/liter; do not adjust pH).

d. 3 M Na acetate, pH 5.2 (3 M NaAc). To 40 g NaAc add dH$_2$O to about 70 ml. Add glacial acetic acid to pH 5.2 and water to 100 ml. Autoclave.

2. *Linearization of template DNA*

a. Method

    (i) Linearize template DNA: mix 10 μg of plasmid DNA, 10 μl 10X restriction buffer, and dH$_2$O to a volume of 100 μl. Add 20 U of a restriction enzyme that generates blunt ends or a 5' overhang and cuts at the 5' end of the piece of DNA to be copied (preferably 700–1000 bp).

Note: Typically, riboprobes are transcribed from the 3' untranslated region of the gene to avoid hybridization of the probe with the mRNA of related genes. If the signal is weak, however, it may be necessary to synthesize a probe corresponding to the 5' end of the gene or to combine 5' and 3' probes. In our experience, probes including the homeodomain do not always hybridize to mRNAs for related genes and can therefore sometimes be used.

    (ii) Incubate 2 hr at the temperature appropriate for the enzyme.

    (iii) Check 10 μl of the digest on a 1% agarose gel in 0.5X TBE including 1 μl/10 ml of a 0.5% ethidium bromide stock solution to make sure the DNA is linear. For a gel loading buffer use 5 μl 70% glycerol containing Orange G or bromophenol blue. (*Note:* Linear DNA migrates more slowly than closed circular plasmid DNA.)

    (iv) Add 10 μl 3 M NaAc, pH 5.2.

    (v) Extract twice with an equal volume of a 1:1 mixture of phenol:chloroform equilibrated with 0.3 M NaAc, pH 5.2, and once with chloroform.

    (vi) Precipitate by adding 2 volumes 100% ethanol, freeze on dry ice, microfuge for 20 min at top speed, wash pellet with 1 ml 100% ethanol. Dry in a SpeedVac (Savant Inc., Farmingdale, NY) and resuspend at 100 ng/μl in DEPC H$_2$O. Store frozen at −20°C. Keeps indefinitely.

3. *Preparation of embryos and larvae*

a. Materials

    (i) MOPS-EGTA fix: 4% paraformaldehyde, 0.1 M MOPS (3[*N*-morpholino]propanesulfonic acid) pH 7.5, 2 mM MgSO$_4$, 1 mM EGTA [ethylene glycol-bis(β-amino-ethyl ether) *N,N,N',N'*-tetraacetic acid], 0.5 M NaCl. Make fresh. (*Note:* Omit NaCl for freshwater larvae.)

Note: To dissolve paraformaldehyde, add 0.4 g to 1 ml 1 N NaOH, heat to 60°C to dissolve, add 9 ml NaPBS, add 1 ml 1 N HCl. Check pH.

    (ii) Acetic acid–ethanol fix: 10% acetic acid, 70% ethanol, made fresh.

b. Method

    (i) Fix embryos for 30 min at room temperature or overnight at 4°C in MOPS-EGTA fix. (*Note:* A 15-min fix in acetic acid–ethanol may give a lower background. It has been suggested that it may not preserve mRNA as well as the MOPS-EGTA fix, but for moderately abundant messages, there is no discernable difference in mRNA preservation.)

(ii) After fixation, transfer embryos to 70% ethanol. After an hour or more change the 70% ethanol. Store at $-20°C$. (*Note:* With prolonged storage, the amount of mRNA may decrease.)

4. *Synthesis of riboprobe*

Note: RNases are everywhere! Wear gloves and decontaminate gel apparatus and all glassware thoroughly before use with RNase AWAY™ (Molecular Bioproducts Inc., San Diego, CA) or 3% H_2O_2 or soak in 1% SDS overnight. Rinse in DEPC-treated H_2O.

a. Materials
   (i) RNA labeling kit, nonradioactive (Genius™4 from Boehringer-Mannheim Corp., Indianapolis, IN). (*Note:* Reagents can be bought separately; this kit comes with SP6 and T7 RNA polymerases; buy T3 RNA polymerase separately.)

b. Method
   (i) Mix 1.5 μl linear DNA template; 12.5 μl DEPC H_2O; 2 μl 10X buffer from kit [400 mM Tris-HCl pH 8.0, 60 mM $MgCl_2$, 100 mM dithiothreitol (DTT), 20 mM spermidine HCl, 100 mM NaCl]; 2 μl digoxigenin RNA labeling mix; 0.5 μl RNase inhibitor (20 U/μl); 2 μl appropriate RNA polymerase (20 U/μl).
   (ii) Incubate at 37°C for 2 hr.
   (iii) Run 1 μl on a 1% agarose gel in 0.5X TBE (include 1 μl/10 ml of a stock solution of 0.5% ethidium bromide in the gel) with appropriate molecular weight standards (e.g., plasmid PBR322 cut with the restriction enzyme HinfI). You should see the plasmid DNA band (about 0.3 kb plus the length of the insert) and an equal-intensity band or smear of RNA. (*Note:* If insufficient probe has been synthesized, more enzyme can be added and the incubation continued. Sometimes the RNA polymerase falls off the DNA prematurely. Carrying out the synthesis at 30°C rather than 37°C may help. However, it may be necessary to use a longer piece of starting DNA even if it includes a very conserved portion of the gene at the 5' end. The polymerase will probably fall off before it reaches that region.)
   (iv) Add 2 μl DNaseI (RNase free) at 10 U/μl.
   (v) Incubate at 37°C for 15 min.
   (vi) Add 100 μl TE, 10 μl 4 M LiCl, 300 μl ethanol.
   (vii) Mix and put at -20°C for 30 min.
   (viii) Spin in a microfuge at top speed for 20 min, wash pellet with 1 ml 100% ethanol. Dry in SpeedVac.

(ix) Resuspend in 25–50 μl DEPC H$_2$O depending on how intense the band was. Store at −20°C.

5. *In Situ hybridization*

Note: RNases are everywhere! Wear gloves. Use newly opened bags of pipette tips and other plastic ware, decontaminate all glassware thoroughly before use with RNase AWAY™ (Molecular Bioproducts Inc., San Diego, CA) or 3% H$_2$O$_2$ or soak in 1% SDS overnight. Rinse in DEPC-treated H$_2$O.

a. Materials

(i) Dishes holding 1–2 ml. Sterile Nunc catalog #176740, or small glass dishes.

(ii) NaPBS (0.9% NaCl, 20 mM sodium phosphate buffer pH 7.4)—treat with DEPC and autoclave.

(iii) Fresh 4% paraformaldehyde (pfa) in NaPBS. (*Note:* See step 3 for dissolving paraformaldehyde.) After use, freeze remainder of solution for Day 3.

(iv) 0.1 M triethanolamine pH 8.0 in autoclaved DEPC-treated H$_2$O. Store frozen.

(v) Acetic anhydride.

(vi) NaPBSTw (NaPBS plus 0.1% Tween-20); store frozen.

(vii) Proteinase K (10 mg/ml stock) store frozen.

(viii) 10% glycine in DEPC water; store frozen.

(ix) 2 mg/ml glycine in NaPBSTw.

(x) 20X SSC [175 g NaCl, 88.5 g trisodium citrate, H$_2$O to 1 liter, pH 7.4 (DEPC treated and autoclaved)]. Store frozen in 50-ml aliquots.

(xi) Deionized formamide (Fluka is best); buy ultrapure or deionize by shaking 20 min with 50 g/liter mixed-bed resin (e.g., Amberlite MB3) and filter through Whatman No. 1 paper. Store frozen.

(xii) Total RNA 10 mg/ml stock (buy cheapest Sigma grade available); dissolve at 50 mg/ml in 10 mM Tris, 1 mM EDTA (TE) pH 7.6, 1% SDS. Digest with 100 μg/ml proteinase K, 3 hr at 37°C, extract once with phenol equilibrated with TE, once with phenol:chloroform 1:1, add $^1/_{10}$ volume 10 M LiCl, 2 volumes 100% ethanol, freeze on dry ice, microfuge at top speed for 15 min, wash pellet with 80% ethanol, dry in SpeedVac. Resuspend to 10 mg/ml and store frozen.

(xiii) 100X Denhardt's solution (0.2 g Ficoll type 8000; 0.2 g polyvinyl pyrrolidone, 0.2 g BSA, dH$_2$O to 10 ml). Pass through a 1.0-μm filter to sterilize.

(xiv) Hybridization buffer: 50% deionized formamide, 100 μg/ml heparin, 5X SSC, 0.1% Tween-20, 5 mM EDTA, 1X Denhardt's, 1 mg/ml total RNA; stored frozen in 1-ml aliquots.

(xv) Hybridization oven with rotating or rocking platform set at 60°C.

(xvi) Waterbath set at 70°C.

(xvii) Variable speed rotating platform.

(xviii) Wash solution 1 (50% deionized formamide, sterile 5X SSC, 1% SDS; made up in DEPC-treated sterile water. Store frozen in 50-ml aliquots).

(xix) Wash solution 2 (50% deionized formamide, sterile 2X SSC, 1% SDS; made up in DEPC-treated sterile water; store frozen in 50-ml aliquots).

(xx) Wash solution 3 (sterile 2X SSC, 0.1% Tween-20; made up in DEPC-treated sterile water; store frozen in 50-ml aliquots).

(xxi) Wash solution 4 (sterile 0.2X SSC, 0.1% Tween-20; made up in DEPC-treated sterile water; store frozen in 50-ml aliquots).

(xxii) Wash solution 5 (NaPBS, 0.1% Tween-20, 2 mg/ml bovine serum albumin (BSA); store frozen in 50-ml aliquots).

(xxiii) RNase stocks (10 mg/ml RNase A in TE; and 10,000 u/ml RNase T1 in 0.1 M NaAcetate pH 5.5; both stocks preboiled 10 min and stored frozen separately from all other *in situ* solutions).

(xxiv) Sheep serum. Pretreat at 55°C for 30 min. Store frozen in 1-ml aliquots.

(xxv) Amphioxus powder: grind frozen or fresh adults in smallest possible volume of NaPBS; add 4 volumes acetone, let sit on ice 30 min. Centrifuge at $10,000g$, wash precipitate twice in acetone. Air dry and grind with a mortar and pestle. Store frozen.

(xxvi) 1:3000 Anti-digoxigenin antibody coupled to alkaline phosphatase (Boehringer-Mannheim Corp.). (To 1.5 mg amphioxus powder add 400 μl NaPBS/0.1% Triton X-100. Heat to 70°C for 30 min. Add 50 μl 20 mg/ml BSA, 50 μl pretreated sheep serum, 0.5 μl Boehringer anti-digoxigenin antibody. Mix and incubate on rotator at room temperature for 1 hr or more, or overnight at 4°C. Add 1 ml NaPBS containing 0.1% Triton X-100, 2 mg/ml BSA. Add 50 μl sheep serum. Store frozen in 200-μl aliquots. Can be reused.

(xxvii) 5 M NaCl (autoclaved).

(xxviii) 1 M Tris pH 9.6 (made in DEPC-treated water).

(xxix) 1 M $MgCl_2$ DEPC-treated.

(xxx) Levamisole (Sigma Chemical Co., St. Louis, MO).

(xxxi) NBT = nitro blue tetrazolium chloride: 75 mg/ml in 70% dimethylformamide.

(xxxii) BCIP = 5-bromo-4-chloro-3-indolylphosphate *p*-toluidine salt (also called X-phosphate); 50 mg/ml in 100% dimethylformamide. (*Note:* Store NBT and BCIP frozen. Keep from light.)

b. Methods
Day 1:

(*Note:* RNases are everywhere! Wear gloves)

(i) Select embryos and larvae—usually five or more of each stage (three if large). Peel fertilization layers off prehatch embryos with sharpened insect pins. (*Note:* Reagents may not penetrate well into amphioxus larvae over 1 week of age; penetration can be facilitated by nicking the larvae with a razor blade. This technique has allowed small adult amphioxus 5 mm long to be labeled as whole mounts.)

(ii) 2–3, 5-min washes with NaPBSTw on rotating platform. To change solutions, use pipetter (with barrel previously cleaned with 1 N NaOH or RNase Away™) and fine tips. Put waste into a second dish, examine under microscope, and return any larvae pipetted over by mistake.

(iii) Change for 7.5 μg/ml proteinase K. Rotate gently. (*Note:* To make 1 ml, add 1 μl 10 mg/ml proteinase K to 100 μl NaPBSTw, then add 75 μl of this to 1 ml NaPBSTw. Add 200 μl to each dish.) (*Note:* For amphioxus larvae 24 hr old or less digest *exactly* 10 min; up to 1 week larvae digest 20 min; over 1 week digest 30 min. Time carefully!)

(iv) After 10 min, add 4 μl 10% glycine stock to each dish with 200 μl proteinase K solution.

(v) Quickly change solution for 2 mg/ml glycine in NaPBSTw; onto rotator for 5 min.

(vi) Change solution for 4% pfa in NaPBS for 1 hr at room temperature. Do not rotate.

(vii) Change solution for 0.1 M triethanolamine; 1 min.

(viii) Change solution for 0.1 M triethanolamine; 5 min shaking.

(ix) Remove supernatant from larvae; quickly add 2.5 μl acetic anhydride to sterile microfuge tube containing 1 ml 0.1 M triethanolamine; vortex briefly and add 300 μl per dish. Let sit 5 min. Do not rotate.

(x) Add 5 μl acetic anhydride to sterile microfuge tube containing 1 ml 0.1 M triethanolamine; vortex and add an extra 300 μl per dish. Let sit 5 min. Do not rotate.

(xi) Change solution for NaPBSTw; place on rotator for 1 min.

(xii) Change solution for NaPBSTw; place on rotator for 5 min; heat an aliquot of hybridization buffer to between 37 and 60°C.

(xiii) Change solution for 100 μl warm hybridization buffer, place on rotator for 1 min at room temperature. (*Note:* Be careful; embryos go transparent in this and are easy to lose.)

(xiv) Change for fresh 200 μl hybridization buffer; rock in hybridization oven at 60°C for 1 hr or longer.

(xv) Thaw probe. If at 100 ng/µl, add 2 µl probe to 200 µl prewarmed fresh hybridization buffer in sterile microfuge tube (amount is for 1 dish, scale appropriately). Vortex and microfuge at top speed for 5 min. (*Note:* If background is high, try decreasing probe concentration.)

(xvi) Warm probe to 70°C. Carefully remove most of the hybridization buffer from the embryos and add the prewarmed probe. Cover dish and place in 70°C waterbath for 5 min (2 min if plastic dishes are used).

(xvii) Move dish of embryos to hybridization oven; rock gently at 60°C overnight.

Day 2:

(i) Prewarm wash solution 1 to 60°C; thaw wash solution 2–4. Remove most of the hybridization solution from the embryos. Save, can be reused.

(ii) Add 0.8 ml wash solution 1; 60°C on rotator 5 min.

(iii) Three washes in wash solution 1 at 60°C for 5 min; 15 min; 15 min.

(iv) Change for wash solution 2; 60°C for 5 min (time carefully) move to room temperature for 10 min.

(v) Change for fresh wash solution 2; room temperature for 15 min (time carefully).

(vi) Change for wash solution 3, remove immediately.

(vii) Change for fresh wash solution 3; room temperature for 5 min.

(viii) Thaw out RNase stocks. Add 2 µl RNase A stock (10 mg/ml) and 1 µl RNase T1 stock (10,000 u/ml) to 1 ml wash solution 3 in a sterile microfuge tube.

(ix) Remove last wash from all embryos and add the RNase solution to all wells. Place in 37°C incubator for 20 min.

(x) Two washes 20 min each in wash solution 3; at room temperature, rotating.

(xi) Change for wash solution 4; 20 min at room temperature, rotating.

(xii) Thaw wash solution 5. Wash in wash solution 5 for 5 min; meanwhile, in microfuge tube, make blocking solution: mix 1 ml wash solution 5 with 100 µl sheep serum.

(xiii) Remove wash and add 200 µl blocking solution (see step xii above) to dish. Rotate at room temperature for 1 hr or more.

(xiv) Replace blocking solution with 100–200 µl antibody (= 1:3000 dilution of anti-digoxigenin antibody preabsorbed with adult powder). Seal dish and put on rotator at 4°C overnight.

Day 3:

(i) Carefully remove the antibody from the embryos; can be refrozen and reused.

(ii) Wash embryos in NaPBSTw, 4 × 20 min at room temperature, rotating.

(iii) During last wash, make up alkaline phosphatase (AP) buffer (mix together

8.3 ml sterile water, 1 ml 1 M Tris pH 9.6, 500 µl 1 M $MgCl_2$, 200 µl 5 M NaCl, 10 µl Tween-20).

(iv) Take off as much of last NaPBSTw wash as possible. Add 500 µl AP buffer, swirl, and remove. A precipitate may form.

(v) Wash 3–4 times in AP buffer, 10 min each on rotator. (*Note:* Don't stint.)

(vi) Add 2.4 mg levamisole into a 5-ml tube, dissolve in 5 ml AP buffer. For 1 ml stain. Transfer 1 ml of this to a microfuge tube and add 2.5 µl NBT and 3.5 µl BCIP. Keep in the dark.

Note: The levamisole may not be necessary if embryos lack endogenous phosphatases.

(vii) Exchange AP buffer on embryos for the staining buffer. Put in the dark for the color to develop. Signal should come up in 20 min to 2 hr; but for some transcription factors, it may be necessary to leave embryos for 3–4 days.

(viii) Transfer embryos into NaPBS, then 4% pfa in NaPBS. Fix for 1 hr.

(ix) After fixation transfer to NaPBS/0.1% Na azide or clear in 80% glycerol in NaPBS/0.1% azide.

(x) Mount on slides in 80% glycerol in NaPBS/0.1% azide. Support coverslip with coverslip fragments or pieces of tape. (*Note:* Whole mounts can be stored in 80% glycerol at $-20°C$.)

6. *For a permanent record*

a. Photograph with tungsten Ektachrome film under DIC optics.

b. If signal is strong, embryos can be transferred to 1% Ponceau S (CI 27195) in 1% acetic acid to counterstain, dehydrated in an ethanol series, embedded in Spurr's resin and sectioned with glass knives on an ultramicrotome at 1–3 µm. When sections are dry, mount coverslips in a drop of immersion oil—do not use Permount! (*Note:* Sectioning is often essential to determine what anatomical structures are labeling and to reveal label in central structures like the notochord when the myotomes are heavily labeled.)

JAGUS AND PLACE.1

PRODUCTION OF RECOMBINANT PROTEINS *IN VITRO* USING COUPLED TRANSCRIPTION/TRANSLATION REACTIONS

Purpose

To provide an easy *in vitro* system for the production of a desired protein product using cDNA cloned in plasmids containing bacteriophage promoters (e.g., T7, T3, or Sp6).

Principle

Transcription of cDNA clones by bacteriophage RNA polymerase is coupled to translation *in vitro* using the micrococcal nuclease-treated rabbit reticulocyte lysate in a single reaction of coupled transcription/translation. This single tube reaction obviates the need for three prior reactions (i) linearization of DNA, (ii) *in vitro* transcription, and (iii) 5' methylation and purification of mRNA.

Optimum conditions for the coupled reaction are closer to those required for *in vitro* translation than transcription. Under the conditions used, transcription is relatively inefficient, but sufficient amounts of mRNA are produced to saturate the protein synthetic capacity of the system. In addition to the standard message-dependent rabbit reticulocyte lysate (MDL) components, the coupled transcription/translation reaction requires DNA, the relevant RNA polymerase, and modified buffers. In most cases, the coupled transcription/translation reactions produce significantly more protein (two- to sixfold) in a 1-hr reaction than do standard *in vitro* rabbit reticulocyte lysate translations using RNA templates (R. Jagus, 1993 Promega Notes PN042, p. 17). Multiple proteins can be expressed from different promoters in the same reaction by using multiple RNA polymerases. In this way protein–protein interactions can be studied.

Source

This is basically the same as the published method for coupled transcription/translation reactions (Craig et al., 1992). A kit with all reagents is marketed by Promega Inc., and described in Promega Notes PN035, PN038, and PN042. It is based on the micrococcal nuclease-treated rabbit reticulocyte translation system of Pelham and Jackson (1976) designed to destroy endogenous mRNA and results in a minimum background of translation. Production of the reticulocyte lysate is described in detail (Jagus, 1987).

Other Applications

This protocol is very useful for confirmation of gene product identity and reading frame and, if used in conjunction with expression-PCR (Kain et al., 1991), can be used for rapid screening and analysis of site-directed mutations.

Procedure for Coupled Transcription/Translation: For laboratories that do not routinely prepare rabbit reticulocyte lysate, the coupled transcription/translation system provided by Promega, their reticulocyte TnT system, is the recommended choice. For laboratories that produce their own reticulocyte lysate, essentially by the method described (Jagus, 1987), the following protocol may be used. It can be modified to use only nonradioactive amino acids for functional studies. Similarly, other radioactive amino acids can be used. Smaller reactions may be used by adjusting the volumes of all reagents proportionally.

Coupled transcription/translation reaction using ^{35}S-methionine

2 M KCl/10 mM MgAcetate	2.5 μl
Amino acid mixture minus methionine, 1 mM	2.0 μl
25X Reaction buffer	2.0 μl
[35]S-methionine (1000 Ci/mmol) at 10 mCi/ml	4.0 μl
RNasin ribonuclease inhibitor (40 U/μl)	0.35 μl
Phosphocreatinine (1 M, freshly prepared)	0.5 μl
Creatine phosphokinase (200 U/ml, CPK, in 50% glycerol)	0.5 μl
Message-dependent lysate (MDL)	35 μl
Plasmid DNA	0.5–2 μl (to give ~0.5 μg)
Nuclease-free H$_2$O to final volume of	50 μl

Add appropriate RNA polymerase to give 600–1000 units/ml.

1. Remove the reagents from storage, placing all reagents on ice. Rapidly thaw the MDL by hand warming and place on ice.
2. Following the example above, assemble the reaction components in a 0.5-ml or 1.5-ml sterile microcentrifuge tube, on ice, in the order given above. After addition of all the components, gently mix the lysate by vortexing. If necessary, microcentrifuge at 4°C for 5 sec to return the reaction mix to the bottom of the tube.
3. Start the reaction by transfer to water bath. Incubate at 30°C for 30–60 min.
4. Take 2 × 2.5 μl for TCA precipitation and incorporation analysis. Take 2.5 μl for SDS-PAGE/fluorography.

Preparation of 25X reaction buffer

Amount of stock solution	Concentration	Concentration in reaction
110 μl 1M HEPES-KOH, pH 7.2	110 mM	4.4 mM
40 μl 1 M Mg acetate	40 mM	1.6 mM
50 μl 0.1 M spermidine	5 mM	0.2 mM
200 μl 50 mM ATP	10 mM	0.4 mM
200 μl 50 mM CTP	10 mM	0.4 mM
200 μl 50 mM GTP	10 mM	0.4 mM
200 μl 50 mM UTP	10 mM	0.4 mM

Total volume: 1.00 ml

The following radiolabeled amino acids have also been recommended for use with coupled transcription/translation reactions:

Amino acid	Final concentration in reaction
³H-Leucine (100–200 Ci/mmol)	0.5 mCi/ml
¹⁴C-Leucine (300 mCi/mmol)	5 μCi/ml
³⁵S-Cysteine (1200Ci/mmol)	0.3 mCi/ml

Notes:

(1) Avoid adding calcium to the translation reaction, otherwise, the micrococcal nuclease used to destroy the endogenous RNA in the lysate may be reactivated and destroy DNA and/or RNA templates.

(2) Ensure that all ethanol is removed from the DNA before adding to the reaction.

(3) Use capped plastic vials to avoid changes in reaction volume.

(4) Note that the reaction mix suggested above is for MDL made by lysing 1 volume of packed reticulocytes with 2 volumes nuclease-free water. Many protocols (including those used by vendors) use a 1:1 lysis ratio. In this case, only half the amount of MDL should be used. Typically, the protein content of the lysate is around 100 (1:2)–200 (1:1) mg/ml.

(5) All handling, except for the actual reaction, should be done at 4°C. Avoid freezing/thawing the lysate more than two times.

(6) If synthesizing a recombinant protein for functional analysis, substitute 1 mM nonradioactive methionine and use 2.5 μl/50 μl reaction. It is a good idea to do a reaction with ³⁵S-methionine at the same time as a check on your reagents.

(7) DNA template considerations:

    (a) Optimal results are found with DNA from which all contaminating RNAs are removed by a method that does not involve RNase. However, for fast screening, "mini-prep" DNA may be used. In our hands, the Promega Wizard Prep technique has given the best results.

    (b) Optimal results appear to be obtained with ~1 μg of plasmid DNA/100 μl reaction volume, although we have used from 0.1 to 2 μg of DNA template and obtained adequate levels of translation. The optimal level may change with DNA purification method. In our hands, levels above 1 μg DNA results in no further amounts of protein produced.

    (c) Circular plasmid DNA gives the best translation results although linear DNA templates can be translated.

    (d) The presence of a poly(A)+ sequence downstream of the gene of interest enhances translation.

Quantification and Analysis

Incorporation of radioactivity into trichloroacetic-acid precipitable material

The incorporation of radioactive amino acids into TCA-precipitable material is determined by diluting small aliquots, 2–5 μl, into 1 ml ice-cold water, to which is added 3 ml 10% TCA. The mixture is heated at 90°C for 15 min to discharge aminoacyl tRNAs and then cooled in ice for 10 min. Precipitated radioactive protein is collected by vacuum filtration using glass fiber filters (Whatman GF/C or equivalent). After rinsing with 5% TCA, the filters are dried under an infrared lamp and assayed in a nonaqueous scintillant. Because of quenching due to precipitated hemoglobin, radioactivity should be measured using a wide window.

Although many protocols suggest measuring the percentage incorporation of radioactivity into TCA-precipitable material, this really tells you little except that you remembered to add the radioactive amino acid and that you added more than enough to support translation. If the number of methionines (or other amino acid used) in the translation product is known, it is possible to calculate from the specific activity the total mass of translated protein. This is extremely important if the translation product is assayed for enzymatic activity and in order to determine the product's turnover number. It should be remembered that the reticulocyte lysate contains endogenous amino acids at various concentrations. The concentration of methionine, for instance, may be from 5 to 15 μM. The endogenous concentration of an amino acid may be measured most simply by measuring incorporation, with short incubation times, using two or three levels of radiolabeled amino acid. As described (Safer et al., 1978), the endogenous pool can be calculated from a comparison of incorporated radioactivity at different input levels.

Denaturing gel analysis of translation products

1. Once the 50-μl translation reaction is complete (or at any desired timepoint), remove a 2.5-μl aliquot and add it to 50 μl of SDS sample buffer. The remainder of the reaction may be stored at −20°C. Or, if you will be assaying for activity, store at liquid nitrogen temperatures.

2. Cap the tube and heat at 95°C for 2 min to denature proteins.

3. A small aliquot (5–20 μl) can be applied to the SDS–polyacrylamide gel. It is not necessary to remove the free amino acids by acetone precipitation of the polypeptides.

4. Run the gel at 15 mA until the sample has entered the running gel and then increase the current to 30 mA. Because the dye front contains the free labeled amino acids, it may be easier for disposal if the tracking dye is not run off the bottom of the gel.

5. In order to visualize the incorporated label, we prefer fluorography to autoradiography. The increased detection sensitivity of fluorography is obtained by infusing an organic scintillant (e.g., Enhance by Du Pont) into the gel. Alternatively, the fixed gel can be exposed to a phosphoimaging screen, such as available from Molecular Dynamics. These systems provide greater sensitivity, speed, and the ability to quantitate the radioactive bands.

6. Dry the gel under a vacuum.

7. Expose the gel on X-Omat AR film (Kodak) for 1–6 hr at −70°C (with fluorography), or 6–15 hr at room temperature (with autoradiography).

JAGUS AND PLACE.2

PRODUCTION OF FUNCTIONAL RECOMBINANT INTEGRAL MEMBRANE PROTEINS USING COUPLED TRANSCRIPTION/TRANSLATION REACTIONS

Purpose

To produce functional recombinant integral membrane proteins *in vitro*. All the advantages of the reticulocyte coupled transcription/translation system can be used in conjunction with such processing events as signal peptide cleavage, membrane insertion, translocation, and core glycosylation, by supplementation of the reaction with microsomal membranes. In many cases, such as the terrapin aromatase, where a sufficiently sensitive assay is available, the processed translation product can be assayed for functional activity.

Principle

The coupled transcription/translation system is used, supplemented with dog pancreatic microsomes for cotranslational and initial post-translational processing of proteins, specifically, signal peptide processing, glycosylation, and membrane insertion.

Origin

Walter and Blobel (1983). See also Promega Notes, PN11 and PN038.

General Protocol for Translation with Microsomal Membranes: This is basically the same protocol, with the same considerations, as in Jagus and Place.1. Most investigators will find it most convenient to use Promega's reticulocyte TnT kit, supplemented with Promega's canine pancreatic microsomes.

1. Remove the reagents from the freezer and allow them to thaw on ice.
2. Mix the following components on ice, in the order given, in a sterile 1.5-ml centrifuge tube:

Coupled transcription/translation reaction using canine pancreatic microsomes

2 M KCl/10 mM MgAcetate	2.5 μl
Amino acid mixture minus methionine, 1 mM	2.0 μl
25X Reaction buffer	2.0 μl
^{35}S-methionine (1000 Ci/mmol) at 10 mCi/ml	4.0 μl
RNasin ribonuclease inhibitor (40 U/μl)	0.35 μl
Phosphocreatinine (1 M, freshly prepared)	0.5 μl

Creatine phosphokinase (200 U/ml, CPK, in 50% glycerol)	0.5 μl
MDL	35 μl
Plasmid DNA	0.5–2 μl (to give ~0.5 μg)
Canine microsomal membranes (~2 equivalents, see Note 1)	5 μl
Nuclease-free H$_2$O	to final volume of 50 μl

Add appropriate RNA polymerase (to give 600–1000 units/ml)

3. Incubate at 30°C for 90 min.
4. Analyze the results of translation and processing using TCA precipitation and SDS-PAGE, as described in Jagus and Place.1. Analyze recombinant protein activity if applicable.

Notes:

(1) The activity of the membranes is defined in equivalents. One equivalent is that amount of membranes required to cleave the signal sequence of preprolactin from 50% of the translation products, as measured by a shift in mobility on SDS–polyacrylamide gels. Two and one-half microliters should be sufficient to process over 90% of the preprolactin. The storage buffer for the microsome membranes is 50 mM triethanolamine, 2 mM DTT, 250 mM sucrose. Promega supplies two control mRNAs with its microsomal membrane preparation, that for β-lactamase and the precursor for yeast α-mating factor. These two mRNAs allow for the assay of signal processing and glycosylation, respectively.

(2) While these reaction conditions are suitable for most applications, the efficiency of processing using other membranes (e.g., oviduct microsomes) may vary. Thus reaction parameters may have to be altered to suit individual requirements. In general, increasing the amount of membranes in the reaction increases the proportion of polypeptides translocated into vesicles but reduces the total amount of polypeptide synthesized.

(3) The amount of protein made will be less than that obtained with the coupled transcription/translation system alone. Depending on the construct used, translation efficiency can drop to between 10% and 50% in the presence of microsomal membranes. In our case, the efficiency drops to 10–20%, depending on the quantity of microsomes used.

(4) In some cases it may be difficult to determine by gel analysis alone if glycosylation has occurred. For example, cleavage of a signal peptide and concomitant glycosylation may result in no shift in SDS-PAGE mobility. It may be necessary to treat the product with endoglycosidases (e.g., Endo H or PNGase F) to unmask signal peptide cleavage.

(5) If synthesizing a recombinant protein for functional analysis, substitute 1 mM nonradioactive methionine and use 2.5 μl/50 μl reaction. It is a good idea to do a reaction with ^{35}S-methionine at the same time as a check on your reagents.

JAGUS AND PLACE.3

TRANSIENT EXPRESSION OF RECOMBINANT INTEGRAL MEMBRANE PROTEINS IN COS CELLS

Purpose

The use of heterologous expression systems to study structural and functional aspects of various forms of cytochrome P450 has found wide acceptance. A key criterion for choice of cell type is that the expression system have low endogenous P450 activities but still express the P450 reductase. The COS cell lines have been found to meet these criteria. Transformation of African green monkey cells (CV-1) with an origin-defective mutant of simian virus 40 (SV40) viral DNA resulted in the integration of a single copy of the complete early region of SV40 DNA into the CV-1 genome, establishing the COS-1 cell line. Since COS-1 cells produce the SV40 antigen, any transiently transfected plasmid DNA that contains the SV40 origin of replication will replicate in transfected cells. Common vectors used that provide efficient promoters in mammalian cells are pCD, pSVL, and pCMV.

Principle

Lipid-mediated delivery of DNA is a more efficient alternative to calcium phosphate precipitation and DEAE dextran-mediated delivery for the transfection of eukaryotic cells (Felgner and Ringold, 1989; Duzgures and Felgner, 1993). These reagents are a mixture of a neutral and a cationic lipid that form liposomes that can interact with anionic macromolecules such as DNA. The lipid–DNA complexes are taken up efficiently by cells. The different commercially available cationic lipids differ in their cationic lipid content. LipofectAMINE shows the highest transfection efficiencies with COS-1 and COS-7 cells as well as with most mammalian cell lines (Ciccarone et al., 1993).

Origin

This is basically the same as the published method (Clark and Waterman, 1991) except using the cationic lipid, LipofectAMINE (Life Technologies) as the delivery vehicle. A more comprehensive guide for the use of cationic lipids is *Guide to Eukaryotic Transfections with Cationic Lipids,* Life Technologies, 1994.

Procedure for Transient Expression in COS Cells: The procedure below contains suggested ranges of amounts of DNA and LipofectAMINE suitable for COS-1 or COS-7 cells. If other cell types are used, optimization studies should be carried out. Commercially available cationic lipids are marketed with transfection protocols, which are usually specified for 35-mm dishes. It is important to scale up all cell numbers and volumes linearly with respect to surface area if larger plates of cells are used. For instance, for measuring *in vivo* aromatase activities using [³H] androgens, we usually use 60-mm tissue culture plates; for measuring *in vitro* aromatase activities from prepared microsomes we use 100-mm plates. The cell numbers plated, amount of DNA and LipofectAMINE, and amount of medium used, are multiplied by 3 and by 8, respectively, compared with the amounts specified for 35-mm plates. The following protocol is for use with 60-mm tissue culture plates.

1. For 60-mm tissue culture plates seed ~6 × 10^5 cells per plate in 3 ml Dulbecco's Modified Minimal Essential Medium (DMEM) containing 10% fetal bovine serum (or newborn calf serum is as effective) and nonessential amino acids.

2. Incubate the cells at 37°C in a CO_2 incubator until the cells are 70–80% confluent. This will usually take 18–24 hr. (*Note:* Since transfection efficiency is sensitive to culture confluence, it is important to maintain a standard seeding protocol from experiment to experiment.)

3. Prepare the following solutions in 12 × 75 mm sterile tubes.
   Solution A: For each transfection, dilute 6 μg DNA (plasmid) in 1.125 ml serum-free, antibiotic-free OPTI-*MEM* (Life Technologies). (*Note:* OPTI-*MEM* is a modification of Eagle's Minimal Essential Medium, buffered with HEPES-KOH and sodium bicarbonate and supplemented with hypoxanthine, thymidine, sodium pyruvate, L-glutamine, trace elements, and growth factors. Cells grown in this medium exhibit reduced serum requirements. This characteristic, along with the improved buffering capacity and a reduced phenol red content, makes this medium ideal for use during cationic lipid transfections. Some investigators have used HEPES-buffered Dulbecco's MEM without serum and antibiotics and found equally effective transfection with cationic lipids.)
   Solution B: For each transfection, dilute 6–45 μl LipofectAMINE Reagent in 1.125 ml serum-free, antibiotic-free OPTI-*MEM*. Peak activity should be at about 18 μl, which is the amount we use for expressing terrapin aromatase. (*Note:* It is critical that the lipid–DNA complexes are formed in the absence of serum.)

4. Combine the two solutions, mix gently, and incubate at room temperature for 15–45 min (we use 30 min). The solution may appear cloudy; however, this will not impede transfection.

5. Wash the cells once with 3 ml serum-free, antibiotic-free OPTI-*MEM*.

6. For each transfection, add 2.25 ml serum-free, antibiotic-free OPTI-*MEM* to each tube containing the lipid–DNA complexes. Mix gently and overlay the diluted complex solution onto the washed cells. (*Note:* Do not use antibiotics during transfection because lipids make cells more permeable to these compounds.)

7. Incubate the cells for 5 h at 37°C in a CO_2 incubator.

8. Remove the transfection mixture and replace with the normal growth medium, DMEM with 10% fetal bovine serum.

9. Replace medium at 18–24 h following start of transfection.

10. Assay cell extracts for gene activity 24–72 h after the start of transfection, depending on cell type and promoter activity. For our aromatase construct, we use 72 h.

MCCARTHY.1

IN SITU HYBRIDIZATION WITH SSDNA PROBES GENERATED BY PCR

Purpose

To localize and quantify specific mRNAs at the cellular level.

Principle

Radiolabeled single-strand DNA probes hybridize selectively to target mRNAs in thin tissue slices. Signal is localized by autoradiography of liquid emulsion directly over the tissue.

Origins

This protocol is based on one established by Brooks et al. (1993).

1. *Preparation of slides*

a. All slide racks and dishes to be used should be autoclaved.

b. Load precleaned glass slides into metal racks wearing gloves.

c. Soak in 0.2 N HCl in 95% ethanol for 30 min or more.

d. Rinse thoroughly in deionized water.

e. Dry slides thoroughly either by letting stand overnight or in an oven.

f. Sub slides by dipping in a fresh solution of 2% 3-aminopropyltriethoxysilane (Sigma #A-3648) in dry acetone for 1 min.

g. Rinse in sterile H_2O twice, emphasize length and agitation.

h. Dry overnight at 42°C.

2. *Tissue preparation*

a. Fresh frozen tissue stored at -70°C should be sectioned on a cryostat at 8–10 μm.

b. Tissue sections should be thaw mounted onto organosilane-subbed slides, briefly heated to 37°C and then either stored at -70°C or immediately postfixed (meaning the same day) according to the following procedure.

3. *Tissue fixation*

Note: Quantities can be approximated by beaker markings except for SSC. All beakers and slide racks must be baked or autoclaved. All buffers must be made with DEPC-treated autoclaved water in order to destroy any RNase. See buffer recipes below.

a. Immerse slides in 4% paraformaldehyde in 100 mM PBS for 5 min.

b. Immerse slides in 100 mM PBS for 5 min.

c. Immerse slides in 100 mM PBS plus 5 mM dithiothreitol for 5 min.

d. Immerse slides in 50% EtOH for 2 min.

e. Immerse slides in 70% EtOH for 2 min.

f. Immerse slides in 95% EtOH for 2 min.

g. Immerse slides in 100% EtOH for 2 min.

Store slides in desiccator under vacuum either overnight or for a couple of hours.

4. *Radiolabeled ssDNA probe preparation*

a. Perform a standard PCR reaction for your desired probe. Product should be in the range of 200–700 bp. Purify the PCR product using any quick spin columns such as Quiagen PCR purification kit.

b. Take about 200 ng (1 μl) of the previously generated PCR template and prepare a PCR reaction in a final volume of 30 μl. Exclude dCTP from the nucleotide mix (or add 0.2 mM dCTP if you want a low specific activity probe) and only add one primer (downstream for antisense, upstream for sense control).

c. Add an appropriate volume of ^{33}P-dCTP to the incubation mixture (3.0–12.0 μl).

d. Run the reaction in the thermocycler using the same annealing conditions as those used to generate the template but run for 50–75 cycles.

Note: This technique can NOT be used to generate ^{35}S-labeled probes because this nucleotide should not be used in the thermocycler as it is highly volatile and will contaminate your machine.

e. Take a 1-μl aliquot to dilute 1:9 and then perform the TCA precipitation assay to determine specific activity of your probe (see RNase protection protocol).

f. Run the remainder of the reaction through a G50 spin column (Boerhinger-Mannheim).

5. *Prehybridization and hybridization*

a. Number slides sequentially and record assay number, original slide number (cryostat section number), tissue type, probe, and stringency in a master table.

b. One-milliliter aliquots of prehybridization buffer (see recipe below) should be stored in the freezer. Calculate an approximately 200-μl volume of buffer/slide. Remove the appropriate number of vials from the freezer to give you half the total volume needed. It is always advisable to overestimate the total volume needed as these buffers are very viscous and you will lose a lot in the pipetting procedure.

c. Add 10 μl of calf thymus DNA (10 mg/ml) per ml prehybridization buffer. Denature by boiling for 10 min. Cool in an ice-water bath, add 10 μl/ml of 100 mM DTT. Add 100% formamide in a 1:1 ratio.

d. Hybridization should be done in Nalgene-style boxes with removable lids and lined with filter paper. Slides are elevated above the filter paper by placing them on tongue depressors or sterile wooden sticks.

e. Pipet approximately 200 μl total (~35 μl/section of tissue) prehybridization buffer onto slides. Spread buffer to completely cover tissue with a glass spreader made from a Pasteur pipette. Prehybridize at room temperature for 3–4 hr.

f. Rinse individual slides in a beaker containing 2X SSC (see recipe below) prior to hybridization. Wipe excess buffer off the backs and sides of the glass slides.

g. Prepare hybridization buffer (see recipe below) in the same manner as prehybridization buffer.

h. Add the labeled probe to the 100% formamide and then add this in a 1:1 ratio to the denatured hybridization buffer. Initial assays should include a range of cpm/tissue section. For example, different slides should be hybridized with 250,000, 500,000, 1,000,000, and 2,000,000 cpm. The objective is to find the concentration at which the signal is no longer increasing (i.e., saturation) but there is not a marked increase in background.

i. When all the hybridization buffer is placed on the slides, soak the filter paper with a 1:1 mix of formamide:2X SSC. This will maintain a moist environment in the chamber but will not cause excess condensation on the lid of the box. If there is some concern for condensation dripping onto the slides or the slides drying out, you can use parafilm coverslips placed on each slide. However, at this hybridization temperature that is usually not necessary.

j. Hybridize overnight at 42°C in a humid oven.

6. *Posthybridization washes*

a. Wash slides twice in 1X SSC at room temperature. The first wash is just a dipping of individual slides in a small beaker of buffer to remove excess radioactivity. Buffer from this wash is radioactive and must be transferred to liquid radioactive waste. Transfer the slides to metal slide racks and, for the second wash, place the slides in a beaker containing ~800 ml 1X SSC for about 60 min. This buffer is not radioactive waste.

b. Transfer slides in metal slide rack to a new beaker containing ~800 ml of 0.1X SSC for overnight at room temperature.

c. Repeat the next morning.

d. In afternoon, wash slides in fresh 0.1X SSC at 60°C for 1 hr, by placing the beakers in a water bath until temperature is equilibrated. Transfer slides in at this time. After 1 hr, remove the 60°C slides from the bath and transfer to a beaker containing fresh 0.1X SSC at room temperature for an additional 1–3 hr.

e. Dehydrate slides in alcohol and ammonium acetate before placing on film overnight (preferably with an enhancing screen at −70°C).

7. *Dehydration protocol*

a. Mix 20.8 g ammonium acetate in 900 ml of autoclaved water. Titrate pH to 5.5 with acetic acid. Using 100% EtOH, set up 4 beakers as below and then dip slides in each for 1 min.

400 ml Ammonium acetate   400 ml EtOH

240 ml Ammonium acetate	560 ml EtOH
80 ml Ammonium acetate	720 ml EtOH
	800 ml EtOH

b. Desiccate either overnight or for several hours before dipping. If placing on film you don't need to worry about this but dry off excess moisture with kim wipes from bottom of slide and around edges.

c. Develop film the next day and if it looks good proceed to emulsion dipping the slides. The purpose of exposing the slides to film is to see if they are in fact worth dipping (emulsion is expensive) and also to give an indication of the intensity and specificity of the signal. It is also possible to perform quantitative densitometry on the films.

8. *Emulsion autoradiography*

a. Photographic emulsion can be purchased from Kodak (NTB-2 or NTB-3) or from Ilford Chemicals. Some investigators dilute the emulsion with H_2O by $1/2$ to $1/3$. This not only essentially gives you more emulsion, it also makes a thinner coating of emulsion on the slides. However, we have generally found that using Kodak NTB-3 undiluted gives a smooth, even coating of emulsion on the slides and avoids the problems that can occur when trying to dilute the emulsion (i.e., spillage, light exposure, contamination).

b. Remove liquid emulsion from 4°C and allow to equilibrate to room temperature for ~1 hr.

c. Place room temperature emulsion in the heated water bath at 44.5–45°C for ~1 hr (work in safety light conditions only but keep the lid on the emulsion at this point).

d. Place glass slides in plastic slide grippers for dipping. Put the frosted end of the slide in the gripper, each one holds up to 5 slides.

e. Remove the lid from the emulsion and vertically dip each set of slides into the emulsion. After dipping, hold slides over the emulsion for a few seconds to allow excess to drip back into the container. Then vigorously shake the excess emulsion off the slides with 3–4 hard jerks of your wrist. Keep the amount of time dipping, draining, and shaking constant between each set of slides.

f. Dry the slides by laying on the thin edge, still in the gripper, for ~1 hr, then invert the gripper so that slides are standing and dry for an additional 1–3 hr.

g. When slides are completely dry, remove from the grippers and place slides in small black slide boxes containing 3–4 humicaps as desiccant. Tape the outside of the boxes with black electrical tape, wrap in aluminum foil, label, and store desiccated at 4°C. It is always a good idea to set aside 1–2 boxes of extra slides as "test boxes" that can be developed at intervals to decide when the experimental slides are sufficiently exposed.

(*Note:* All of these procedures can be done under safe light conditions but emulsion is extremely sensitive and so exposure should be kept to a minimum. DO

NOT use a sodium-based (orange glow) safety light, as emulsion is sensitive to sodium-based emissions.)

h. Depending on the specific activity of the probe and abundance of the targeted message, test slides should be developed beginning at 1 week.

9. Slide developing

a. Remove box of test slides from 4°C and allow to equilibrate to room temperature.
b. Place slides in metal slide racks under safe light conditions.
c. Filter previously prepared Kodak Dektol Developer 1:1 with dH₂O into an appropriate sized container. Filter Kodak Rapid Fixer (do not dilute). Prepare three containers containing only dH₂O.
d. Bring all solutions to 15°C by incubating on ice.
e. Develop slides according to the following schema:
   (1) 4 min in Dektol
   (2) 1 min in dH₂O (this is the stop bath, do not use commercial stop bath)
   (3) 10 min in Rapid Fixer
   (4) 10 min in dH₂O
   (5) 10 min in dH₂O
   Continuing to rinse the slides under a stream of dH₂O at this point helps to reduce generalized crud before counterstaining and cover-slipping. Also, while slides are in dH₂O you can scrape the emulsion off the backs of the slides with a razor blade (be sure it's the back!) to reduce the amount of counterstain taken up by the emulsion.
f. Counterstain the slides with cresyl violet or hematoxylin and eosin according to standard histological protocol. In general, it will take longer to stain and differentiate the tissue but you should keep the staining very light so that the silver grains can easily be visualized over the cells.
g. Cover-slip and you're done!

10. Buffers and solutions

a. Prehybridization buffer

Stock solution	Amount of stock to tube
(1) NaCl, 5.0 M = 14.61 g/50 ml	12 ml
(2) Tris 7.6 = 2.98 g/20 ml	1 ml
(3) BSA 1.6% = 600 mg/10 ml	167 µl
(4) EDTA 250 mM = 1.681 g/20 ml (place in hot water to dissolve if necessary)	400 µl
(5) Na pyrophosphate 5% = 2 g/40 ml	500 µl
(6) Ficoll 6% = 600 mg/10 ml	167 µl

(7) PVP 6% = 600 mg/10 ml (polyvinyl
    pyrrolidone, catalog# P5288, Sigma)   330 μl

(8) Yeast tRNA, 50 mg/ml   100 μl

(9) Salmon testis DNA 10 mg/ml   5000 μl

   (shear in syringe several times prior to adding to the buffer using up to a
   25-gauge needle)

(10) Add autoclaved double-distilled water to make 50 ml volume.

(11) Aliquot 1 ml portions to labeled 1.5-ml microfuge tubes and freeze.

b. Hybridization buffer: All ingredients are the same as above with the addition of:

   (1) Dextran sulfate = 20%   10 g

   (2) Reduce Salmon testis DNA to   1000 μl

   (3) Add autoclaved double-distilled water to make 50 ml volume.

   (4) Aliquot 1 ml to labeled 1.5-ml microfuge tubes and freeze. Pipet slowly to
       ensure fairly accurate 1-ml aliquots. Buffer may need to be heated to get all
       of the dextran sulfate into solution before pipetting.

   (5) At the time of the assay add 10 μl/ml of 5 mM DTT and 10 μl/ml of calf
       thymus DNA (10 mg/ml).

   (6) Denature hybridization buffer by heating to 100°C for 10 min and then
       cooling in an ice-water slush. Dilute with 100% formamide 1:1 at time of
       assay. The radiolabeled probe should be added to the formamide prior to
       adding it to the hybridization buffer.

c. Paraformaldehyde stock solution (8%) in 1X PBS: Weigh 160 g of parafor-
    maldehyde and place in hood covered. Measure 500 ml 4X PBS and 1.5 liters of
    autoclaved water into flask, mark 2-liter line, and remove and reserve 500 ml in
    autoclaved graduated cylinder. Use large stir bar. Heat no higher than 50–60°C
    (do NOT autoclave) while adding paraformaldehyde gradually in small quan-
    tities via wide funnel. After all paraformaldehyde is in solution, add reserved 1X
    PBS to 2-liter mark. Clear with NaOH (pellets) and store at 4°C; a plastic
    container can be used.

d. PBS stock solution (400 mM or 4XPBS): Dissolve 30.92 g of $NaH_2PO_4 \cdot H_2O$
    (monobasic), 195.4 g Na_2HPO_4 (dibasic), and 144 g NaCl into 4 liters of auto-
    claved water.

e. SSC stock solution (20X): Dissolve 87.7 g of NaCl and 44.1 g of sodium citrate
    in 500 ml autoclaved water.

MCCARTHY.2

LYSATE RIBONUCLEASE PROTECTION ASSAY

Purpose

To quantify specific mRNAs in extremely small tissue samples (~100–200 μg
protein) without prior extraction of RNA.

Principle

Radiolabeled antisense RNA probes are allowed to hybridize with target mRNA present in tissue. The resultant RNA:RNA hybrids are resistant to digestion by RNase. Enzymatic digestion of all unhybridized RNAs results in "protected" RNA fragments of predicted size (based on probe size) that are separated by electrophoresis and visualized by autoradiography. The experimental signal is quantified by standardization to a control gene also probed for in each sample. By hybridizing the probes directly in tissue, the lysate version of the RNase protection assay avoids the variance inherent in RNA extraction and recovery.

Origin

This protocol is based on procedures established by Strauss and Jakobowitz (1993) and the Ambion Inc. (Austin, TX) *Direct Protect Lysate Ribonuclease Protection Assay Kit™* (catalog #1420).

1. *Tissue preparation*

a. Microdissect fresh frozen tissue and place directly into Hybridization Buffer (see recipe below) in a 1:10 (w/v) ratio.

b. Sonicate sample, either with an internal probe sonicator that has to be cleaned between each sample or preferably with an external water-based sonicator. Wear ear plugs!

c. Samples can either be stored at $-70°C$ for up to 6 months or radiolabeled riboprobe is immediately added and hybridization is allowed to proceed overnight at $37°C$.

2. *Probe preparation*

a. Mix the following ingredients together in a microfuge tube:

5X Transcription Buffer (comes with the enzyme)	6 μl
DTT, 100 mM	2 μl
RNasin (40 units)	1 μl
rATP (10 mM)	1.5 μl
rCTP (10 mM)	1.5 μl
rGTP (10 mM)	1.5 μl
rUTP (0.2 mM)	1.5 μl
DNA (linearized plasmid or PCR product; 0.2–0.8 μg)	1.0 μl
^{32}P-rUTP (10 mCi/ml, 800 Ci/mmol)	5–13 μl
RNA polymerase (10–20 Units/ml)	1.5 μl
H$_2$O bring up to total of	**30 μl**

b. Incubate at RT for 1–2 hr.

Note: If using SP6 polymerase you will greatly increase yield by incubating at

37°C; however, this may also increase the probability of getting incomplete transcripts, which may produce extra protected bands on your gel. In general, a lower template concentration and shorter incubation period at room temperature will increase the fidelity of your probe. Also, the specific activity of the probe can be manipulated by varying the amount of cold rUTP, but you should always have some or the efficiency of the reaction is greatly reduced.

c. Add 1.5 μl of RNase-free DNase (15 units) and incubate for 30 min at 37°C.

d. Purify radiolabeled probe by G50 spin column (Boerhinger-Mannheim). It helps to increase the volume to 50–100 μl just before pipetting it onto the column.

Note: Many people consider a gel-purified probe to be far superior in that it ensures your probe is all of one length. We have not found that necessary so far.

e. Add appropriate amount of probe (0.1–10 ng in less than 10 μl volume) to tubes containing tissue and hybridization buffer. The radiolabeled probe cannot be stored for more than a day or two as it will rapidly undergo radiolysis.

f. Hybridize overnight at 37°C.

3. *RNase digestion*

a. Add 500 μl of RNase digestion buffer (see recipe below) to each tube. Vortex, briefly spin, and incubate at 37°C for 1 hr.

b. At completion of RNase digestion, a 30-μl aliquot of sample can be removed and used for subsequent protein quantification as an additional control.

c. After incubation, add 20 μl of 10% lauryl sarcosyl and 10 μl of proteinase K (10 mg/ml) to each tube. The lauryl sarcosyl will inactivate the RNase before it and the tissue proteins are degraded by the proteinase K. Mix, spin, and incubate at 37°C for 1 hr.

d. Add 500 μl of isopropyl alcohol to precipitate the RNA; mix well.

e. Place tubes at −20°C for ∼1 hr. It is not necessary to add carrier during this precipitation.

4. *Separation and detection of protected fragments*

a. During the RNase digestion, prepare a 6% polyacrylamide gel (0.75 mm) with a 20-tooth comb. For two gels this requires : 35 ml dH$_2$O, 5 ml 10X TBE, 10 ml 30% acrylamide:bisacrylamide (29:1), 300 μl fresh 10% ammonium persulfate, and 30 μl TEMED. Pour immediately into vertical gel apparatus (we use Hoefer Scientific SE400 sandwich gel and with two apparatuses can run four gels and up to 75 samples). Note that you do not need to use a denaturing gel to get good band resolution, but run the gel at ∼150–175 volts or 20 amperes.

b. Remove tubes from freezer and microfuge at 4°C for 30 min at 10,000g.

c. Pour supernatant off being careful not to dislodge the pellet. Remember that this liquid is radioactive waste!

d. Re-spin tubes for an additional 5 min and now carefully pipette off the remaining supernatant. This can be done with a glass pipette on which the tip has been drawn out very thin or we find the flattened tips used for gel loading to be useful for this purpose.

Note: It is absolutely critical to remove all the supernatant at this point to avoid aberrant migration of bands in the gel. Do NOT air dry the samples as this will leave residual salts. An alternative approach has been used by Strauss and Jakobowitz (1993). Following the proteinase K digestion, the samples are first phenol and then chloroform extracted, precipitated with 100% ethanol and the pellet washed once with 95% ethanol prior to air drying. This substantially increases the amount of work but we have recently found a reduction in variability between samples by incorporating these additional steps.

e. Add 10–15 µl gel loading buffer to each sample. Vortex vigorously and microfuge before loading on the gel. Do NOT heat the samples if you are running a nondenaturing gel.

f. Electrophorese at 175–200 volts using 0.5X TBE running buffer until the bromophenol blue exits the bottom of the gel (3–5 hr).

g. Remove the gel from the apparatus, after removing one glass plate to expose the gel, fix it by gently immersing in a 7% acetic acid solution for 5 min. Rinse in dH_2O for 2 min, mark the gel by notching one corner, then dry in a gel dryer. We find that the Hoefer Scientific Easy Breeze Gel Dryer gives a uniformly smooth dry gel that is not overly brittle.

h. Expose the dried gels to X-Omat radiographic film for 2–24 hr with an enhancing screen if necessary.

5. *Quantification and analysis*

a. Films can be analyzed by densitometry and signal standardized to the housekeeping gene band present in each lane.

b. In addition, by lining up the gel with the autoradiogram, the bands can be cut from the gel and quantified in a scintillation counter. We have found this to be a reliable method of quantification. Using a light box to align the autoradiogram and the gel, and scotch taping the back of the gel before cutting out the bands, greatly facilitates the procedure. The experimental signal is again standardized to the housekeeping band present in each lane.

c. Quantification of protein in the 30-µl aliquot removed prior to proteinase K digestion allows for a third possible standardization between signals.

d. In order to make quantitative comparisons it is essential that your probe be at saturating concentrations in relation to your target mRNA. This needs to be determined empirically by increasing amounts of tissue in the presence of excess probe. A linear increase in signal indicates a saturating level of probe. Probes of relatively high specific activity (5×10^8 to 5×10^9 cpm/µg) are useful for establishing the appropriate amount of probe. If the target mRNA is in abundance, the specific activity (but not the total amount) of the probe can be reduced in subsequent experiments.

6. *Calculation of riboprobe-specific activity*

Below are sample calculations based on using ^{32}P-rUTP, 800 Ci/mmol, 10 mCi/ml.

a. Determine proportion of ^{32}P-rUTP incorporated by TCA precipitation as follows.

 (1) Remove a 1-μl aliquot from the incubation mixture prior to G50 column purification. Dilute 1:9 and then spot 1 μl on each side of a Whatman GC filter. Allow to air dry or place under a heat lamp.

 (2) Cut the filter in half. One side becomes the total, the other side is washed 3–4 times in ~300 ml of ice-cold TCA buffer (20 mM sodium pyrophosphate, 5% trichloroacetic acid). Wash the filters for 2 min each time with swirling.

 (3) Immerse the filters briefly in ice-cold 70% ethanol. Allow to dry at room temperature or under a heat lamp.

 (4) Place each filter in a scintillation vial and count.

  If prewash sample = 2,300,000 cpm and postwash sample = 1,500,000 cpm, then 1,500,000/2,300,000 = 65% of ^{32}P-rUTP was incorporated.

b. Determine the number of moles of ^{32}P-UTP in the reaction.

 If added 9 μl to the reaction, then:

 9 μl (^{32}P) $\times$ 0.01 mCi/μl = 0.09 mCi

 0.00009 Ci/800 Ci/mmol = 1.13 $\times$ 10^{-7} mmol

 = 1.13 $\times$ 10^{-10} mol = 0.113 nmol in the reaction.

c. Determine the number of moles of unlabeled rUTP in the reaction.

 If added 1.5 μl of 0.2mM, then:

 (200 μmol/1000 ml) $\times$ 1.5 μl $\times$ (1 ml/1000 μl) = 300 $\times$ 10^{-6} mmol

 = 0.3 $\times$ 10^{-9} mol or 0.3 nmol

d. Determine the total amount of UTP in the reaction.

 0.113 nmol + 0.30 nmol = 0.413 nmol total UTP

e. Determine how much total UTP was incorporated into the riboprobe

 0.413 nmol $\times$ 65% incorporation = 0.269 nmol

f. Determine how much riboprobe was synthesized.

 Assuming equimolar amounts incorporated of all four nucleotides, then incorporated 0.269 nmol of each. Sum of MW of the ribonucleotides is ~1320 daltons, so multiply by 0.269 = 355.1 ng synthesized.

g. Determine how many cpm were incorporated into riboprobe.

 Final reaction volume = 31.5 μl, TCA precipitation = 0.1 μl, so 31.5 μl/0.1 μl $\times$ 1,500,000 = 472,500,000 cpm in RNA

h. Calculate the specific activity of the probe.

 472,500,000 cpm/355.1 ng $\times$ 1000 ng/μg = 13.3 $\times$ 10^8 cpm/μg

i. Calculate the mass of probe per sample.

 Added 4,000,000 cpm/sample

 divided by 13.3 $\times$ 10^8 cpm/μg $\times$ 1000 ng/μg = **3.01 ng probe/sample**

7. *Controls and characteristics of the probes*

a. You must have a suitable housekeeping gene that can be used as an internal standard in each lane. Templates for synthesis of various genes can be purchased commercially and include species-specific (rat, mouse, human) templates for β-actin, GAPDH, 18S, and 28S RNA. Of course, it is important to use a control probe that is not regulated by the same stimuli as you are attempting to quantify. There are no "perfect" control genes though, so you just do the best you can.

b. The signal for the housekeeping gene is then used to standardize that for the experimental signals, such that the experimental signal is expressed as a ratio of that for the housekeeping gene.

c. Each assay should include a probe control in which the probes are exposed to RNase digestion in the absence of any tissue. This should result in no signal remaining. If there is substantial signal in this lane it suggests you have DNA template contamination, a self-hybridizing (and therefore self-protecting) probe, or insufficient RNase.

d. Specificity of the probe can further be demonstrated by constructing a standard curve of unlabeled sense probe that is then hybridized to the radiolabeled antisense probe.

e. For sufficient resolution between bands, construct riboprobes that differ in size by 200–300 nt, but don't exceed 900 nt. With increasing size of the probe, integrity of the mRNA becomes more critical.

For a thorough and excellent discussion of other aspects of the Lysate RPA assay, see the instruction manual for the Direct Protect™ kit from Ambion (catalog #1420).

8. *Buffer recipes*

a. Hybridization/lysis buffer

Guanidine thiocyanate @ 4 M	47.3 g
1% β-Mercaptoethanol	1 ml
dH$_2$O	bring volume to 100 ml

b. 50X RNase cocktail (store at −20°C)

RNase A @ 1 mg/ml	10 mg
RNase T1 @ 20,000 U/ml	200 µl
Hepes @ 10 mM	24 mg
NaCl @ 20 mM	12 mg
Triton X-100 @ 0.1%	10 µl
EDTA @ 1 mM	4 mg
Glycerol @ 50%	5 ml
Sterile H$_2$O	bring volume to 10 ml

c. RNase digestion buffer (store at RT)

Tris-HCl @ 10 mM	1.57 g

NaCi @ 300 mM	19.48 g
EDTA @ 5 mM	1.86 g
dH$_2$O	bring volume to 1 liter

MCCARTHY.3

GENE KNOCKOUT BY *IN VIVO* ANTISENSE OLIGODEOXYNUCLEOTIDES

Purpose

To temporarily and selectively reduce the synthesis of a particular protein.

Principle

Synthetic DNA designed to be complementary to a targeted mRNA will selectively hybridize to it and prevent its translation into protein, thereby blocking gene expression.

Origins

This protocol is based on my personal experience as well as the published literature. More extensive reviews of the technique can be found in McCarthy et al. (1993a) and McCarthy (1994b).

1. *Selection of the antisense oligonucleotide*

a. Presumably you have first done your homework and established that your protein is a reasonable candidate for knockout by antisense oligonucleotides and the only thing left to do is try it. Criteria for being a "reasonable" protein include: (1) does not have an excessively long half-life or is extremely abundant, (2) does not share a high degree of homology (at the mRNA level) with other closely related proteins, (3) is not constitutively expressed at a baseline level that will be difficult to reduce (in general, it helps to have a gene product for which you know the temporal and stimulus constraints of its expression), (4) its expression is discretely enough localized that you can reasonably target it, and (5) you have a method to quantify any changes in its levels.

b. Design an antisense oligonucleotide that is 15–22 nucleotides long. First make sure that you have the proper orientation. One way to check is that if your antisense spans an AUG translation start codon, it should read CAT when in the 5' to 3' orientation, which is the direction you want for ordering or synthesis.

c. Establish that your oligo does not have significant homology to other known mRNAs than the one you are targeting. This can be accomplished by using the BLAST program from NCBI available via the internet (blast-@ncbi.nlm.nih.gov). To receive a complete copy of the manual send an initial message of "help" to this address. After completing a search, if you have 10–12 contiguous bases that are homologous to another mRNA, this probably should be rejected as your antisense oligonucleotide unless the homology is strictly in the 3′ untranslated region of the other mRNA.

d. Control oligonucleotides should be at least two of three varieties: (1) *Scrambled oligonucleotides* (sometimes called *nonsense oligonucleotides*) that consist of the same nucleotide content as the antisense oligonucleotide but in a random order. The purpose of this oligonucleotide is to control for nucleotide content, but it does not control for secondary structure or nonspecific sequence effects. This oligonucleotide also has to be checked for sequence homology to other known mRNAs. (2) *Missense oligonucleotide* in which only a single nucleotide is altered from the antisense oligonucleotide. This will usually control for nonspecific sequence or structural effects but not nonspecific hybridizations. (3) *Sense oligonucleotide*, which is simply the same sequence as the mRNA and therefore not complementary to it. This will control for structural features such as palindromes and stem-loops but does not maintain composition.

e. Once you have the sequences you need, the oligonucleotides can be ordered from any number of commercial companies or synthesized at the core facility available at your university. The most important issues are to make sure to obtain sufficient quantities (make clear to the manufacturer that you are ordering oligos for antisense experiments as opposed to PCR, etc.) and purity. The oligonucleotide MUST be purified before being used in biological systems, preferably by HPLC although gel purification can suffice. If purification is done by HPLC, take steps to ensure that all nitryl residues are removed.

f. If rapid degradation by endogenous nucleases is of particular concern (which it should not be in the brain) you can order modified oligonucleotides that are resistant to DNase digestions. The most popular is the phosphorothioate modification in which a sulfur group is substituted for a phosphate on the phosphodiester backbone. However, at lengths of 15 or more, phosphorothioate oligos become extremely toxic. To protect against this you can have your oligonucleotide synthesized as a chimera in which the 3–4 bases on the 5′ and 3′ ends are phosphorothioate but the internal nucleotides are unmodified. This will offer protection against the endonucleases, which are the most serious threat to oligonucleotide integrity, without significant toxicity.

2. *Administration of the oligonucleotide*

a. Depending on your target organ, you can administer the oligonucleotide by injection, osmotic mini-pump, or just about any way you can think of that will give a sustained release. In general, infusion into large peripheral structures is going to be problematic. Targets that work well include the brain, the testis, and very localized subcutaneous infusions (Whitesell et al., 1991).

b. Timing and frequency of oligonucleotide infusions are critical issues. If the onset of gene induction is established or can be manipulated (i.e., light, steroids, behavior), this offers considerable advantage. Administration of oligo 6–12 hr prior to gene expression onset and again 6–12 post onset is a good way of bracketing the time of maximum mRNA synthesis. In the brain, once or twice daily infusions have proved effective (Wahlestedt et al., 1993; McCarthy et al., 1994a, 1994b).

c. Vehicles can also be varied to suit your needs. Most investigators use physiological saline or artificial CSF. However, sesame oil has also been used with some success when the number of infusions is limited to one or two. The oligonucleotides are not soluble in oil and must be suspended by heating and sonication but indirect evidence suggests that this may prolong the action of the oligonucleotides by slowing their release.

3. *Quantification of gene knockout*

a. First, you should establish that your antisense oligonucleotide treatment has had the predicted behavioral or physiological effect but that your control oligonucleotides have not.

b. Second, you MUST establish that the concentration of your targeted protein has been reduced. Furthermore, you must establish that a closely related but untargeted protein has not been reduced; in other words, you have not nonspecifically reduced all protein synthesis with your antisense oligonucleotide.

c. Third, quantification of the targeted mRNA is an option but not a necessity. Reduction in the mRNA requires the activity of RNase H as a mechanism in antisense oligonucleotide effects and it has not been established that this is the case for all tissues or cell systems.

Note: When dealing with antisense oligonucleotide studies, expect anything. As the number of studies using this approach proliferates, so does the number of unexpected, unexplained, but undeniable effects of antisense oligonucleotides.

PALUMBI.1

SEQUENCING SINGLE-STRAND ASYMMETRIC PCR PRODUCTS

When single-strand PCR products are consistently obtained, single-strand sequencing is the preferred method of sequencing. It is the easiest, quickest, and least expensive of all the sequencing methods we generally use. However, and here's the rub, obtaining the proper amount of single sequencing template consistently is tricky (see Single-Strand DNA Amplifications). We suggest that if you begin with this method of sequencing you should move on to other methods if you are unable to produce readable sequences consistently.

1. *Asymmetric PCR:* Use conditions that are similar to those that are successful for double-strand PCR for your particular template. Adjust concentration of primers so that one of them is at about 1/100th normal. This is the "limiting" primer (see

below). Asymmetric PCR is discussed in detail by Gyllensten and Erlich (1988) and Palumbi (1995b).

2. ***Purification and concentration of template:*** Prior to sequencing, the PCR buffer, excess primer(s) and nucleotides left over from the PCR reaction mix must be removed, and the DNA concentrated. We accomplish this in two ways: propanol precipitation or Centricon-30 tubes (Amicon, Inc.). Both methods work equally well. We initially used Centricon tubes but moved to propanol precipitation because it was much less expensive.

a. Propanol precipitation of PCR products

    (i) For 100 μl of PCR reaction add:

        50 μl of 7.5 M NH$_4$OAc

        150 μl of 2-propanol

    (ii) Mix well and incubate at room temperature for at least 10 min.

    (iii) Microcentrifuge for 8–10 min.

    (iv) Remove supernatant. Wash with 500 μl of 70% EtOH.

    (v) Microcentrifuge for 2 min. Dry pellet.

    (vi) Resuspend in 8–20 μl of 0.1X TE buffer.

b. Ultrafiltration purification

    (i) Place 2.0 ml of 2.5 M NH$_4$OAc in top half of a Centricon-30 tube; add the PCR product, avoiding mineral oil, which will cling to the plastic pipet tip (before dispensing the product wipe the pipet tip with a Kimwipe). For lower capacity Millipore tubes, reduce the amount of NH$_4$OAc.

    (ii) Centrifuge it at 1500g for 15–20 min (Centricon) or 5–10 min (Millipore). For Millipore tubes, do a pilot experiment the first time you use a new centrifuge. Over-spinning Millipore tubes can result in sample attachment to the filtration membrane. Centricon tubes are designed to prevent this problem.

    (iii) Repeat this 2 more times with dH$_2$O, removing water from bottom half of the filter assembly when needed.

    (iv) For Centricon tubes, invert tube and spin 20 sec to bring concentrated sample into tip of conical tube. This should yield about 20–40 μl of sample. For Millipore tubes, remove 20–40 μl with a micropipette.

    (v) Store sample at −20°C.

3. ***Primer annealing:*** Once the PCR product is purified and concentrated you are ready to anneal the primer to the template to initiate sequencing. When you are generating single-strand template via asymmetric amplification, you must use the primer that was limiting or absent in the amplification. The primer that was more abundant in the PCR reactions will produce the most DNA. This DNA will remain single strand because there is not an equal amount of the other strand of DNA to complement with it. The "limiting" primer complements with this excess single-strand DNA and therefore must be used in the sequencing reaction.

a. Use 7 μl of the purified (e.g., precipitated or filtered) ssNDA template.

b. To this, add 1 μl of 10–100 μM primer and 2 μl of 5X Sequenase® buffer.

c. Incubate at 65°C for 2–5 min and then let cool to room temperature in the heating block (for 30 min or so). Many people shorten the annealing step by incubating the annealing reaction for 5 min at 65°C and simply moving to the sequencing steps.

4. *Sequencing:* We use the Sequenase® System from US Biochemical.

a. Add 2.5 μl of ddGTP (labeled Mix G), ddATP (labeled Mix A), ddTTP (labeled Mix T), and ddCTP (labeled Mix C) to separate tubes (or racks). These are the termination mixes. For each template there should be four separate tubes labeled G, A, T, and C.

b. Make up Labeling cocktail. For each sample add:

  (i) 1.0 μl DTT (dithiothreitol, 0.1 M)

  (ii) 2.25 μl dH₂O.

  (iii) 1.75 μl 1X TE (we do not use the Sequenase® dilution mix; we have gotten better results with TE in side-by-side comparisons).

  (iv) 0.25 μl ³⁵S-ATP (US Biochemical recommends 0.5).

  (v) 0.25 μl Sequenase® (US Biochemical recommends 0.5). (*Note:* Using this procedure, we do not normally use the Labeling mix provided by US Biochemical. There seems to be enough nucleotides left over from the purification step to enhance sequencing without the labeling mix step.)

c. Add 5.5 μl of the Labeling cocktail to the annealed templates.

d. Pipette 3.5 μl of this mix into each of the four ddNTP tubes and incubate for 2–5 min at room temperature. (Do not let this reaction go longer than 5 min.)

e. Add 4 μl of Stop Solution to each ddNTP mix.

f. Store reactions in the freezer until ready to load onto acrylamide gel.

PALUMBI.2

SEQUENCING DOUBLE-STRAND PCR PRODUCTS

In the past, we have had trouble with this method of sequencing. Gene-cleaning the double-strand product coupled with the addition of a nonionic detergent during the primer-annealing step (Bachmann et al., 1990) seems to have cleared up most of the previous problems. Recently, we have been consistently generating sequences in excess of 400 base pairs using this method.

Unfortunately, this method is especially sensitive to "junk" left over from the PCR reaction. As a result, double-strand templates must be prepared very carefully. We generally cut the DNA product out of a 1% TAE gel and Geneclean (Bio101, Inc.) this slice. While this initial "work" is tedious, the final pure, concentrated template is generally sufficient for four or more sequencing reactions. In addition, because both DNA strands are present it is possible to sequence in either direction. Thus there is the potential to double the amount of data that can be collected with each PCR amplification. The details of this procedure were worked out by Grace Tang, University of Hawaii, Kewalo Marine Laboratory.

1. *Purification and concentration of template*

a. Precipitation of DNA

    (i) Following PCR, remove oil by adding 100 μl of chloroform to a 100-μl PCR reaction, vortex, and transfer the aqueous layer to a clean vial.

    (ii) Concentrate by precipitating DNA using linear polyacrylamide as a carrier (Gaillard and Strauss, 1990). We typically mix:

    80 μl of PCR product

    20 μl of 0.5 M KC1–0.125% polyacrylamide

    250 μl 95% EtOH (−20°C).

    (iii) Incubate at −70°C for 30 min or −20°C overnight.

    (iv) Microcentrifuge at high speed for 10 min.

    (v) Wash with 70% EtOH (−20°C).

    (vi) Resuspend in 10 μl of 1X TE buffer.

    (vii) Electrophorese in a 1% TAE gel (see Sambrook et al., 1989 for buffers and conditions).

    (viii) Stain gel in EtBr (to visualize PCR products). Locate and excise band.

    (ix) Place the band in a 1.5-μl eppendorf vial.

b. Geneclean (Bio101, Inc.)

    (i) Add 0.5 μl NaI solution to each vial.

    (ii) Incubate at 50°C for 5 min.

    (iii) Add 3–5 μl of glassmilk to absorb DNA. Mix well on ice for 5 min.

    (iv) Wash glassmilk with 0.3 ml NEW wash. Spin this solution quickly (20–25 sec). The glassmilk will pellet easily. Pour off the supernatant. Add more NEW wash and resuspend the glassmilk. Repeat this step 2 more times. The final spin should run for about 2 min.

    (v) Add 7–40 μl of TE. The amount of TE you add depends on how much DNA you started with. For bright double strands we normally add about 28 μl. This is enough for four sequencing reactions.

    (vi) Let this solution soak at 50°C for 5 min to 1 hr.

    (vii) Spin for 2 min and gather supernatant. This is your template.

2. *Primer annealing:* Once the PCR product is purified and concentrated you are ready to anneal the primer for sequencing. You can use either of the original PCR primers for sequencing.

a. Mix together:

    7 μl DNA template

2 μl Sequencing buffer

1–2 μl 5% NP-40

1 μl 10 or 100 μM primer

b. Boil mixture for 3 min.

c. Plunge into liquid nitrogen. 95% EtOH at −70°C also works well.

d. Warm slowly to room temperature in a metal block initially chilled to −20°C.

3. *Sequencing*

a. Add 2.5 μl of ddGTP (labeled Mix G), ddATP (labeled Mix A), ddTTP (labeled Mix T), and ddCTP (labeled Mix C) to separate tubes (or racks). These are the termination mixes. For each template there should be four separate tubes or positions labeled G, A, T, and C.

b. Make up Labeling cocktail. For each sample add:

1.0 μl DTT (dithiothreitol, 0.1M)

0.4 μl 5X Labeling mix

1.1 μl H_2O

1.0 μl 5% NP-40

0.25 μl ^{35}S-ATP

0.5 μl Mn buffer (emphasizes the region close to the primer and tends to reduce shadowing problem)

1.5 μl Sequenase® dilution buffer

0.25 μl Sequenase®

c. Add 5.5 μl of the Labeling cocktail to each annealed template. *Do not let labeling reaction proceed longer than 5 min.*

d. Pipette 3.5 μl of this mix to each of the four ddNTP tubes. Incubate for 2–5 min at room temperature. If you get a lot of nonspecific termination in the sequencing reactions, try doing this step at 37–45°C.

e. Add 4 μl of Stop Solution to each ddNTP mix.

f. Store reactions in the freezer until ready to load onto acrylamide gel.

PALUMBI.3

SOLID PHASE SEQUENCING AND STRIPPING WITH BIOTINYLATED PRIMERS

1. *Attaching PCR product to streptavidin/magnetic beads*

a. Wash streptavidin-linked magnetic beads (Dynabeads M-280, Dynal, Inc.) three times in 1 volume of sterile 2 M NaCl, 1 mM EDTA, 10 mM Tris pH 7.5 (2X Bead and Wash buffer). Use 40 μl of beads per sequencing reaction. Resuspend in the original volume of 2X BW buffer. (*Note:* A quantity of 20–30 μl of beads also works, although sequencing reactions tend to be faint.)

b. Add 40 µl of streptavidin-linked magnetic beads to 45–95 µl of the double-strand PCR product. Be sure not to transfer any mineral oil to the bead/PCR mixture.

c. Mix well. Allow the biotin to bind to the streptavidin by rotating this solution slowly for 30–120 min at room temperature.

d. Pull beads to bottom of tube with magnet and remove supernatant. You can assay 5 µl of this for unbound PCR products, but be aware that the high salt content will make an agarose gel run abnormally unless it is run slowly.

e. Wash the beads by resuspending them in 200 µl BW buffer, pulling them to the bottom with the magnet, and removing the fluid again. Do this washing step twice.

2. *Denaturing attached DNA*

a. NaOH method. This is the method recommended by the bead manufacturer. It works well in preparation of the bound strand for solid phase sequencing, but preparation of the stripped strand requires careful neutralization and/or precipitation.

   (i) Denature the DNA by adding 60 µl of 0.2 M NaOH (fresh! who knows why?) and rotating slowly for 10 min. (*Note:* The biotinylated strand remains fixed to the magnetic beads.)

  (ii) Pull beads to bottom with magnet, and collect 70 µl of supernatant. (Don't be greedy, leave a few microliters behind rather than get beads in the sample.)

 (iii) Wash the beads in 200 µl BW buffer twice, and save the beads for solid phase sequencing.

 (iv) Propanol ppt the supernatant with:

    70 µl of 5 M ammonium acetate (pH 6.8) [Note this is lower than the normal pH because of the NaOH.]

    40 µl of isopropanol

    1 µl of 10 µg/µl tRNA (The tRNA driver is necessary! Without it sequences are very faint.)

  (v) Incubate at −20°C for at least 1 hr.

  (vi) Spin at 4°C for 10 min in a microfuge.

 (vii) Wash with 70% EtOH.

(viii) Dry pellet, and resuspend in 7 µl of 1X TE buffer.

b. Heat denaturation method. This method uses heat to denature the DNA and speed to separate the two dissociated strands. The method works well to prepare both bound and stripped strands and does not rely on extra steps to neutralize or precipitate DNA. The method is especially simply since it combines denaturation with primer annealing in a single step. It is thus faster and less subject to mistakes than the NaOH method. However, high heat probably disrupts the biotin–streptavidin bond, and so the timing of these steps is unusually important. Don't go off to lunch.

(i) To the washed and rinsed beads, add 16 µl of *stripped strand annealing denaturing cocktail*. The stripped strand primer is the same as the biotinylated primer (a biotinylated or nonbiotinylated form is OK), or it can be another primer that anneals to the same strand. The cocktail is made up as:

3.2 µl 5X sequencing buffer

1.6 µl 10 µM primer

11.2 µl water

for each sample to be sequenced.

(ii) Boil each tube for 2 min, then IMMEDIATELY summon beads with a magnet and carefully remove 14 µl of buffer (containing stripped strands, annealed to the sequencing primer and NO BEADS).

(iii) IMMEDIATELY wash the beads with 150 µl BW buffer to remove extra stripped strands and prevent reannealing to the bound strand.

(iv) Add 150 µl of BW buffer. Beads can be stored in this state for sequencing later.

(v) Stripped, single strand from step ii is ready to sequence. You can assay 4 µl of this on an agarose gel to ascertain how well the denaturation has worked (single-stranded DNAs are generally fuzzy and faint), leaving 10 µl left for sequencing.

(vi) To prepare the beads from step iii for sequencing, rinse them with TE, and add 16 µl of bead cocktail:

3.2 µl 5X sequencing buffer

1.6 µl 10 µM primer

11.2 µl water

In this case the primer is the one opposite to the biotinylated primer.

(vii) Place beads and cocktail in a heating block at 65°C for 2 min, and let the block cool slowly to room temperature (over 15–30 min). Alternatively, use the PCR machine to ramp from 65°C to 37°C over 15 min.

(viii) Proceed with labeling and sequencing reactions for both beaded and stripped strands. Follow normal procedures except that you must make sure the beads are in suspension during the extension phase of the sequencing reaction. Also DON'T put magnetic beads onto sequencing gel. ZAP!

POTTS.1

PCR DIRECTLY FROM PLASMID-CONTAINING BACTERIAL COLONIES

Purpose

Rapid PCR-based screening for colonies containing plasmids with the appropriate size insert.

Protocol

1. Number colonies to be picked on the bottom of the original plate.
2. Prepare an LB/Amp/X-Gal grid plate (or other appropriate selective medium) and number the grid such that it reflects colonies to be picked from the original plate.
3. Pick a numbered colony with a sterile toothpick and gently streak or touch the toothpick to the appropriate numbered cell on the plate in step 2. Place the toothpick in an eppendorf or PCR tube containing 50 μl of sterile water and swirl.
4. Repeat step 3 for all selected colonies.
5. Place the samples in boiling water for 15 min.
6. These boiled samples can now be used as template in PCR (1 μl boiled product per 25 μl PCR).
7. Boiled samples can be stored at −20°C for future PCRs.
8. Clones to be developed can be rescued from grid plate in steps 2 and 3.

POTTS.2

ETHIDIUM BROMIDE STAINING OF SEQUENCING SIZE GELS

Purpose

To stain sequencing size acrylamide gels with ethidium bromide in order to visualize DNA fragments.

Solutions

10 mg/ml ethidium bromide, 1X TBE

Protocol

1. After gel has run desired distance, turn off power pack, drain upper buffer chamber, remove comb and remove gel from electrophoresis apparatus.
2. Place gel with the siliconized plate uppermost and remove spacers.
3. Insert spatula between glass plates at the upper edge of the gel and pry spatula upward. The two plates will separate with the gel sticking to the lower plate.
4. To 70 ml of 1X TBE, add 8 μl of ethidium bromide, mix and apply evenly to area of gel where DNA is expected to be. Let sit for 3–8 min.
5. With 50 ml 1X TBE, rinse ethidium bromide off gel and into tray. Dispose of liquid into ethidium bromide waste container.
6. Take gel on glass plate to sink. Keeping the plate close to horizontal, rinse gel gently for 3 min with distilled water. Let gel drain.
7. Take plate back to benchtop and, with scalpel blade, excise unwanted portion of gel and dispose.

8. The gel can now be exposed to UV light for visualization of DNA. A photo-graph can be taken through the plate if the quantity of DNA per fragment is sufficient (approximately 30 ng). (*Note:* It has been our experience that the glass plates transmit UV light in only one direction, so this must be determined before pouring gel.) If DNA quantities are not sufficiently high, the gel must be removed onto plastic wrap for photodocumentation.

9. To transfer to plastic wrap, cover desired portion of gel with plastic wrap and smooth out wrinkles using a test tube.

10. Flip plate over so glass plate rests on top of gel and plastic wrap. Hold plastic wrap to benchtop and slowly lift one side of the glass plate away from the plastic wrap. The gel should drop smoothly off the plate, sticking to the plastic wrap.

11. Transfer gel/plastic wrap to tray and visualize with UV illumination.

POTTS.3

DENATURING GRADIENT GEL ELECTROPHORESIS (DGGE)

Purpose

To detect variants based on DNA molecules that differ in sequence by at least one base and therefore have slightly different melting properties and subsequent differential electrophoretic mobilities.

Equipment

Gradient maker (Hoefer, SG 50 gravitational gradient marker), Heating Circulator (Fisher Scientific, Isotemp Immersion Circ., Model 730), acrylamide gel holder and plates, peristaltic pump, and buffer chamber. We currently use a custom made plexiglass buffer chamber (46 cm × 29 cm × 28 cm, 20-liter volume) and gel holder and plates (18 cm × 23 cm), but a complete DGGE apparatus is commercially available from Bio-Rad.

Stock Solutions

1. 40% Acrylamide:
   100 g acrylamide
   5 g bis-acrylamide
   bring up to 250 ml—ddH$_2$O
2. 80% Denaturant:
   84 g urea
   80 ml formamide
   12.5 ml 20X TAE

37.5 ml 40% acrylamide

bring up to 250 ml—ddH$_2$O

3. 0% Denaturant:

12.5 ml 20X TAE

37.5 ml 40% acrylamide

bring up to 250 ml—ddH$_2$O

Table for Making Various Percentage Denaturation Solutions

	Percent Denaturant Desired								
	0%	10%	20%	30%	40%	50%	60%	70%	80%
ml of 0% dnt	17.5	15.3	13.1	10.9	8.75	6.6	4.4	2.2	0
ml of 80% dnt	0	2.2	4.4	6.6	8.75	10.9	13.1	15.3	17.5

Protocol

Pouring Gels

1. Set magnetic mixer ~1 ft above bench.
2. Tape and clamp gel plates (19 × 23 cm).
3. Make sure chambers of gradient maker are closed.
4. Put one micro-stirbar (10 mm × 3 mm) in each chamber.
5. Add appropriate amount of 0% denaturant to left chamber of gradient marker and open left chamber to let bubble pass through to right chamber. CLOSE QUICKLY once bubble is through. (*Note:* Left chamber holds lower % denaturant, right chamber holds higher % denaturant.)
6. Add remaining amounts of denaturant to each chamber and put gradient maker on mixer—turn mixer on.
7. Attach draining hose with needle to gradient maker; with gel standing upright, tape needle to top of gel plates with tip of needle inserted between plates.
8. To each chamber add: 45 μl 20% APS (ammonium persulfate) and 15 μl TEMED.
9. Once gradients mix well (~10 sec), open right chamber to let higher % denaturant drain to plates slowly (~45° angle). When denaturant reaches the needle, open the left and right chambers completely.
10. Make sure gradient is mixing and a steady stream of acrylamide is pouring.
11. Fill to top of plates, place comb in gel, and lay flat.
12. Cover gel with plastic wrap and clamp to allow gel to polymerize.
13. Let gel polymerize at least an hour before running.

Running Gels

1. 1X TAE buffer should fill buffer chamber.
2. Turn on heating circulator about 1 hr before desired loading time to allow buffer to reach running temperature of 60°C.
3. When buffer chamber is heated, take tape off gel and put in gel holder!
4. Remove comb and fill gel holder buffer chamber with dH_2O.
5. Flush out each well with syringe and straighten wells with gel loading tip.
6. Remove dH_2O from gel holder buffer chamber.
7. Put gel holder in buffer chamber and fill top of gel holder with buffer. (Gel holder should be parallel to electrode in chamber.)
8. Flush out each well again. Wells tend to fill up with urea causing sample to float out of wells.
9. Just before loading, flush wells again.
10. Load 10–20 μl of sample, depending on sample concentration.
11. Place outflow hose of peristaltic pump in gel holder buffer chamber and inflow hose in primary buffer chamber; turn pump on. Insure flow rate of pump is slow enough that the samples in the wells are not disturbed.
12. Attach negative electrode onto gel holder and positive electrode on buffer chamber.
13. Run at 250 V for 3–6 hr. Time must be optimized for each locus.
14. Stain gel with EtBr to visualize DNA (see Potts.2).

POTTS.4

SINGLE-STRAND CONFORMATION POLYMORPHISM (SSCP) ANALYSIS

Purpose

To detect variants based on single-strand DNA molecules that take on specific sequence-based secondary structures under nondenaturing conditions and subsequently migrate at different rates in a polyacrylamide gel.

Protocol

1. Denature DNA by one of the following methods:
    (a) Boil PCR product for 5 min, immediately place on ice.
        Keep samples on ice until they are loaded on the gel.
    (b) Add 1 μl denaturant (0.5 M NaOH, 10 mM EDTA).
2. Mix 5 μl of denatured PCR product with an equal volume of formamide loading

dye (95% formamide, 20 mM EDTA, 0.05% bromophenol blue, and 0.05% xylene cyanol).

3. Load onto a 5%–10% polyacrylamide gel (see Sambrook et al., 1989, p. 6.39).
4. Run gel at low voltage to avoid heat-induced conformational changes. Running time for the gel will be longer not only because the gel is run at a low voltage but also because single-strand DNA runs slower than double-strand DNA.
5. Stain gel with EtBr to visualize DNA (see Potts.2).

Note: Several parameters such as DNA concentration, polyacrylamide concentration, and denaturants affect the degree of strand separation during electrophoresis. Therefore each locus must be optimized for parameters such as amount of PCR product, temperature at which the gel is run (room temperature or 4°C), and addition of 5%–10% (V/V) glycerol (which can act as a weak denaturant).

POTTS.5

HETERODUPLEX ANALYSIS

Purpose

To detect variants based on DNA sequence mismatches in heteroduplex (double-strand) DNA. This technique may be used to detect genetic variants within a PCR sample or to detect variation among different PCR samples.

Protocol

1. Boil PCR product or mix of PCR products (equal volumes) for 5 min.
2. Allow boiled PCR product to cool to room temperature over a period of an hour. This allows for the single-strand DNA to anneal into both homoduplex and heteroduplex conformations.
3. Add equal volume of formamide loading dye (95% formamide, 20 mM EDTA, 0.05% bromophenol blue, and 0.05% xylene cyanol) to 5 μl of PCR product.
4. Load on a 5% polyacrylamide gel (see Sambrook et al., 1989, p. 6.39).
5. Run gel. Remember single-strand DNA migrate slower than double-strand DNA so running times are usually overnight.
6. Stain the gel with EtBr to visualize DNA (see Potts.2).

POWERS.1

ALLELE-SPECIFIC PCR

Allele-specific PCR provides a rapid method for screening populations for the presence of known alleles. The method is much simpler than sequencing and can be used for detecting variation that cannot be observed by protein electrophoresis, restriction enzyme analysis, or Southern blotting with allele-specific oligo-nucleotides (Powell et al., 1992) See Figures 6, 7, and 8.

1. Design PCR primers with their 3' nucleotide at the position that differs between alleles. The length of the primers should be adjusted to give a predicted annealing temperature of 44°C (predicted annealing temperature = 4°C per G/C bp + 2°C per A/T bp). As a result, these primers are likely to be relatively short (12–14 bp).

2. The primers can then be used in conventional PCR to detect the presence or absence of specific alleles. We use PCR conditions as follows: 94°C, 1 min; 50°C, 2 min; 72°C, 3 min; 30 cycles; and then one cycle of 72°C for 7 min. Reaction conditions are as recommended by Perkin Elmer; [MgCl$_2$] 1.5 mM final. Primer concentration is 300 ng per 50 μl reaction.

3. All possible combinations of specific primers should be used for each template. If an individual is homozygous for a particular allele, only one reaction should give a product. Heterozygous individuals should yield bands in two reactions.

Note: PCRs must be performed at high "stringency" (i.e., at high annealing temperatures) to ensure that the primers yield product only when they are perfectly matched to the template. [With primers designed to LDH-B in *Fundulus heteroclitus* (T$_m$ = 44°C), we found that annealing temperatures below 50°C caused low levels of amplification for mismatched primers.] To avoid problems associated with uneven heating of the thermal cycler block, relatively long annealing plateau times should be used (e.g., 2 min at 50°C).

POWERS.2

PREPARATION OF ISOLATED NUCLEI

Most techniques developed for the investigation of DNA–protein interactions and transcription rate assume that cells in culture are available. For many organisms or questions of interest, cultured cells are not readily available or appropriate. We have developed a simplified method for the isolation of intact nuclei that is useful for a variety of organisms. Nuclei prepared in this way can then be used in techniques such as *in vivo* footprinting or nuclear run-on assays.

Solutions

Buffer A: 0.3 M sucrose, 10 mM Tris pH 7.5, 1 mM DTT, 1 mM EDTA, 50 mM KCl, 1 μg/ml pepstatin, 10 μg/ml leupeptin, 100 μg/ml chymostatin, 0.5 mM spermidine.

Buffer DD: 1 mM DTT, 10 mM MgCl$_2$, 12.5% sucrose, 5% glycerol, 50 mM KCl.

Nuclear Isolation

Note: All steps should be performed quickly and in the cold.

1. Rapidly sacrifice the organism and excise desired organ (e.g., liver).
2. Immediately homogenize on ice in 4.5 ml buffer A using 25 passes with a Dounce homogenizer (pestle B) (*Note:* These volumes are appropriate for tissues from 50 to 500 mg and should yield sufficient material for at least six footprints or several run-on assays. If larger amounts of tissue are available, these volumes can be scaled up as appropriate.)
3. Filter lysate through a 5-cc syringe partly filled with silanized glass wool to remove particulate matter.
4. Mix with an equal volume of buffer B (equivalent to buffer A plus 2 M sucrose).
5. Layer over a cushion of 4 ml of buffer B in an ultracentrifuge tube.
6. Pellet nuclei by ultracentrifugation at 15,600 rpm for 45 min in an SW-41 rotor at 4°C.
7. Resuspend pelleted nuclei in 1.6 ml buffer DD; these nuclei can be used immediately for *in vivo* footprinting or be frozen at −80°C and used for nuclear run-on assays.

POWERS.3

IN VIVO FOOTPRINTING

Footprinting is a widely used technique for determining the location of DNA–protein interactions. Modifications of this approach have been developed that allow the probing of these contacts *in vivo* in isolated intact nuclei or cultured cells. The *in vivo* footprinting technique (Mueller and Wold, 1989; 1991) on which this protocol is based was originally designed for use with cells in culture and uses the highly toxic chemical dimethyl sulfate (DMS) as an agent for chemical cleavage of genomic DNA. The approach outlined here modifies this technique for use with isolated intact nuclei and uses DNase I as an agent for enzymatic DNA cleavage.

Solutions

5X First Strand Synthesis Buffer: 200 mM NaCl; 50 mM Tris•Cl pH 8.9; 25 mM $MgSO_4$; 0.05% gelatin.

First Strand Synthesis Mix: 6 μl 5X first strand buffer; 0.3 μl 1 pmol/μl primer 1; 18.21 μl H_2O; 0.25 μl 2 U/μl Vent DNA polymerase (New England Biolabs); 0.24 μl 25 mM 4dNTP mix.

Linker Mix: 20 mM primer LMPCR.1; 20 mM primer LMPCR.2; 250 mM Tris•Cl, pH 7.7. (See Mueller and Wold, 1989, for suggested primer sequences.)

Ligase Dilution Solution: 110 mM Tris•Cl pH 7.5; 17.5 mM MgCl$_2$; 50 mM DTT; 125 μg/ml BSA.

Ligase Mix: 10 mM MgCl$_2$; 20 mM DTT; 3 mM ATP; 50 μg/ml BSA; plus for each 25 μl reaction—5 μl linker mix; 1 μl T4 DNA ligase (3 units).

1. Prepare digested experimental and control DNA.
    a. Incubate 0.25-ml aliquots of isolated nuclei (see Powers.2) with 0, 1.0, 2.0, and 5.0 μg of DNase I at 20°C for 3 min to prepare digested experimental DNA.
    b. Terminate reaction by addition of EDTA to 50 mM final concentration.
    c. Isolate DNA by proteinase K digestion (150 mg/ml, 2 hr, 55°C) Phenol extract 1X, chloroform extract 1X. Ethanol precipitate.
    d. The remaining undigested nuclei (approximately 0.5 ml) can be used to provide control ("naked") DNA. These nuclei should be deproteinated (by phenol:chloroform extraction and ethanol precipitation) prior to DNase I treatment.
    e. The appropriate amount of DNase to use must be determined empirically to result in approximately the same length of footprinting ladder as does the experimental DNA. We find that approximately 0.003–0.25 μg DNase I per μg DNA provides the appropriate range.

2. Perform ligation mediated PCR (LMPCR; Mueller and Wold, 1989).

    Note: LMPCR requires the design of three overlapping gene-specific primers for the region to be footprinted. They should be oriented such that primer 1 is the most distant from the region to be footprinted and T$_m$ primer 3 > T$_m$ primer 2 > T$_m$ primer 1. (See Ausubel et al., 1987, supplement 20, or Mueller and Wold, 1991, for a detailed discussion of important considerations in primer design.)

    a. Prepare first strand synthesis.
        (i) Transfer 5 μl (approximately 2 μg) digested DNA to a 0.5-ml micro-centrifuge tube. Chill in an ice water bath.
        (ii) Prepare the first strand synthesis mix (with primer 1) *on ice.* Add 25 μl to DNA sample.
        (iii) Denature DNA for 5 min at 95°C. Anneal primer for 30 min at 5°C above calculated T$_m$. Extend for 10 min at 76°C.
    b. Prepare ligations.
        (i) Thaw linker mix in ice water bath and prepare ligase dilution solution and ligase mix (without adding linker mix or ligase) and chill in ice water.
        (ii) Immediately before setting up ligations, add linker mix and ligase to ligase mix (on ice).
        (iii) Add 20 μl ligase dilution solution to DNA sample. Keep on ice.
        (iv) Add 25 μl ligase mix, mix with pipettor, and incubate overnight at 17°C.

(v) Prepare and chill precipitation salt mix (2.7 M sodium acetate, 1 mg/ml yeast tRNA).

(vi) Add 9.4 μl and 200 μl ice-cold 100% ethanol to sample.

(vii) Mix thoroughly and chill for 2–8 hr at −20°C.

c. Prepare PCRs.

(i) Precipitate DNA by centrifugation and resuspend pellet in 50 μl water.

(ii) Set up 100-μl PCRs with Taq DNA polymerase under standard conditions. (Use 10 pmol each primer 2 and LMPCR.2), cover with mineral oil, and microcentrifuge briefly.

(iii) Carry out 18 cycles of PCR as follows: 1 min at 95°C, 2 min at 2°C above calculated T_m primer 2; 3 min at 73°C. For each cycle add an extra 5 sec to the extension step. Allow final extension to proceed 10 min.

d. End label primer 3.

(i) End label primer 3 with ^{33}P by using T4 DNA kinase and standard techniques. [*Note:* We use ^{33}P rather than ^{32}P because it does not require extensive shielding. Because it is a lower energy emitter, slightly longer exposure times may be required than when using ^{32}P (1–2 days vs. 6 hr). However, substantially sharper bands are produced when ^{33}P is used, which helps in the resolution of doublet bands.]

(ii) Purify labeled primer with a spin column (e.g., Sephadex G-25).

e. End label PCR reactions.

(i) Prepare a 1X PCR mix containing labeled primer 3 and add 5 μl to sample.

(ii) Carry out two rounds of PCR to label the DNA. First denaturation 3–4 min at 94°C; second denaturation 1 min; anneal end labeled primer 3 for 2 min at 2°C above calculated T_m. Extend 10 min at 73°C.

(iii) Add 295 μl polymerase stop solution. Ethanol precipitate.

(iv) Resuspend pellet in 7 μl loading buffer.

(v) Denature samples for 5 min at 85–90°C. Load contents of tube onto 6% sequencing gel. After completion of run, fix and dry the gel and autoradiograph for 24–48 hr without intensifying screen.

POWERS.4

NUCLEAR RUN-ON ASSAYS

Nuclear run-on assays (Ausubel et al., 1987, supplement 9 with modifications) are designed to measure gene transcription rate by identification and quantification of newly transcribed RNA. In these assays, RNA transcripts do not appear to be initiated, but transcripts that have already been initiated are elongated. This method therefore gives a reasonably accurate estimate of the level of transcription occurring in the cell at the time of nuclear isolation.

Solutions

DNase I Buffer: 20 mM Hepes, pH 7.5; 5 mM $MgCl_2$; 1 mM $CaCl_2$.

Digestion Buffer: 0.5 M NaCl; 50 mM $MgCl_2$; 2 mM $CaCl_2$; 10 mM Tris•Cl, pH 7.4.

2X Reaction Buffer Minus Nucleotides: 10 mM Tris•Cl pH 8; 5 mM $MgCl_2$; 0.3 M KCl.

2X Reaction Buffer with Nucleotides: 1 ml 2X reaction buffer minus nucleotides; 10 μl 100 mM each ATP, CTP, and GTP; 5 μl 1 M DTT.

TES Solution: 10 mM TES, pH 7.4; 10 mM EDTA; 0.2% SDS.

Elution Buffer: 1% SDS; 10 mM Tris•Cl, pH 7.5; 5 mM EDTA.

1. Prepare test filters (plasmids bound to nitrocellulose).
   a. Linearize plasmids containing the cDNAs of the gene whose transcription rate will be assessed. Typically, this will consist of one or more cDNAs of interest plus actin and/or tubulin as a control for generalized (i.e., non-specific) increases in cellular transcription rate.
   b. Denature the linearized DNA (200 μg in 40 μl) by incubating with 49 μl of 1 M NaOH for 30 min at room temperature.
   c. Neutralize by addition of 4.9 μl of 6X SSC and place on ice.
   d. Spot the linearized DNA onto nitrocellulose filters using a dot blot or slot blot apparatus (20 μg of each cDNA). Rinse with 500 μl of 6X SSC.
   e. Air dry the nitrocellulose filters overnight. Bake for 2 hr at 80°C under vacuum and store in a vacuum desiccator.

2. Perform nuclear run-on transcription assays.
   a. Thaw 200 μl frozen nuclei (see Powers.2) at room temperature and transfer to a polypropylene tube. Immediately add 200 μl of 2X reaction buffer (with nucleotides) plus 10 μl [α-^{32}P] UTP (10 μCi/μl). React 30 min at 30°C with shaking.
   b. Mix 40 μl of 1 mg/ml RNase-free DNase I and 1 ml digestion buffer. Add 0.6 ml of this solution to the labeled nuclei. Incubate for 5 min at 30°C.
   c. Extract sample with 1 ml phenol:chloroform:isoamyl alcohol. Save aqueous layer.
   d. Add to aqueous layer: 2 ml water, 3 ml 10% TCA: 60 mM sodium pyrophosphate, and 10 μl of 10 mg/ml E. coli tRNA carrier.

e. Filter TCA precipitate onto Millipore-type HA (0.45 μm) filters. Wash filters 3 times with 10 ml of 5% TCA:30 mM sodium pyrophosphate.

f. Transfer the filters to a glass scintillation vial. Incubate with 1.5 ml DNase buffer and 37.5 μl of 1 mg/ml RNase-free DNase I for 30 min at 37°C. Stop the reaction by adding 45 μl of 0.5 M EDTA and 68 μl of 20% SDS.

g. Heat the samples to 65°C for 10 min to elute RNA. Remove supernatant and save. Add 1.5 ml elution buffer to filters. Incubate for 10 min at 65°C. Remove supernatant and combine with first.

h. Add 4.5 μl of 20 mg/ml proteinase K to supernatant. Incubate 30 min at 37°C.

i. Extract once with 3 ml buffered phenol:chloroform:isoamyl alcohol.

j. Transfer aqueous phase to a silanized 30-ml Corex tube. Add 0.75 ml of 1 M NaOH. Leave on ice 10 min. Stop reaction by adding 1.5 ml of 1 M Hepes (free acid).

k. Precipitate RNA by adding 0.53 ml of 3 M sodium acetate and 14.5 ml ethanol. Chill overnight at −20°C.

l. Centrifuge for 30 min at 10,000g. Remove ethanol and resuspend in 1 ml TES solution. Shake for 30 min at room temperature to dissolve.

m. Count 5-μl aliquots of each sample in duplicate by spotting onto Whatman GF/F filters.

n. Mix 1 ml of RNA solution with 2 ml TES solution.

3. Hybridize the labeled RNA to the test filters.

a. Hybridize a total volume of 3 ml RNA solution (approximately 10^6 to 2×10^7 dpm) to cDNA test filters for 36 hr at 65°C in 5-ml glass scintillation vials.

b. After hybridization, wash each filter twice in 25 ml 2X SSC for 1 hr at 65°C.

c. Transfer filter to a scintillation vial containing 0.4 ml of 0.1 N NaOH.

d. Incubate for 15 min at 55°C.

e. Add scintillation cocktail and count.

Note: Some important controls: To ensure that the amount of RNA added does not saturate the filters, samples can be spiked with labeled cRNA (RNA produced *in vitro* from an Sp6-containing vector). Negative controls should also be included on the filters. Plasmid DNAs or cloned cDNAs from distantly related organisms are appropriate negative controls for nonspecific hybridization.

POWERS.5

SITE-DIRECTED MUTAGENESIS

Site-directed mutagenesis is a technique that is very widely used in molecular biology to change DNA sequences of interest to examine their function. Many different approaches to site-directed mutagenesis are available. This technique is a modification of that of Kunkel (Kunkel et al., 1987).

Solutions

Polymerization Ligation Buffer: 400 mM Tris pH 7.5, 100 mM Mg Cl$_2$, 50 mM DTT, 500 mM NaCl.

1. Obtain sequence to be mutagenized.
    a. Clone the sequence to be modified into a phagemid (a plasmid containing the f1 origin of replication; e.g., pBluescript KS, Stratagene) using standard techniques.
    b. Transform into competent CJ 235 cells (Bio Rad).

2. Isolate single-strand DNA.
    a. Inoculate a single colony into 10 ml 2X YT (with 100 μg/ml ampicillin).
    b. Grow for 1–2 hr at 37°C with good aeration.
    c. Add kanamycin to 70 μg/ml; grow with shaking at 37°C for 8–16 hr.
    d. Centrifuge at 4000g for 15 min to pellet cells.
    e. Carefully remove approximately 9 ml of the supernatant. Add to 2 ml of 2.5 M NaCl, 20% PEG solution. Mix well and place on ice for 30 min to pellet phage particles.
    f. Centrifuge at 8000–10,000g for 30 min.
    g. Discard supernatant and dry the inside of the tube as much as possible.
    h. Resuspend phage pellet in 250 μl TE and incubate with 10 μl DNase-free RNase (10 mg/ml) for 15 min at room temperature.
    i. Phenol/chloroform extract until no white interface is visible; extract twice with water saturated ether.
    j. Pass the extracted DNA through a Sephadex G-50 spin column.
    k. Ethanol precipitate the eluate.
    l. Resuspend pellet in 25–50 μl TE and check integrity and concentration by running on a 1% agarose gel versus single-strand M13 of known concentration.

3. Mutagenize the sequence.
    a. Synthesize an oligonucleotide that contains the modified sequence.
    b. End label the mutagenic oligonucleotide with ^{32}P using standard techniques.
    c. Anneal this primer to the single-strand DNA template obtained above as follows:
        (i) Combine primer with 1 μg single-strand DNA at a 15:1 molar ratio in a total volume of 20 μl dH$_2$O.
        (ii) Add 1.3 μl of 20X SSC.
        (iii) Heat to 70°C and allow to cool to room temperature slowly (45–60 min) (*Note:* This can be conveniently done in a PCR machine.)
        (iv) Place on ice.

    d. Add 10 μl polymerization–ligation buffer, 10 μl 10 mM ATP, 5 mM dNTPs, 45 μl dH$_2$O, 1 μl Sequenase (Stratagene), and 2 μl (4 units) DNA ligase.

    e. Keep on ice for 5 min then transfer to room temperature for 5 min.

    f. Incubate at 37°C for 1–2 hr.

    g. Terminate the reaction by adding 3 μl of 500 mM EDTA.

    h. Transform into competent cells (e.g., XL-1 blue).

    i. Screen colonies for mutants by PCR or colony hybridization.

    j. The entire gene should be sequenced to ensure that no other random mutations occurred.

SOMMER.1

CULTURING FREE-LIVING NEMATODES

1. *Caenorhabditis elegans* and other free-living nematodes are maintained on NG agar plates (5 cm for normal culture, 9 cm for mass culture and DNA preparation). NG Agar (Brenner, 1974):

NaCl   3 g

Agar   17 g

Peptone   2.5 g

Cholesterol (5 mg/ml in EtOH)   1 ml

Water   975 ml

Autoclave, then add:
CaCl$_2$, 1 M   1 ml
MgSO$_4$, 1 M   1 ml
K$_3$PO$_4$, 1 M, pH 6.0   25 ml

2. Spread plates with a lawn of OP50, a uracil-requiring strain of *E. coli* (Brenner, 1974). This strain can be obtained from the *C. elegans* Genetic Center (address given below). Keep plates at room temperature for at least 3 days before use, so that excess moisture can evaporate.

3. Transfer nematodes by means of a platinum wire pick, whose end has been flattened (~32 gauge, sealed into the end of a pasteur pipette). Coat the pick with bacteria from the transfer plate in order to transfer worms. Worms should be active directly after transfer.

4. Hermaphroditic species can be cultured by the transfer of one or several hermaphrodites. Male–female species should be transferred by using at least two individuals per sex.

5. The *C. elegans* wild-type strain N2 can be grown between 12°C and 25°C; other species have different temperature requirements.

6. The *C. elegans* and some other species can be obtained from:
   Caenorhabditis Genetics Center
   University of Minnesota
   1445 Gortner Avenue
   St. Paul, Minnesota 55108-1095

SOMMER.2

CELL LINEAGE AND CELL ABLATION OF VULVAL CELLS

1. Prepare a 0.4-mm-thick pad of agar by flattening a drop of a hot 5% agar solution (Noble agar, DIFCO) on a microscope slide with the help of a second slide. The latter is supported by two spacer slides raised 0.4 mm from the bench by two pieces of adhesive tape.
2. When set, remove the upper slide by sliding and add a small drop (2–3 ml) of buffer, using a drawn capillary and a mouth tube.
3. Precoat the center of a coverslip with a very thin layer of bacteria. Bacteria are needed for worms to survive on the slide.
4. After mounting worms, remove excess agar with a razor blade, and seal the edges of the agar pad with silicone grease (only for long-term observation, greater than 1 hr). Worms having enough bacteria around can survive for several hours.
5. Particular cell lineages (e.g., the vulval cell lineage) can be observed using a 100× objective and Nomarski interference contrast optics. Cell lineages are determined by continuous observation of nuclei as they divide.
6. For laser microsurgery immobilize worms by anesthetizing them with 10–15 mM sodium azide added freshly to the agar of step 1 of the protocol.
7. A pulsed dye laser is arranged as an epi-illuminator (for details see Avery and Horvitz, 1987).
8. Specimens are recovered from the slide using spit tubing and a pulled capillary tube immediately after laser microsurgery and are transferred individually to a fresh plate.
9. Observe worm several hours after the ablation to confirm the death of the ablated cell.
10. Incubate worms until cell lineage of the remaining cells of interest can be observed.

SOMMER.3

PCR CLONING OF GENES FROM NEMATODES

1. Single worms (transferred into a PCR tube by a drawn capillary and a mouth tube) or previously prepared genomic DNA (grow worms according to Som-

mer. 1 on 9-cm plates and isolate genomic DNA according to standard techniques described in Sambrook et al., 1989) can be used for PCR. Use one or several intact worms or between 10 and 100 ng genomic DNA.

2. PCR reactions are performed in 20-μl reaction volume using standard buffer [50 mM KCl; 10 mM Tris, pH 8.2; 2 mM MgCl$_2$; and 0.5 units Taq polymerase (Cetus)]. For the cloning of heterologous genes use degenerate primers, which consider the codon redundancy of the different amino acids. Use 5 mM of each primer in the PCR reaction.

3. Subject the reaction mixture to 30 cycles of 1 min at 94°C, 2 min at 50°C, and 2 min at 72°C and a final extension of 10 min at 72°C.

4. Separate PCR products (5 μl) on an agarose gel. If no DNA bands are observed, blot the gel and hybridize under low stringency (2X SSPE at 65°C) with the homologous fragment of an organism where the gene of interest is already cloned (in our case, *Caenorhabditis*).

5. If a signal is seen, isolate the appropriate region from a duplicate gel and subject it to another round of PCR amplification (30 cycles).

6. The second round yields normally enough DNA for cloning into the M13 vector. For cloning incubate the PCR reaction with 2.5 units Klenow for 15 min at 37°C to produce blunt ends.

7. Use standard techniques for ligation and transformation (Sambrook et al., 1989).

8. Positive clones after transformation can be obtained by low-stringency hybridization (see step 4) and can subsequently be sequenced using standard techniques.

SOMMER.4

DNA-MEDIATED TRANSFORMATION IN *CAENORHABDITIS ELEGANS*

1. Worms are immobilized on a dried pad of agarose (2% agarose) under an oil layer (voltalef). Pads are prepared on 48 × 60-mm diameter micro cover glasses (Thomas Scientific) by placing a molten drop of agar on a cover glass, laying a second cover glass on top using gentle pressure. The upper cover glass is removed and the lower cover glass with the agarose pad is dried at 100°C for 15 min.

2. Worms are picked onto the oil-covered pad. They are viable for up to 30 min. Thus, just a few worms (1–10) can be injected simultaneously.

3. Injection needles with a tip of less than 1 μm can be drawn by using a conventional needle puller system. Edge needles with HF (1:2 in H$_2$O) to increase the opening of the needle. During the course of injection the needles break frequently but can readily be used with bore sizes up to 3 μm. Fill 1 μl of DNA solution into the needle using a capillary tube that has been drawn to a long bore by hand.

4. Injections are done using a Zeiss IM microscope or a corresponding system with a Plan-40 or Plan-Neofluar-40 lens and a long working distance condenser. Different micromanipulation systems are available. We have used the Eppendorf Micromanipulator (Eppendorf, Hamburg, Germany).

5. For injection, press needle against the cuticle of the worm so it depresses cuticle to the site of injection and then gently hit the manipulator to drive the needle into the worm.

6. Recover worms using a recovery buffer (4% glucose; 2.4 mM KCl; 66 mM NaCl; 3 mM $MgCl_2$; 3 mM $CaCl_2$; 3 mM Hepes pH 7.2).

7. Worms are transferred to individual petri plates (see Sommer.1).

8. The use of this DNA-mediated transformation system with respect to particular DNA fragments is described in detail in Sommer.5.

SOMMER.5

CLONING AND CHARACTERIZATION OF GENES USING THE DNA-MEDIATED TRANSFORMATION IN *CAENORHABDITIS ELEGANS*

1. A major consideration for a DNA-mediated transformation system is the assay for the presence of injected DNA. In *C. elegans* the *rol-6* transformation marker is used very frequently. In this case the DNA of a dominant mutation in the *rol-6* gene is injected (pRF4). Successful transformants will produce roller progeny, which are easy to distinguish from wild-type.

2. In *C. elegans,* the Genome Sequencing Project provides genomic libraries in YAC, a cosmid, or lambda vectors.

Genome Sequencing Project
MRC Laboratory of Molecular Biology
Hills Road, Cambridge CB2 2QH, UK

Department of Genetics
Box 8232
Washington University School of Medicine
4566 Scott Avenue
St. Louis, Missouri 63110, USA

In most cases, cloning a gene in *C. elegans* is based on genetic analysis of a previously isolated and analyzed mutation or an already existing mutation. Mutations in *C. elegans* are maintained by the *Caenorhabditis* Genetics Center (see Sommer.1).

3. Genetic mapping analysis usually provides information about the physical map position of the gene of interest in the genome. Usually, YAC clones covering the corresponding genomic area are injected into mutant worms together with pRF4-DNA. Transformed F_1 animals that show the roller phenotype are picked indi-

vidually to new petri plates. The most commonly used transformation is extra-chromosomal. Stable transformation is indicated by the presence of the transformation marker in the F_2 generation. Thus F_2 progeny are analyzed for the roller phenotype. If the coinjected genomic DNA rescues the mutation of inter-est, the F_2 "roller" animals should be wild-type fate phenotype of interest beside the roller phenotype.

4. Starting from a YAC rescue, the procedure is repeated with cosmids covering the area of the particular YAC clone (as mentioned above, the Genome Sequencing Projects usually can provide the YAC and cosmid clones of interest).

5. After obtaining rescue of mutant phenotypes with a cosmid, subclones of the cosmid-DNA have to be made to localize the gene further. Once rescue has been obtained with a DNA fragment of ~10 kb (typical gene size for *C. elegans*), this fragment is molecular characterized using standard technology, including se-quencing analysis.

6. Once the gene is cloned, the DNA transformation system can be used for two additional purposes. First, the expression pattern can be analyzed by making a fusion construct containing the promoter of the gene of interest fused to the β-Gal gene of *E. coli* (Fire, 1992; Mello et al., 1995).

7. Second, misexpression of the gene of interest can be studied by making a construct using a human heat shock promoter fused to the coding sequence of the gene of interest (Stringham et al., 1992; Mello et al., 1995).

STRASSMANN.1

GENOMIC DNA EXTRACTION FOR PCR OF MICROSATELLITES
Modified from a protocol obtained from M. Antolin originally based on a protocol in Bender et al., 1983.

We use this protocol to process insects weighing 30–50 mg that are processed in 1.5-ml microfuge tubes. It can also be used on tissue removed from insect thoraces to avoid cuticular compounds that might inhibit PCR. For smaller insects, we reduce volumes and use 0.5-ml microfuge tubes. Very hard animals might be sliced in two before proceeding. Those with copious gut contents could be eviscerated. A key characteristic of the first two grinding techniques is that the grinding surface is never reused, thereby avoiding contamination. The third grinding technique is useful for producing high-grade genomic DNA for library construction, but it is too tedious for PCR and does not use a fresh grinding surface. We centrifuge at room tempera-ture, but a refrigerated centrifuge could be used, if available. Two sets of labeled 1.5-ml microfuge tubes are needed (three if the tissues are very fatty and step 2f. is included).

Grinding Buffer

(filter sterilize and store at 4°C one week or less)

0.1 M NaCl
0.2 M Sucrose
0.1 M Tris-HCl (pH 9.1)
0.05 M EDTA
0.05% SDS (sodium dodecyl sulfate)

1. Homogenize Tissues: Use one of the following techniques (the third is only for libraries).

Tissue Grinding Using a Minibeadbeater: This works very well for hard material such as whole adult insects and is very fast. A minibeadbeater that grinds 6 samples at once is available.

    a. Prepare 1.5-ml eppendorf tubes to be used in the grinding process by adding 0.5 ml of 2-mm diameter glass grinding beads. Add 0.2 ml of 0.5-mm diameter glass grinding beads.

    b. Put 150 µl grinding buffer in a 1.5-ml tube and then add the insect or other tissue (30–50 mg).

    c. Grind for 15–30 sec or longer in a MiniBeadBeater (Biospec Products, P.O. Box 722, Bartlesville, OK 74005) at top speed. No big pieces of tissue should remain at this point. Repeat grinding, if necessary. Find the slowest speed and shortest time necessary to homogenize your tissue to keep DNA shearing to a minimum.

    d. Spin briefly at low speed to bring tissue and beads to the bottom of the tube. Do not pellet the tissue.

    e. Add another 300 µl of grinding buffer, mix well, and tap down.

Tissue Grinding Using Disposable Minipestles: This method is good for small, soft tissues weighing under 10 mg. For example, using this technique, DNA can be extracted from insect eggs or the head capsule of late instar larvae.

    a. Prepare a set of minipestles by slightly melting the tip of a 200-µl pipette tip and inserting it all the way to the bottom of a 0.5-ml eppendorf tube to mold it.

    b. Place the egg or head capsule at the bottom of a 0.5-ml eppendorf tube.

    c. Add 50 µl of grinding buffer.

    d. Grind tissue by hand with minipestle attached to pipetman to get a better grip.

    e. Wash the minipestle into sample tube with 200 µl of grinding buffer, and discard minipestle.

Tissue Grinding in Liquid Nitrogen for Library Grade Genomic DNA

a. Place the entire mortar and pestle in a styrofoam cooler, and add liquid nitrogen to cover. Wait 10 min until the liquid nitrogen has mostly evaporated but leave the mortar in the styrofoam container sitting in the residual liquid nitrogen.

b. Grind 40 mg of tissue (does not all have to be from same individual for libraries). Wear heavy gloves!

c. With a cold spatula, transfer tissue to a 1.5-ml microfuge tube containing 300 μl cold grinding buffer.

2. DNA Extraction From Homogenized Tissue

a. Incubate homogenized tissue in grinding buffer for 30 min in a 65°C water bath.

b. While tube is still warm, add 86 μl of 8 M potassium acetate (if tissue is under 10 mg, use 70 μl) and mix well by inverting tubes, then tap them to bring contents to the bottom.

c. Incubate at 4°C for 30 min. This should precipitate the salts and SDS.

d. Spin at 10,000 rpm for 5 min.

e. Transfer approximately 200 μl (or as much as possible without taking up tissue parts) to a new set of labeled 1.5-ml tubes. Go straight to step g unless tissues are very fatty.

f. Optional step for high lipid content tissue: Add an equal volume (i.e., about 200 μl) of chloroform isoamyl alcohol 24:1 and mix gently for 10 min. Spin at 10,000 rpm for 5 min and transfer the upper phase to a new set of 1.5-ml tubes.

g. Add 500 μl of cold (stored at −20°C) 100% ethanol, mix well, and incubate at −20°C overnight. The DNA can be left to precipitate for a shorter time if it is visibly precipitating out. Even 15 min on ice is sometimes sufficient.

h. Spin at 10,000 rpm for 5 min. Some spin longer but this can heat the DNA excessively if the centrifuge is not refrigerated.

i. Aspirate off ethanol from pellet. Be very careful not to lose the DNA pellet, which sometimes looks like a smear on the side of the tube.

j. Optional ethanol wash to remove residual salts: Carefully add 750 μl of cold 70% ethanol without disturbing the pellet. Spin at 10,000 rpm for 5 min. aspirate off ethanol from pellet.

k. Dry DNA in speed vac (no heat) or by air.

l. Resuspend DNA in 50 μl of sterile ultrapure water. Do not resuspend in TE since the EDTA will interfere with PCR unless the DNA is substantially diluted with water after this step.

3. Check the Quality and Quantity of DNA

a. Add 2 µl of tracking dye (0.25% bromophenol blue, 0.25% xylene cyanol FF, 30% glycerol, Sambrook 6.12) to 2 µl of sample. We line up the dots of tracking dye on a piece of parafilm to which we add the genomic DNA samples. Then we load them in the same order into the gel. If DNA was extracted from eggs or small tissues, we skip running it unamplified on a gel since there will be too little DNA to show up, though plenty for PCR.

b. Run samples in a 0.7–1% agarose gel using a TBE buffer (0.089 M Tris, 0.089 M boric acid, 0.002 M EDTA) at 70 volts for about an hour. Run 0.1 and 0.2 µg uncut lambda as a quantity standard, and a lambda/HindIII digest for a size standard.

c. Visualize the nucleic acids by bathing the gel 20 min in 100 ml dH$_2$O to which 2 µl of 10 mg/ml ethidium bromide solution has been added. Store ethidium bromide in the dark at 4°C, and wear gloves since it is mutagenic and dispose of waste properly. We add about 6 tablespoons of activated carbon per gallon of waste, stir for an hour, then filter and dispose of the filter and contents in a special waste container.

d. Wearing protective goggles, look at the gel under UV light and take a picture with a Polaroid camera (107C b/w film). The DNA will appear as a high molecular weight band or as a smear from the size of the uncut lambda (about 48 kb) to the fluffy looking puff of RNA at the smaller sizes (smaller runs faster), depending on the preparation used. The smear is fine for PCR, but for a library, the DNA should be mostly very high molecular weight. From this photograph, you can also estimate the amount of DNA you have by comparison with the lambda standards.

e. Dilute an aliquot in sterile ultrapure water as necessary for PCR. In general, a 1/10 to 1/100 dilution is necessary for microsatellite amplification, though DNA extracted from insect eggs should be used undiluted. Though somewhat variable depending on genome size, approximately 1 ng of DNA is sufficient for a 10-µl PCR reaction. Avoid using too much DNA, which can cause multiple bands.

STRASSMANN.2

DNA PREPARATION FOR PCR USING THE DTAB/CTAB METHOD
Modified from Gustincich et al., 1991

Although Strassmann.1 is our standard protocol, we sometimes use the following protocol to process insect thoraces weighing 30–50 mg and use 1.5-ml microcentrifuge tubes. The protocol can be halved for smaller tubes. This protocol can be tried if Strassmann.1 fails to yield DNA that works for the PCR since the protein-denaturing (DTAB) and DNA-precipitating (CTAB) detergents may remove impurities not removed in Strassmann.1. Here, we give instructions for using the mini-beadbeater to homogenize tissues, but the disposable minipestles or grinding in liquid nitrogen described in Strassmann.1 can also be used with the first three ingredients of step 1.

1. To numbered, autoclaved 1.5-ml tubes add:

    130 μl of grinding buffer (0.1 M NaCl; 0.05 M Tris•Cl 8.0; 0.01 M EDTA)

    15 μl of 0.5 M EDTA

    5 μl of 10 mg/ml RNase

    0.5 ml 2-mm diameter glass beads

    0.2 ml 0.5-mm diameter glass beads

2. Place insect or tissue (30–50 mg) in tubes on top of beads and grind for 30 sec in minibeadbeater.

3. Spin briefly to get debris and beads to bottom of tube.

4. Incubate in water bath at 65°C for 5 min.

5. Remove from bath, add the following, then mix well:

    300 μl 8% dodecyltrimethylammonium bromide (DTAB)

    1.5 M NaCl

    100 mM Tris·Cl pH 8.0

    50 mM EDTA

6. Incubate at 65°C for 5 min.

7. Under hood, add 450 μl chloroform:isoamyl alcohol (24:1).

8. Mix for 5 min (tip covered rack back and forth by hand).

9. Spin at 10,000 rpm for 5 min.

10. Number another set of autoclaved 1.5-ml tubes.

11. Recover supernatant into clean tubes; avoid picking up the white layer at the interface between both phases (use clean pipette tip for each sample; set pipette to 100 μl).

12. Add: 400 μl H_2O and 50 μl 5% cetyltrimethylammonium bromide (CTAB)/0.4 M NaCl.

13. Incubate at 4°C overnight.

14. Spin 15 min.

15. Using aspirator, carefully aspirate off supernatant. The DNA pellet may be nearly invisible, so take care. It may help to mark the side of the tube where the DNA should precipitate after spinning, which can easily be done by aligning the caps in the same orientation in the centrifuge.

16. Resuspend in 120 μl 1.2 M NaCl.

17. Add 250 μl cold (stored at −20°C) 100% ethanol.

18. Spin at 10,000 rpm for 15 min.

19. Aspirate off ethanol from pellet. Be very careful not to lose the DNA pellet, which sometimes looks like a smear on the side of the tube.

20. Optional ethanol wash that would remove residual salts: Carefully add 750 µl of cold 70% ethanol without disturbing the pellet. Spin at 10,000 rpm for 5 min. Aspirate off ethanol from pellet.
21. Dry DNA in speed vac (no heat) or by air.
22. Resuspend DNA in 50 µl of sterile ultrapure water. Do not resuspend in TE since the EDTA will interfere with PCR.
23. Check the quality and quantity of DNA following step 3 in Strassmann.1.

STRASSMANN.3

INSECT SPERM DNA EXTRACTION AND PCR

From Peters et al., 1995

This protocol, which has been tested on wasps and ants, describes sperm extraction from an insect spermatheca and subsequent cell lysis and PCR. We do not precipitate the DNA since the sample is so small. This protocol may also be appropriate for other very small tissue samples, where DNA cannot actually be precipitated. Before PCR, it is important to spin down, then agitate, these samples to disperse the DNA.

Solutions

Promega PCR Reaction Buffer B (10X): 20 mM Tris-HCl pH 8.0, 100 mM KCl, 0.1 mM EDTA, 1 mM DTT, 50% glycerol, 0.5% Nonidet-P-40, 0.5% Tween-20.

Sperm DNA Solution (5X): 150 mM KCl (from KOH and HCl), 50 mM Tris-HCl pH 9.0, 15 mM DTT.

Sperm PCR Buffer (10X): 200 mM KCl, 1% Triton X-100.

1. Under a dissecting microscope at 25×, dissect insect in 10% NaCl and remove spermatheca.
2. Transfer spermatheca to a clean microscope slide with small drops of 10% NaCl.
3. Use fine forceps and insect pins to separate sperm ball from maternal tissue. In wasps, the sperm are surrounded by an inner membrane that facilitates their removal from the rest of the spermatheca. Crush this inner sac but do not attempt to separate sperm. If sperm must be picked up from the surrounding fluid, use acid etched, then hooked tungsten wires or very fine micropipettes (make friends with a neurobiologist who will introduce you to these wonderful tools).
4. Pick up sperm ball with insect pins and transfer it to a 0.5-ml tube with 20 µl 50 mM DTT (dithiothreitol). Do not let sperm ball dry out at any point.
5. Inspect the pin at 25× magnification to verify transfer of the sperm ball. Keep

sperm on ice (but do not freeze) while dissecting more insects. Wash and wipe forceps, and so on, between samples with H_2O and EtOH. Use fresh pins for each wasp.

6. Centrifuge for 30 sec.

7. Add 20 μl 0.5 M KOH. Add solutions to the side of the tube not the bottom, as undigested sperm balls may bind completely and irremovably to plastic pipette tips.

8. Spin for 5 sec.

9. Incubate for 5 min in a water bath at 65°C.

10. Vortex for 10 sec.

11. Incubate for another 5 min in a water bath at 65°C.

12. Spin for 10 sec.

13. Add 20 μl 0.5 M HCl. This must neutralize the KOH. Test solutions before use and alter HCl concentration (not volumes) used, if necessary.

14. Add 6.66 μl 0.5 M Tris-HCl pH 9.0. The pH must be exact as it establishes the pH for the PCR reaction.

15. Quick-spin for 10 sec.

16. Flick to mix.

17. Quick-spin for 10 sec and store at 4°C until used (do not freeze).

18. Set up a set of numbered 0.5-ml eppendorf tubes in a rack. To minimize errors of DNA transfer, each tube should be placed directly in front of the tube containing genomic DNA to be transferred into it.

19. Use 2.0 μl of the sperm cocktail for a 10-μl PCR reaction. Before setting up PCR, quick-spin sperm DNA for 30 sec, then vortex for 5 sec.

20. Prepare a PCR master mix that will supply each tube with:

    0.2 μl 10 mM dNTP mix

    1 μl sperm buffer

    3.84 μl dH_2O

    0.085 μl *Taq* polymerase (5 units/μl)

    0.16 μl ^{35}S-dATP (12.5 μCi/μl)

    0.7 μl 25 mM $MgCl_2$

    [If using Promega *Taq*, buy *Taq* in storage buffer B (not A, see solutions above).] If you are preparing PCR with a single primer pair, you may include the primer in the master mix by adding 2 μl of primer mix (2.5 μM each) to it.

21. Pipette 6 μl of master mix to the bottom of each sample tube, or 8 μl if the mix contains the primers too. Since no sperm DNA is yet present, the same pipette tip can be used for all tubes, thereby avoiding contaminating a large number of tips with radiation.

22. To one side of the tube add 2 μl 2.5 μM primer mix. Omit this step if the primers are included in the master mix.

23. Pipette 2 μl of sperm DNA template solution into the bottom of the tube. Use a fresh pipette tip for every sample.

24. Add a drop of mineral oil to keep the sample from evaporating.

25. Centrifuge the sample tubes briefly to mix all components in the bottom of the tube.

26. Cycle the samples between denaturing, annealing and extending temperatures. For some species (determined empirically), a double amplification is necessary in which the second amplification basically serves to add more Taq for additional cycles. The first run is for 23 cycles beginning with denaturing at 92°C for 60 sec, annealing (usually around 50°C) for 60 sec, then extending at 72°C for 45 sec. (If this doesn't work, try a 5-min denaturing at 94°C initially and annealing for 2 min for the first five cycles.) After the completion of all cycles, a long extension at 72°C for 10 min or so fills in incomplete strands.

27. If a second amplification is necessary, make a master mix that will supply each tube with 0.5 μl 10X buffer:

    4.1 μl dH$_2$O

    0.085 μl *Taq* polymerase (5 units/μl)

    0.3 μl 25 mM MgCl$_2$

    Then run the thermal cycle in the same way as above for 40 more cycles.

28. After the PCR is finished, store samples in the refrigerator at 4°C or in the freezer at −20°C until they are used. Add 5 μl denaturing tracking dye—identical to sequencing stop solution (95% formamide, 20 mM EDTA, 0.05% bromophenol blue, 0.05% xylene cyanol FF)—before storing the samples.

STRASSMANN.4

PCR USING 35S-LABELED dATP
Modified from Sambrook 14.18 and Ausubel 15.1.1

This is the crucial step of amplifying microsatellite length polymorphisms so they can be visualized on a denaturing acrylamide gel. Along with genomic DNA extraction and visualization, it is one of the most heavily used techniques in a microsatellite lab. Wipe test the thermocycler frequently and keep in a vented hood if radiation appears to be escaping (Trentmann et al., 1995). If you are only interested in whether or not the primers work, run an unlabeled PCR out on a 2% agarose gel (Strassmann.1 step 3). Visualize the PCR product with ethidium bromide, which is adequate for determining the success of the reaction, but not for scoring variation in length of product among individuals. The only change in the following protocol for an unlabeled PCR is that the 35S-dATP is omitted.

1. Set up a group of 0.5-ml eppendorf tubes in a rack. Tubes are labeled with sample numbers. To minimize errors of DNA transfer, each tube should be placed directly in front of the tube containing genomic DNA to be transferred into it. (We now use 96 well plates loaded with an octapipettor.)

2. Prepare a PCR master mix containing the following quantities for each reaction sample to be prepared:

0.1 μl 10 mM dNTP mix

1 μl 10X buffer (provided with *Taq*)

4.08 μl dH$_2$O

0.62 μl 25 mM MgCl$_2$

0.05 μl *Taq* polymerase (5 units/μl)

0.15 μl ^{35}S-dATP (12.5 μCi/μl)

The total working PCR volume will be 10 μl, including the genomic DNA template. As a general rule, the master mix should have enough for 1 extra sample for every 10 working samples to allow for losses due to pipetting errors (e.g., for 20 samples make enough for 22). If you are preparing PCR with a single primer pair, you may include the primer in the master mix by adding 2 μl of primer mix (2.5 μM each) to it. Leave it separate if some of the tubes you are setting up will have different primers.

3. To the bottom of the tube add 2 μl primer mix. Omit this step if the primers are included in the master mix (which is often the case).

4. Pipette 6 μl of master mix to the side of each sample tube; 8 μl if mix contains the primers too. The same pipette tip can be used for all tubes since no genomic DNA is yet present, thereby avoiding contaminating a lot of tips with radiation.

5. Pipette 2 μl of genomic DNA template solution by placing it to another side of the tube. Use a fresh pipette tip for every sample.

6. Add a drop of mineral oil to keep the sample from evaporating.

7. Centrifuge the sample tubes briefly to mix all components in the bottom of the tube.

8. Cycle the samples between denaturing, annealing and extending temperatures. For most applications we use the same basic program. We do 30–35 cycles of 92°C denaturing for 60 sec, a 60-sec annealing (at a temperature optimized for the primers used: usually between 45°C and 60°C) and 45-sec extension at 72°C. After that, an extra 5 min at 72°C are added to allow for the completion of the extension. PCR times vary considerably among thermal cyclers. Our thermal cycler takes between 2-1/2 and 3 hr. Different procedures could shorten this time.

9. After the PCR is finished, samples are stored in the refrigerator at 4°C or in the freezer at −20°C until they are used. Add 5 μl denaturing tracking dye—identical to sequencing stop solution (95% formamide, 20 mM EDTA, 0.05% bromophenol blue, 0.05% xylene cyanol FF)—before loading the samples. Samples with loading buffer can be stored at 4°C for as much as a month although each time they are boiled for loading the DNA seems to degrade some.

STRASSMANN.5

VISUALIZING ³⁵S-LABELED MICROSATELLITE LENGTH POLYMORPHISMS ON SEQUENCING GELS

From Ausubel et al. 7.4 and BioRad's Sequi-Gen nucleic acid sequencing cell instruction manual.

This technique visualizes the length polymorphisms in PCR products generated in Strassmann.4 by running ³⁵S-containing PCR products out on a denaturing sequencing gel next to an M13 sequencing reaction that is used as a size standard. This technique has the advantages of generating a large size permanent record of microsatellite lengths and the size standard.

We prefer a wide sequencing apparatus that does not require taping and use the BioRad Sequi-Gen II apparatus with 38 cm × 50 cm plates. The sequence for our size standard, M13mp18, appears in the Sequenase™ Version 2.0 DNA Sequencing Kit Step-By-Step Protocols (Amersham/US Biochemical) and is also in GenBank. The numbers in the manual do not match actual fragment sizes produced by the −40 primer used to produce the sequence: 41 must be added to the numbers written above the sequences in the US Biochemical (USB) manual. For example, the protocols list C's at positions 79 and 80. In the sequence produced by the −40 primer, these actually correspond to fragment sizes of 120 and 121 base pairs. We now use an octapipettor to load gels directly from 96 well plates.

1. Clean gel rigs thoroughly using Alconox™ powdered detergent. Detergent residues may cause poor microsatellite resolution and some detergents leave worse residues than others. Rinse the gel rig very well in distilled water. Cleaning a gel rig well takes about 20 min!

2. Squirt acetone on the side of each glass plate that will touch the gel and rub hard all over with a large kimwipe. This should take several minutes.

3. Squirt 100% ethanol on the same side of the glass plates and rub hard all over with a large kimwipe, again taking several minutes.

4. Every 3 or 4 gels, polish the integrated plate chamber (IPC) by squirting a small amount of Rain-X (available at auto supply stores) on a kimwipe and polishing it over the surface. Clean this surface with dH₂O. Though cheaper and less hazardous than silane-based products, it is critical that not too much Rain-X be used.

5. Place 0.4 mm gel spacers along either side of the IPC. Carefully align and place the front plate on top. Clamp the two plates together, making sure the clamps are oriented properly.

6. Place a rubber strip in the bottom of the casting tray with a piece of filter paper on top. The small amount of acrylamide poured into the casting tray and sucked up between the plates serves as the bottom seal of the gel.

7. In a small beaker, place 30 ml of sequencing gel solution [6% acrylamide/methylene bisacrylamide (19:1, available as a 40% stock solution)], 7 M urea and TBE (0.089 M Tris, 0.089 M boric acid, 0.002 M EDTA), 100 µl

25% ammonium persulfate (made fresh weekly), 100 μl TEMED; swirl well and pour into casting tray.

8. Place gel rig in casting tray and watch to see that the seal is sucked into the gel rig about 1 cm. Allow to set for 3 min.

9. Lay gel rig IPC down on corks so one upper corner is highest.

10. In a small beaker place 100 ml sequencing gel solution, 100 μl ammonium persulfate, and 100 μl TEMED and stir well with a spatula.

11. Pour this solution slowly and carefully between the gel plates starting at the lower open corner. Take great care to avoid bubbles. If, despite careful pouring, bubbles appear, this is usually a sign of dirty plates or that the bottom seal failed.

12. Insert combs in top slot of gel. If using sharktooth combs, put the combs in upside down (i.e., smooth side toward gel). If using well-forming combs, put them in teeth down. We use combs with 60–65 lanes on the 38-cm-wide gel.

13. Clamp plates together at the origin with binder clamps and let polymerize overnight. However, polymerization for as little as an hour or as long as 2 days has no obvious adverse effects on results.

14. When ready to run the gel, remove the clamps and the casting tray and place the gel rig properly in the buffer tray.

15. Make 2 liters of 1X TBE buffer and pour 300 ml in the tray and the rest in the IPC, reserving a little to add, if needed, during the run.

16. Wipe the urea off the front plate with a kimwipe, remove the combs, connect electrodes, and turn on the power and set limits (we use 2100 volts, 150 mA, 250 watts). Run at 95 watts for 45 min to 1 hr to preheat gel to 50°C.

17. Denature DNA samples for 5 min in a heating block with cells filled with water at 88°C. Keep careful track of sample order. Samples should be numbered consecutively.

18. In preparation for loading the gel, set sharktooth combs so the tips barely penetrate the top of the gel. If block combs are used, be sure wells are clearly defined. Using a pasteur pipette, squirt out all wells to remove urea (this is very important and may need to be done repeatedly, if you load the gel slowly). With a felt tip pen on the front glass mark the location of any bubbles and other gel imperfections that require that lanes be skipped.

19. Load a size standard in the first two lanes, in the last two, and somewhere in the middle. We use A, G, and T in the first lane and C in the second lane from an M13mp18 sequencing reaction, thus devoting 6 lanes per gel for size standards.

20. Load 4 μl per lane of DNA sample. Flush the tip in the buffer container between samples. As you load, transfer the tubes in order to a tube rack. Remember to flush urea regularly. Check tube order after loading for errors or mixups. Save samples at −20°C until the autorad is developed, since samples can be rerun if there is something wrong with the gel. If you load too slowly the gel may cool below 40°C which could lead to blurry bands. Wrapping the gel rig in 2 or 3 layers of thin plastic will keep it from cooling for roughly 20 min.

However, once the rig has been warmed and loaded remove the plastic to prevent overheating.

21. Reconnect electrodes and run gel for 2–4 hr (depending on product size) at 95 watts, which should keep the gel between 45°C and 55°C. Lower temperatures do not reliably denature the DNA completely, which leads to fuzzy bands, and higher temperatures are likely to crack the gel rig or degrade the gel during extended runs.

22. Dismantle the gel apparatus and remove the IPC from the front plate, and gel.

23. Lay a piece of 3MM Whatman paper on top of the gel and pick it up carefully from the front glass. We do not fix the gel with 10% methanol/10% acetic acid, as fixing does not appear to improve band resolution and makes lifting the gel from the plate more difficult.

24. Cover the gel with Saran Wrap and put it in a gel drier with vacuum to dry at 80°C for 45 min to an hour. A low cost alternative is to clamp the gel/filter paper by its edges to a glass plate and dry for 20–25 min with a hair dryer.

25. When dry, place the gel in an autoradiography cassette with a sheet of x-ray film. Films are made by Amersham, Du Pont, Fuji, or Kodak and differ appreciably in price, sensitivity, and background color, so several should be tried to find what works best for you. Film exposure times vary between 1 and 4 days, depending on film sensitivity, locus, and initial amount of target DNA in the PCR that originated the fragments. We use a geiger counter to estimate the correct exposure time.

STRASSMANN.6

HIGH MOLECULAR WEIGHT GENOMIC DNA ISOLATION USING GEL BARRIER TUBES
Modified from Thomas et al., 1989.

This is the main technique we use for obtaining library grade genomic DNA. It uses SST tubes (SST tubes are used for blood serum separation and are available from Becton Dickinson, Rutherford, NJ), which eliminate the need for phase separation between additions of solvents. The clinical centrifuge must have a swinging bucket rotor that allows the tubes to be perfectly horizontal so the gel stays equally thick all around. Strassmann.1 can also yield library grade genomic DNA if the tissue is ground in liquid nitrogen.

Initial Buffer: 1.8 ml 0.1 M NaCl, 0.05 M Tris-HCl (pH 8.0), 10 mM EDTA.

1. Add the following to an SST tube and place at 4°C:
   Initial buffer
   150 ml 20 mg/ml proteinase K
   100 ml 10% SDS

2. Grind 500 mg or less (no more!) of tissue or whole insects, using a mortar and pestle that has been chilled by immersion in liquid nitrogen. Place the entire mortar and pestle in a styrofoam cooler, add liquid nitrogen to cover, and begin grinding (leaving the mortar and pestle in the cooler) when most of the liquid nitrogen has evaporated. Wear heavy gloves!

3. Using a small, cold, metal spatula, transfer the ground tissue to the SST tube. Invert several times to mix and then incubate at 58°C for 30 min. Do this quickly so the DNA is not attacked by enzymes!

4. Add 1 ml each of phenol (saturated with Tris to pH 8.0) and chloroform/isoamyl alcohol (24:1).

5. Mix gently for 5 min by hand (hold lid on!).

6. Centrifuge at 2700 rpm in a clinical centrifuge for 10 min. After centrifuging, the organic solvents and proteins end up below the gel barrier, and so fresh solvents can be added without removing the old ones.

7. Add another 1 ml each of phenol (saturated with Tris to pH 8.0) and chloroform/isoamyl alcohol (24:1).

8. Mix gently for 5 min by hand (hold lid on!).

9. Centrifuge at 2700 rpm in a clinical centrifuge for 10 min.

10. Add a third dose of 1 ml each of phenol (saturated with Tris to pH 8.0) and chloroform/isoamyl alcohol (24:1).

11. Mix gently for 5 min by hand (hold lid on!).

12. Centrifuge at 2700 rpm in a clinical centrifuge for 10 min.

13. Extract a final time with 2 ml chloroform/isoamyl alcohol (24:1).

14. Mix gently by hand for 8 min.

15. Centrifuge at 2700 rpm in a clinical centrifuge for 10 min.

16. Decant supernatant to a fresh, empty, plastic centrifuge tube (all solvents are below gel barrier) and add:

    250 ml 3 M NaOAc (pH 5.2–5.6)

    5 ml 100% ethanol

17. Precipitate DNA for at least 2 hr (overnight is better) at room temperature.

18. Spin in clinical centrifuge at 2700 rpm for 10 min. Pour off supernatant.

19. Wash pellet carefully by adding 4 ml 70% ethanol to tube, then pouring or aspirating it off without dislodging the pellet. The ethanol removes residual salts, chloroform, and phenol.

20. Dry in speed vac.

21. Resuspend pellet in 250 µl of TE (0.01 M Tris, 0.001 M EDTA). Check quality on 1% agarose/TBE gel (see Strassmann.1 step 3a). Yield should be about 100 µg of DNA, if initial tissue was 500 mg. DNA from wasps after this extraction is of very high molecular weight and is purple colored. Though this contaminant does not inhibit restriction enzyme digestion or ligation, it may inhibit PCR.

STRASSMANN.7

DIGESTION AND PURIFICATION OF DNA FOR PLASMID LIBRARIES
Modified from Sambrook 1.53 and Ausubel 3.10.1 and 3.14

In this protocol the plasmid is cut open and dephosphorylated so that the genomic DNA insert can be ligated in. The DNA to be inserted is prepared by digesting with a restriction enzyme. Then it is run out on an agarose gel and the size range of insert desired is selected. The DNA is extracted from the agarose and then ligated into the plasmid (Strassmann.8). [Use high-quality agarose (e.g., Seakem/Seaplaque) as poor grade agaroses may inhibit subsequent ligations.] This is a sticky-end ligation so the restriction enzymes have to leave complementary ends. It is also important that there are restriction enzyme recognition sites to cut the insert out again on either side of the insertion site, and that they work under the same conditions. The protocol describes using a Bluescript phagemid (Stratagene) and the 6 cutter BamHI on the plasmid and the 4 cutter SAU3A on the genomic DNA. Plan carefully to avoid predawn laboratory visits.

1. *Plasmid digestion*

a. To a 0.5-ml autoclaved tube add:

    7 μl 10X Buffer (provided with restriction enzyme)

    1 μl BSA (bovine serum albumin)

    7 μl 50 mM spermidine

    3 μl—60 units BamHI

    50 μl—10 μg Bluescript SK + (Stratagene) in TE (if needed, dilute with TE, not water)

b. Incubate at 37°C at least 5 hr (overnight preferable).

c. Add:

    8 μl 10X CIP buffer (stored in refrigerator)

    3 μl CIP (75 u calf intestinal phosphatase)

d. Incubate at 37°C for 3–6 hr in a waterbath

e. Add:

    2 μl BAP (250 units bovine alkaline phosphatase)

f. Incubate at 62°C for 2.5 hr.

g. Cool slowly to room temperature.

h. Load on 1% Seakem/TBE gel. Use TAE (0.4 M Tris acetate, 0.001 M EDTA), if extracting DNA with Geneclean II. If the digestion was complete (run an aliquot on a gel), then running on a gel is unnecessary (so skip steps h and i). Simply precipitate the plasmid DNA (with 1/10 volume 2 M NaCl and 2.5 volumes 100% ethanol), resuspend it in 100 μl TE, and quantify an aliquot on a gel against a lambda standard.

i. Run 12 hr at 45 V, though some people just run it out in an hour or so at 70 V.

Since you will have about 80 μl of sample to run out, use the gel combs that produce a large slot, not a narrow lane. In the narrow lines on either side, run out a size standard such as lambda/HindIII.

2. *Genomic DNA digestion*

a. To an 0.5-ml autoclaved tube add:

138 μl H_2O

35 μl 10X Buffer

3.5 μl BSA

35 μl 50 mM spermidine

8 μl—10 mg/ml RNase

120 μl—~20 μg genomic DNA in TE

10 μl—~100 units SAU3A

b. Incubate at 37°C for 8–12 hr.

c. Run on 1% Seakem/TBE gel (if extracting DNA from agarose using Geneclean II use TAE) for about 12 hr at 35 V (some do this at 70 V for about an hour) in large slot comb, with lambda/HindIII as a size standard.

3. *Isolation:* During this process, view the DNA as briefly as possible under UV light, making nicks as to where to cut out the appropriate size bands. It is preferable to cut out and remove the DNA band of interest, then to take a photograph of the remainder for reference, should any question about what was chosen arise subsequently.

a. Using the lambda/HindIII ladder, cut out pBluescript band immediately above 2031 bp lambda/HindIII band. This should be the linear, dephosphorylated Bluescript. It will be a tight, very bright band.

b. Cut out digested, genomic DNA from 300 to 600 bp in length. This will result in a larger amount of agarose from which the DNA must be extracted.

c. Isolate the DNA from the agarose. There are many ways of doing this, none entirely satisfactory. Geneclean II (Bio101) works moderately well, though this is at the small end of its DNA size class, and the gel buffer needs to be TAE not TBE. A potentially hazardous technique, which we have not tried, is advocated by some (Bruford et al., 1992). Freezing and thawing a few times breaks down the gel matrix, which simplifies the following procedure.

d. For each type of DNA to be purified: Make a small drainage hole in each of five 0.5-ml tubes. Plug with siliconized glass wool. Put the 0.5-ml tubes into 1.5-ml tubes.

e. Fill each tube with small gel slices and spin at 10,000 rpm for 1 min.

f. Collect eluate and continue filtering until all the gel has been filtered (each tube may need to be loaded several times).

g. Rinse by adding 50 μl of TE to the residue in the top of each filter. Spin through and collect.

h. Filter eluate through 1 Ultrafree 0.45-μm filter (Millipore); do not exceed 12,000 rpm and collect new eluate.

i. Wash with 200 μl TE.

j. Filter eluate through Ultrafree 30,000 NMWL filter (do not exceed 8000 rpm) to concentrate. Wash filter with 2X 200 μl TE.

k. Resuspend sample in 100 μl TE. You should have between 4 and 8 μg of DNA.

l. Run out 2 μl of both plasmid and genomic DNA and size and quantify standards on a 1% agarose gel to check quality and quantity of DNA. There should be a single plasmid band, and the genomic DNA should be between 300 and 600 bp.

STRASSMANN.8

LIGATION OF TARGET DNA INTO PLASMID
Modified from Ausubel 3.16

We have found that the best ratios for ligation are two to four times as many sticky ends of insert to plasmid. Since the plasmid (Bluescript) is 2900 bases, and the insert averages 400 bases, this means that ratios of about twice as much DNA of plasmid relative to insert are required. You may want to try ratios of plasmid:insert of sticky ends 4:1, 2:1, 1:1, 1:2, 1:4, and 1:6 and a control of 1:0 to verify that the plasmid is properly dephosphorylated, which prevents it from closing up on itself when ligase is added.

1. In a 0.5-ml microfuge tube put:

   5 μl vector (100 ng)

   Target DNA (none in control, about 50 ng in test ligations depending on ratio used)

   1 μl T4 ligase (6 Weiss units/μl)

   2.5 μl 10X buffer that came with the T4 ligase

   Ultrapure water to bring total volume to 25 μl (depends on dilution of target DNA)

2. Incubate at 14°C in a circulating cold water bath for 10 hr.

STRASSMANN.9

TRANSFORMATION OF COMPETENT CELLS
Modified from Hanahan 1983, Sambrook 1.76

SOC

2% Tryptone

0.5% Yeast extract

10 mM NaCl

2.5 mM KCl

10 mM MgCl$_2$

10 mM MgSO$_4$

2 mM Glucose

1. Dilute the ligation reaction with 75 μl of TE, giving 100 μl in all.

2. Thaw competent cells (e.g., XL1-Blue, Stratagene) on ice for 5 min. Use one 1.5-ml tube for each ligation reaction and 100–200 μl of competent cells per tube.

3. Add 15 μl ligation reaction and mix gently either by sucking in and out with a pipette tip that has been cut so it has a larger opening at the tip, or by tapping gently on the tube.

4. Incubate on crushed ice for 8 min (or as long as 30 min).

5. Heat shock at 42°C for 1 min in a water bath.

6. Ice down for 2 min.

7. Add 900 μl SOC that is at room temperature.

8. Incubate at 37°C, shaking 240 rpm for 45 min (no longer so cells cannot duplicate!).

STRASSMANN.10

TRIAL PLATING OUT OF TRANSFORMED CELLS

1. In a sterile work area set out one 82-mm LB/amp (50 ng/ml) plate for each tube of transformed cells.

2. Pipette 50 μl of transformed cells on each plate and spread them out with a glass rod (bent from a pasteur pipette or glass rod so it is thin and will cool quickly) that has been sterilized by flaming after dipping in alcohol, and cooled. Touch the rod first to a blank area of the plate to be sure it is cool and then spread the cells gently.

3. Incubate plates in an oven at 37°C for about 12 hr until discrete colonies are visible on the plates.

4. Determine which is the best ligation ratio, and whether or not there is any background. The best ligation ratio will have the most colonies on the agar plate. Because of the ampicillin in the agar, only cells with plasmids survive. The background plate indicates how many colonies have recircularized plasmids lacking inserts. On this plate are distributed cells that were transformed with dephosphorylated plasmids to which ligase but no insert was added. If all the plasmids were dephosphorylated, there should be no background at all. However, since you will only be selecting clones with microsatellites, some background is tolerable. It is best if the background is below about 20%, though a

higher percentage will only mean a larger library must be plated out to get sufficient positives.

STRASSMANN.11

PLATE OUT LIBRARY ON NYLON MEMBRANES
Modified from Ausubel 6.2.1

Twenty large (132-mm) nylon membranes with about 1000 colonies each will yield a library of 20,000 colonies. When probed with three or four trinucleotide repeats, this size library should yield our usual goal of 10 highly polymorphic, easily scored trinucleotide microsatellites. For taxa where microsatellites are rare, it is advisable to either plate out a larger library, enrich the inserts for microsatellites (Fleischer, Chapter 7 in this volume), or plate out a very dense phage library with positives that can be converted to plasmids, using a Lambda Zap II system (Stratagene; C. Hughes, *in preparation*).

1. Calculate how much of the transformation should be plated on each membrane to yield 1000 colonies. In our experience, only about half as many colonies from a given transformation grow on nylon as grow directly on agar plates, so double the number of cells plated on nylon, as compared to directly on agarose. Dilute the transformation with LB/amp so that you will plate between 1 and 3 ml per plate for even coverage.
2. Squirt a little alcohol into the Buchner funnel to sterilize it and be sure the vacuum is working. Let it dry.
3. Place 3 pieces of Whatman 3MM filter paper in the Buchner funnel. Saturate with LB/amp.
4. Write an identifying number in pencil at the margin of a nylon membrane. By convention, in our laboratory, the number is always on the same side as the colonies.
5. Place the labeled nylon membrane on the damp filter paper (caution—these membranes are delicate—handle carefully with forceps). It should become moist from the lower filter paper. If not, moisten with more LB/amp but be sure there are no pools. The vacuum should be weak but working well.
6. Using a 1-ml pipetman with a tip cut to a larger aperture, slowly squirt the cells all over the membrane. Avoid the edges of the membrane, and try to get very even coverage.
7. With forceps, carefully transfer the nylon membrane to a prepared large LB/amp plate.
8. Repeat for the other 19 plates.
9. Grow in an incubator at 37°C for 8–14 hr. Check periodically to catch when colonies are clearly raised up but still very small (roughly the size of medium to

large insect pinheads). At this point, transfer membranes on their plates to a 4°C refrigerator or proceed with making replicate membranes.

STRASSMANN.12

REPLICATING MEMBRANES
Modified from Ausubel 6.2.2

Normally one copy is denatured and probed while the other is the source of live colonies. Another replicate can be made in the same way and the colonies archived in a −80°C freezer (Dreyer et al., 1991). To do this, filter paper the size of the petri plate is cut and moistened with a 50% glycerol 50% LB/amp solution. One piece is put in the lid and one in the floor of the petri plate. One membrane can be put on the floor and another in the lid, colony side toward the middle. The plate is then sealed with parafilm and frozen.

1. Number fresh petri plates to match numbers on original plates.
2. Take membrane with colonies and place on filter paper. Take a fresh membrane and number it in pencil. Place pencil side down carefully on original membrane. Cover with filter paper, then piece of glass. Press hard (put most of your weight on it).
3. Heat dissecting needle in flame. Poke 4 or 5 asymmetric holes through both membranes for subsequent alignment.
4. Using forceps, carefully separate membranes and place each on its petri plate. Allow to grow for several hours until colonies are again raised up. Then store at 4°C until you are ready to proceed.

STRASSMANN.13

PROBING A LIBRARY OR A SOUTHERN WITH A SHORT OLIGONUCLEOTIDE
Modified from Ausubel 3.6, 5.4.1 and Rosenberg et al., 1990

This protocol takes your denatured library membranes and probes them with a microsatellite oligo to identify clones containing inserts with repeat regions. You can probe for oligos with the same melting temperatures at the same time. This means that although you must probe for AAT alone, you can probe simultaneously for AAC, AAG, TAG, and CAT, though we label each probe separately with ^{32}P. CAC, CAG, GAG, and GAC can also be probed for together. We usually found the most repeats with AAT, AAC, AAG, TAG, and CAT (J.E. Strassmann et al., *in preparation*). There is only one other trinucleotide repeat: CCG. We usually synthesize oligonucleotides that are ten repeats (30 bases) long.

Hybridizing Solution

5X SSC

0.5% SDS

5X Denhardt's reagent [50X Denhardt's reagent contains 5g Ficoll (Type 400, Pharmacia), 5 g polyvinylpyrrolidone, 5 g BSA in dH_2O to 500 ml—and is filter sterilized and stored in aliquots at $-20°C$, Sambrook 0.49]

1. Denature, neutralize, and bind DNA to membrane or strip off previous probe.

New membranes:

a. Denature with 1.5 M NaCl, 0.5 M NaOH for about 20 min. (Southerns can be bathed in this solution. Libraries need to be laid, colony side up, on saturated filter paper. Take GREAT care not to wet the colonies on top because they might wash off. Periodically, add more solution to the filter paper with a pasteur pipette to keep it wet.)

b. Neutralize with 1.5 M NaCl, 0.5 M Tris-HCl pH 7.2, 0.001 M EDTA for 10–20 min. (Same procedure as above.)

c. While still damp, put in Stratalinker at 1200 times 100 microjoules for 1 min, or bake 3 hr at 65°C.

Previously probed membranes:

a. Strip by submerging membranes in 1.5 liters boiling 0.1% SDS and let return to room temperature slowly. (Boil water without SDS, add SDS to pan in oven, add boiling water over membranes, and stir occasionally.)

2. Prepare probe. This protocol uses TDT and alpha labeled ^{32}P, which we prefer to the alternative of using kinase and gamma labeled ^{32}P. Use all ^{32}P precautions when making this! Try to add ingredients carefully so tube does not have to be centrifuged.

3. Add the following ingredients to a 0.5-ml microfuge tube. Add ^{32}P-labeled dCTP last.

	Quantity per hybridizing tube
3.6 µM oligo	6 µl
TdT 5X buffer	3 µl
TdT enzyme	1 µl
H_2O	2 µl
α-^{32}P-dCTP	3 µl (30 µCi)

4. Incubate in 37°C water bath for 3–6 hr.

5. Prehybridize membranes with 30 ml of hybridizing solution for at least an hour at 40–60°C (depending on the melting temperature of the oligo with which you are probing), using the same temperature as for hybridizing. Membranes can be rolled together 10 deep in the hybridizing tube that fits your oven (we use a

Robbins Scientific Hybridization Oven) but be sure that they lie smoothly against each other without wrinkles or folds. Use a tube for every 10 or so large round filters. Place membranes carefully in hybridizing tubes, putting DNA side toward tube interior. Turn on rotisserie motor and let membranes prehybridize at least an hour.

6. Hybridize membranes. Take hybridizing tubes out of hybridizing oven, pour out prehybridizing solution and set tubes up in rack behind shield.

7. Add 1.5 ml hybridizing solution to tubes.

8. Take probe out of waterbath (shield tube) and add 300 μl hybridizing solution to it. Add the necessary volume of probe to each tube, trying to hit the bottom of the tube with it and not the membranes.

9. Screw lids on tubes firmly. Put in balanced positions in oven and turn on motor. Leave for 10–48 hr at a temperature determined by oligo binding.

10. Wash membranes. Pour hybridizing solution into ^{32}P waste and begin washing with 5X SSC, 0.1% SDS either made fresh or warmed slightly to dissolve crystals. Do not warm above the hybridizing temperature. Use 100 ml for first 20-min wash, then dump in ^{32}P liquid radioactive trash.

11. Use 100 ml for next 20-min wash, and dispose of according to your radiation safety license. Keep washing until a geiger counter held 1 inch from a single membrane reads 20–50 counts per second.

12. You may need to use a more stringent wash such as 100 ml of 2X SSC, 0.1% SDS for 20 min.

13. Wrap membranes in Saran Wrap, place in film cassette, add film (in darkroom), and expose for 10–24 hr.

STRASSMANN.14

PICKING UP POSITIVES FROM A LIBRARY
Modified from Ausubel 6.6

By matching the autorads from colony probing with microsatellites with your live library membranes, you can identify and select those colonies most likely to have genomic DNA inserts containing microsatellites. We find it most efficient to identify all positives, select them, grow them up, and freeze away an aliquot (in a microfuge tube with equal amounts of plasmid grown in LB/amp and glycerol). Sequencing of positives can then proceed uninterrupted with probing processes that require ^{32}P.

1. Clean the working area well with alcohol.

2. When it is dry, light a bunsen burner and work very close to it. (Be careful because gloves can melt!)

3. Place the best autorad on the light box.

4. Carefully pick up the first membrane with sterile forceps. Take care never to breathe on the membrane. Place the membrane on the autorad and line up the holes in the membrane with the dots or x on the autorad. You should be able to see dark spots on the autorad through the colony membrane. These spots correspond either to colonies or to dirt. It is very difficult to distinguish between the two, and there are no good rules, so, if there is a dark spot that lines up with a colony, it is worth picking up. If there is no colony under a dark spot, then it is probably dirt. Sometimes, it will be hard to tell to which of several discrete colonies the dark spot on the autorad corresponds. In these cases, pick up separate colonies that might be producing the positive signal and put their numbers on the autorad. Sometimes, the colonies are so close together that you are not sure which of the colonies are the positive and you cannot pick up just one colony. In these cases, be sure to touch one or several sticks to all the candidate colonies, and put each stick in a culture tube. Be sure to note this on the datasheet, because, if these colonies turn out positive on the Southern, they will have to be streaked out again to permit picking up of single colonies. There may be quite a number like this because the library is quite dense. It is best to get as few colonies as is possible per tube.

5. To pick up a colony, touch it firmly with a wooden stick (held in forceps).

6. Open a numbered culture tube and drop the stick in, holding it as near the flame as possible. Flame the opening of the tube before closing it.

7. Record the colony number on the autorad, then repeat type on the culture tube, and both on the datasheet.

8. Turn the shaker on to 140–240 rpm and 35°C and let grow up overnight. Plasmid DNA can now be extracted from these cultures for further analysis. Aliquots of LB/amp containing the plasmid can be combined with an equal volume of glycerol and frozen at −70°C for later analysis.

SWALLA.1

SUBTRACTIVE HYBRIDIZATION

From Swalla et al., 1993.

Initial PCR

1. PCR of different cDNA libraries (N = number of reactions)

2. Prepare master mix:

Volume	Component
10 μl	10X PCR buffer (final concentration 1.5 mM $MgCl_2$)
8 μl	10 mM dNTPs (final concentration 200 μM each)
8 μl	10 pM/μl T3 primer (5′-ATTAACCCTCACTAAAGGGA-3′)

8 μl   10 pM/μl T7 primer (5′-GCGTAATACGACTCACTATA-3′)

1 μl   Amplitaq

<u>65 μl</u>   cDNA library 10^5 plaque forming units (pfu)/μl

100 μl/reaction

3. Heat the cDNA libraries to 95°C for 5–10 min.

4. Add master mix and 40 μl (2 drops) mineral oil. Spin briefly.

5. Cycles:

	Temperature	Time
Denature	94°C	1 min
Anneal	55°C	2 min
Extend	72°C	3 min

Number of cycles = 15

6. Prepare second master mix:

Volume	Component
	_____ N reactions
	_____ N ddH$_2$O (add for total volume of 100 μl/reaction)
_____	5 N 10X PCR buffer (final concentration 1.5 mM MgCl$_2$)
_____	N 10 mM dNTPs (final concentration 200 μM each)
_____	2 N 10 μM T3 primer
_____	2 N 10 μM T7 primer
_____	0.5 N Amplitaq
<u>20 μl</u>	Amplified inserts from step 5 above

100 μl/reaction

7. Cycles:

	Temperature	Time
Denature	94°C	1 min
Anneal	55°C	2 min
Extend	72°C	3 min

Number of cycles = 4

8. Cycles:

	Temperature	Time
Anneal	55°C	2 min
Extend	72°C	3 min

Number of cycles = 2

9. Delay   72°C   10 min

10. Soak   4°C   soak

cDNA for Subtraction (Library of Interest)

1. Phenol-chloroform extract PCR products 2X.
   Add 100 µl of 50:49:1, phenol:chloroform:isoamyl alcohol.
   Vortex for 1 min. Spin for 5 min at top speed in microfuge.
   Remove top layer and transfer to a clean tube. Repeat.
2. Chloroform extract 1X. Repeat step 1 with 50 µl chloroform.
3. Ethanol precipitate. Add ¹/₁₀ volume of 3 M Na acetate, then 2.2 volumes 100% ethanol; leave at −20°C for at least 30 min.
4. Spin down one-quarter of the DNA at top speed in microfuge. Dry in a speed vac, and resuspend:
5. 7 µl DEPC-treated water (up to 24 µl total)
   5 µl 5X T7 transcription buffer (supplied with the enzyme)
   3 µl rNTP mixture (ATP, CTP, GTP, UTP)
   <u>1 µl 750 mM DTT</u>
   16 µl total
6. Mix well. Spin briefly in microfuge at top speed. Then add:
   4 µl T7 RNA Polymerase 20 u/µl (80 *u* total)
7. Mix gently. Incubate at 37°C for 2 hr.
8. Add 1 unit of RQ1 DNase (1 µl) to digest the DNA template. Leave at 37°C for 15 min.
9. Ethanol precipitate. Add ¹/₁₀ volume of 3 M Na acetate, then 2.2 volumes 100% ethanol. Leave at −20°C for at least 30 min.
10. Convert the RNA to cDNA by using the T3 primer and reverse transcriptase. A commercial kit is the easiest way to do this. The RNA is resuspended in reverse transcriptase buffer with the T3 primer and dNTPs according to the directions in the kit. After the RNA is converted to cDNA, the RNA template is hydrolyzed by treatment with 1 N NaOH, 0.5% SDS, and the cDNA is ethanol precipitated as in step 9.

RNA for Subtraction

1. Phenol-chloroform extract PCR products from Initial PCR 2X.
   Add 100 µl of 50:49:1, phenol:chloroform:isoamyl alcohol.
   Vortex for 1 min. Spin for 5 min at top speed in microfuge.
   Remove top layer and transfer to a clean tube. Repeat.
2. Chloroform extract 1X. Repeat step 1 with 50 µl chloroform.
3. Ethanol precipitate. Add ¹/₁₀ volume of 3 M Na acetate, then 2.2 volumes 100% ethanol. Leave at −20°C for at least 30 min.
4. Spin down one-quarter of the DNA at top speed in microfuge. Dry in a speed vac, and resuspend:

5. 7 µl DEPC-treated water (up to 24 µl total)

   5 µl 5X T7 transcription buffer (usually supplied with T7 polymerase)

   3 µl rNTP mixture (ATP, CTP, GTP, UTP; each 10 mM in H$_2$O)

   <u>1 µl 750 mM DTT</u>

   16 µl total

6. Mix well. Spin briefly in microfuge at top speed. Then add:

   4 µl T7 RNA polymerase 20 u/µl (80 u total)

7. Mix gently. Incubate at 37°C for 2 hr.

8. Add 1 unit of RQ1 DNase (1 µl) to digest the DNA template. Leave at 37°C for 15 min.

9. Ethanol precipitate. Add $^{1}/_{10}$ volume of 3 M Na acetate, then 2.2 volumes 100% ethanol. Leave at −20°C for at least 30 min.

10. Label the subtracting library RNA with photoactivatible biotin (PAB; Clontech Laboratories, Inc., Palo Alto, CA) by use of a PAB sunlamp.

11. Resuspend RNA in 20 µl of DEPC-treated water.

12. Put into tube and leave under sunlamp on ice for 15 min (no caps on tubes).

    10 µl of RNA

    50 µl of PAB (1 µg/µl)

13. Add 60 µl 100 mM Tris-Cl, pH 9

    40 µl TE 10:1

14. Butanol extract 2X

    Add 100 µl of TE-saturated butanol.

    Vortex for 1 min. Spin for 5 min at top speed in microfuge.

    Remove bottom layer and transfer to a clean tube. Repeat.

15. Phenol-chloroform extract.

    Add 100 µl of 50:49:1, phenol:chloroform:isoamyl alcohol.

    Vortex for 1 min. Spin for 5 min at top speed in microfuge.

    Remove top layer and transfer to a clean tube.

16. Chloroform extract 1X. Repeat step 15 with 50 µl chloroform.

17. Ethanol precipitate. Add $^{1}/_{10}$ volume of 3 M Na acetate, then 2.2 volumes 100% ethanol. Leave at −70°C for at least 30 min.

18. Repeat steps 12 through 17.

Subtractive Hybridization

1. Mix single-strand 3 µl cDNA with 10-fold excess biotin-labeled RNA.

   Add 1 µl 250 mM Hepes, pH 7.6, 1% SDS, 10 mM EDTA, 1 µl 2.5 M NaCl.

   Overlay with mineral oil.

2. Put in PCR machine. Use 95°C for 2 min, then 65°C for 24 hr.

3. Add 5 µg of streptavidin to precipitate biotin-labeled transcripts.

4. Phenol extract to remove the biotin-labeled transcripts 2X.

   Add 50 μl of 50:49:1, phenol:chloroform:isoamyl alcohol.

   Vortex for 1 min. Spin for 5 min at top speed in microfuge.

   Remove top layer and transfer to a clean tube. Repeat.

5. Repeat subtraction 2X by repeating steps 1–4.

6. Label subtracted cDNA (92 ng) with α-^{32}P-dCTP by random primer labeling and use as a probe to screen the *M. oculata* cDNA library.

Screening cDNA Libraries

1. XL-1 Blue Culture grown in TB broth + 10 mM MgSO$_4$ + 0.2% maltose at 37°C to O.D.600 < 1.0 (overnight).

   Pellet for 10 min at high speed (1000g) in clinical centrifuge.

   Resuspend in 10 mM MgSO$_4$ to O.D.600 = 0.5.

2. NZYM plates dried and warmed at 37°C.

3. Top agar melted, aliquoted to 3 ml, cooled to 50°C.

4. Dilute cDNA library 1:1000 in SM Buffer (final concentration 1×10^6 pfu/μl)

5. Shake 40 μl diluted phage with 200 μl XL-1 Blue (O.D.600 = 0.5) for 15 min at 37°C (1×10^4 per 150×15 mm plate; 3×10^3 per 100×15 mm plate).

6. Pour plates. Incubate at 37°C for 5–6 hr.

7. Refrigerate at 4°C for at least 2 hr.

8. Lifts:

   A—2 min for the first filter; B—5 min for the second filter.

   Submerge in denaturing solution (1.5 M NaCl, 0.5 M NaOH) for 2 min.

   Submerge in neutralizing solution (1.5 M NaCl, 0.5 M Tris-Cl, pH 8) for 5 min.

   Rinse (0.2 M Tris-Cl, pH 7.5, 2X SSC) for 1–5 min.

9. Air dry (DNA side up).

10. Bake for 2 hr at 80°C, under vacuum.

Filter Hybridization

1. Prewash for 1 hr at room temperature

Prewash	(Store at 4°C; warm to room temperature before use)
50 mM Tris, pH 8	12.5 ml of 1 M Tris-Cl, pH 8
1 M NaCl	62.5 ml of 4 M NaCl
1 mM EDTA	0.5 ml of 500 mM EDTA, pH 8
0.1% SDS	2.5 ml of 10% SDS
	178.25 ml of sterile DDH$_2$O
	250 ml (put into sterile jar—do not autoclave)

2. Prehybridize 3–6 hr in 60 ml Hybridization Buffer, 42°C

Hybridization buffer	(Store at −20°C; warm to 42°C before use)
50% formamide	50 ml formamide
5X Denhardt's	10 ml 50X Denhardt's
5X SSPE	25 ml 20X SSPE
0.5% SDS	5 ml of 10% SDS
100 μg/ml DNA	1 ml of 10 mg/ml salmon sperm DNA
	13 ml of sterile DDH$_2$O
	100 ml (put into sterile jar—do not autoclave)

3. Add probe (use isolated insert–random prime to label) and hybridize 12–24 hr.
4. Wash in 2X SSC, 0.2% SDS, room temperature; 4X, 15 min each (until few counts come off in the wash—check with geiger counter).
5. Put filters under Saran Wrap (DNA side up) and expose to film at −70°C.
6. Pick plaques with sterile 6-in. sterile glass disposable pipet and put in 500 μl SM Buffer. Add 20 μl chloroform, vortex vigorously, leave in 4°C overnight before rescreening.

TERWILLIGER AND DURSTEWITZ.1

REVERSE TRANSCRIPTION (RT) AND PCR AMPLIFICATION OF HEMOCYANIN mRNA

1. *Reverse transcription of total RNA: First strand synthesis.*

a. Dilute 1 μl total RNA (1 μg/μl) with 10.65 μl autoclaved H$_2$O and add 0.75 μl of oligo-dT primer (0.27 μg/μl).
b. Incubate at 65°C for 3 min.
c. Cool slowly to room temperature and spin briefly in a microcentrifuge.
d. Add 4 μl 5X RT-buffer (250 mM Tris-HCl pH 8.5, 200 mM KCl, 30 mM MgCl), 1 μl 20 mM DTT, 1 μl 25 mM dNTPs, 1 μl RNAsin (10 units/μl), and 0.6 μl AMV reverse transcriptase (17 units/μl).
e. Vortex, spin briefly, and incubate at 42°C for 1.5 hr.
f. Dilute to a total volume of 500 μl with autoclaved water and store at −20°C.

2. *PCR amplification of first strand cDNA.*

a. Add 10 μl of cDNA from the RT reaction described above to 18.5 μl H$_2$O, 5 μl 10X PCR-buffer (670 mM Tris-HCl), 4 μl 2.5 mM dNTPs, 5 μl 10X BSA (bovine serum albumin 1 μg/μl), 1 μl of each primer (0.2 μg/μl), 0.5 μl Taq polymerase (5 units/μl), and 5 μl 40 mM MgCl$_2$.
b. Mix well, spin, and overlay with a drop of PCR-oil.

c. Carry out PCR reaction using the following protocol:

Denature   94°C for 40 sec

Anneal   55°C for 40 sec

Polymerize   72°C for 1 min

Repeat 35 cycles, then 5 min at 72°C and hold at 4°C.

d. Analyze 10-μl aliquots of each reaction on 1.2% agarose TAE-minigels.

TERWILLIGER AND DURSTEWITZ.2

NORTHERN BLOTS USING TOTAL RNA FROM DIFFERENT TISSUES AND DEVELOPMENTAL STAGES OF *CANCER mAGISTER*

1. To 0.3 μg RNA from each tissue or developmental stage (in a volume of up to 15 μl H_2O) add RNA sample buffer (75% formamide, 7.8% formaldehyde, 15 mM MOPS, 6 mM NaOAc, 0.75 mM EDTA) to a total volume of 20 μl.

2. Denature for 5 min at 65°C, then chill on ice.

3. Spin briefly, then add 2 μl 10X RNA loading dye (25 mg xylene cyanol FF in 6 ml glycerol/2 mM EDTA).

4. Electrophorese samples for 2.5 hr at 160 V on a 1.2% agarose formaldehyde gel (6% HCHO, Sambrook et al., 1989).

5. Soak gel for 40 min in 20X SSC (Sambrook et al., 1989).

6. Pressure blot gel onto a nitrocellulose or nylon membrane (Hybond, Amersham) for 1 hr at 75 mm Hg.

7. Crosslink for 2 min with UV (Stratalinker, Stratagene).

8. Bake for 1 hr at 80°C.

9. Prehybridize blots with agitation for 2 hr at 42°C in 50% formamide, 5X SSPE, 2X Denhardt's, 0.1% SDS (Sambrook et al., 1989).

10. Hybridize with agitation overnight at 42°C with a ^{32}P random prime labeled 750 bp 5′ probe that had previously been amplified by PCR (described above).

11. Wash 4× for 20 min at 45°C in 2X SSC, 0.1% SDS. Monitor activity of blot with hand-held geiger counter.

12. Evaluate by autoradiography with intensifying screen overnight.

TERWILLIGER AND DURSTEWITZ.3

CREATING NESTED DNASE I DELETIONS IN A HEMOCYANIN cDNA

Based on Lin et al., 1985, and Christine and Bernard Thisse, personal communication.

1. Precipitate 50 μg plasmid DNA (e.g., 1800 bp hemocyanin cDNA cloned into a Bluescript SK vector) with 2 volumes ethanol and 0.1 volume 3 M NaOAc.

2. Resuspend in 300 μl DNase I/Mn^{2+} buffer (0.2 M Tris-HCl, 10 mM MnCl$_2$, 1 mg/ml BSA).

3. Divide into 5 aliquots and digest each for 5 min with 4 μl of the following serial dilutions of DNase I: 0.2 ng/μl, 0.1 ng/μl, 0.05 ng/μl, 0.02 ng/μl, and 0.01 ng/μl.

4. After 5 min terminate digests by phenol/chloroform extraction (Sambrook et al., 1989).

5. Run 6-μl aliquots of each reaction on a 0.6% agarose gel. Choose the reaction showing the best linearization (no supercoiled or nicked DNA, no smear); discard the others.

6. Ethanol precipitate DNA (see above) and spin, wash, and dry pellet.

7. Resuspend in 150 μl TE (pH 8.0) and digest with 20–50 units of enzyme 1 (SmaI) in a total volume of 200 μl at 37°C for 2 hr.

8. Add 200 μl PEG (13% PEG$_{8000}$ in 1.6 M NaCl); mix and store on ice for 90 min.

9. Spin for 15 min at 16,000g at 4°C in microcentrifuge. Resuspend pellet in 400 μl TE (pH 8.0).

10. Extract with phenol, phenol/chloroform, and chloroform.

11. Ethanol precipitate (see above).

12. Spin, wash, and dry pellet. To repair ends of fragments, add 10 μl 0.25 mM dNTPs, 5 μl (0.1 M Tris-HCl pH 7.8, 0.1 M MgCl$_2$), 34 μl H$_2$O, and 1 μl (5 units) Klenow fragment. Incubate for 15 min at room temperature.

13. Stop reaction by incubating for 15 min at 68°C.

14. To religate fragments, add 40 μl 5X blunt end ligation buffer (250 mM Tris-HCl pH 7.5, 50 mM MgCl$_2$, 25% PEG$_{8000}$, 5 mM ATP, 5 mM DTT), 1 μl T4 DNA ligase (1:10 dilution, 40 NEB units) (20,000 NEB units equals 300 Weiss Units), and 109 μl H$_2$O. Incubate overnight at 16°C.

15. Stop ligation by incubation at 68°C for 15 min.

16. Digest DNA with 40 units enzyme 2 (EcoRI) in a total volume of 400 μl. Incubate at 37°C for 2 hr. Ethanol precipitate and resuspend in 50 μl TE.

17. Transform 200 μl competent *E. coli* XL-1 Blue cells with 1 and 10 μl of the mixture (Sambrook et al., 1989) and plate out on LB/Amp (50 μg/ml) plates. Grow overnight at 37°C.

18. Select 60 positive clones and isolate plasmid DNA by alkaline lysis miniprep (Sambrook et al., 1989).

19. Determine insert size by restriction analysis with enzymes 3 and 4 (KpnI/XbaI; Sambrook et al., 1989).

20. Select 15 clones with useful insert sizes (150 to 1800 bp) and sequence by dideoxy method (Sequenase V. 2.0, US Biochemical) according to manufacturer's instructions.

Protocols by Author

Application codes are:
- C – Cloning and Cell Preparation
- D – DNA Preparation and Manipulation
- G – Gel Electrophoresis
- H – Hybridization
- M – Molecular Characterization of Macromolecules
- P – PCR
- R – RNA, Expression and Transfection
- S – Sequencing

Protocols by Application Codes*

Cloning and Cell Preparation (C)

FERRARIS.1	Genomic DNA Lambda Library Screen
FLEISCHER.2	Construction and Screening of Microsatellite-Enriched Genomic Libraries
HOLLAND.2	Purifying DNA from Recombinant Bacteriophage Lambda
POTTS.1	PCR Directly from Plasmid Containing Bacterial Colonies
POWERS.2	Preparation of Isolated Nuclei
SOMMER.1	Culturing Free-Living Nematodes
SOMMER.2	Cell Lineage and Cell Ablation of Nematodes
STRASSMANN.7	Preparing Vector and Genomic DNA for Ligation
STRASSMANN.8	Sticky-end Ligation of Genomic DNA into Vector
STRASSMANN.9	Transformation of Competent Cells
STRASSMANN.10	Plating of Transformed Cells
STRASSMANN.11	Plating onto Nylon Membranes
STRASSMANN.12	Replicating Membranes
STRASSMANN.13	Probing a Library or Southern with a Labelled Oligonucleotide
STRASSMANN.14	Picking Positives
SWALLA.1	Subtractive Hybridization
TERWILLIGER.3	Creating Nested DNaseI Deletions

DNA Preparation and Manipulation (D)

FLEISCHER.1	Multilocus DNA Fingerprinting (see Loew and Fleischer)
FLEISCHER.2	Construction and Screening of Microsatellite-Enriched Genomic Libraries
GROSBERG.1	Simple Extraction and RAPD-PCR Protocols

*(most protocols appear in 2 or more categories)

Gel Electrophoresis (G)

Hybridization (H)

Molecular Characterization of Macromolecules (M)

Polymerase Chain Reaction (P)

GOLDEN.2	Proteinase K Digestion of Cells Prior to PCR
GOLDEN.3	Amplification of DNA Using Nested PCR
GROSBERG.1	Simple Extraction and RAPD-PCR Protocols
MCCARTHY.1	*In Situ* Hybridization with ss-DNA Probes Generated by PCR
PALUMBI.1	Sequencing Single Strand Asymmetric PCR Products
PALUMBI.2	Sequencing Double Strand PCR Products
PALUMBI.3	Solid Phase Sequencing From Magnetic Beads
POTTS.1	PCR Directly from Plasmid Containing Bacterial Colonies
POTTS.2	Ethidium Bromide Staining of Large Acrylamide Gels
POTTS.3	Denaturing Gradient Gel Electrophoresis (DGGE)
POTTS.4	Single-Strand Conformation Polymorphism (SSCP) Denaturation and Electrophoresis
POTTS.5	DNA Heteroduplex Analysis
POWERS.1	Allele Specific PCR
POWERS.3	*In Vivo* Footprinting
SOMMER.3	PCR Cloning of Genes from Nematodes
STRASSMANN.3	Insect Sperm DNA Extraction and PCR
STRASSMANN.4	Microsatellite PCR Using ^{35}S-dATP
STRASSMANN.5	Visualizing ^{35}S Labelled PCR Products on Sequencing Gels
SWALLA.1	Subtractive Hybridization
TERWILLIGER.1	Reverse Transcription (RT) and PCR from RNA

RNA, Expression and Transfection (R)

BURKE.1	Whole Mount *In Situ* Hybridization
CAMERON.1	Whole Mount *In Situ* Hybridization
FERRARIS.2	Transcription Start Site Determination by Dideoxy-Termination Sequencing of RNA
FERRARIS.3	Transcription Start Site Determination Using a PCR-Enhanced Method for Determining 5′ End Sequence
FERRARIS.4	Transient Transfection of Cultured cells
HOLLAND.3	Whole Mount *In Situ* Hybridization Applicable to *Amphioxus* and Other Small Larvae
JAGUS.1	Production of Recombinant Proteins *In Vitro* Using Coupled Transcription/Translation Reactions
JAGUS.2	Production of Functional Recombinant Integral Membrane Proteins Using Coupled Transcription/Translation Reactions
JAGUS.3	Transient Expression of Recombinant Integral Membrane Proteins in COS Cells